THE STRUCTURE
AND FUNCTION OF MUSCLE

SECOND EDITION

VOLUME III

Physiology and Biochemistry

CONTRIBUTORS

RICHARD H. ADRIAN

JOHN V. BASMAJIAN

NANCY A. CURTIN

R. E. DAVIES

SETSURO EBASHI

E. J. DE HAAN

G. S. P. GROOT

MARION HINES

DAVID NACHMANSOHN

D. M. NEEDHAM

YOSHIAKI NONOMURA

LEE D. PEACHEY

SUNE ROSELL

BENGT SALTIN

H. R. SCHOLTE

J. M. TAGER

E. M. WIT-PEETERS

KENNETH L. ZIERLER

THE STRUCTURE AND FUNCTION OF MUSCLE

Second Edition

VOLUME III

Physiology and Biochemistry

Edited by

Geoffrey H. Bourne

Yerkes Regional Primate Research Center
Emory University
Atlanta, Georgia

1973

ACADEMIC PRESS   New York   San Francisco   London
A Subsidiary of Harcourt Brace Jovanovich, Publishers

ACADEMIC PRESS, INC.
111 Fifth Avenue, New York, New York 10003

United Kingdom Edition published by
ACADEMIC PRESS, INC. (LONDON) LTD.
24/28 Oval Road, London NW1

LIBRARY OF CONGRESS CATALOG CARD NUMBER: 72-154373

PRINTED IN THE UNITED STATES OF AMERICA

CONTENTS

1. Electrical Properties of the Transverse Tubular System

Lee D. Peachey and Richard H. Adrian

2. The Neuromuscular Junction—The Role of Acetylcholine in Excitable Membranes

David Nachmansohn

3. Some Aspects of the Biophysics of Muscle

Kenneth L. Zierler

4. Energy Need, Delivery, and Utilization in Muscular Exercise

Sune Rosell and Bengt Saltin

5. The Control of Muscular Activity by the Central Nervous System

Marion Hines

6. Electromyography

John V. Basmajian

7. Proteins of the Myofibril

Setsuro Ebashi and Yoshiaki Nonomura

8. Biochemistry of Muscle

D. M. Needham

9. Biochemistry of Muscle Mitochondria

E. J. De Haan, G. S. P. Groot, H. R. Scholte,
J. M. Tager, and E. M. Wit-Peeters

10. ATP Breakdown Following Activation of Muscle

Nancy A. Curtin and R. E. Davies

LIST OF CONTRIBUTORS

Numbers in parentheses indicate the pages on which the authors' contributions begin.

RICHARD H. ADRIAN, *Physiological Laboratories, Cambridge University, Cambridge, England* (1)

JOHN V. BASMAJIAN, *Emory University Regional Rehabilitation Research and Training Center, Georgia Mental Health Institute, Atlanta, Georgia* (273)

NANCY A. CURTIN, *Graduate Group on Molecular Biology, University of Pennsylvania, Philadelphia, Pennsylvania* (471)

R. E. DAVIES, *Graduate Group on Molecular Biology and School of Veterinary Medicine, University of Pennsylvania, Philadelphia, Pennsylvania* (471)

SETSURO EBASHI, *Department of Pharmacology, Faculty of Medicine, University of Tokyo Japan, Bunkyo-ku, Tokyo, Japan* (285)

E. J. DE HAAN, *Laboratory of Biochemistry, B.C.P. Jansen Institute, University of Amsterdam, Amsterdam, The Netherlands* (417)

G. S. P. GROOT, *Laboratory of Biochemistry, B.C.P. Jansen Institute, University of Amsterdam, Amsterdam, The Netherlands* (417)

MARION HINES, *Department of Anatomy, Johns Hopkins University, Baltimore, Maryland* (223)

DAVID NACHMANSOHN, *Departments of Neurology and Biochemistry, College of Physicians and Surgeons, Columbia University, New York, New York* (31)

D. M. NEEDHAM, *Department of Biochemistry, University of Cambridge, Cambridge, England* (363)

YOSHIAKI NONOMURA, *Department of Pharmacology, Faculty of Medicine, University of Tokyo Japan, Bunkyo-ku, Tokyo, Japan* (285)

LEE D. PEACHEY, *Department of Biology, University of Pennsylvania, Philadelphia, Pennsylvania (1)*

SUNE ROSELL, *Department of Pharmacology, Karolinska Institutet, Stockholm, Sweden (185)*

BENGT SALTIN, *Department of Physiology, Gymsiology, Gymnastik- och idrottshogskolan, Lidingovagen, Stockholm, Sweden (185)*

H. R. SCHOLTE, *Department of Biochemistry, Medical Faculty, University of Rotterdam, Rotterdam, The Netherlands (417)*

J. M. TAGER, *Laboratory of Biochemistry, Medical Faculty, University of Amsterdam, Amsterdam, The Netherlands (417)*

E. M. WIT-PEETERS, *Plesmanlaan, Amsterdam, The Netherlands (417)*

KENNETH L. ZIERLER, *Clayton Laboratory for Study of Control of Cell Function in Health and Disease, Department of Medicine and Department of Physiology, The Johns Hopkins University School of Medicine, Baltimore, Maryland (117)*

PREFACE

In the years elapsed since the first edition of this work was published in 1960, studies on muscle have advanced to such a degree that a second edition has long been overdue. Although the original three volumes have grown to four, we have covered only a fraction of the new developments that have taken place since that time. It is not surprising that these advances have not been uniform, and in this new edition not only have earlier chapters been updated but also areas in which there was only limited knowledge before. Examples are the development of our knowledge of crustacean muscle (172 of 213 references in the reference list for this chapter are dated since the first edition appeared) and arthropod muscle (205 of 233 references are dated since the last edition). Obliquely striated muscle, described in 1869, had to wait until the electron microscope was focused on it in the 1960's before it began to yield the secrets of its structure, and 33 of 43 references dated after 1960 in this chapter show that the findings described are the result of recent research. There has also been a great increase in knowledge in some areas in which considerable advances had been made by the time the first edition appeared. As an example, in Dr. Hugh Huxley's chapter on "Molecular Basis of Contraction in Cross-Striated Muscles," 76 of his 126 references are dated after 1960.

The first volume of this new edition deals primarily with structure and considers muscles from the macroscopic, embryonic, histological, and molecular points of view. The other volumes deal with further aspects of structure, with the physiology and biochemistry of muscle, and with some aspects of muscle disease.

We have been fortunate in that many of our original authors agreed to revise their chapters from the first edition, and it has also been our good fortune to find other distinguished authors to write the new chapters included in this second edition.

To all authors I must express my indebtedness for their hard work and patience, and to the staff of Academic Press I can only renew my confidence in their handling of this publication.

Geoffrey H. Bourne

PREFACE
TO THE FIRST EDITION

Muscle is unique among tissues in demonstrating to the eye even of the lay person the convertibility of chemical into kinetic energy.

The precise manner in which this is done is a problem, the solution of which has been pursued for many years by workers in many different disciplines; yet only in the last 15 or 20 years have the critical findings been obtained which have enabled us to build up some sort of general picture of the way in which this transformation of energy may take place. In some cases the studies which produced such rich results were carried out directly on muscle tissue. In others, collateral studies on other tissues were shown to have direct application to the study of muscular contraction.

Prior to 1930 our knowledge of muscle was largely restricted to the macroscopic appearance and distribution of various muscles in different animals, to their microscopical structure, to the classic studies of the electro- and other physiologists and to some basic chemical and biochemical properties. Some of the latter studies go back a number of years and might perhaps be considered to have started with the classic researches of Fletcher and Hopkins in 1907, who demonstrated the accumulation of lactic acid in contracting frog muscle. This led very shortly afterward to the demonstration by Meyerhof that the lactic acid so formed is derived from glycogen under anaerobic conditions. The lactic acid formed is quantitatively related to the glycogen hydrolyzed. However, it took until nearly 1930 before it was established that the energy required for the contraction of a muscle was derived from the transformation of glycogen to lactic acid.

This was followed by the isolation of creatine phosphate and its establishment as an energy source for contraction. The isolation of ADP and ATP and their relation with creatine phosphate as expressed in the Lohmann reaction, were studies carried out in the thirties. What might be described as a spectacular claim was made by Engelhart and

Lubimova, who in the 1940's said that the myosin of the muscle fiber had ATPase activity. The identification of actin and relationship of actin and myosin to muscular contraction, the advent of the electron microscope and its application with other physical techniques to the study of the general morphology and ultrastructure of the muscle fibers were events in the 1940's which greatly developed our knowledge of this complex and most mobile of tissues.

In the 1950s the technique of differential centrifugation extended the knowledge obtained during previous years of observation by muscle cytologists and electron microscopists to show the differential localization of metabolic activity in the muscle fiber. The Krebs cycle and the rest of the complex of aerobic metabolism was shown to be present in the sarcosomes—the muscle mitochondria.

This is only a minute fraction of the story of muscle in the last 50 years. Many types of discipline have contributed to it. The secret of the muscle fiber has been probed by biochemists, physiologists, histologists and cytologists, electron microscopists and biophysicists, pathologists, and clinicians. Pharmacologists have insulted skeletal, heart, and smooth muscle with a variety of drugs, *in vitro, in vivo,* and *in extenso;* nutritionists have peered at the muscle fiber after vitamin and other nutritional deficiencies; endocrinologists have eyed the metabolic process through hormonal glasses. Even the humble histochemist has had the temerity to apply his techniques to the muscle fiber and describe results which were interesting but not as yet very illuminating—but who knows where knowledge will lead. Such a ferment of interest (a statement probably felicitously applied to muscle) in this unique tissue has produced thousands of papers and many distinguished workers, many of whom we are honored to have as authors in this compendium.

Originally we thought, the publishers and I, to have *a book* on muscle which would contain a fairly comprehensive account of various aspects of modern research. As we began to consider the subjects to be treated it became obvious that two volumes would be required. This rapidly grew to three volumes, and even so we have dealt lightly or not at all with many important aspects of muscle research. Nevertheless, we feel that we have brought together a considerable wealth of material which was hitherto available only in widely scattered publications. As with all treatises of this type, there is some overlap, and it is perhaps unnecessary to mention that to a certain extent this is desirable. It is, however, necessary to point out that most of the overlap was planned, and that which was not planned was thought to be worthwhile and was thus not deleted.

We believe that a comprehensive work of this nature will find favor with all those who work with muscle, whatever their disciplines, and that although the division of subject matter is such that various categories of workers may need only to buy the volume which is especially apposite to their specialty, they will nevertheless feel a need to have the other volumes as well.

The Editor wishes to express his special appreciation of the willing collaboration of the international group of distinguished persons who made this treatise possible. To them and to the publishers his heartfelt thanks are due for their help, their patience, and their understanding.

Emory University, Atlanta, Georgia   GEOFFREY H. BOURNE
October 1, 1959

CONTENTS OF OTHER VOLUMES

Volume IV: Pharmacology and Disease

1

ELECTRICAL PROPERTIES OF THE TRANSVERSE TUBULAR SYSTEM

LEE D. PEACHEY and RICHARD H. ADRIAN

I. Introduction

The transverse tubular system or T system of striated muscle was discovered by electron microscopy in the 1950s. The most important paper of that period (Porter and Palade, 1957) described the structure we now call the T system as rows of vesicles in the space between the terminal cisternae of the sarcoplasmic reticulum (SR) and termed it the "intermediary vesicles." This analysis, in effect, included the T system as part of the SR. Since that time, through further comparative studies and the use of improved preparation methods, electron microscopists have demonstrated convincingly that the T system is a branched network of tubules derived from and remaining attached to the surface plasma membrane of the muscle fiber. This has fostered a general feeling that the T system should not be considered as a part of the SR, but as part of the surface membrane complex of the cell. To be sure, it is found deep in the fiber, and it associates closely and in a specific way with the intracellular SR, but in an important respect it is part of the fiber surface membrane.

Of greatest interest to us are the function and physiological properties of the T system. Several recent physiological and morphological studies have been directed to this problem, and work in this area continues actively at present. Our aim in this chapter will be to review the present state of our knowledge, and to indicate some of the uncertainties and possible future directions for advance. We also hope to resolve some uncertainties and apparent discrepancies in earlier papers by presenting an analysis of the T system as a cable network with special properties at the surface of the fiber and with nonlinear properties in its membrane.

II. Mathematical Analysis

A. Cable Analysis as a Network

The electrical properties of nerve and muscle have been analyzed in terms of the ionic and capacity currents flowing across the cell surface membrane. The basis of this approach is a set of equations that has become known as "cable theory" because of its application to transmission cables. If we are to understand the operation of the T system as a passive electrical network with linear properties, or as a structure generating its own active potential changes, it is necessary as a first step to extend the equations of one-dimensional cable theory to two- and even three-dimensional cable networks representing the T system structure.

Morphological evidence suggests that we can represent the T system in frog twitch striated muscle fibers as a set of regular two-dimensional tubular networks in which the tubular diameter is small compared to the mesh and in which the mesh itself is small compared to the fiber diameter, as done by Adrian et al. (1969a). Tubules are thought to be present between all the fibrils throughout the cross section of the fiber. This is probably the general pattern in vertebrate striated fibers, although there is evidence of scanty longitudinal elements of the T system in some fiber types, for example, frog twitch fibers (Eisenberg and Eisenberg, 1968). In frog slow fibers (Page, 1965; Flitney, 1971), the longitudinally oriented T tubules may make up a substantial fraction of the whole T system. In this case, it is probable that the T system approximates more closely to a three-dimensional network than to a two-dimensional network. For generality, therefore, we shall derive a T system cable equation in three dimensions. Cases of two-dimensional networks, with or without radial symmetry, are special cases of the general equation and may be treated by setting appropriate terms in the general equation to zero.

Since the muscle fiber is cylindrical, the networks to be considered will be limited by the surface of a cylinder, and for T systems with negligible longitudinal connections, the network is in the shape of a circular disk with the same radius as the fiber (Fig. 1). Cylindrical and polar coordinate systems are therefore appropriate for the cable equations.

Several authors have considered this problem (Falk and Fatt, 1964; Falk, 1968; Adrian *et al.*, 1969a; Schneider, 1970), and what follows is based on their approach. In particular, we shall use the treatment of Adrian *et al.* and define the following properties of the tubular network: G_L, specific conductivity of the tubular lumen (mho cm^{-1}), G_w, conductance per unit area of tubular membrane (mho cm^{-2}), C_w, capacitance per unit area of tubular membrane (F cm^{-2}). From these basic constants, a set of practical constants is derived. These practical constants are the properties of the tubular system referred to the volume of the muscle fiber rather than to the tubule, and they involve the fraction of the fiber occupied by the tubular system (ρ), the volume to surface ratio of the tubules (ζ), and a geometrical factor (σ), which is $\frac{1}{2}$ for certain regular two-dimensional networks and $\frac{1}{3}$ for certain regular three-dimensional networks. The practical constants are:

$$\bar{G}_L = G_L \rho \sigma \quad \text{mho cm}^{-1}$$
$$\bar{G}_w = G_w \rho / \zeta \quad \text{mho cm}^{-3}$$
$$\bar{C}_w = C_w \rho / \zeta \quad \text{F cm}^{-3}$$

The currents across the walls of the tubular system are considered to be sufficiently small so that they produce a negligible potential gradient in the extratubular sarcoplasm. We can therefore write

$$I_w = \bar{G}_w u + \bar{C}_w \frac{\partial u}{\partial t} \tag{1}$$

where I_w is the current across the tubular wall in unit volume of fiber and u is the displacement of the potential difference across the tubular membrane from its resting value in the sense sarcoplasmic potential minus potential in the tubular lumen (t represents time).

In a three-dimensional regular network, the current within the network at any point is

$$\tilde{G}_L \text{ grad } u$$

and the current entering or leaving the network across the tubular walls is

$$\tilde{G}_L \text{ div grad } u$$

therefore

$$\frac{\tilde{G}_L}{\tilde{C}_w} \nabla^2 u - \frac{\tilde{G}_w}{\tilde{C}_w} u = \frac{\partial u}{\partial t} \tag{2}$$

or

$$\nabla^2 u - \nu^2 u = \frac{1}{\kappa} \frac{\partial u}{\partial t} \tag{3}$$

where $\nu = 1/\lambda_T = (\overline{G}_w/\overline{G}_L)^{1/2}$ and $\kappa = \overline{G}_L/\overline{C}_w$; κ may be called the propagation constant of the system. It has the dimensions of a diffusion constant but is numerically much greater; ν is the reciprocal of the tubular length constant λ_T. [*Note:* Adrian *et al.* (1969) define ν differently, as a dimensionless quantity $\nu = a/\lambda_T$, where a is the fiber radius.] In a cylindrical coordinate system (r,θ,z), Eq. (3) becomes

$$\frac{1}{r} \frac{\partial}{\partial r} \left(r \frac{\partial u}{\partial r} \right) + \frac{1}{r^2} \frac{\partial^2 u}{\partial \theta^2} + \frac{\partial^2 u}{\partial z^2} - \nu^2 u = \frac{1}{\kappa} \frac{\partial u}{\partial t} \tag{4}$$

and the equivalent equation for a two-dimensional network in polar coordinates (r,θ) is

$$\frac{1}{r} \frac{\partial}{\partial r} \left(r \frac{\partial u}{\partial r} \right) + \frac{1}{r^2} \frac{\partial^2 u}{\partial \theta^2} - \nu^2 u = \frac{1}{\kappa} \frac{\partial u}{\partial t} \tag{5}$$

For a two dimensional network with radially symmetrical boundary conditions (i.e., where $\partial u/\partial \theta = 0$, for example, a source of current at $r = 0$), or for a circumferentially uniform boundary condition at the surface of the fiber $(r = a)$, Eq. (5) becomes

$$\frac{1}{r} \frac{\partial}{\partial r} \left(r \frac{\partial u}{\partial r} \right) - \nu^2 u = \frac{1}{\kappa} \frac{\partial u}{\partial t} \tag{6}$$

which is equivalent to Eq. (9) in Adrian *et al.* (1969a). Equation (3) has the same form as the differential equation that describes the diffusion of a substance that is simultaneously involved in an irreversible first order chemical reaction. Solutions to it may be obtained by the methods described in Carslaw and Jaeger (1959), Crank (1956), and Danckwerts (1951). Particular solutions depend on the boundary conditions, and these in turn must be related to the particular experimental method used to change the potential within the tubular system. In a general way, one may distinguish point sources of current which require solutions of Eqs. (4) or (5) and radially symmetrical situations where Eq. (6) is appropriate. The former correspond to the experiments of Huxley and Taylor (1958) and Huxley and Peachey (1964) who applied point sources of current to the tubular network in the form of local externally applied micro pipettes, and the latter to the experiments of Adrian *et al.* (1969a,b), who altered the potential of the membrane of the fiber surface by current from an internal microelectrode. For the latter type of experiment current flow within the tubular network can, as a reasonable approximation, be treated as if it were only radial. An analytic solution of Eq. (6) has been given by Adrian *et al.* (1969a). They wished to know how rapidly the tubular capacity was charged when the potential between the sacoplasm and the external fluid was changed suddenly. In their experiments and those of Adrian *et al.* (1969b), the potential across the surface of the fiber was altered suddenly using a voltage clamping system with two internal microelectrodes. In terms of the variables of Eq. (6) the initial and boundary conditions are that at $r = a$,

$$\begin{aligned} u_a &= 0 && \text{for} \quad t < 0 \\ u_a &= V_s && \text{for} \quad t \geq 0 \end{aligned} \tag{7}$$

The appropriate solution of Eq. (6) is

$$\frac{u}{V_s} = \frac{I_0(\nu r)}{I_0(\nu a)} - 2 \sum_{n=1}^{\infty} \frac{\beta_n \exp[-(\nu^2 a^2 + \beta_n^2)(\kappa t/a^2)]}{(\nu^2 a^2 + \beta_n^2)} \frac{J_0(\beta_n r/a)}{J_1(\beta_n)} \tag{8}$$

where $I_0(\ )$ is the hyperbolic Bessel function of zero order. $J_0(\ )$ and $J_1(\ )$ are Bessel functions of the first kind, and $\beta_1, \beta_2, \ldots, \beta_n$ are the positive roots of $J_0(\beta) = 0$. In the steady state, the distribution of potential across the wall of the network is Eq. (8) with the time-dependent term equal to zero, which gives

$$\frac{u}{V_s} = \frac{I_0(\nu r)}{I_0(\nu a)} \tag{9}$$

B. Introduction of an Access Resistance

The boundary condition used to obtain Eqs. (8) and (9) supposes that the properties of the T system are the same up to and including the surface of the fiber. It is perhaps more realistic to allow for some special access resistance which will represent an extra resistance of the tubular network at or just under the surface of the fiber. This might

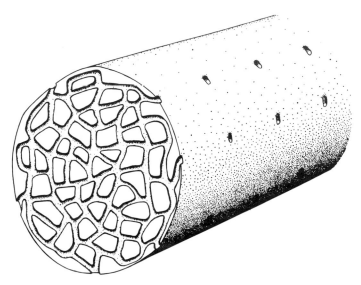

Fig. 1. Drawing of a muscle fiber cut across through one T system network. The openings of other T system networks are seen along the length of the muscle fiber. Not all tubules near the surface of the fiber are shown opening to the outside, a situation that could give rise to an access resistance, as discussed in the text.

arise, for example, if not all tubules were connected to the surface or open to the outside (Fig. 1). The apparent rather wide circumferential separation of active spots (Huxley and Taylor, 1958) supports this notion. The scanning electron micrographs of McCallister and Hadek (1970) also indicate that tubular openings of frog toe muscle are spaced about twice as far apart in the circumferential direction as in the longitudinal direction. This is several times farther apart than one would see if a tubule joined the surface between each pair of adjacent fibrils.

Another source of an access resistance could be a constriction of the tubular lumen at the fiber surface or a tortuosity of the tubular path near the surface. Either increased tubular length or a reduced luminal caliber over a significant length would increase radial resistance near the fiber surface equivalent to the access resistance proposed. Such a peripheral T system tortuosity or constriction could help to explain the failure of electron microscopists to demonstrate surface connections of the T-system in sections of frog muscle, as is discussed in Chapter 9 of Volume II.

Let this access resistance be called R_s (Ω cm^2). The current (i_a) across the access resistance entering the T system will be

$$i_a = \frac{1}{R_s} (V_s - u_a) = \bar{G}_L \left(\frac{\partial u}{\partial r} \right)_{r=a} \tag{10}$$

or at $r = a$, $\partial u/\partial r + h(u_a - V_s) = 0$, where $h = 1/G_L R_s$. The solution of Eq. (6) that satisfies the boundary condition for a suddenly applied potential V_s at the mouths of the tubules (that is outside the access resistance R_s) is

$$\frac{u}{V_s} = \frac{I_0(\nu r)}{I_0(\nu a) + (\nu/h)I_1(\nu a)} - 2ah \sum_{n=1}^{\infty} \frac{\beta_n^2 \exp[-(\nu^2 a^2 + \beta_n^2)\kappa t/a^2]J_0(\beta_n r/a)}{(h^2 a^2 + \beta_n^2)(\nu^2 a^2 + \beta_n^2)J_0(\beta_n)} \quad (11)$$

where $\beta_1, \beta_2, \ldots, \beta_n$ are the positive roots of

$$\beta J_1(\beta) - ahJ_0(\beta) = 0$$

As before, the steady state potential distribution across the wall of the T system is obtained by setting the time-dependent term equal to zero.

$$\frac{u}{V_s} = \frac{I_0(\nu r)}{I_0(\nu a) + (\nu/h)I_1(\nu a)} \quad (12)$$

The lines in Figs. 2 and 3 show the potential distribution along the radius of a fiber given by Eqs. (8) and (11) at different times following a sudden displacement of the potential at the surface of the fiber. The numbers against the curves are times in milliseconds after the beginning of a pulse at the fiber surface for a fiber of 40 μ radius with a propagation constant $\kappa = 5 \times 10^{-3}$ cm^2 sec^{-1} and $\lambda_t = 100$ μ. In terms of the electrical properties of the T system, this implies $C_w = 1$ μF cm^{-2}, $G_w = 0.5 \times 10^{-4}$ mho cm^{-2}, $G_L = 10^{-2}$ mho cm^{-1}, with $\rho = 0.003$, $\sigma = \frac{1}{2}$ and $\zeta = 10^{-6}$ cm. For Eq. (11) and Fig. 3, $R_s = 100$ Ω cm^2.

The general conclusion from calculations using Eqs. (8) and (11) is that for rapid changes of potential at the surface of the fiber, it is R_s and the propagation constant ($\kappa = G_L \sigma \zeta / C_w$) that mainly determine how rapidly the potential in the network changes. The steady state distribution is determined by R_s and the tubular length constant, $\lambda_T = (G_L \sigma \zeta / G_w)^{1/2}$.

C. Tubular Length Constant and Resistance*

Adrian *et al.* (1969b) imposed step depolarization at a point on a fiber and observed the resultant shortening of the myofibrils across the diameter of the fiber at that point. The fibers were in the presence of tetrodotoxin to prevent the occurrence of action potentials. In some experiments, the potential steps were long enough to allow the potential to reach its steady state distribution. With threshold depolarizing steps, only the most superficial myofibrils shortened. Increasing the depolariz-

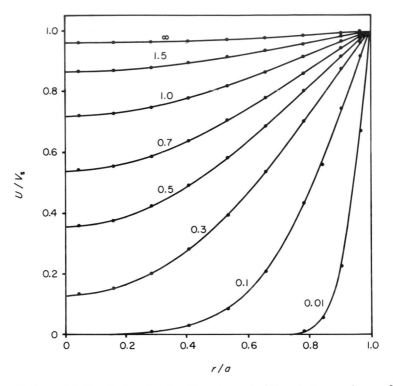

Fig. 2. Potential distribution in the T system at different times after sudden displacement of the potential at the surface of the fiber. The surface potential displacement is V_s, and the potential throughout the T system is u. The abscissa is radial distance from the center of the fiber expressed as a fraction of the radius of the fiber. The numbers adjacent to the curves are times, in milliseconds, after the displacement of the surface potential. The curves are obtained from Eq. (8), and the points are obtained using the numerical approximation method described in the text (curves redrawn with permission from Adrian *et al.* 1969a). For both calculations, $a = 40 \mu$, $\kappa = 5 \times 10^{-3}$ cm^2 sec^{-1}, and $\lambda_T = 100 \mu$.

ing step by a few millivolts (2–6 mV) made axial as well as superficial myofibrils shorten. In terms of the potential distributions in the T system shown in Figs. 2 and 3, this small potential span represents the difference between the steady-state potential across the wall of the T system at the periphery and at the fiber axis. From it and Eq. (9) or (12), λ_T may be calculated. The effective resistance (R_T) of the T system may also be estimated, provided G_L is known.

$$R_T = V_s/\bar{G}_L \left(\frac{\partial u}{\partial r}\right)_{r=a}$$

For the tubular system without access resistance

$$R_T = \frac{\lambda_T I_0(a/\lambda_T)}{\bar{G}_L I_1(a/\lambda_T)} \tag{13}$$

For a tubular system with an access resistance R_s

$$R_T = \frac{\lambda_T I_0(a/\lambda_T)}{\bar{G}_L I_1(a/\lambda_T)} + R_s \tag{14}$$

Estimates of λ_T and R_T from the experimentally measured span between the depolarizations at which the superficial and axial myofibrils first shorten depend critically upon the condition assumed to hold at the tubular mouths. Adrian *et al.* (1969b) assumed that there was no access resistance. They used Eqs. (9) and (13) to estimate λ_T and to compare the measured fiber resistance with the calculated resistance of the T system. For fibers with a radius of 40 μ, they estimated λ_T to be 60 μ

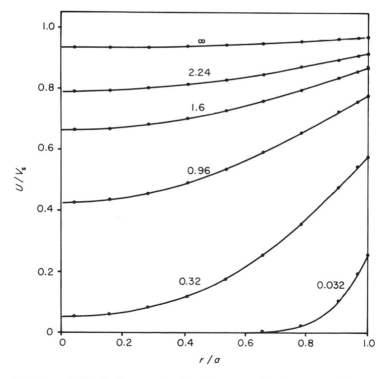

Fig. 3. Potential distribution, as in Fig. 2, except for the case with an access resistance, $R_s = 100$ Ω cm². Equation (11) was used for the curves, and the numerical approximation method for the points. Other parameters are the same as for Fig. 2.

and $R_T = 1400\ \Omega$ cm^2. The measured membrane resistance of these fibers was very nearly equal to the estimated R_T, suggesting that little or no current passed across the surface of the muscle fiber in the conditions of these experiments. Stated another way, this suggests that all or most of the conductance into the fiber is due to the T system. This conclusion is improbable because it is not consistent with the evidence about tubular and surface resistance from comparison of normal and detubulated fibers (Gage and Eisenberg, 1969). These authors estimated that the tubules contribute only one-third to one-half of the total fiber conductance.

The presence of a tubular access resistance can remove this difficulty if we say that the observed potential span includes the potential drop across the access resistance. This would require that the most superficial myofibrils can be activated by a potential change across the surface membrane of the fiber or at least outside most of the access resistance. Figure 4, which shows terminal cisternae in relation to a tubule and the surface membrane at the mouth of a tubule, gives morphological plausibility to this assumption.

Using Eqs. (12) and (14) to recalculate the results of Adrian et al. (1969b) for a fiber with a 40 μ radius and $R_s = 100\ \Omega$ cm^2 increases the estimate of λ_T to 85 μ and of R_T to 2580 Ω cm^2. Since the measured fiber resistance was about 1400 Ω cm^2, the recalculated results suggest that only about half the conductance of the fiber is attributable to the T system, which is consitent with the results of Gage and Eisenberg (1969).

D. Numerical Method of Analysis

When the coefficients in the T system differential equations (Eq. 2 or 4) are not constants but are functions of time and/or potential, the equations are said to be nonlinear and cannot as easily be solved in closed form. Therefore, we have employed a numerical approximation method for obtaining the spatial distribution of potential in the T system at various times during and after the application of a square potential change at the surface of the fiber. After briefly describing the method, we will show that it closely duplicates results obtained with the closed solution for the linear case and then present some results of nonlinear numerical solutions of interest in relation to the experimental results of Costantin (1970).

1. LINEAR METHOD

The disk-shaped network ($r = a$) representing the T system is divided into n concentric cylindrical shells of thickness $\Delta r = a/n$. For one centi-

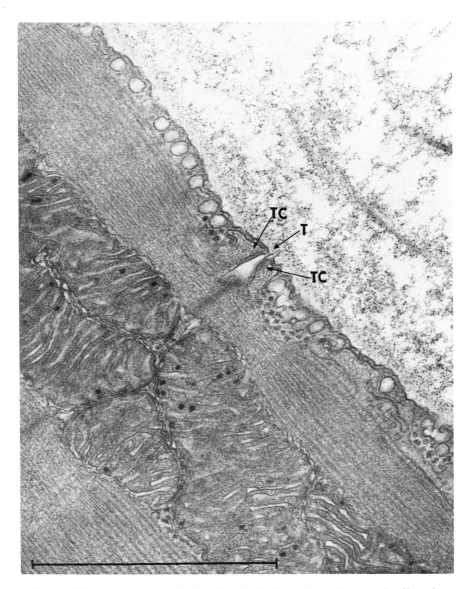

Fig. 4. Electron micrograph of a longitudinal section of a muscle fiber from the semitendinosus muscle of a frog (*Rana temporaria*). Two terminal cisternae (TC) of the SR are seen adjacent to a transverse tubule (T) forming a triad. The constricted mouth of this transverse tubule suggests one possible source of an access resistance, as discussed in the text. The close apposition of peripheral terminal cisternae to the surface membrane of the muscle fiber suggests the possibility that calcium release near the surface of the fiber is triggered by potential changes outside this access resistance. The line indicates 1 μ.

meter length of fiber, the following parameters are calculated for each
shell, using the same geometric and electrical constants as given earlier:
(1) tubular volume (cm^2); (2) tubular wall capacitance (F cm^{-1});
(3) tubular wall conductance (mho cm^{-1}); (4) the effective radius,
which divides each shell into two half-shells of equal volume (cm);
(5) the resistance between effective radii for each pair of adjacent shells,
and the resistance from the radius of the nth shell to its outer surface
(mho cm^{-1}). The equivalent circuit is given in Fig. 5.

Initial conditions are introduced by setting the potential in each shell
equal to the resting potential and time equal to zero. For a square pulse
of magnitude V_s (volts) and duration T_p (sec), the potential at the
outer surface of the outermost shell is set at V_s plus the resting potential
for $0 \leq T \leq T_p$ and at the resting potential for $T_p < T$.

The numerical calculation procedes in time steps of ΔT, during which
the potential in each shell is considered to be constant and currents
flow through the radial and tubular wall resistance according to Ohm's
Law. The net charge transferred into each shell is calculated as the
net current into the tubular lumen in that shell times ΔT. When this
has been done for all shells, each charge is applied to the wall capacity

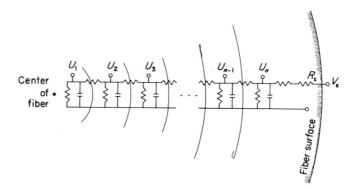

Fig. 5. Equivalent circuit used for the numerical method of analysis of time-dependent potential changes in the T system. The fiber is considered to be divided
into n shells of equal thickness. The potential u across the wall of the T system
is considered to be uniform within each shell. Each shell contains a capacitance
and a resistance representing the wall capacitance and wall resistance of the T
system contained in that shell. Each shell is connected to radially adjacent shells
by a resistance representing the radial component of the tubular luminal resistance.
An additional access resistance R_s is in series with the outermost radial resistance.
For the linear method, the wall conductance in each shell is held constant. To
introduce a nonlinearity representing a sodium conductance change in the T system
membrane, each wall conductance is varied as a function of potential in the shell
and time.

of its shell to obtain a potential change for that shell. The result is a new potential distribution in the shells for the time ΔT. Repeating this process gives the distribution for the time $2 \Delta T$, and for m repetitions, the potential distribution at $T = m \Delta T$ is obtained.

For the method to be numerically stable, the quantity $\Delta T/(\Delta r)^2$ times the maximum radial conductance between shells must be less than $\frac{1}{2}$. In practice we kept it substantially smaller than this. Spatial accuracy is determined by the choice of n which sets Δr. We found that the results were not changed appreciably in the third significant figure by choices of n greater than about 12. We regularly used 16 shells, which conveniently gives shells at $\frac{1}{2}$, $\frac{1}{4}$, etc. of the fiber radius. Choosing $\Delta T = 10^{-6}$ sec gave satisfactorily stable results, but all runs chosen for publication here have also been run at $\Delta T = 10^{-7}$ sec for safety. If significant differences had been obtained for the two choices of ΔT, it would have indicated instability using the larger ΔT. They never were.

To check the numerical method, the calculations in Figs. 2 and 3 were computed using the same electrical parameters as used for the exact solution and using as numerical parameters $n = 16$ and $\Delta T = 10^{-7}$ sec. For the case with an access resistance (Fig. 3), an appropriate additional resistance was inserted in series between the fiber surface and the electrical center of the outermost shell (see Fig. 5). The results are presented as the circles in Figs. 2 and 3. It can be seen that the results agree satisfactorily with the exact solution (solid lines), indicating that the method is valid for the linear case with or without an access resistance.

2. Nonlinear Method

To introduce a nonlinearity representing a time- and potential-dependent sodium conductance into the numerical calculations, it was necessary only to add an additional current across the tubular wall calculated for each shell at each time interval. The form of the sodium conductance we chose was a simple one. A conductance was turned on instantaneously for all potentials above a threshold, chosen for the examples presented here at −50 mV. This instantaneous conductance increased linearly with potential above this threshold, saturating abruptly at a potential of −20 mV. At each time interval, the existing sodium conductance for each shell was reduced by a fractional amount, which gave the equivalent of an exponential inactivation with a 1 msec time constant.

Using this nonlinearity and the numerical calculation method, we investigated near threshold square pulses of brief duration as a model for the voltage-clamp experiments of Costantin (1970). The results are discussed next.

3. Results of Nonlinear Calculations

Figures 6–10 show time-dependent changes in potential in the innermost shell ($r = 0.044a$) for numerical calculations with 16 shells and $\Delta T = 10^{-7}$ sec. In each case, the surface potential was a square wave depolarization of 3 msec duration. Electrical and geometric parameters are the same as used for Fig. 3 and are given in the legend. There was no sodium conductance for Fig. 6. The sodium conductance for Fig. 7 had a threshold of −50 mV, a slope of 0.5 mho/cm² V (referred to the area of the tubular membrane) and saturated at −20 mV. The calculation for Fig. 7 assumed an external and tubular sodium concentration of 120 mM and a sodium equilibrium potential of 35 mV. For Figs. 8 and 9, the external and tubular sodium concentrations were reduced to 60 mM and 30 mM, respectively. The sodium equilibrium potential was reduced by 17.5 mV for each twofold reduction in sodium concentration. Internal sodium concentration was assumed constant and

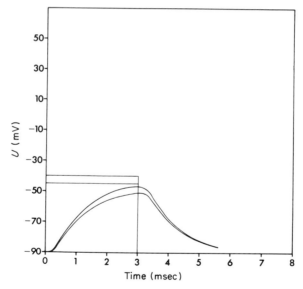

Fig. 6. Time dependence of potential across the T system network calculated by the numerical method. The surface potential change is shown as a square wave of 3 msec duration and with two amplitudes, 45 and 50 mV. The curved lines show the potential near the center of the fiber for the case with no sodium conductance change (linear method). Parameters used in the calculation are: $a = 40$ μ, $n = 16$, $\Delta T = 10^{-7}$ sec, $G_L = 10^{-2}$ mho cm⁻¹, $G_w = 5 \times 10^{-5}$ mho cm⁻², $R_s = 100$ Ω cm, $C_w = 10^{-6}$ F cm⁻², $\rho = 3 \times 10^{-3}$, $\sigma = 0.5$, $\zeta = 10^{-6}$ cm. This gives values of $\kappa = 5 \times 10^{-3}$ cm² sec⁻¹ and $\lambda_T = 10^{-2}$ cm. Resting potential of the fiber surface and all parts of the T system is −90 mV.

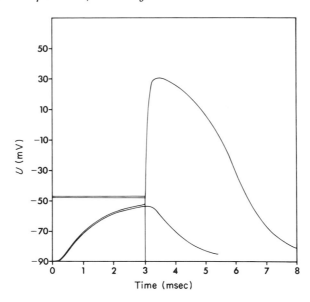

Fig. 7. Numerically calculated T system potentials as in Fig. 6, except for the inclusion of a sodium conductance calculated for each shell starting at −50 mV, increasing with a slope of 0.5 mho cm^{-2} V^{-1} (referred to area of T system membrane) and saturating at −20 mV and inactivating with a 1 msec time constant. The sodium equilibrium potential was assumed to be +35 mV, corresponding to a sodium concentration of 120 mM in the tubular lumen and of 29.9 mM in the sarcoplasm. The 3 msec square waves at the fiber surface have amplitudes of 42 and 43 mV. For the case of 42 mV pulse amplitude, the T system potential inside R_s never gets as high as −50 mV, and the T system remains passive. When the surface pulse is 43 mV, however, the T system reaches the threshold for sodium activation and moves quickly toward the sodium equilibrium potential, even though the fiber surface is held down to −47 mV.

equal to 29.9 mM. For Figs. 8 and 9, the slope of the sodium conductance was reduced in proportion to the tubular sodium concentration, to represent a reduction in carrier concentration.

Figure 6 shows the pulse form of the potential across the T system membrane near the center of a fiber for 3 msec square waves of two heights at the surface in the case with no sodium conductance. This corresponds to Eq. 8, the equation for a passive T system with an access resistance. Figure 7 shows the effect of including a sodium conductance and normal sodium concentrations. A 1 mV increase in the size of the surface depolarization triggers a regenerative depolarization of the T system at the center of the fiber. Reduction of the sodium concentration to one-half normal gives a similar result, although the potential in the T-system overshoots less due to the lowered sodium equilibrium potential and the threshold is slightly raised (Fig. 8). A further reduction

of sodium concentration to one-fourth normal raises the threshold further
and reduces the size of the regenerative response in the T system (Fig.
9).

Figure 10 shows the numerically calculated profiles of potential across
the T system wall through the diameter of the fiber at various times
during and after 3 msec depolarizing pulses at the surface of the fiber.
Figure 10A corresponds to the 43 mV subthreshold pulse in Fig. 7.
Figure 10B corresponds to the 44 mV pulse which triggers a regenerative
response in the T system. For the subthreshold pulse, the potential across
the T system near the center of the fiber sags below the surface potential
during the pulse (up to 3 msec) and follows the surface potential
downward after the end of the pulse. For the larger pulse (Fig. 10B),
however, a dramatically different potential profile is seen. During the
last millisecond of the pulse, a wave of depolarization exceeding the
surface depolarization spreads inward toward the center of the fiber.

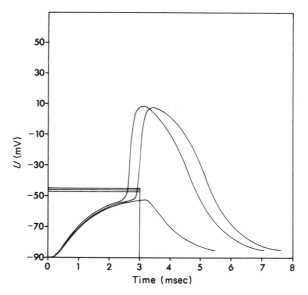

Fig. 8. Same as Fig. 7, except that the external sodium concentration and the
slope of the sodium conductance increase were reduced by one-half. This gives
a sodium equilibrium potential of +17.5 mV. The surface pulses have amplitudes
of 43, 44, and 45 mV. The 43 mV pulse now is not sufficient to cause a regenerative
response in the T system, even though the potential across the T system membranes
near the fiber surface, but inside, R_s gets slightly above −50 mV (not shown
in figure). The two larger pulses do, however, trigger a regenerative response,
which rises and falls sooner in the case of the larger pulse. The highest potential
reached is less than in the case of Fig. 7, largely because of the reduced sodium
equilibrium potential.

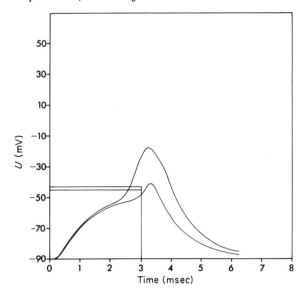

Fig. 9. Same as Figs. 7 and 8, except that the external sodium concentration and sodium conductance slope were reduced by one-half again (external sodium = 30 mM, conductance slope = 0.125 mho cm^{-2} V^{-1}, sodium equilibrium potential = 0). The surface pulses are 45 and 47 mV. The active response in the T system is clearly seen but is small.

At 4 msec, 1 msec after the end of the surface pulse, the center of the fiber is still more depolarized than the surface was during the pulse. As seen in the middle curve of Fig. 7, which shows the time-dependence of the potential across the T system at the center of the fiber for the same case as Fig. 10B, this inversion of potential profile lasts for several milliseconds after the pulse ends.

4. Comparison with Experimental Results

These results fit rather well the experimental results of Costantin (1970), who found contraction spreading across the entire cross section of the muscle fiber for pulses 0–2 mV greater than the threshold for surface contraction, in fibers bathed in Ringer's fluid with one-half the normal sodium concentration (in the absence of tetrodotoxin). A greater reduction in external sodium concentration led to a reduction in effectiveness of spread along the T system, as noted here in Figs. 9 and 10. Furthermore, Costantin often noted stronger contractions near the center of the fiber than near the surface, implying the sort of reversal of potential gradient shown in Fig. 10B.

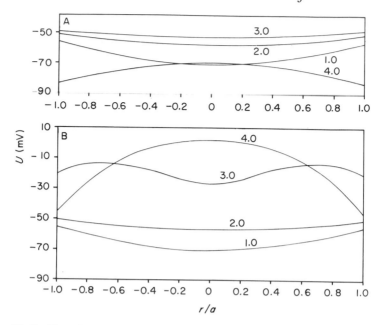

Fig. 10. Profiles of T system potential across the diameter of the fiber at various times after the application of the pulses shown in Fig. 7. The numbers against the curves are times, in milliseconds, after the beginning of a surface pulse of 3 msec duration. Figure 10A is for the 43 mV pulse, which does not trigger a regenerative T system response. Figure 10B is for the 44 mV pulse, which triggers a response that spreads inward at around 3 msec and results in an inverted potential profile at 4 msec.

That this simple sodium activation and inactivation system in the T system can reproduce the two most important observations in the Costantin (1970) experiments supports the idea that the T system does act with a propagated signal based on a regenerative sodium activation and inactivation. Obviously these calculations should be extended to a model with conductance characteristics of various sorts, perhaps including some more like those of the surface membrane, before drawing any conclusions on the real nature of the sodium system in the T system.

III. Tubular Capacity—Lumped or Distributed?

It is widely agreed that there is good experimental evidence that the walls of the transverse tubules are the site of the greater part of the membrane capacity of muscle fibers. Gage and Eisenberg (1969) have shown that fibers treated with a Ringer's fluid made hypertonic with glycerol and then returned to Ringer's fluid lose the ability to admit into the T tubules large molecules such as ferritin and peroxidase. At the same time, the membrane capacity is much reduced. The existence of two separable capacities, one at the fiber surface and one in the transverse tubules, raises the question of how the tubular capacity should be included in the equivalent circuit for the muscle fiber. The models discussed so far in this chapter assume that the tubular capacity is uniformly distributed throughout the cross section of the fiber and that the resistance of the tubular lumen is sufficient to make λ_T comparable in magnitude to the fiber diameter.

In their important analysis of the AC impedance of frog twitch skeletal muscle fibers measured with two intracellular electrodes, Falk and Fatt (1964) showed that the measured impedence locus in the complex plane deviates over a wide range of frequencies from that expected for a cable made up simply of a resistance (R_m) and a capacitance (C_m) in parallel (Fig. 11A). A reasonable agreement between the experimental points and the calculated impedance locus could be achieved when the elements of the linear equivalent circuit representing the muscle fiber contained in addition to the parallel components ($R_m = 3100 \ \Omega \ cm^2$, $C_m = 2.6 \ \mu F \ cm^{-2}$) a second pathway made up of a resistance $R_e = 330 \ \Omega \ cm^2$ in series with a capacitance $C_e = 4.1 \ \mu F \ cm^{-2}$ (Fig. 11B).

In the transitional range of frequencies, the measured impedance showed some deviation from the calculated locus (see Fig. 8, Falk and Fatt, 1964). The deviation was of the kind that might be expected if the capacity C_e were distributed throughout the cross section of the fiber and R_e represented the luminal resistance of the tubules. Falk and Fatt considered this more complex model, but concluded that the difference between their experimental results and the locus based on the model with undistributed C_e was not great enough to justify the conclusion that C_e was distributed. They therefore suggested that C_e is the tubular

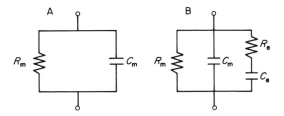

Fig. 11. Two equivalent circuits for skeletal muscle fibers considered by Falk and Fatt (1964) and others. A shows a simple parallel capacity and resistance, as used to describe the nerve membrane. In B, an extra current path consisting of a capacitance in series with a resistance is added to represent the T system.

capacity and that R_e is the resistance of a structure between the tubular wall and the sarcoplasm, possibly the membrane of the terminal cistern. It is implicit in this suggestion that the luminal resistance of the tubules is sufficiently small to make λ_T effectively infinite. It is difficult to assess how firmly based this conclusion can be considered. Schneider (1970), who has repeated some of the measurements, considers that his results can be explained in terms of a distributed capacity model similar to the model proposed by Adrian *et al.* (1969a).

Recently, Nakajima and Hodgkin (1970) have measured the membrane capacity of isolated frog muscle fibers in a number of ways. Since the capacity per unit length of fiber of the tubular walls will increase as the square of the diameter, the sum of the tubular and surface capacities, calculated for a square centimeter of fiber surface, will increase linearly with the radius. In the model proposed by Falk and Fatt, therefore, the capacity measured with rectangular current pulses (= low frequency capacity) should increase linearly with the fiber radius. In a distributed capacity model, the capacity measured in this way will also increase with the radius, but the increase will fall below the linear relation as the fiber radius a increases, because when a becomes comparable to λ_T, the steady state potential across the tubular capacity at the center of the fiber is less than at the periphery. Hodgkin and Nakajima also obtained estimates of the high frequency capacity from the exponential rise of the foot of the action potential and the propagation velocity. As is to be expected, this is substantially less than the low frequency capacity derived from rectangular current pulse measurements, but it is still much larger than the capacity that Nakajima and Hodgkin measured on fibers detubulated by the method of Gage and Eisenberg (1969).

If we assume that the wall of the T system and the surface of the fiber have the same capacity per unit area (C_w) one can make a number of assumptions that will lead to different relationships between the fiber diameter and the measured high and low frequency capacities. The simplest case is the equivalent circuit of Falk and Fatt (see Fig. 11B). In this case, the capacity is lumped. In terms of the parameters of the Adrian *et al.* (1969a) model, R_e can be made the equivalent of R_s. If $G_w = 0$ ($\lambda_T = \infty$), the measured low frequency capacity (C_m) will be

$$C_m = C_w[1 + (\rho a/2\zeta)] \tag{15}$$

The capacity (C_f) measured from the foot of an action potential that rises exponentially with a time constant τ_f is

$$C_f = C_w\left[1 + \frac{\tau_f}{R_s C_w + (2\tau_f \zeta/\rho a)}\right] \tag{16}$$

If we assume that $G_w \neq 0$ but that $\lambda_t = \infty$ (i.e., that we can treat the tubular capacity as lumped), then

$$C_m = C_w \left(1 + \frac{\rho a}{G_w R_s \rho a + 2\zeta} \right) \tag{17}$$

and

$$C_f = C_w \left(1 + \frac{\tau_f}{R_s C_w + 2\tau_f \zeta / \rho a + \tau_f R_s G_w} \right) \tag{18}$$

These relations are plotted in Fig. 12B for $\rho = 0.003$, $\zeta = 10^{-6}$ cm, $C_w = 1$ μF cm^{-2}, $G_w = 0.5 \times 10^{-4}$ mho cm^{-2}, $R_s = 100$ Ω cm^2, $\tau_f = 0.13$ msec. Since $\tau_f R_s G_w \ll R_s C_w$, the curves for Eqs. (16) and (18) are practically identical.

If λ_T is finite ($\nu > 0$), the capacity (C_m) measured by rectangular current pulses and assuming no access resistance ($R_s = 0$) is:

$$C_m = C_w \left[1 + \frac{2\rho a}{\zeta} \sum_{n=1}^{\infty} \frac{\beta_n^2}{(a^2 \nu^2 + \beta_n^2)^2} \right] \quad {}^*(19)$$

where $\beta_1, \beta_2, \ldots, \beta_n$ are the positive roots of $J_0(\beta) = 0$. This equation is the equivalent of equation 22 in Adrian *et al.* (1969a). The capacity measured by the foot of the action potential for this case will be (Nakajima and Hodgkin, 1970):

$$C_f = C_w \left[1 + \rho \left(\frac{\tau_f G_L}{2\zeta C_w} \right)^{1/2} \frac{I_1(\mu a)}{I_0(\mu a)} \right] \tag{20}$$

where $\mu = (\bar{C}_w / \tau_f \bar{G}_L)^{1/2} = (2C_w / \tau_f G_L \zeta)^{1/2}$. If $R_s > 0$ and λ_T is finite, then:

$$C_m = C_w \left(1 + \frac{2\rho a}{\zeta} \sum_{n=1}^{\infty} \frac{a^2 h^2 \beta_n^2}{(a^2 h^2 + \beta_n^2)(a^2 \nu^2 + \beta_n^2)^2} \right) \quad {}^*(21)$$

* Hodgkin and Nakajima (1972) have given a closed form for the infinite series of Eq. (19).

$$\sum_{n=1}^{\infty} \frac{4\beta_n^2}{(a^2 \nu^2 + \beta_n^2)^2} = 1 - \frac{I_1^2(\nu a)}{I_0^2(\nu a)}$$

Similarly the infinite series in Eq. (21) can be expressed in the closed form

$$\sum_{n=1}^{\infty} \frac{4a^2 h^2 \beta_n^2}{(a^2 h^2 + \beta_n^2)(a^2 \nu^2 + \beta_n^2)^2} = \frac{I_0^2(\nu a) - I_1^2(\nu a)}{\{I_0(\nu a) + (\nu/h) I_1(\nu a)\}^2}$$

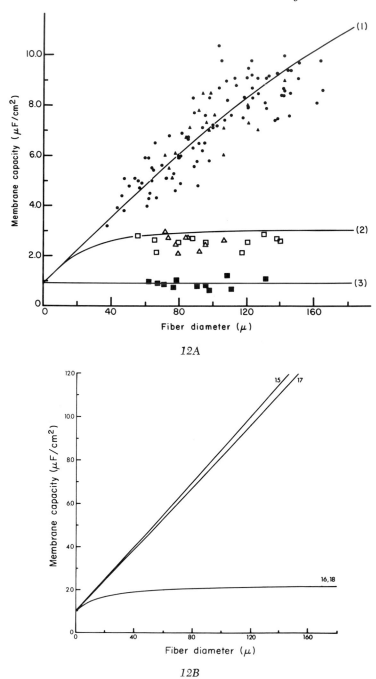

12A

12B

Fig. 12. See facing page for legend.

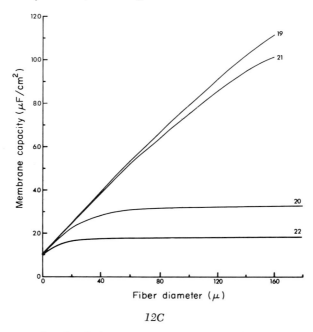

12C

Fig. 12. Measured and calculated capacities of muscle fibers plotted against fiber diameter. Capacity is expressed as microfarads per square centimeter referred to the area of the fiber surface, not including the T system. A is from Nakajima and Hodgkin (1970), who measured capacity of frog muscle fibers both using retangular pulses (low frequency capacity) and by estimating capacity from the time constant of the exponentially rising foot of a propagated action potential and the conduction velocity (high frequency capacity). The points around curve 1 are low frequency capacity measured with normal fibers. The measured high frequency capacity values are shown for normal fibers as the open symbols around curve 2. The low frequency capacities for detubulated fibers are the closed symbols around curve 3. Curve 1 is plotted from Eq. (19), curve 2 from Eq. 20, and curve 3 is C_w. B and C are plots of Eqs. (15)–(22), as indicated by the numbers against the curves.

where β_1, β_2, . . . , β_n are the positive roots of $\beta J_1(\beta) - ah J_0(\beta) = 0$ and

$$C_f = C_w \left(1 + \rho \left(\frac{\tau_f G_L}{2 \zeta C_w} \right)^{1/2} \frac{I_1(\mu a)}{I_0(\mu a) + (\mu/h) I_1(\mu a)} \right) \qquad (22)$$

where $\mu = (\bar{C}_w / \tau_f \bar{G}_L)^{1/2}$ and τ_f is the time constant of the exponentially rising part of the action potential. Figure 12C plots Eqs. (19)–(22) for $C_w = 1$ μF cm^{-2}, $G_w = 0.5 \times 10^{-4}$ mho cm^{-2}, $R_s = 100$ Ω cm^2, $\tau_f = 0.13$ msec, $G_L = 10^{-2}$ mho cm^{-1}, $\rho = 0.003$, $\zeta = 10^{-6}$ cm, and $\sigma = \frac{1}{2}$.

Figure 12A, which is reproduced from Nakajima and Hodgkin (1970), shows the measurements of capacity and plots Eqs. (19) and (20) with $C_w = 0.9 \ \mu\text{F}/\text{cm}^2$, $G_L = 10^{-2}$ mho cm^{-1}, $\rho = 0.03$, and $\zeta = 10^{-6}$ cm. Curve 1 and the associated points are Eq. (19) and capacity measurements with rectangular current pulses. Curve 2 is Eq. (20), and the open symbols are capacity measured from the foot of the action potential. Line 3 is C_w, and the points on it are capacities of detubulated fibers measured from the foot of the action potential. Figure 12C compares the curves for the two models with distributed capacity with and without access resistance. The effect of an access resistance is to reduce the apparent capacity, and the reduction is greater for the high frequency capacity measured from the foot of the action potential than for the low frequency capacity.

The scatter of the capacity measurements does not allow an unequivocal choice of the model that gives the best fit, but there is certainly a suggestion in these results that the low frequency capacity is not linearly related to the diameter. An advantage (already mentioned) of the model with distributed capacity and an access resistance (R_s) is that it can reconcile the measurements of membrane resistance with the measurements of λ_T from visual observation of radial spread of contraction (Adrian et al., 1969b). Furthermore, both Falk and Fatt (1964) and Adrian et al. (1969a) satisfactorily used equivalent circuits with lumped capacities, though with series resistances $R_e = 330 \ \Omega$ cm^2 and $R_s = 150 \ \Omega$ cm^2, respectively, which were substantially greater than the estimated effective radial resistance of the T system [67 Ω cm^2 (Adrian et al., 1969a)]. Likewise, Schneider (1970) estimates the specific resistance of the tubular lumen to be 300 Ω cm, which is substantially greater than would be expected if the content of the tubules was similar to the extracellular fluid. The discrepancy is partially resolved if one supposes that their R_e and R_s are made up of an access resistance of 100 Ω cm^2 and an effective radial T system resistance of 67 Ω cm^2. If the model with access resistance represents the true situation in striated muscle, the tubular capacity is neither entirely lumped nor entirely distributed, and it is not surprising that the impedance measurements (Falk and Fatt, 1964; Schneider, 1970) could be interpreted in terms of models with either lumped or distributed capacity.

IV. Excitation–Contraction Coupling

If the model for the electrical behavior of the T system presented in the previous section is on the right lines, what does it allow one to say about the contraction–excitation process? We shall focus on the first stages of this process, which lead to the release of Ca^{2+} ions. At rest, these ions seem to be stored in the sarcoplasmic reticulum, most probably in the terminal cisternae, and at least partly in a bound form. If the sarcoplasmic Ca^{2+} concentration rises during a twitch to between 1 and 3×10^{-5} M and troponin takes up 0.07 μmole gm^{-1} muscle fiber (Ebashi *et al.*, 1969), we shall not be far wrong if we estimate that a single action potential leads to the release of 10^{-7} mole of Ca^{2+} from the sarcoplasmic reticulum in 1 cm^3 of muscle. This movement of Ca^{2+} is substantially larger than the entry of Na^+ into the fiber in the course of an action potential (6×10^{-9} moles cm^{-3} impulse^{-1} for a fiber with a diameter of 100 μ).

One can consider the electrical consequences of this movement of Ca^{2+} by relating its magnitude either to the capacity of the electrical double layer at the surface of the membrane separating the sarcoplasmic reticulum from the sarcoplasm (SR membrane) or to the capacity of the SR membrane itself. These two capacities are relevant physical quantities for two mechanisms that have been suggested for the release of Ca^{2+} by the SR; release of Ca^{2+} bound to the surface of the SR membrane, or alternatively, transfer of Ca^{2+}—some of which may be bound within the reticulum—across the SR membrane. These are both possible mechanisms. The evidence, which does not exclude binding at the surface, but certainly suggests that Ca^{2+} is bound within the reticulum, has been reviewed by Ebashi and Endo (1968) and by Ebashi *et al.* (1969).

Peachey (1965) estimates that the surface area of the SR membrane is $27 \times 10^3 \times a$ cm^2 for each square centimeter of fiber surface, where a is the fiber radius in centimeters. For a fiber with a diameter of 100 μ, the surface area of the SR is 135 cm^2 for each square centimeter of fiber surface, or 54×10^3 cm^2 cm^{-3} of muscle fiber. If the activating Ca^{2+} were uniformly bound to the surface of the SR membrane, unbinding 10^{-7} moles cm^{-3} of Ca^{2+} corresponds to an alteration in the surface charge density of 0.36 μC cm^{-2}, which would produce a change in the surface potential of the SR membrane of not more than 15 mV (Chandler *et al.*, 1965).

To calculate the consequences of a transfer of Ca^{2+} across the SR membrane, we need to know the membrane capacity per unit area. There is no morphological reason to suppose that the capacity of the SR membrane is very different from other cellular membranes, and we shall therefore assume that it is 1 μF cm^{-2}. Under this assumption, the capacity of the SR membrane related to unit volume of muscle fiber is 54×10^3 μF cm^{-3}. Transfer of 0.0193 coulombs of charge (10^{-7} mole

Ca^{2+}) across the capacity would leave the potential of the inside of the reticulum more negative by 360 mV. It seems most unlikely that the potential across the SR membrane does in fact change by 360 mV. This would imply a resting potential in the SR of $+324$ mV, since the calcium equilibrium potential during contraction may be estimated to be -36 mV. This value for E_{Ca} assumes that the sarcoplasmic Ca^{2+} concentration rises to 2×10^{-5} M during a twitch and that the ionized calcium concentration in the reticulum is 0.35×10^{-3} M (Ogawa, quoted in Ebashi et al., 1969). It seems much more likely that there is an SR action potential involving an increase in Ca^{2+} permeability and a change of internal SR potential from a value somewhat positive to the sarcoplasm toward the Ca^{2+} equilibrium potential. A large Ca^{2+} movement would then be made possible by means of a compensating movement of charge carried into the SR by some other ion or ions. In much the same way, the movement of Na$^+$ ions during an action potential in nerve or muscle is larger than the minimum necessary movement calculated from the potential change and the membrane capacity; for an interval of time during the spike, the inward Na$^+$ and outward K$^+$ currents take place simultaneously.

If we suppose that the Ca^{2+} release takes place over a period of 25 msec (Jöbsis and O'Connor, 1966), a movement of ions in the opposite direction equivalent to a current of 0.77 A cm^{-3} must take place during or very shortly after the period of Ca^{2+} release. In a 100 μ fiber this is a current of 1.93 mA cm^{-2} of fiber surface. For obvious reasons, the Ca^{2+} movement must be across the SR membrane, but there is an alternative pathway for the compensatory movement of charge, which might enter the SR via the T tubules and the apposed membranes of the triad. The fact that there is no recordable potential change across the surface membrane of the fiber during the period when Ca^{2+} is leaving the SR, suggests that the non-Ca^{2+} currents do not enter the SR from the T tubules, but flow across the SR membrane. Furthermore, one can say that during the release of Ca^{2+}, the resistance of the SR membrane to ions other than Ca^{2+} is small (probably $< \frac{1}{100}$) compared to the resistance of the fiber surface and SR–T system junctions in series. If we postulate a transient driving force of 100 mV for the non-Ca^{2+} current, the resistance of the SR membrane during the release of calcium would need to be as low as 50 Ω cm^2 related to the fiber surface. This resistance calculated per unit area of SR membrane is 6750 Ω cm^2. The resistance to the noncalcium current might fall to this value as part of the Ca^{2+}-releasing action potential; alternatively it might have this value at rest as well as during calcium release, in which case Ca^{2+} release could depend on a permeability change to Ca^{2+} alone. Perhaps in favor of the latter hypothesis is the fact that a resistance of 6750 Ω cm^2 is not unlike other excitable membranes at rest.

If local currents via the T system are to be the means of stimulating the Ca^{2+} action potential by changing the potential across the SR membrane directly, the SR resistance and capacity deduced above mean that remarkably small changes of potential would have to initiate a change in calcium permeability. Nor is it much help to suggest a larger resting resistance for the SR membrane, because this would make the brief action potential at the fiber surface an ineffective stimulus for the SR membrane.

Previous authors have called attention to the similarities (Fahrenbach, 1965; Peachey, 1965) and differences (Franzini-Armstrong, 1970) between the membrane contact morphology in the triad and the morphology of the low-resistance intercellular bridge between cells of the salivary gland (Wiener *et al.*, 1964). The similarities have led to the suggestion that there is a low resistance between the tubular lumen and the interior of the SR, and that therefore the component of the fiber resistance attributed to the tubular membrane (Gage and Eisenberg, 1969) should be located in the SR membrane. If this were the case, the resting resistance of 1 cm^2 of SR membrane would be very high indeed. Moreover, measurements of low frequency capacity should reveal the capacity of the SR membrane. Experimentally measured capacities of muscle (e.g., 6–8 μF cm^{-2} fiber surface) seem to be much more consistent with measured specific capacities of cell membranes (e.g., 1 μF cm^{-2}) and with the area of the T system (e.g., 7 cm^2 per square centimeter of fiber surface) than with the area of the SR (e.g., 135 cm^2 per square centimeter of fiber surface). To suggest that the greater part of the low frequency capacity is attributable to the SR membrane would involve assuming a very low capacity per unit area of that membrane.

The conclusion seems to be that whatever the resting resistance of the SR membrane, it will be difficult for local currents flowing across the tubular membrane and SR membrane in series to trigger calcium release by means of the potential produced across the SR membrane. The difficulty applies equally to the release of membrane-bound Ca^{2+} and to Ca^{2+} release depending on a permeability change. The argument above depends crucially on the value assumed for the capacity of the SR membrane, and the difficulty could be resolved by assuming that the effective capacity of the SR membrane was substantially less than 135 μF cm^{-2} of fiber surface. In this connection, it should be mentioned that Falk and Fatt (1964) thought that there might be a capacity of 1.1 μF cm^{-2} across R_e (C_{re}; related to fiber surface area). In terms of their suggested morphological equivalents, C_{re} would represent the capacity of the SR membrane, but the very normal morphological appearance of this membrane makes such a small value seem unlikely. An alternative way out of the difficulty is to suppose that the local currents of the action potential do not enter the SR vesicles, but only flow across the membrane of the T tubules. Such an assumption removes the difficulties involved in supposing that the SR membrane and the tubular membrane are in series; it makes it necessary to suppose that current can flow to and from the sarcoplasm between the tubule and the terminal cistern where they appear to be in contact; a mechanism is required that allows a potential change across the wall of the tubule to affect the permeability or binding properties of the SR membrane. Not much can be said about such a mechanism at the present stage. Falk and Fatt (1964) suggested that some substance (possibly Ca^{2+}) is absorbed or bound to the sarcoplasmic surface of the T tubules and that this is released by a change in potential across the tubular wall. Such an activator might diffuse into the myofibrils to activate contraction directly, or it might move to the SR vesicles in the triad and activate release of Ca^{2+} from there. Both mechanisms are possible.

Direct activation would require one twitch to release a calcium ion from each 500 $Å^2$ of tubular surface. A two-stage process requiring the movement to the SR of an activator for the Ca^{2+} releasing process is attractive for a number of reasons. The quantity of activator released from the tubular membrane would not need to be large. The main Ca^{2+} release would be from a cellular component known to store Ca^{2+}, the SR. Adrian *et al.* (1969a) examined the strength duration relation between step depolarization of the surface membrane in the presence of TTX (tetrodotoxin) and just visible contraction, finding that the results could be fitted by a model involving movement of a charged particle in a constant field from a store to a site of action. Though the model they proposed is by no means an unique explanation of the results, it serves as a framework for quantitative description of the early steps of the activation process. They also showed that a just subrheobasic depolarization could not achieve an activator concentration of more than about 40% of the threshold. A possible explanation of this apparently regenerative step was given in terms of an activator of contraction whose release from and uptake into a store were both potential-dependent, and whose uptake rate was inversely related to its concentration over a certain subtreshold range of concentration. Equally likely, the regenerative increase of activator concentration could be due to a release rate which was an appropriate nonlinear function of the activator concentration. Alternatively, the finding of a regenerative step in the early stages of contraction activation might be explained by supposing that a change in transtubular potential uncovers or provides carriers in the SR membrane which promote a regenerative Ca^{2+} current. In one such model one could think of the transtubular potential as removing an inactivating particle from a calcium transporting mechanism in the SR membrane with an activation variable that depended on the potential across the SR membrane. Intermittent continuity of the T system and SR membranes would be required. The calcium action potential of the SR membrane of the triad would spread to other parts of the SR membrane by the usual local circuit current flow. There is no clear evidence that allows a choice between these and several other possible mechanisms.

ACKNOWLEDGMENT

Dr. Peachey's contribution is supported by a grant from the National Science Foundation (GB-6975X). Computations were done in the Mathematical Laboratories, Cambridge University, and the University of Pennsylvania Medical School Computing Facility. The latter is supported by the National Institutes of Health (No. RR-15).

REFERENCES

Adrian, R. H., Chandler, K. C., and Hodgkin, A. L. (1969a). *J. Physiol. (London)* **204**, 207.
Adrian, R. H., Costantin, L. L., and Peachey, L. D. (1969b). *J. Physiol. (London)* **204**, 231.
Carslaw, H. S., and Jaeger, J. C. (1959). "Conduction of Heat in Solids," 2nd ed. Oxford Univ. Press, London and New York.
Chandler, W. K., Hodgkin, A. L., and Meves, H. (1965). *J. Physiol. (London)* **180**, 821.
Costantin, L. L. (1970). *J. Gen. Physiol.* **55**, 703.
Crank, J. (1956). "The Mathematics of Diffusion." Oxford Univ. Press (Clarendon), London and New York.
Danckwerts, P. V. (1951). *Trans. Faraday Soc.* **47**, 1014.
Ebashi, S., and Endo, M. (1968). *Progr. Biophys. Mol. Biol.* **18**, 123.
Ebashi, S., Endo, M., and Ohtsuki, I. (1969). *Quart. Rev. Biophys.* **2**, 351.
Eisenberg, B., and Eisenberg, R. S. (1968). *J. Cell Biol.* **39**, 451.
Fahrenbach, W. (1965). *Science* **147**, 1308.
Falk, G. (1968). *Biophys. J.* **8**, 608.
Falk, G., and Fatt, P. (1964). *Proc. Roy. Soc., Ser. B* **160**, 69.
Flitney, E. (1971). *J. Physiol. (London)* **217**, 243.
Franzini-Armstrong, C. (1970). *J. Cell Biol.* **47**, 488.
Gage, P. W., and Eisenberg, R. S. (1969). *J. Gen. Physiol.* **53**, 298.
Hodgkin, A. L., and Nakajima, S. (1972). *J. Physiol. (London)*, in press.
Huxley, A. F., and Peachey, L. D. (1964). *J. Cell Biol.* **23**, 107A.
Huxley, A. F., and Taylor, R. E. (1958). *J. Physiol. (London)* **144**, 426.
Jöbsis, F. F., and O'Connor, M. J. (1936). *Biochem. Biophys. Res. Commun.* **25**, 246.

McCallister, L. P., and Hadek, R. (1970). *J. Ultrastruct. Res.* **33**, 360.
Nakajima, S., and Hodgkin, A. L. (1970). *Nature (London)* **227**, 1053.
Page, S. (1965). *J. Cell Biol.* **26**, 477.
Peachey, L. D. (1965). *J. Cell Biol.* **25**, 209.
Porter, K. R., and Palade, G. E. (1957). *J. Biophys. Biochem. Cytol.* 3, 269.
Schneider, M. F. (1970). *J. Gen. Physiol.* **56**, 640.
Wiener, J., Spiro, D., and Loewenstein, W. R. (1964). *J. Cell Biol.* **22**, 587.

_____ ② _____

THE NEUROMUSCULAR JUNCTION—THE ROLE OF ACETYLCHOLINE IN EXCITABLE MEMBRANES

DAVID NACHMANSOHN

I. Excitable Membranes

Bioelectricity was first discovered in the latter part of the eighteenth century, when it was demonstrated that the powerful shock delivered by certain fish is an electrical discharge. The question was immediately raised as to the mechanism by which living cells produce electricity. The fundamental importance of the problem for biology in general became apparent when it was firmly established during the nineteenth century that nerve impulses are propagated by electrical currents. Thus, the understanding of the nervous system, one of the most vital functions of the organism, became intrinsically linked to the knowledge of the mechanism by which living cells generate electricity.

At the turn of the century, two notions were widely accepted. First, in a fluid system such as the living cell, ions must be the carriers of electrical currents. Since it was known that Na^+ ions are highly concentrated in the outer environment of cells, whereas in the cell interior a high K^+ concentration prevails, Overton (1902) proposed that during activity, Na^+ ions move into the cell interior and K^+ ions flow to the outside. Overton's assumption was borne out by many subsequent investigations, particularly when, after World War II, the availability of radioactive ions facilitated measurements of ion movements. The second notion was concerned with the control of these ion movements. It was postulated that excitable membranes, i.e., those surrounding nerve and muscle fibers, must be able to change their permeability to ions during electrical activity. Since cell membranes were not yet demonstrated, but only postulated on the basis of a variety of observations, the mechanism of such permeability changes was obscure and not even subject to speculations.

A. General Properties of Cell Membranes

During the last decade, however, the properties and function of membranes have been one of the most actively explored fields in biological sciences. Much information has accumulated owing to electron microscopy in combination with biochemical and biophysical methods. The notion of "unit membrane," proposed by Robertson (1960a,b) and based essentially on the Danielli–Davson model, assumed a structure 80 Å thick and formed by a bimolecular leaflet of phospholipids to which proteins are attached on the inside and outside by ionic forces. While this notion appeared at first attractive to many investigators, it soon proved to be inadequate to account for a great variety of biochemical, biophysical, and electron microscope observations (e.g., Sjoestrand, 1963; Elbers, 1964; Korn, 1966; Green and Perdue, 1966; Green and McLennan, 1969; Sjöstrand and Barajas, 1968). Membranes appear to be a mosaic of functional units (Palade, 1963). The units are formed by lipoprotein complexes. But in spite of the many efforts of elucidating the molecular structure, in spite of the application of an ever-increasing range of highly refined chemical and physical methods, the topography and the precise molecular arrangements of phospholipids and proteins are at present still under lively discussion. Quite a few interesting models have been proposed in the last few years in an effort to integrate the available information (see, e.g., Lenard and Singer, 1966; Vanderkooi and Green, 1971; Benson, 1968; Blasie and Worthington, 1969). The models are based on data obtained with a variety of methods and applied to different types of membrane, such as, e.g., mitochondria, chloroplasts, outer segments of retinal rods, myelin, etc. It appears likely that there are marked differences of molecular structure between these membranes. Although the phospholipids and proteins are the main components, the wide variations in their ratio support the assumption of important dissimilarities.

However, the most pertinent feature of recent developments is the conceptual change with respect to function. As is now widely recognized, membranes are highly dynamic and well-organized structures, the site of a great number of different proteins and enzymes, and of intensive chemical reactions. This is illustrated by the well-explored mitochondrial as well as by various other membranes (Green and McLennan, 1969; Racker, 1965, 1967, 1969; Korn, 1969; Loewenstein, 1966; Membrane Proteins, 1969). The large number and heterogeneity of the proteins in membranes is today widely recognized. George Palade (1970) estimates the number of different proteins in the membranes of the endoplasmic reticulum of liver cells to be about 50 (personal communication). Whereas the importance of phospholipids is not contested, e.g., in their role as barriers, the emphasis in the present notions of functional activities of membranes, whatever the structural arrangements may be, has shifted to the central role of proteins and enzymes. Proteins account more readily than phospholipids for the great diversity of the function of different cell membranes, their high degree of specificity, and their remarkable efficiency.

The notion of the crucial role of proteins in membranes induced Sjöstrand and Barajas (1968) to initiate the development of new procedures aimed at preserving the conformation of proteins in their native state when specimens are prepared for examination by electron microscopy. These authors found that the image of the unit membrane is created by extensive denaturation. When proteins are stabilized by inter-molecular cross-linking with glutaraldehyde and dehydrated by ethylene glycol instead of by acetone or alcohol used in the standard procedures, denaturation seems markedly reduced. The electron micrographs of the mitochondrial membranes obtained by the authors are quite different from those resulting from standard procedures; for instance, the thickness of the membrane is 150–300 Å, and there is evidence for the presence of globular structures. The new approach appears to be a beginning in the right direction. It may stimulate efforts resulting in much more realistic pictures of cellular and subcellular structures than those obtained with currently used methods.

On the basis of their electron microscopic observations, Sjöstrand and Barajas (1970) have recently designed a model which tries to describe the molecular structure of mitochondrial membranes as revealed by electron microscopy in combination with the information available from the large number of biochemical studies and especially those of the protein and enzyme complexes. Although quite obviously the authors are obliged to make assumptions of a hypothetical character, the models seem to the writer to present the most successful attempt of a picture which is close to the extraordinary complexity of membranes suggested by all available data. They offer a more realistic and integrated starting point for probing structure and function of membranes than the more schematized models previously proposed. They are especially in line with present conceptual ideas as to the dynamic character of membranes considered to be highly intricate biochemical machines.

With their new methods, the authors had found the membranes to consist of a great variety of particles ranging from below 40 to 100 A in diameter. In view of the average thickness of about 150 A, the authors suggest that the membrane particles are arranged in a staggered 3-dimensional fashion and not spread out in the usually assumed strictly 2-dimensional array. The membrane structure is conceived as being maintained primarily through hydrophobic interactions between proteins, lipids and proteins, and lipids. A predominantly hydrophobic milieu would be found in the interior of the membrane, and both nonpolar amino acid side chains exposed at certain regions of the surfaces of the protein molecules and the hydrocarbon tails of phospholipids would contribute to the nonpolar character of the membrane interior. Figure 1 shows a photograph and Fig. 2 the drawing of the proposed model of the molecular structure of a mitochondrial membrane.

The model is based on the information available on mitochondrial membranes. Since the functions of various types of membranes differ, it appears likely that their chemical composition and structural arrange-

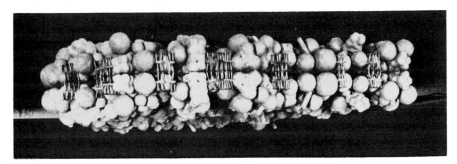

Fig. 1. Photograph of the molecular structure of the proposed model of a mitochondrial membrane element. Most of the molecules and molecular subunits are represented by spheres. Lipid bilayer sections are indicated by nails. (From Sjöstrand and Barajas, 1970.)

ments will show many differences from those of mitochondrial membranes. However, it appears likely that there will also exist many basic similarities in the overall principles which dominate the structural elements and the functional aspects of cell membranes.

B. *Special Features of Excitable Membranes*

The ability of excitable membranes of generating and propagating electrical currents, generally attributed to rapid and reversible changes in permeability to ions, raises the question as to the specific molecular events controlling these ion movements across membranes. This problem must be considered in the light of the new knowledge about cell membranes in general. In the concept of Hodgkin a simple diffusion process following the concentration gradient is postulated (Hodgkin, 1951, 1964). The assumption of chemical reactions involved in this process is explicitly

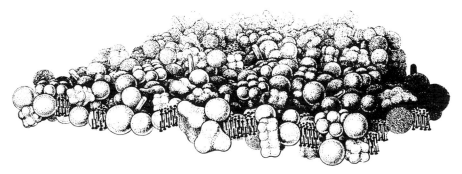

Fig. 2. Drawing of the model showing the lack of any simple repeat pattern. (From Sjöstrand and Barajas, 1970.)

rejected (see, e.g., Baker *et al.*, 1962; Keynes and Aubert, 1964); they are only accepted for the recovery process in which Na⁺ ions are extruded to the exterior and K⁺ ions enter the interior *against* their concentration gradients. The notion of a simple diffusion process is difficult to reconcile with present notions of membranes in general and with a variety of specific experimental facts. Drastic modifications of ion composition both in the interior of the axon and its outer environment have for a considerable length of time no effect on the electrical parameters contrary to the predictions of the theory (Tasaki, 1968).

In a review article on electrodiffusion models of the excitable membrane of squid axons Cole (1965), one of the pioneers in the experimental and theoretical analysis of the physical properties and the bioelectric phenomena of the excitable membrane, discusses the model for passive ion flow, especially in respect to the squid giant axons, in the light of theories on ion movements, as originally described by Nernst (1888, 1889) and further developed by Planck (1890a,b). In view of the penetrating inquiry performed in this article the final conclusions of the author may be quoted in full.

A general explanation of the voltage-clamp characteristics in terms of the Planck several-ion model has not been found, and this model appears to be unable to account for all the relaxations required by the membrane. The sodium ion phenomena in the membrane seem to be almost entirely contrary to the properties of a single-ion model. The single-ion theory has long been an attractive explanation for the potassium behavior in the membrane but further consideration suggests that here also the model is so inadequate as to make it an improbable explanation.

The contrasts between the various calculations and the array of experimental facts give a basis to conclude that the simple process of electrodiffusion is not a principal factor in the behavior of the squid axon membrane.

Another serious difficulty for the assumption of a simple diffusion process arises from the more recent measurements of heat production and absorption associated with electrical activity, as was found by Hill and his associates (Abbott *et al.*, 1958). The authors find it hard to believe that the drastic changes of permeability in a material like the excitable membrane could occur without the intervention of work or chemical reaction. In a subsequent lecture, Hill (1960), discussing various factors which may contribute to the heat production observed, again stressed as the only reasonable assumption that the early heat produced and absorbed is due to chemical reactions associated with the permeability cycle during electrical activity. In still more recent measurements the heat production coinciding with electrical activity was found to be diphasic—heat is produced during the rising phase of the action current and absorbed during its falling phase (Howarth *et al.*, 1968).

The heat production and absorption are not attributed by the authors to chemical reactions. Their interpretation is open to question and will be discussed later in the light of the chemical theory proposed by the writer more than two decades ago and to be presented in this chapter.

II. Role of Acetylcholine in Nerve Activity

A. Neurohumoral Transmission

Whereas electrical currents are generally assumed to propagate nerve impulses along nerve and muscle fibers, the well known theory of neurohumoral transmission proposes that at junctions (synapses) between nerve and nerve or nerve and muscle, chemical compounds (acetylcholine (AcCh) or others) are released from the nerve ending and, crossing the junctional gap, transmit the impulse to the second cell, acting as chemical mediators. This idea was never accepted by many prominent neurobiologists. Their main objection was the similarity of many electrical properties of the membranes at junctions and in axons. This made it difficult, in their view, to interpret the observations reported in terms of neurohumoral transmission and to assume two basically different mechanisms of nerve impulse propagation, a purely physical one in the axonal membranes and a chemical one between the junctional membranes (see, e.g., Erlanger, 1939). The theory of neurohumoral transmission proposed in the 1930s was based on observations with classic methods of physiology and pharmacology. These methods are essential in many biological studies. They are inadequate for the analysis of the molecular mechanism responsible for the permeability changes controlling ion movements, events which take place in microseconds in a membrane of about 100 Å thickness. Thus a new approach to the controversial problem of the role of AcCh appeared imperative.

B. Biochemical Approach

Attending frequently vigorous controversies between proponents and opponents of neurohumoral transmission at the meetings of the English Physiological Society in 1935 and 1936, the writer was struck by the total lack of any biochemical approach to the problem. Having been trained in biochemical laboratories in which proteins and enzymes played a central role in the ideas about cell function, it seemed to him imperative to investigate a variety of aspects of the proteins associated with the function of AcCh. Such information seemed likely to provide pertinent clues to the controversial problem of the role of AcCh. Further-more, certain basic notions developed by Meyerhof in the analysis of the physical and chemical events during muscular contraction should be applicable to other cell functions, such as nerve impulse conduction, even if many specific aspects are fundamentally different. If one accepts, for example, the notion that the reactions in a living cell are chemically and energetically coupled, which was one of the most fundamental con-cepts resulting from Meyerhof's muscle work, it appeared necessary to integrate the formation and hydrolysis of AcCh into the intermediary metabolism of the nerve cell, to study the sequence of energy transforma-tions during nerve activity, and to correlate the biochemical data ob-tained with the physical events of conduction.

An approach, based on principles of bioenergetics and protein chemis-try, was initiated more than three decades ago. Three factors were deci-sive in the progress achieved in this period—(1) The spectacular growth of enzyme and protein chemistry, and of biopolymers in general, culmi-nating in the exploration of tridimensional structures; (2) the extraordi-nary development of highly refined methods of instruments for an analy-sis of cellular function on cellular, subcellular, and molecular levels; and (3) the information about biomembranes mentioned above.

Membranes form an extremely small fraction of the cell mass. Even with all the refined methods and techniques available, studies of the chemical reactions and especially the analysis of the proteins in mem-branes of many tissues offer great difficulties. It is, therefore, fortunate that a material is available which is particularly favorable for studying the proteins and enzymes located in excitable membranes and associated with bioelectricity—the electric organs of fish. These organs are the most powerful bioelectric generators created by nature. The electroplax, the single cell of these organs, has electrical parameters similar to those in nerve and muscle fibers. It is only their arrangement in series as in a voltaic pile that is responsible for the power of the discharge. In *Electrophorus*, the electric eel, there are 5000–6000 cells arranged in series and the discharge amounts to 600 volts. However, the most important feature of electric organs is their high specialization for their main function, bioelectrogenesis. The excitable membrane of electroplax is the site of the most active metabolic reactions in contrast to the rather inert mass of the rest of the cell. When the writer became interested in AcCh esterase and investigated its occurrence, distribution and con-centration in a great variety of excitable tissues—nerve, muscle, brain ganglia, etc.—in many species, he came across, in 1937, the electric organs of *Torpedo marmorata*. Determinations of the concentrations of AcCh esterase in the electric tissue showed that 1 kg of tissue (fresh weight) hydrolyzes 3–4 kg of AcCh per hour. Similar values were obtained in 1938 with the electric tissue of *Electrophorus*. This extraordinarily high concentration of an enzyme is particularly striking in a specialized tissue which is formed by 92% water and only 3% protein. It was therefore immediately apparent that this material was uniquely suitable for a biochemist interested in the study of the enzymes and proteins associated with AcCh function. The use of this material was instrumental during the last 30 years in the isolation, identification, and characterization of the proteins associated with the action of AcCh and their relationship to the function of the excitable membrane.

In this chapter, a few aspects will be discussed as an illustration of how the use of bioenergetics and of protein chemistry for the elucidation of the role of AcCh in nerve activity resulted in new information and suggested a new concept entirely different from that obtained by the application of classic methods of physiology and pharmacology.

C. Chemical Theory of the Function of AcCh in Excitable Membranes

A vast amount of biochemical data has accumulated in the last 30 years, and the evidence is continuously increasing, in favor of the assumption the AcCh is not a transmitter between two cells. It is never released from nerve terminals; its appearance in the outside fluid is an artifact, as will be fully discussed in Section V. AcCh acts as a signal *within* excitable membranes; it triggers off a series of molecular changes which result in increased permeability. AcCh is intrinsically associated with excitability; its action is an integral part of the processes generating and propagating electrical currents. The role of AcCh is fundamentally similar in the membranes of nerve and muscle fibers and in the pre- and postsynaptic membranes of junctions.

The following picture of the role of AcCh in the membrane fits best the available information. In the resting condition, AcCh is present in the membrane in bound form, probably associated with a protein or a lipoprotein. It is released by excitation and acts as a signal recognized by a receptor protein located at a distance of a few angstroms within the membrane. The reaction induces a conformational change of the protein. The change may possibly release by allosteric action Ca^{2+} ions bound to carboxyl groups of the proteins, possibly also to $P—O^-$ groups of phospholipids. Calcium ions have long been assumed to be involved in the excitability of nerve and muscle fibers (for a summary, see Brink, 1954). The Ca^{2+} ions released may act on a cell component(s) referred to by Changeux as "ionophore" (Changeux *et al.*, 1969a) and forming a barrier to the movements of ions. Calcium ions may induce further conformational changes of phospholipids and other polyelectrolytes. The end results of the sequence of chemical reactions is the change of permeability to ions permitting the movements across the membrane of many thousands of ions, possibly as many as 20,000–40,000 in each direction, per molecule of AcCh released. While Ca^{2+} ions may be the main force in effecting the permeability changes, a specific reaction is necessary to control this activity. Among the components of a cell, only proteins have the ability of recognizing a specific ligand such as AcCh and thus respond to a signal. The series of reactions described thus act as typical amplifiers of the signal released by excitation. AcCh esterase, by rapid hydrolysis of AcCh, permits the return of the receptor to its original conformation; the barrier to the ion movements is thereby reestablished. A schematic presentation of this process is shown in Fig. 3, comparing the membrane in resting and active form.

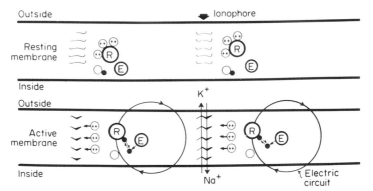

Fig. 3. Schematic presentation of the role of AcCh in excitable membranes. AcCh is postulated to act as a signal initiating a series of reactions that result in increased permeability to ions. In the resting condition of the axonal membrane, AcCh (●) is bound to a storage form. On excitation (electric current, K^+, H_3O^+ ions?) AcCh is released and acts on the receptor protein inducing a conformational change. Ca^{2+} (⊕) ions bound to carboxyl groups of the protein may thereby be released as a result of allosteric action; these ions act on the "ionophore," inducing conformational changes of phospholipids or other polyelectrolytes and thus permitting accelerated ion movements. The end result is an amplification of the signal initiating a new electric circuit. The process is repeated at successive points of the membrane and the impulse is thus propagated along the axon. The AcCh–receptor complex is in a dynamic equilibrium with the free ester and the receptor. The free ester will be susceptible to attack by AcCh-esterase. The rapid hydrolysis will permit the receptor to return to its resting condition. The barrier to the rapid ion movements is reestablished, but in the meantime the electric circuit generated has activated the adjacent point. The protein assembly effecting these processes must be structurally well organized within the membrane, as known for other assemblies and multienzyme systems. The structural organization permits the high speed, precision and efficiency.

A brief comment on the properties of Ca^{2+} ions may be indicated in the context of their proposed role. One of the characteristic features of these ions is their osmotic coefficient φ_p. Tested with alginic acid, Ca^{2+} ions were found to have a φ_p *of* 0.01 as compared with 0.15 for Mg^{2+} ions or 0.3–0.4 for the monovalent Na^+ and K^+ ions, indicating that only 1% is osmotically active and 99% are bound (Katchalsky *et al.*, 1961). The strong degree of ion binding suggests the possibility of an association with polymers, presumably localized at definite sites along the polymer chain. Out of the one hundred carboxylic groups of bovine serum albumin, eight seem to form true complexes with Ca^{2+} ions, whereas the others form loose combinations of the ordinary electrostatic type (Carr, 1953). Another important feature of φ_p is its independence of the molecular weight of the polymer. The strong and specific combination of Ca^{2+} ions with polymers and other features discussed by Katchalsky (1964) may be responsible for some of the remarkable biological properties and their well known important role in·muscular contraction or in the permeability control of the cell walls of adjacent cells (see, e.g., Loewenstein, 1967). Although many observations support the assumption of an essential role of Ca^{2+} ions in the permeability changes of excitable membranes, more information on a molecular level is needed for a precise and satisfactory picture.

The evidence on which the concept of the role of AcCh in excitable membranes is based has been repeatedly described. The earlier data may be found in a monograph (Nachmansohn, 1959), the later developments in several reviews (Nachmansohn, 1963a,b, 1964, 1966a,b,c, 1968, 1969, 1970, 1971). In this chapter, a few more recent observations will be presented.

III. Enzymes Hydrolyzing and Forming Acetylcholine

A. AcCh Esterase

1. LOCALIZATION

AcCh and the two enzymes which hydrolyze and form the ester, AcCh esterase and choline *O*-acetyl transferase (choline acetylase), have been shown to be present in a large variety of types of nerve and muscle fibers—motor and sensory, cholinergic and adrenergic, peripheral and central, etc.—and in a large variety of species, vertebrates and invertebrates (for a summary, see Nachmansohn, 1959). Three decades ago, it was suggested that AcCh esterase is localized at or near the cell surface, i.e., in or near the site of the presumable localization of the plasma membrane. This assumption was based on a variety of indirect biochemical data. During the last decade a great number of investigators have demonstrated, by using electron microscopy combined with histochemical staining techniques, that AcCh esterase is indeed a membrane-bound enzyme. It is present in the conducting membranes of axons and muscle fibers as well as in pre- and postsynaptic membranes of junctions.

These discoveries are illustrated in Figs. 4–6. Figure 4 is an electron micrograph of unmyelinated squid axons prepared by Dr. Virginia Tennyson. It demonstrates the exclusive localization of the enzyme in the plasma membranes of the three axons visible in the picture. In unmyelinated fibers, the findings were irregular. Sometimes the enzyme was present, but frequently it was absent, in spite of the presence of enzyme activity when tested by chemical determinations. Even sections of such fibers only 500–1000 Å thick are rich in lipid, and this may prevent an adequate access of substrate to the enzyme. Therefore, Brzin (1966) treated sections of a single myelinated fiber of frog sciatic nerve with the detergent Triton 100 X and found the enzyme to be regularly present in the plasma membrane (Fig. 5). Figure 6 shows an electron micrograph prepared by Lundin and Hellström (1968) from a muscle fiber of plaice. Again, the enzyme is localized exclusively in the excitable membrane of the muscle fiber, the sarcolemma. When the preparation was first exposed to low concentrations (10^{-6} M) of diisopropylphosphorofluoridate (DFP), a strong enzyme inhibitor, no enzyme activity was detectable.

In view of the central role which the electric tissue played in the development, recent studies of Changeux and his associates on the localization of AcCh esterase in the electroplax of *Electrophorus* are of particular interest. When the excitable and inexcitable membranes of an electroplax were separated mechanically, virtually all of the enzyme was found to be localized in the former, very little in the latter (Changeux *et al.*, 1969b). The data are in agreement with unpublished observations in this laboratory by Drs. M. Brzin and W.-D. Dettbarn several years ago. When the isolated membranes were stained with the usual procedures applied for the localization of AcCh esterase and using Triton 100 X, the enzyme was shown to be present in the excitable membrane (Fig. 7); in the inexcitable membrane, no enzyme was detectable. The enzyme is uniformly distributed throughout the excitable membrane. This is in agreement with previous data of Bloom and Barrnett (1966), who found the enzyme to be present in the membranes surrounding the nerve terminal and the postsynaptic and conducting membranes.

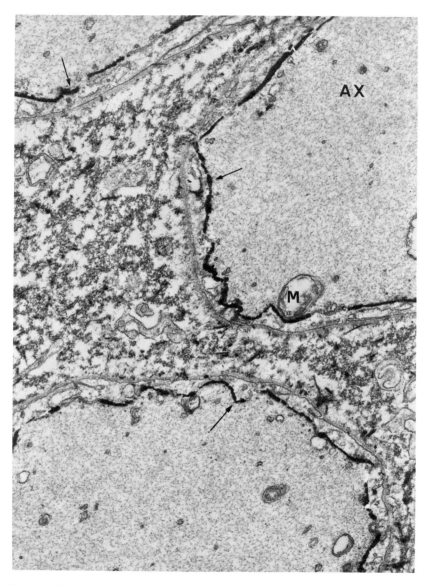

Fig. 4. Electron micrograph showing the exclusive localization of AcCh esterase (arrows) in the plasma membrane of small nerve fibers (AX) in the stellar nerve of squid. A mitochondrion (M) is located close to the axolemmal surface. A basement membrane surrounds the Schwann cell, separating it from the fibrillar material of the intercellular matrix. Incubated in acetylthiocholine without inhibitors, ×20,500. (V. Tennyson.)

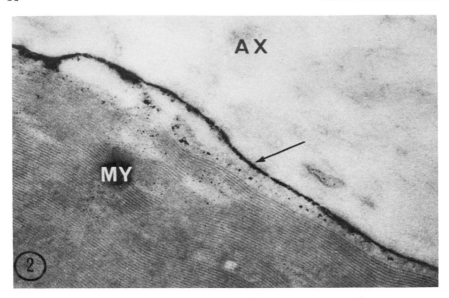

Fig. 5. Large myelinated (MY) ventral root axon (AX) taken from a frog sciatic nerve. The fiber was treated with Triton 100× before the incubation for testing acetylcholinesterase activity, in the same way as in Fig. 4. Dense end product is present on the axolemmal (plasma) membrane (arrow) (×3200). Figure reprinted by permission of Editor, *Proc. Nat. Acad. Sci. U.S.* 56, 1560 (1966).

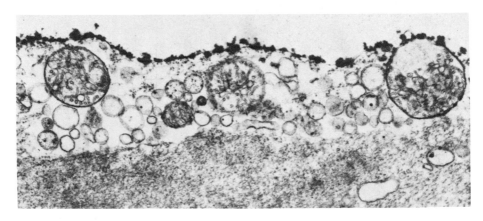

Fig. 6. AcCh esterase activity located in the plasma membrane (sarcolemma) of a plaice muscle. Section of a cell kept at 0°C in isotonic sucrose solution showing black precipitates along the sarcolemma as a result of cholinesterase activity. Substrate: Acetylcholine for 45 min at 0°C, ×21,000. (From Lundin and Hellström, 1968.)

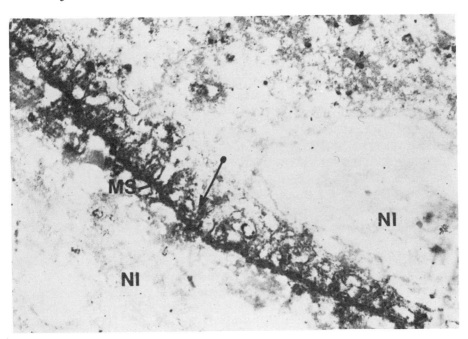

Fig. 7. Electron micrograph of an isolated excitable membrane fragment of the electroplax of *Electrophorus*. The picture shows the striking uniformity of the distribution of AcCh esterase at the innervated surface (MS) of the membrane. The separation of the electroplax membranes has been achieved by differential centrifugation and by means of discontinuous saccharose gradients. In the usual cytochemical staining procedure for demonstrating AcCh esterase activity, acetylthiocholine was used as the substrate. No staining was found in the noninnervated (NI) membrane. (×8860) (From Changeux *et al.*, 1969b.)

These results require several comments. As mentioned before, the writer has stressed for many years the extraordinary feature of the finding that the highly specialized organ may hydrolyze 3–4 kg of AcCh per hour per kilogram of tissue (fresh weight). Since it is now evident that almost all of the enzyme is localized in the excitable membrane, and since the membrane forms at most 10^{-4} of the cell mass, the concentration of the enzyme in electric tissue becomes even more impressive; a more correct estimate is probably that 1 gm of excitable membrane may hydrolyze 30 or more kg of AcCh per hour. It is difficult to estimate the surface area of the membrane because of the presence of many deep invaginations. Moreover, the thickness of the membrane is unknown; it may be greater than the usually assumed 100 Å [see the findings of Sjöstrand and Barajas (1968) mentioned above]. But with all reservations in view of some uncertainties, the fraction of the membrane volume formed by the enzyme must be considerable. Tentative estimates of Professor Aharon Katchalsky and the writer made on the basis of enzyme activity and molecular weight suggest that the enzyme may form about 5% of the total membrane volume.

Another pertinent aspect which should be stressed is the uniformity of the enzyme distribution. Although an electroplax may have $2-4 \times 10^{-4}$ synaptic junctions, they form at most 5–10% of the total surface area of the excitable membrane; more than 90% of the membrane is conducting. In view of the uniform distribution, it is evident that the conducting membrane contains by far the largest fraction of the enzyme. As will be described later, the AcCh receptor protein is also present in the conducting membrane, although AcCh, curare, and other quaternary ammonium derivatives reach the receptor and react with the protein at the level of synaptic junctions only. In contrast, congeners of AcCh that have tertiary ammonium groups and are therefore more lipid soluble, such as physostigmine, local anesthetics, etc., act on both types of membrane. This pharmacological difference is observed with most axonal and junctional membranes and will be discussed in great detail in Section V.

2. PROPERTIES OF THE ENZYME

When Meyerhof isolated the glycolytic enzymes from muscle, his success opened the way for the studies of intermediary reactions in glycolysis and led to the elucidation of the Meyerhof–Embden pathway in the 1920s and 1930s. These developments were an exciting and impressive period for all those who spent some of these years in Meyerhof's laboratory. The high concentration of AcCh esterase found in the electric organ of *Torpedo* in 1937 suggested to the writer that this tissue may be a most suitable material for isolating the enzyme biologically associated with the hydrolysis of AcCh. Studies on the properties of the isolated enzyme would almost certainly contribute valuable information about its function. In the relatively few studies on cholinesterase reported at that time, serum cholinesterase was used. This enzyme, as we know today, is distinctly different from that biologically responsible for the hydrolysis of the ester, which is now generally referred to as acetylcholinesterase (Augustinsson and Nachmansohn, 1949a) and which has a number of special features.

AcCh esterase was extracted from electric tissue of *Torpedo* in 1938 (Nachmansohn and Lederer, 1939). That was the first time that this enzyme was obtained in solution and in rather active form. In the early 1940s, the enzyme, extracted from electric tissue of either *Torpedo* or

Electrophorus, was purified about 400 to 500-fold. The specific activity of the best preparation was about 8 mmoles of AcCh hydrolyzed per milligram of protein per minute (Rothenberg and Nachmansohn, 1947). Large amounts of tissue were at that time difficult to procure; therefore, the protein amounts with high specific activity finally obtained were small. However, partially purified preparations were suitable for kinetic studies and for the analysis of the hydrolytic mechanism. AcCh esterase was one of the first ester-splitting enzymes which were used for the analysis of the molecular groups in the active site (Nachmansohn and Wilson, 1951; Nachmansohn, 1959; Wilson, 1960; and others). The effects of many competitive inhibitors used in neuromuscular disorders and in medicine were interpreted in terms of their reactions with the molecular groups in the active site.

Of particular theoretical as well as practical interest were the studies with organophosphates, potent inhibitors of AcCh esterase and other ester-splitting enzymes. A few of these compounds are volatile and potential chemical warfare agents; many are widely used as insecticides. The fatal effects are due to the reaction with AcCh esterase because of the vital function of this enzyme. It is well known today that the phosphorus atom of organophosphates forms a covalent bond with an oxygen atom of the serine residue in the active site of ester-splitting enzymes. The phosphorylated enzyme is much more stable than the acetyl enzyme. The reaction was first thought to be irreversible, but this notion has undergone considerable modification (e.g., Hobbiger, 1963; Nachmansohn, 1971).

The explanation of the reaction mechanism of organophosphates with AcCh esterase permitted the development of an antidote against poisoning by organophosphate insecticides and even some of the nerve gases, namely pyridine-2-aldoxime methiodide (PAM), synthesized by Wilson and Ginsburg (1955). This compound rapidly and quite specifically reactivates the enzyme; the nucleophilic oxygen of the aldoxime attacks the phosphorus atom of the phosphoryl enzyme and displaces the phosphoryl group from the oxygen of the serine residue, thereby restoring the enzyme activity. The specific biochemical lesion is repaired. The efficiency of PAM is greatly increased when it is given in combination with atropine. The latter does not reverse the enzyme inhibition, the primary lesion, but it protects against excess of AcCh, a secondary effect of organophosphate poisoning. Mice which had received a ten-to twenty-fold sure lethal dose of Paraoxon survived when treated with PAM in combination with atropine (Kewitz *et al.*, 1956). Subsequently, even more spectacular successes were reported by many investigators. Japanese investigators were the first to apply PAM on a large scale to agricultural workers as antidote against poisoning by the organophosphate insecticide Parathion (e.g., Namba and Hiraki, 1958). PAM is today the key antidote in organophosphate poisoning (in the commercially available product, the iodide is replaced by chloride). It is unfortunate that many physicians still use in poisoning by organophosphate insecticides only atropine, although this compound is toxic in milligram quantities, whereas PAM is harmless to humans in doses of 1–2 gm. Treatment with atropine alone is completely inadequate. Several deaths were recently reported in newspapers of victims of parathion poisoning treated with atropine only; the lives of these victims could have been saved almost certainly by the use of PAM.

In the last decade a vast literature has appeared about this problem, which cannot be discussed here. But it has been briefly mentioned for the following reason. The present trend against stable insecticides, such as DDT, has increased the interest in other unstable insecticides. Organophosphate insecticides are unstable, but the available compounds have a toxicity which is higher than is desirable. Many hundreds of organophosphates have been and many more could be synthesized. There is a good possibility of developing compounds which are much more specific for certain insects and less toxic for men and animals. These efforts would be useful in the fight against pollution of the environment by insecticides; their use is necessary in the fight against hunger and disease. However, the development of more efficient and less toxic organophosphate insecticides requires great efforts and systematic studies of their toxicology about which the present information is entirely inadequate and actually nonexistent. The problem has been discussed in some detail in a recent review article (Nachmansohn, 1971).

The developments of protein chemistry during the last decade, which permit an analysis of molecular properties and have even led to the exploration of the tridimensional structure of several proteins and enzymes, made it desirable to obtain large amounts of homogeneous enzyme protein and possibly to crystallize it. Multi-step procedures, using ion exchange and exclusion chromatography were used for the purification of electric eel tissue by Kremzner and Wilson (1963). These procedures were applied, with some modifications, to a large scale purification by Leuzinger and Baker (1967). Using 10 kg of toluene-treated electric tissue as starting material, about 60 mg protein formed by a single component were obtained with a specific activity estimated to be about 12 mmoles of AcCh hydrolyzed per milligram of protein per minute. The preparation led to the crystallization of AcCh esterase (Leuzinger *et al.*, 1968). The crystals have the form of hexagonal prisms (Fig. 8). The molecular weight of the enzyme has been measured by equilibrium centrifugation according to the methods described by Van

Fig. 8. Crystals of AcCh esterase. (A) Regular hexagonal prisms; (B) possibly pyramidal termination. According to Dr. Barbara Low, the most common form of growth observed, A is compatible with true hexagonal symmetry; B may imply, however, a crystal system of lower order. The crystals exhibit low birefringence so that it is not possible at present to determine whether they are uniaxial or biaxial. (From Leuzinger *et al.*, 1968.)

Holde and Baldwin (1958) and Yphantis 1964) and was found to be 260,000, in good agreement with the previously reported values of 240,000 (Lawler, 1961) and 230,000 (Kremzner and Wilson, 1963). The protein seems to be formed by two different polypeptide chains, on the basis of the determinations of C-terminal groups with two different methods (Leuzinger *et al.*, 1969). However, the chromatographic procedures used have several drawbacks. They are quite laborious since seven steps are involved, some of them not easily reproducible. The most serious difficulty is the very poor yield; 5–10% of the initial enzyme activity is obtained as homogeneous protein. In view of the difficulties in procuring the material and the extremely high costs involved, such a yield is prohibitively small for studies aimed at elucidating the structure and properties of the enzyme protein. Thus new and more efficient purification procedures became imperative.

The introduction of affinity chromatography by Cuatrecasas *et al.* (1968), used in a variety of protein purification systems, fortunately offered the promise of a greatly improved method at this critical moment. The procedure was applied to the purification of AcCh esterase by Kalderon *et al.* (1970), using toluene-treated electric tissue which was prepared in this laboratory in the usual way. The affinity ligand contained phenyltrimethylammonium (PTA), a specific reversible inhibitor of AcCh esterase which was used as the affinity group, and 6-amino-caproic acid, which extended the affinity group from the resin matrix. The ligand with a free amino group was attached to Sepharose 4B by means of the coupling agent cyanogen bromide. The chromatographic column selectively retained the enzyme which was eluted with a salt gradient. The extent to which the target protein is retained on affinity columns was found by Cuatrecasas (1970) to depend upon the length of the attachment arm which links the specific inhibitor to the resin. Rosenberry et al. (1972) therefore prepared a series of affinity columns, using several specific inhibitors of AcCh esterase with attachment arms of varying length. These affinity columns are currently still under investigation in this laboratory, but one colunm proved so far to be particularly efficient and produced higher yields than either the multistep procedure or the affinity chromatography described previously. The inhibitor used was PTA attached by a di-(6-aminocaproyl) linkage to Sepharose 4B: [*N*-(6-aminocaproyl-6'-aminocaproyl)-*p*-aminophenyl]trimethylammonium bromide hydrobromide [di-(6-aminocaproyl)-PTA].

$$H_3C\overset{\oplus}{-}\underset{\underset{CH_3}{|}}{\overset{\overset{CH_3}{|}}{N}}\!-\!\!\left\langle\right\rangle\!-\!\!\left[NH-\overset{\overset{O}{\|}}{C}\!+\!(CH_2)_5\!\right]_{\!2}\!-\overset{\oplus}{NH_3}\quad 2\,Br^{\ominus}$$

During the course of this work a report by Berman and Young (1971) appeared in which the purification of AcCh esterase by affinity chromatography was described and the retention properties of several inhibitor linked agarose gels were compared. The studies of Rosenberry, Chang, and Chen supplement this report in providing information both on somewhat different inhibitor ligands and on the elution profile as well as the purity of the recovered enzyme. Moreover, a critical analysis is given of some of the difficulties which had resulted in previous conflicting reports.

The best results with the di-(6-aminocaproyl)-PTA gel were obtained with a relatively low concentration of ligand (about 1 μmole/ml packed gel) and with the elution of retained protein on a nonlinear potassium chloride gradient. The yield of pure, homogeneous enzyme accomplished in the single step is about 70% of the initial preparation. The specific activity is about 575–600 mmoles of AcCh hydrolyzed/mg protein/hr. The homogeneity of the enzyme was established both by polyacrylamide gel electrophoresis and by immunoelectrophoresis procedures. For estimations of the specific activity an extinction coefficient ($\epsilon^{1\%}_{280\ nm}$) of 18.0 $\pm$ 0.4 was assumed, a value obtained after many careful studies. This value is lower than that reported by Kremzner and Wilson (1963) who found for their preparation an $\epsilon^{1\%}_{280\ nm}$ = 22.9. This difference of the relative UV absorption values apparently accounts for the discrepancy between the estimated specific activities.

In view of the discrepancies between the specific activities reported from various laboratories, Rosenberry et al. (1972) made a number of other thorough and precise studies in order to find possible explanations for the differences. They found several intrinsic difficulties in obtaining values of specific activities for homogeneous enzyme. The stability of highly purified enzyme varies greatly from preparation to preparation. This may be one factor accounting for some of the differences reported. Another inherent difficulty is the significant differences between protein estimates obtained from techniques measuring microquantities of nitrogen as NH_3 or N_2 and those monitoring dry weight and refractive index. It appears that microdeterminations of this protein which measure NH_3 or N_2 and assume that the protein contains 16.0% nitrogen may underestimate the protein content by 10–15%.

ACTIVE SITE TITRATIONS. The number of active sites per molecule was reevaluated with two new techniques: by direct spectrofluorometric titration with the carbamoylating agent N-methyl-(7-dimethylcarbamoyl)quinolinium iodide (M7C), and by equilibrium dialysis with 3,3′-bis{α-([14C]trimethylammonium)methyl}azobenzene dibromide (bis-Q); (see also page 82). The kinetic parameters for M7C were previously defined by Rosenberry and Bernhard (1971). Using this compound for titrating the enzyme, one may calculate both the turnover number for the enzyme-catalyzed hydrolysis and the number of catalytic sites per enzyme molecule, provided the enzyme molarity, molecular weight, and total protein and active enzyme concentration are known. The calculations lead to an average value of 3.29 ± 0.11 active sites per molecule of enzyme. The turnover number was found to be 8×10^5 per active site/min.

Equilibrium titrations of the purified enzyme with bis-Q also indicate 3.3 active sites per molecule. A similar number of active sites has recently been reported by Kasai and Changeux (1971) for decamethonium binding to enzyme in membrane fragments. Previous reports of four active sites per molecule (Kremzner and Wilson, 1964; Rosenberry and Bernhard, 1971) were based on the assumption of specific activity of 660 and 730, values which have not been confirmed in these studies.

The most precise values presently available are 3.2 to 3.4 active sites per molecule. This value raises some problems. Will the number of active sites after further investigations turn out to be three or four? The value of three active sites is not readily reconciled with the assumption of four subunits reported by Leuzinger et al. (1969) and with the view of symmetry considerations which postulates an even number of active sites for oligomeric enzymes (Monod et al., 1965). However, recent reports by Millar and Grafius (1970) suggest six subunits. This is of particular interest for this problem because three active sites per hexamer would indicate higher symmetry than three active sites per tetramer. Studies of the number of molecular subunits are currently underway in this laboratory. More information is needed for a satisfactory answer.

3. Essential Role in Conduction

The evidence of the presence of high concentrations of the enzyme in the various types of excitable membranes, of axons and muscle fibers does not yet indicate its essential role in electrical activity, which is a prerequisite for the postulated role of AcCh in the permeability changes. There are other prerequisites, such as the high speed of enzyme action. Because nerve fibers may conduct up to one thousand impulses or more per second, the signal interacting with the receptor must be removed with a speed compatible with the function proposed. The turnover time of the enzyme is 30–40 μsec (Lawler, 1961; Wilson and Harrison, 1961). Thus, it has the necessary speed of activity.

While these features may be suggestive, the postulated role of AcCh requires the evidence that the enzyme activity is directly associated with electrical activity. Potent inhibitors of AcCh esterase should affect and eventually block electrical activity. This evidence has been obtained with a variety of potent inhibitors; exposure of a great number of different types of nerve and muscle fibers to reversible competitive inhibitors, such as physostigmine, affects electrical activity reversibly, whereas exposure to organophosphates affects it irreversibly. However, the demonstration of the relationship between electrical and enzyme activity offers many complex problems and pitfalls. Many of them have been known for a long time. Many more were entirely unknown two decades ago, in the early period of attempts aimed at establishing the relationship between electrical and enzyme activity. Biologists have become increasingly aware of the intrinsic difficulties of correlating reactions of enzymes studied *in vitro* with special cellular mechanisms that must be tested on the intact cell. The last decade has provided much information about factors which may drastically modify chemical reactions taking place in a structure, as compared with those in solution, thus making extrapolations from observations of enzymes *in vitro* to their behavior in a structure difficult and problematic. In addition to structural barriers, well known for a long time, the importance of such factors as microenvironment (charges, pH, etc.) has become apparent, particularly in structures as highly organized as membranes. Pertinent in this context is also the phenomenon referred to by Schatz and Racker as "allotopy" (Racker, 1967)—ATPase activity in the mitochondrial membranes is sensitive to oligomycin, while resistant in solution. As stressed by Racker, such alterations of the properties of enzymes bound to a membrane may be more general than realized at present. Other factors active in the intact cell and absent in solution, such as control mechanisms, feedback, and regulatory enzymes, form by now an integral part of textbook biochemistry.

The complexity of the problem of a correlation between electrical and AcCh esterase activity in excitable membrane has been discussed in detail in a recent article (Nachmansohn, 1971). Therefore, in this chapter only a few facts will be presented as illustrations. When a frog sciatic nerve is exposed to physostigmine, a potent competitive inhibitor of AcCh esterase with a K_I of 10^{-7} M (Augustinsson and Nachmansohn, 1949b), electrical activity is reversibly blocked (Bullock *et al.*, 1946, 1947). However, the concentration required is about 10^{-2} M, four to five orders of magnitude higher than the K_I, and the time of exposure is 30–60 min. The frog sciatic nerve is formed by several thousands of myelinated axons and is surrounded by a sheath impervious to many compounds. The high concentration of inhibitor required and the slowness of its action was, therefore, attributed to the structural barriers preventing the compound to reach the receptor in the excitable membranes. When in the late 1950s a single fiber preparation was developed, permitting one to measure electrical activity on single nodes of Ranvier, physostigmine was applied to this preparation (Dettbarn, 1960a). Effects of the compound in 5×10^{-5} to 5×10^{-6} M (1.5–15 μg/ml) on electrical activity appear rapidly, in seconds. Significantly, the electrical parameters in the single fiber preparation are modified by the action of the inhibitor in a way that has been postulated by electrophysiologists, if the proposed theory were to be correct. First, a potentiating effect was observed, an increased spike height and a larger descending phase. On longer exposure, the amplitude increased and the prolongation became greater. At a concentration of 2×10^{-4} M (60 μg/ml), the spike height was first reduced and the duration was increased; after exposure of 15–20 min, blockage of conduction occurred. With 10^{-3} M (300 μg/ml) conduction was blocked in 25 sec. These modifications of the parameters could not possibly have been observed, for many reasons, on a preparation formed by many thousands of axons. Dettbarn (1960a) also found the effectiveness of physostigmine to increase with increasing pH. A roughly linear relationship exists between the concentration of the conjugate base in solution and the percentage attenuation of the action potential, consistent with the assumption that the neutral molecule may readily penetrate; no such relationship was observed with the cationic form. It would be difficult to attribute the effects observed on the electrical activity of the axon to some unknown unspecific effect, in view of the high affinity of physostigmine for AcCh esterase and the rapidity, effectiveness, and mode of action. Moreover, when this single fiber preparation is exposed to d-tubocurarine, electrical activity measured at the nodes of Ranvier is blocked rapidly and reversibly, as previously described for the neuromuscular junction (Dettbarn, 1960b). This observation will be further discussed in Section V.

Another series of observations relevant to the problem of functional interdependence between electrical and enzyme activity may be briefly discussed. On exposure of squid giant axons to diisopropylphosphorofluoridate (DFP), irreversible blockage of electrical activity is obtained, but a long period of exposure (30–40 min) and a high concentration (5×10^{-3} M) are required. The concentration is again three to four orders of magnitude higher than that which blocks the enzyme in solution. This discrepancy was considered by many as evidence that the action of DFP was due to a general toxic effect and has nothing to do with the inhibition of AcCh esterase. When, however, the axoplasm of the exposed axon was extruded, it was found that at the time when electrical activity was irreversibly blocked, less than 0.5 μg/gm, i.e., less than $\frac{1}{2000}$ of the outside concentration was present in the interior (Feld *et al.*, 1948). When radioactively labeled DFP was used, it was found that the compound is rapidly hydrolyzed by an enzyme referred to first as DFPase by Mazur (1946), and later as phosphorylphosphatase by Augustinsson and Heimburger (1955); the amounts of DFP found in the axoplasm were the same as reported 20 years earlier (Hoskin *et al.*, 1966). In a tissue which is rich in enzyme inactivating DFP, the meaning of the outside concentration appears questionable. The results certainly do not support the assumption of a general toxic effect.

While it is at present poorly understood why in some types of fibers a relatively high concentration of organophosphate and long exposure time is required for obtaining irreversible block of electrical activity, although in some fibers rapid irreversible effects are obtained, another type of observation appears significant in this context. Pyridine-2-aldoxime methiodide (or related compounds), potent and specific reactivators of phosphorylated AcCh esterase mentioned above, restored in some preparations electrical activity which was irreversibly blocked by organophosphates, in spite of the fact that the compound is a quaternary ammonium derivative and does not readily penetrate into cell fibers (Hinterbuchner and Nachmansohn, 1960; Rosenberg and Dettbarn, 1967). It would be difficult to explain this restoration of electrical activity by a chemical reaction, known to remove specifically the phosphoryl group from the serine in the active site of AcCh esterase, in any other way than by the reactivation of the enzyme activity. These observations thus clearly demonstrate the direct interdependence between electrical and chemical activity.

It would be desirable to measure directly the relationship between electrical and enzyme activity by correlating the effects of inhibitors on both functions and to find out the minimum enzyme concentration compatible with unimpaired conduction. It is impossible to obtain this information with reversible inhibitors, as has been tried in the past, for reasons that have been repeatedly discussed (e.g., Nachmansohn, 1959, 1971). When it was reported, in the 1940s, that organophosphates are irreversible inhibitors of AcCh esterase, it was thought that this type of compounds may permit the examination of the relationship. In the last few years it has become evident that there is at present no possibility of answering this question, because the methods now available are satisfactory for the determination of the enzyme activity in solution, but they do not permit a quantitative evaluation of the exact enzyme concentration in normal tissue. Simple homogenization turned out to be inadequate. Several procedures may markedly increase the activities measured after homogenization; in some tissues the increase may be severalfold. After the tissues have been exposed to organophosphates, the difficulties are compounded by a variety of complicating factors; some of them were known for some time, others became apparent only recently. Thus, for the time being, there is no information available about the enzyme activity in tissues after their exposure to organophosphates and the relationship between this activity and conduction. Reports in the literature about the enzyme activities aimed at correlating them with electrical activity appeared at a time when these difficulties were not yet recognized. These data have lost their validity in the light of our present knowledge (Nachmansohn, 1971). For the topic of this chapter, the most important question is whether the enzyme is essential for electrical activity. This question has been answered in the affirmative by the observations that specific inhibitors of the enzyme affect and may even block electrical activity, an evidence obtained with several preparations and with potent competitive reversible as well as irreversible inhibitors. Once such a relationship has been convincingly demonstrated, even in a few cases only, it must be assumed that it has general validity, in spite of the difficulty of explaining at present some of the findings because of the known complexities of cellular mechanisms and the still unexplored behavior of many compounds within a tissue. The assumption that AcCh esterase present in remarkably high concentration in excitable membranes is sometimes associated with electrical activity and sometimes not, appears less acceptable. Once the role of ATP in the elementary process of muscular contraction, its function in the sliding mechanism of actin and myosin filaments, has been demonstrated in the rabbit muscle, it would be difficult to question this role of ATP on the basis of some unexplained observations on other muscles which are perhaps for some reason less suitable for analysis.

B. Choline O-Acetyltransferase

1. ENERGETIC ASPECTS OF BIOELECTROGENESIS

It is well known that there are two different classes of phosphate compounds. One class, formed by a group of relatively stable phosphate esters, such as α-glycerophosphate of glucose 6-phosphate, is hydrolyzed with a relatively small free energy change of the order of -1500 to -3000 cal per mole. The second class of phosphate derivatives, mostly anhydrides of phosphoric acid, are less stable and hydrolyzed with a large free-energy change, with a ΔF of $-10,000$ to $-12,000$ cal per mole. The free energy available from oxidative reactions is usually trapped by the latter class of phosphate derivatives and used for endergonic life processes.

The notion of the two different classes of phosphate derivatives was developed in Meyerhof's laboratory. When phosphocreatine was discovered, Meyerhof and Suranyi (1927) measured the heat of its hydrolysis and found a high enthalpy in contrast to that of other phosphate esters. This was probably one of the most startling discoveries in modern biochemistry and was the beginning of the notion of phosphate derivatives with a large free energy release on hydrolysis. Later, ATP, phosphoenolpyruvic acid, 1,3-diphosphoglyceric acid, and other phosphate derivatives were found to be hydrolyzed with a large free energy change. Meyerhof (1937) reviewed the importance of the distinction between these two classes of phosphate derivatives for the intermediary metabolism in general, and in particular for the implications for the problem of energy supply and the sequence of energy transformations in muscular contraction. Meyerhof's ideas were further extended and elaborated in the articles of Lipmann (1941) and Kalckar (1942). The history of the development of these early concepts as well as those of more recent years may be found in the book *Biological Phosphorylations* edited by Kalckar (1969).

An analysis of the mechanism of bioelectrogenesis requires the combination of chemistry, energetics, and physics. When the writer first began to suspect, in 1940 and 1941, that the action of AcCh was not that of a neurohumoral transmitter, but an intracellular process associated with bioelectrogenesis, the problem of the energy supply and of the sequence of energy transformations during electrical activity appeared to him to be essential elements for obtaining a satisfactory insight into the function of AcCh and the chemical basis of nerve activity (Nachmansohn, 1972).

The ionic concentration gradients between the outer environment and the interior of conducting fibers are most likely the primary source of energy of the ion movements carrying the electrical currents. If AcCh is the trigger responsible for making the potential source of energy, the ionic concentration gradients, effective by initiating the series of processes leading to increased ion permeability of excitable membranes, and if the ester is rapidly inactivated by hydrolysis, permitting the return of the membrane to its resting condition and reestablishing the barrier to ion fluxes, the question arises what is the reaction that provides the energy for the acetylation of choline in the recovery period? Were phosphate derivatives with a large free energy change involved? The electric organs being the most powerful bioelectric generators known and being highly specialized in their function seemed to offer a most suitable material for testing this question. The electric tissue of *Electrophorus,* although formed to 92% by water, has a remarkably high concentration of phosphocreatine, as high or higher than striated muscle; the concentration of ATP is also high, although slightly lower than in muscle.

When the breakdown of phosphocreatine, roughly coinciding with electrical activity, was measured in the electric tissue of *Electrophorus,* it turned out to be more than adequate to account for the total external and internal electrical energy released (Nachmansohn *et al.,* 1943a; for later reviews, see Nachmansohn, 1959, 1963a). It was assumed that the phosphocreatine breakdown was coupled as in muscle with the hydrolysis of ATP and that it was used for the rephosphorylation of ADP. To the energy released by the breakdown of phosphocreatine had to be added that of lactic acid formation. The total chemical and electrical energy per gram of electric tissue depends on the size of the specimen, since large specimens have fewer cells per gram than small specimens. For a medium-sized eel, the total electrical energy was estimated to be 47 μcal per discharge per gram tissue (Cox *et al.,* 1946) compared to 140 μcal from the two chemical reactions. The computation of the chemical energy was based on the Gibbs free energy, ΔF or ΔG; the energy of ATP hydrolysis was assumed to be $-12,500$ cal per mole, that of lactic acid formed $-25,000$ cal per mole. A detailed description and analysis of the data may be found in the reviews quoted above.

It appears likely that under the conditions of the experiments, part of the chemical energy was released subsequent to electrical activity and was, therefore, used for the restoration of the ionic concentration gradient for which the energy of ATP hydrolysis is required, as was shown many years later in several laboratories for many cells. The process is not specific for nerve and muscle fibers. A direct relationship between ATP hydrolysis and the elementary processes of bioelectrogenesis appeared for many reasons unlikely, especially in view of the low speed of the reaction. But it appeared possible that part of the energy of ATP hydrolysis might be used for the acetylation of choline, the first recovery process following the hydrolysis of AcCh.

2. Discovery of Enzymatic Acetylation

When ATP was added to extracts of brain and electric tissue, AcCh formation was obtained, but it was disappointingly small. Only when sodium fluoride was added for inhibiting ATPase, was a strong and regular synthesis of AcCh observed (Nachmansohn and Machado, 1943).* Soon afterwards, the rapid inactivation of the enzyme system on dialysis was observed and suggested the necessity of a coenzyme in the system which was subsequently demonstrated (Nachmansohn *et al.*, 1943b; Nachmansohn and Berman, 1946). When extracts of acetone-dried brain powder in which ATPase is inactivated were used, the AcCh formation was greatly increased (Nachmansohn and John, 1944).

These observations were the first enzymatic acetylations achieved in a soluble system with ATP hydrolysis as the source of energy. It seemed at that time surprising and difficult to explain that ATP provided the energy for acetylation and not acetylphosphate as described by Lipmann (1940). However, in the following decade the mechanism of acetylation in bacterial and animal cells and the structure and role of the coenzyme, referred to by Lipmann as coenzyme A (CoA), have been fully elucidated in the laboratories of Lipmann, Ochoa, Lynen, and many others and have become an integral part of biochemistry textbooks. Acetylation takes place in two steps. The first is the formation of acetyl-CoA; in most bacterial cells, acetylphosphate is the source of energy of CoA acetylation, catalyzed by an enzyme referred to as phosphotransacetylase (Stadtman *et al.*, 1951; Stadtman, 1952). In animal cells, ATP plus acetate form adenyl acetate plus P—P; the acetate of adenyl acetate is

---

* Colleagues have frequently questioned the writer as to why this discovery, which initiated in the following decade a period of intensive investigations on the mechanism of acetylations and its role in intermediary metabolism, was published in the *Journal of Neurophysiology*, usually unknown to biochemists. The paper was first submitted to *Science*, then to the *Journal of Biological Chemistry*, and finally to the *Proceedings of the Society of Experimental Biology and Medicine*. All three journals refused acceptance. It was only then that the writer appealed to John F. Fulton to publish the paper in his journal.

transferred to CoA (Berg, 1956). The acetyl group of acetyl-CoA is transferred to a receptor, a reaction catalyzed by various enzymes specific for the receptor. Choline acetylase, now generally referred to as choline O-acetyltransferase, was redefined as the enzyme that transfers the acetyl group from acetyl-CoA to choline (Korey *et al.*, 1951).

The properties and function of choline O-acetyltransferase, its partial purification, etc., have been repeatedly reviewed (e.g., Nachmansohn, 1959, 1963b; Davis and Nachmansohn, 1964; Prince, 1967; and others). The reader interested in this enzyme is referred to the literature mentioned. It is of interest that Meyerhof, the pioneer in stressing the biological importance of phosphate derivatives in endergonic life processes, when asked to give the opening address at the first symposium on phosphorus metabolism, held at the Johns Hopkins University in 1951, shortly before his death, quoted among three outstanding examples of the role of phosphates in living cells the discovery of the role of ATP (and phosphocreatine) in bioelectricity and in enzymatic acetylation (Meyerhof, 1951).

IV. AcCh Receptor Protein

Since the beginning of this century it was recognized by Ehrlich, Langley, and many others that effects on cells produced by any chemical compound, drug, toxin, etc., must be caused by the reaction with specific molecules referred to as receptors. Whatever their nature may be, these molecules must have specific sites capable of binding the chemical compound before it produces the characteristic response in the cell. An early attempt at a more quantitative approach to the interaction between drugs and receptors was made by Clark (1937). More recent notions of receptor mechanisms are discussed in the review of Furchgott (1964) and in the book *Molecular Pharmacology* edited by Ariëns (1964). An evaluation in which classic views of receptors are analyzed and integrated with modern concepts and notions of proteins and biopolymers in general may be found in a review by Mautner (1967).

The notion of a receptor reacting with AcCh found increasing attention after World War II, as a result of the widespread investigations of inhibitors of AcCh esterase. It became apparent that some of the effects were not readily explained in terms of enzyme inhibition; they were attributed to the action on a receptor as distinct from the enzyme (e.g., Wescoe and Riker, 1951; Riker, 1953; Werner and Kuperman, 1963; and many others). The notion was essentially an operational term, its precise nature was an open question. The complex nature of the preparations used in electrophysiological and pharmacological studies of the action of AcCh and its congeners did not offer the possibility for an analysis of the chemical nature and the properties of the AcCh receptor. The writer suspected for more than two decades that the receptor is a protein—see, e.g., his Harvey Lecture in 1953 (Nachmansohn, 1955), but this assumption was based more on theoretical considerations than on experimental evidence (see below).

A. Monocellular Electroplax Preparation

A new chapter in the studies on the nature and properties of the receptor was opened by the development of the monocellular electroplax preparation (Schoffeniels and Nachmansohn, 1957; Schoffeniels, 1957, 1959). The two outstanding features of this preparation which make it a uniquely favorable material for the studies of the receptor protein are: (1) it is formed by a single cell, therefore, many obstacles encountered in the interpretation of reactions between ligands and proteins in multicellular preparations are eliminated; and (2) the cell is so highly specialized in its specific function, i.e., bioelectrogenesis, that the excitable membrane, in which the receptor must be located, is the dominating structure, while other cell functions appear to be poorly developed, as indicated by the low protein–high water content, and the slow rate of metabolic reactions of the cell, with the exception of those directly associated with the activity of the membrane. Among other remarkable and useful characteristics may be mentioned (1) the large size of the cell—which is 6–10 mm long, 1–2 mm high, and 0.2 mm thick; (2) the rectangular shape of the excitable membrane; and (3) the presence of both synaptic and conducting parts of the membrane which can be readily distinguished by means of electrical parameters. The cell has many thousands of synaptic junctions, but the surface area, as mentioned before, amounts to about only 5% of the total surface area of the excitable membrane. Finally, the presence of a nonexcitable membrane facilitates comparison between the two functionally different parts of the membrane surrounding the cell.

During the last decade several refinements and techniques, introduced mainly by Higman, Bartels, and Podleski,' have greatly increased the sensitivity of the preparation for the analysis of the reactions between the receptor protein and specific ligands (Higman and Bartels, 1961, 1962; Higman et al., 1963, 1964). The preparation permits precise and reproducible titrations of the dose–response curves and the evaluation of dissociation constants between ligands and the protein.

B. Reactions with the Receptor

1. RECEPTOR ACTIVATORS AND INHIBITORS

According to theory, the primary molecular event initiating the permeability changes in excitable membranes is the conformational change of the AcCh receptor protein induced by AcCh. AcCh and related compounds acting on and depolarizing the membrane are therefore referred to as receptor activators (agonists). Other compounds, also closely related in structure to AcCh, react with the active site of the receptor, but apparently are unable to induce a conformational change; they block the response of the membrane, but do not depolarize. In analogy with enzyme chemistry, they are referred to as receptor inhibitors (antagonists, antimetabolites). Receptor activators have usually one structural feature in common: most of them are methylated quaternary ammonium derivatives. Their tertiary analogs are either poor activators or, more frequently, receptor inhibitors.

The striking uniformity of the distribution of AcCh esterase in the excitable membrane of the electroplax has been described before (Section III). The enzyme is present in the postsynaptic and conducting part of the membrane, as well as in that of the nerve terminals (Bloom and Barrnett, 1966; Changeux et al., 1969b). The presence of the AcCh receptor in both parts of the excitable membrane, the synaptic and conducting parts, and its essential role in electrical activity have been demonstrated in experiments with a variety of specific receptor inhibitors, such as physostigmine, special antimetabolites referred to as local anesthetics (see below), and others. They were shown to block reversibly the direct response, i.e., that of the conducting membrane, independent of the events at the synaptic junctions (Higman and Bartels, 1961; Bartels, 1968; Bartels and Nachmansohn, 1969).

While the two proteins AcCh receptor protein and AcCh esterase are thus present and functional in all parts of the excitable membrane of the electroplax, as in the excitable membranes of nerve and muscle cells, AcCh applied from the outside acts only at the level of the junctions, because only there can the membrane be reached by certain quaternary ammonium derivatives, such as AcCh, curare, and many of their congeners. The same situation applies to most preparations; when applied externally, AcCh initiates the reactions leading to increased permeability only at the pre- and postjunctional membranes, because conducting membranes, with few exceptions, are fully protected against the action of this type of compounds. This limitation of the action of externally applied AcCh cannot be used as an indication that the function

of internally released AcCh differs at junctional and conducting parts
of the membrane, either in those of the electroplax or in others, as
will be discussed in detail in Section V. When AcCh or other activators
are added to the electroplax, the whole membrane is depolarized; although the action is limited to the junctions, a potential difference develops between many thousands of junctions in this cell and the rest
of the membrane and acts as a stimulus that leads to the release of
AcCh within the conducting part of the membrane and thus to their
depolarization. On the other hand, when receptor inhibitors are applied
which are unable to reach the conducting membrane (e.g., curare),
only the junctions are blocked, but the conducting membrane still responds to stimuli. When, in presence of curare receptor inhibitors are
applied which reach the conducting part of the membrane (e.g., physostigmine, local anesthetics) they block the direct response.

2. Evidence for Protein Nature; Active Site

Although the protein nature of the receptor has been postulated by
the writer for two decades because, among other reasons, only proteins
have the special ability of recognizing a small ligand proposed to act
as signal, studies on the monocellular electroplax preparation have provided strong evidence for this assumption; they also brought at the
same time much information about the active site of the receptor. A
great number of different types of ligands, congeners of AcCh, were
tested and the response of the membrane was evaluated by means of
dose–response curves. Among the many ligands used may be mentioned a series of n-alkyltrimethylammonium and of n-aryltrimethylam-
monium ions (Podleski, 1966, 1969). These reactions were compared
with those of purified AcCh esterase obtained from electric eel tissue.
In comparing the reactions of the enzyme inhibitors in solution with
those of the receptor in the intact membrane, only binding forces can
be properly correlated; receptor activators, because of the biological
activity involved as a second factor, are not suitable. On the basis of
these studies it became apparent that there are marked differences in
the active site of these two proteins. This was subsequently substantiated
by experiments with a variety of other types of compounds. There is
in the active sites of both enzyme and receptor an anionic subsite. The
quaternary ammonium ion derivatives are several hundred to a thousand
times more potent in their effects on the receptor than their tertiary
analogs. An anionic site in the enzyme has been established a long
time ago and is generally accepted. However, there is evidence that
there is no equivalent in the active site of the receptor to the esteratic

site of the enzyme. There are some indications that there is a hydrophobic group near the anionic site which contributes to the binding of the ligands.

Pertinent information as to the difference between the active site of the receptor and that of the enzyme was obtained with a series of benzoquinonium and ambenonium derivatives (Webb, 1965). Both compounds and their derivatives are receptor and enzyme inhibitors, and thus the factor of the biological activity of the receptor does not interfere in these experiments. When the dissociation constants between receptor protein and these derivatives were determined and compared with those obtained for the enzyme in solution, close similarities were found with some of these derivatives. Small modifications of the structure of these derivatives, sometimes the substitution of one atom in these rather large molecules, had a strong effect on the dissociation constants with the two proteins, sometimes in the same, sometimes in the opposite direction. While these data support the assumption of the protein nature of the receptor, they also indicate differences in the active site. This does not appear surprising in view of the entirely different biological functions of receptor and enzyme. Recently a definite separation of the two proteins has been achieved by Changeux and his associates (see page 76).

When a receptor inhibitor has an ($\overset{+}{N}\rightarrow OH$) group at an appropriate distance, namely about 5–6 Å (e.g., 1-methylacetoxyquinolinium), it forms a hydrogen bond with an atom in the active site (Podleski and Nachmansohn, 1966). Such a hydrogen bond formation was previously demonstrated with the hydroxyphenyltrimethylammonium ion which has a similar ($N^+\rightarrow OH$) distance on AcCh esterase (Wilson and Quan, 1958). However, it has been shown that blocking the esteratic site by specific irreversible enzyme inhibitors reacting with the oxygen of the serine residue does not interfere with this hydrogen bond formation between receptor inhibitor and the active site of the receptor (Podleski, 1967).

Further support for the protein nature of the receptor was obtained by the evidence of the presence of sulfhydryl and disulfide groups essential for the function of the receptor. The response of the excitable membrane of the electroplax to three receptor activators tested, AcCh, carbamylcholine, and trimethylbutyrylammonium, was found to be inhibited when sulfhydryl groups were blocked by p-chloromercuribenzoate (PCMB) or disulfide bridges were reduced by 1,4-dithiothreitol (DTT), a reagent designed to reduce disulfide bridges with a minimum formation of mixed disulfide (Cleland, 1964). Exposure of the electroplax of 0.5 mM PCMB or to 1 mM DTT for a few minutes reduced the depolarization obtained by the activators by about 30–40%. Extensive washing does not restore the response. However, the inhibition by PCMB is reversed by reducing, that by DTT by oxidizing compounds (Karlin and Bartels, 1966).

3. EFFECT OF QUATERNARY GROUPS. CONFORMATIONAL CHANGES

In spite of all the dramatic advances of protein and enzyme chemistry during the last two decades, the extraordinary catalytic power of enzymes remains one of the most fascinating, but still not fully understood activities in living cells. The key–lock concept of Emil Fisher assumed a rigid protein which absorbs the substrate to specialized catalytic groups. The resulting formation of an enzyme substrate complex seemed, at least to some extent, to account for the specificity of enzymes. However, the progress in the analysis of enzyme mechanisms has made it apparent that the simple lock–key theory did not provide a satisfactory explanation. A notion referred to as induced fit theory was introduced by Koshland and his associates (Koshland, 1960; Koshland *et al.*, 1962; Koshland and Neet, 1968). He postulated on the basis of a vast amount of experimental data, using a variety of enzymes and substrates, that ligands may induce conformational changes in the protein, producing a precise orientation of catalytic groups and thereby permit the reaction. Binding alone is not sufficient without the favorable change in conformation. During the last decade, this notion has found strong support by a variety of developments, such as the exploration of the tridimensional structure of proteins and enzymes in association with kinetic and chemical studies and the introduction of the notion of allosteric effects of the role of subunits and enzymes. Conformation and conformational changes of proteins during activity form today an integral part of biochemical thinking.

The possibility that a molecule such as a protein may not be a rigid structure and that local changes of conformation may take place during activity occurred to the writer some twenty years ago, long before any experimental evidence was available. It appeared striking that quaternary ammonium derivatives have so much more powerful pharmacological actions than their tertiary analogs, as was known for nearly a century. How could one extra methyl group increase so strongly the biological response? As long as this phenomenon was limited to observations of pharmacological effects, the complexity of the system excluded interpretations on a molecular level. However, when the enzymes associated with the hydrolysis and the formation of AcCh became available in purified solutions, it became possible to explore experimentally the problem of the effect of the extra methyl group in quaternary compounds. The results seemed to suggest that some process in the protein may be induced by the extra methyl group. The tetrahedral structure of a methylated quaternary ammonium group is more or less spherical. A direct contact between all four methyl groups of such a structure and the protein surface would not be readily possible, since one of the groups will be located away from the protein. Thus, it appeared possible that such a quaternary compound may induce a conformational change in the protein, possibly a localized one, permitting the macromolecule to have contact with all four groups of the tetrahedral structure and that this conformational change increased the catalytic efficiency.

The effect of the extra methyl group was first analyzed with a purified solution of AcCh esterase. When the inhibitory strength of ammonium and of hydroxyethylammonium ions with an increasing number of methyl groups was tested, it was found that introduction of two or three, respectively, methyl groups increased the strength of binding of the inhibitor to the enzyme about equally; the extra methyl group of the two quaternary compounds hardly increased the binding strength or, as was found later, at least to a much lesser degree (Wilson, 1952). Since the extra methyl group did not contribute significantly to the binding, the strong difference between the rates of hydrolysis of AcCh and its tertiary analog dimethylaminoethyl acetate by AcCh esterase, seemed to support the assumption that a local change of conformation may take place during activity whereby the methyl group of the quaternary form is enveloped, and that this was possibly the explanation of the greater catalytic efficiency.

The possiblity of conformational changes of AcCh esterase during the catalytic activity found some further, although still very indirect, support by studies of the enthalpies and entropies of activation. $\Delta H\ddagger$ and $\Delta S\ddagger$, of the ester of ethanolamine and its methylated derivatives with AcCh esterase from electric tissue (Wilson and Cabib, 1956). Substitution of the first two protons by methyl groups produced little change in the activation energies, but the extra methyl group had a very pronounced effect. The enthalpy of activation, $\Delta H\ddagger$, of the hydrolysis of AcCh is about 14,000 cal, as compared with about 8000 cal for that of the tertiary analog. On the basis of this finding, one could expect that the hydrolysis of the quaternary ester would be less favored than that of the tertiary. But the entropy of activation is extremely favorable for the quaternary compared to that of the tertiary compound. While it is negative with the tertiary analog, it is positive with the quaternary group. The difference amounts to about 30–40 entropy units. This large favorable entropy of activation and the higher rate of hydrolysis of quaternary esters, when catalyzed by the enzyme, appears all the more significant in view of the observations of Chu and Mautner (1966) that the nonenzymatic hydrolysis of the tertiary ester is very much faster than that of the quaternary analog. Thus, the actual difference of efficiency between the enzyme-catalyzed hydrolysis of tertiary and quaternary compounds seems to be much greater than is apparent simply on the basis of comparing the two enzyme-catalyzed reactions.

An even more striking difference of catalytic efficiency between quaternary and tertiary analogs has been observed with choline O-acetyltransferase. The rate of acetylation of choline was found to be about twelve times higher than that of dimethylethanolamine (Berman et al., 1953). On the other hand, the difference between the rates of acetylation of dimethyl- and monomethylethanolamine is small.

By far the strongest difference between AcCh and its tertiary analog is observed when the pharmacological actions are compared. On the basis of the differences of the catalytic efficiencies of the two enzymes and of the pharmacological actions, and on the firm conviction that the receptor on which acetylcholine acts is also a protein, the writer proposed, in his Harvey Lecture in 1953, as a possible explanation for the action of AcCh on the receptor, that it may induce a conformational change of the protein and thereby initiate other changes leading to increased permeability (Nachmansohn, 1955). A quantitative evaluation of the difference in potency between AcCh and its tertiary analog in their reaction with the receptor became possible when the monocellular electroplax preparation became available. By the removal of one methyl group from the quaternary form, the potency of the depolarizing action decreased two hundredfold (Bartels, 1962).

At the time when the suggestion was made that a conformational change of the receptor may initiate the sequence of reactions resulting in increased membrane permeability, the view appeared quite speculative. Today, the notion of conformational changes of protein, as mentioned before, is widely accepted. Although in the case of the three proteins associated with the function of AcCh there is still no direct evidence for this view, many kinetic and chemical data obtained in recent years have lent support to this assumption.

4. ESSENTIAL ROLE IN ELECTRICAL ACTIVITY

According to theory, the primary molecular event initiating the permeability changes of excitable membranes during electrical activity is the action of AcCh on the receptor, thereby inducing a conformational change. A crucial test for the validity of the theory is the ability of specific inhibitors of the AcCh receptor—i.e., compounds analogous in structure to AcCh and capable to reach the receptor in the conducting membrane—to block electrical activity by competing with the internally released AcCh and preventing its reaction with the receptor, which otherwise would induce the conformational change and lead to depolarization. A series of compounds that are analogs of AcCh in structure have long been known as "local anesthetics" (e.g., procaine, tetracaine). The differ from AcCh in two respects—(1) They are receptor inhibitors, i.e., they block electrical activity without depolarizing the membrane; (2) they are able to reach the receptor not only in the membranes of the junctions, but also in the conducting parts of the membrane, due apparently to their greater lipid solubility which permits them to penetrate structural barriers protecting conducting membranes (see Section V). The electroplax preparation permits a quantitative evaluation of the effects of the molecule substitutions that transform AcCh step by step into receptor inhibitors that act on junctional as well as on conducting parts of the membrane.

A systematic analysis of a great number of compounds in which a group of either the quaternary ammonium or the acyl part of AcCh has been substituted, has provided information about the molecular groups that transform one type of compound into the other type (Bartels, 1965; Bartels and Nachmansohn, 1965). Table I gives some of the most characteristic features of the transformation of the biological action in relation to chemical structure. When the methyl group on the carbonyl groups, for example, is substituted by a saturated benzyl ring, the potency of the compound as an activator is decreased about two hundred-fold compared to that of AcCh; but the compound is an activator, and the action is limited to the junction. Even in high concentration it does not act on the receptor of the conducting membrane, apparently because it is unable to reach it. However, such a small modification as a substitution of a saturated by an unsaturated benzyl ring (i.e., benzoylcholine) changes the properties of the molecule. It may act as a receptor activator or as an inhibitor according to the experimental condition used, and in concentrations of 10^{-3} M, benzoylcholine acts also on the conducting membrane. Thus, the compound is a typical transitory form in chemical structure and in biological action between AcCh and local anesthetics. Addition of an amino group to the phenyl ring in para position transforms the analog into a receptor inhibitor and increases the penetrating ability; the molecule has acquired the two features characteristic of a local anesthetic. Both inhibitory potency and ability of penetration may be increased by further substitution in either the quaternary or in the acyl part of the molecule.

These experiments lent further support to the presence of the receptor in the conducting parts of the membrane, since these analogs of AcCh act similarily as receptor inhibitors in both junctional and conducting

TABLE I

Transformation of Acetylcholine by Substitutions[a]

Compound	Concentration (M)		
	Synaptic junctions		Conducting membrane
	Activator	Inhibitor	
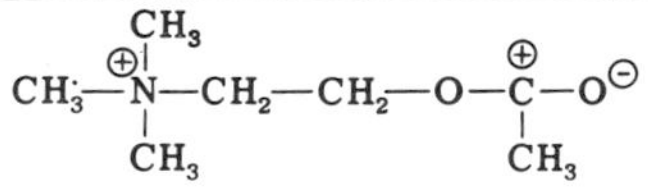 Acetylcholine	2.5×10^{-6}	0	0
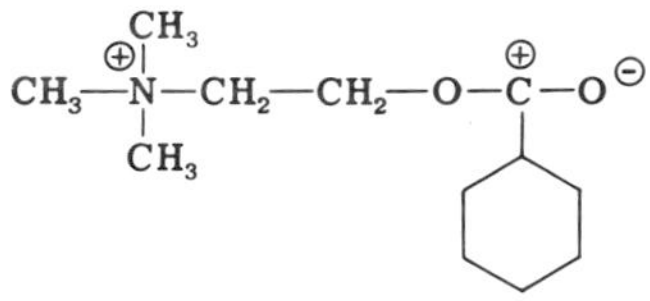 Hexahydrobenzoylcholine	5×10^{-4}	0	0
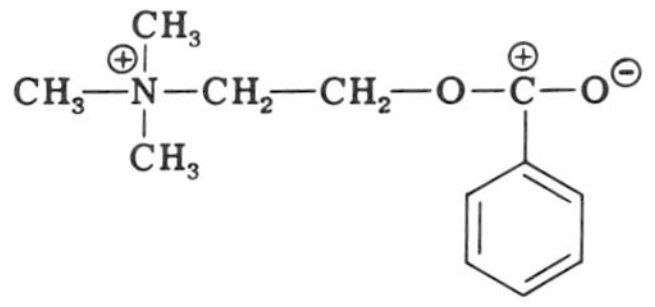 Benzoylcholine	5×10^{-4}	5×10^{-4}	1×10^{-3}
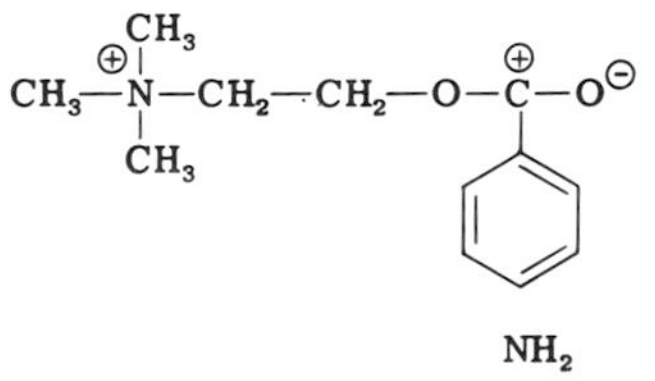 p-Aminobenzoylcholine	0	1×10^{-3}	2.5×10^{-3}
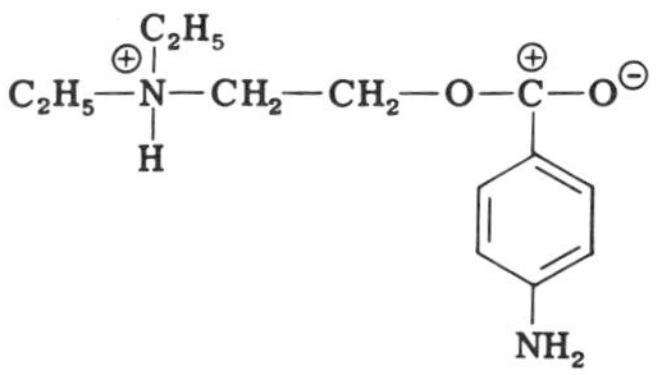 Procaine	0	2.5×10^{-4}	5×10^{-4}

TABLE I (*Continued*)

Compound	Concentration (M)		
	Synaptic junctions		Conducting membrane
	Activator	Inhibitor	
$CH_3-\overset{\oplus}{\underset{CH_3}{N}}-CH_2-CH_2-O-\overset{\oplus}{C}-O^{\ominus}$ (with phenyl ring and C_4H_9-NH substituent) Tetracainemethiodide	0	2×10^{-5}	1×10^{-5}

[a] Substitutions of AcCh gradually transform the molecule from a receptor activator acting on the synaptic junctions only into a receptor inhibitor blocking electrical activity both of junctions and of the conducting membrane ("local anesthetic"). Benzoylcholine is a transitory form, both in structure and in biological activity. For details, see text.

parts of the electroplax membrane. The usual type of competitive action with carbamylcholine or curare can only be demonstrated at the junction, because these quaternary compounds, when applied externally, do not reach the receptor in the conducting part; there the action is limited to the competition with internally released AcCh. Moreover, since it is known that local anesthetics block all electrical activity in all types of conducting fibers, the observations with these typical antimetabolites of AcCh offer evidence that the receptor is not only present but essential for electrical activity. The evidence parallels that obtained for the essential role of the enzyme in conduction with AcCh esterase inhibitors (Section III).

In view of several reports and suggestions that local anesthetics affect electrical activity be reacting with phospholipids and that their effects may be potentiated or antagonized by Ca^{2+} ions, a brief comment on these observations and their interpretations may be appropriate. The ability of phospholipids to react with local anesthetics is obvious. However, for the problem of the basic mechanism by which local anesthetics act on excitable membranes, reducing or blocking excitability, the possibility of their reactions with several cell components does not yet offer a satisfactory explanation. Many classical antimetabolites are capable of reacting with other cell constituents, but their effects are nevertheless attributed to the competition with the physiologically active metabolic analog for the target protein. An illustration of this point in the field of nerve activity is the block of synaptic transmission by d-tubocurarine. The effect is generally attributed to the competition with AcCh for the receptor protein, although d-tubocurarine is well known to react with a number of positively charged macromolecules by Coulombic and Van der Waals' forces; the binding forces are, however, much weaker than those for the receptor. When the question is raised whether local anesthetics, so strikingly analogous in structure to AcCh, affect electrical activity by competing with AcCh for the receptor protein, or in an unspecific way with other cell components, such as phospholipids, a few basic principles must be considered. When, e.g., a competitive reaction between an enzyme in solution and two ligands is postulated, a straight forward test for this assumption is the Lineweaver–Burk plot. When, however, a compound, postulated to act as an antimetabolite (or competitive inhibitor) of a specific protein, is tested on a complex structure, such as the cell membrane, many factors influence the ligand–protein interaction and require a critical analysis before a satisfactory answer can be obtained.

It is well established that AcCh, carbamylcholine, d-tubocurarine, and some related quaternary ammonium derivatives, when applied externally, react with the AcCh-receptor protein at the synaptic junctions only. They act there, as is well known, on both the pre- and the postsynaptic membranes. It is equally well established that the AcCh-receptor protein and AcCh esterase are present and functional in the conducting parts of excitable membranes, as repeatedly discussed in this chapter. But in the conducting parts of the membranes these proteins are protected by structural barriers (Schwann cell or equivalent structures). This problem is fully discussed in Section V. Therefore, the receptor and enzyme cannot be reached there by poorly lipid-soluble compounds applied externally, such as the quaternary derivatives mentioned, except in a few axons where the barrier is apparently poor or after the fiber has been treated by agents reducing the lipid barrier (as shown, e.g., with the squid giant axons, where the Schwann cell becomes more pervious by exposure to phospholipase A or simply to its split product lysolecithin). After such a treatment, AcCh and d-tubocurarine have been shown to penetrate into the interior, so they are capable of reaching the receptor and they produce the expected effects on electrical activity.

The same situation prevails in the electroplax. AcCh and carbamylcholine act at the synaptic junctions only; the depolarization of the whole excitable membrane is an indirect effect produced by the small currents developing between the many thousand depolarized junctions and the polarized conducting membrane. When the competitive nature of the action between local anesthetics and externally applied carbamylcholine is tested on the electroplax, the classical characteristics of competitive action are observed at the junctions only, as described in the original papers mentioned above. In contrast, local anesthetics are lipid soluble and are therefore capable of crossing the structural barrier and reach the receptor protein in the conducting parts of the excitable membrane, which forms more than 90% of the total surface area. In the conducting parts of the membrane, the compounds react with the receptor by competing only with the internally released AcCh and thus reducing or blocking excitability. Therefore, when the action of local anesthetics in the presence of carbamylcholine is tested on the electroplax, some deviation from the classical picture of competitive action must of necessity be expected, even if the receptor protein is present in both parts of the membrane and the reaction of the local anesthetic is a specific reaction with the receptor and the decisive factor responsible for the reduced or blocked excitability. The structural element clearly interferes with the simple relationship prevailing between two competitive compounds in solution. Obviously, reactions between local anesthetics and intramembranous phospholipids may, and probably will, take place and may even influence the effects observed, especially since phospholipids are most likely involved in electrical activity. But it seems difficult to escape the conclusion that chemical structures, so strikingly analogous to AcCh and shown to compete with AcCh for the receptor protein at the junction, must also act on the receptor protein in the conducting parts of the membrane. The assumption that an unspecific reaction of local anesthetics, such as the binding to phospholipids fully accounts for block of electrical activity, is, however, readily compatible with the view that only ions and phospholipids—but not proteins—are involved in the mechanism generating bioelectric currents.

Local anesthetics are frequently referred to as "stabilizers" of excitable membranes and compared to the "stabilizing" action of Ca^{++} ions. Therefore they have been assumed to share a common site of action. Blaustein and Goldman (1966) have shown, e.g., that some electrical parameters affected by procaine are antagonized by an increased external Ca^{2+} ion concentration and potentiated by a reduced one. Assumptions such as a "stabilization" of a complex structure are a rather vague description of a phenomenon and appears unsatisfactory for a bliochemical explanation. An interpretation would become more meaningful when the effects would be associated with specific chemical events, especially since Ca^{2+} ions have a great diversity of potent actions. It is therefore of interest to know that many observations on electrical parameters indicate definite differences between the actions of Ca^{2+} ions and local anesthetics; in some cases even opposite effects have been described. Without going into the details of the large number of electrophysiological studies, it may be mentioned that these differences led several investigators (see, e.g., Aceves and Machne, 1963) to the suggestion that Ca^{2+} ions and local anesthetics do not compete for identical, but different binding sites, although they are most likely involved in a mutually interacting system. The writer's view that the action of AcCh inducing a conformational change of the receptor may release Ca^{2+} bound to the protein by allosteric action is fully compatible with this suggestion; it provides, moreover, a basis for experimental tests of the chemical reactions underlying Ca^{2+} effects.

C. Cooperativity and Allosteric Action

Regulatory enzymes act at critical metabolic steps and have specific functions in regulation and coordination. They are activated or inhibited by metabolic effectors that are not substrates catalyzed by the enzymatically active site; they act on different sites. Monod et al. (1963) introduced the notion of allosteric effectors and referred to their binding sites as allosteric sites. The effectors do not act directly on the catalytic reaction, but produce an alteration of the molecular structure of the protein referred to as an allosteric transition which modifies the properties of the active site and changes kinetic parameters. When the reaction velocity of such enzymes is plotted against substrate concentration, the curve is not the usual hyperbola corresponding to a Langmuir isotherm, as is to be expected from a simple first order reaction, but has a sigmoid form indicating a second order reaction. There is a cooperative effect in the binding of more than one molecule to the enzyme. Cooperativity and allosteric systems have been in recent years extensively studied in many laboratories. The symmetry model of Monod et al. (1963, 1965) has found support in studies of a number of enzymes. Koshland and his associates analyzed cooperative properties on the basis of the induced fit theory and suggested a modified and more flexible interaction between ligands and protein which may induce a new conformation of the subunit (Kirtley and Koshland, 1967; Koshland and Neet, 1968). The prediction of their "sequential" model was borne out in experiments with several enzymes which do not fit the symmetry model.

Changeux and his associates were the first to recognize certain common properties of regulatory enzyme systems and excitable membranes. The biological activities of both systems depend upon threshold concentration of the regulatory ligand, and both systems exhibit cooperative phenomena and other similarities of behavior (Changeux et al., 1967).

The dose–response curve of the electroplax membrane to receptor activators has a sigmoid shape (Higman *et al.*, 1963). On the basis of the notion that biological membranes are an ordered collection of repeating globular lipoprotein units organized into a two-dimensional crystalline lattice, Changeux *et al.* (1967) interpreted the sigmoid shape of the response of the membrane in terms of allosteric systems and cooperativity (see also Changeux and Thiery, 1968). Similar ideas were discussed by Karlin (1967). A detailed analysis of the response of the electroplax membrane to AcCh and its congeners was performed by Changeux and Podleski (1968). Dose-response curves were obtained with several different activators within a large range of concentration. In all cases, the characteristic sigmoid shape was obtained. The Hill coefficient was 1.7; this is widely accepted as an indication of the cooperative character of the response. The maximal responses of three activators tested—decamethonium, carbamylcholine, and phenyltrimethylammonium—were different. The maximal response seems to be directly related to the structure of the ligand. Changes of the ionic environment, either low Na⁺ of high K⁺ medium, as compared to the usual medium of the Ringer's solution, did affect the absolute values of the maximum depolarization, but the relative differences between the different activators remained the same. Thus, the differential amplitude of the maximal response is determined by the elementary interaction between the receptor activator and the receptor protein in the membrane. This observation would be difficult to interpret in terms of the ionic theory, considering the ionic concentration gradients as the only determining factor of electrical parameters.

D. Special Ligands

1. Affinity Labeling

Compounds forming a covalent bond with a molecular group in the active site of an enzyme have long been important tools in the analyses of these sites. Organophosphates, for example, were used in the studies of AcCh esterase and other ester-splitting enzymes. A different type of compounds forming covalent bonds, referred to as affinity labeling, has been introduced by Wofsy *et al.* (1962). The reagent, because of its steric complementarity to the active site, first combines specifically and reversibly with the site with which it forms a reversible complex; then a small and reactive group reacts with one or more amino acid residues to form irreversible covalent bonds. Affinity labeling of the AcCh receptor was first attempted by Changeux *et al.* (1968) with *p*-(trimethylammonium)benzenediazonium fluoroborate (TDF) applied to the electroplax membrane. TDF is a structural analog of phenyltrimethylammonium (PTA), a potent activator of the receptor. It was therefore reasonable to assume that it would form a reversible complex with the anionic site of the receptor; the diazonium would form a covalent bond with an amino acid residue in or near the active site. Exposure of the electroplax produced an irreversible block of the response to receptor activators, although the concentration required appeared high, 10^{-4} *M*. A competitive action for the receptor was demonstrated with reversible receptor inhibitors, such as *d*-tubocurarine and others.

While these observations supported indeed the assumption of a typical affinity labeling, more recent observations on the reaction of TDF and related compounds with the electroplax membrane suggest a modification of the reaction mechanism proposed on the basis of earlier experiments. When phenyltrimethylammonium was replaced by an uncharged *p*-nitro group, the effect on the electroplax response was just as strong as that of TDF (Mautner and Bartels, 1970). When the *p*-nitro group is replaced by an acetoxy group, which attracts electrons less strong than either the *p*-nitro or the trimethylammonium groups and thereby decreases the positive charge of the diazonium, the blocking strength decreased by an order of magnitude. The experiments suggest that it is the positively charged diazonium group which is attracted to the anionic subsite of the receptor.

Another more potent affinity labeling was obtained in a two-step procedure (Karlin and Winnik, 1968). As mentioned before, the response of the electroplax to AcCh and its congeners is blocked when disulfide bridges are reduced by DTT; the block is reversed by oxidizing compounds. When *N*-ethylmaleimide (NEM) is applied after DTT, the reversal by oxidizing compounds is prevented. Presumably NEM forms a covalent bond with the exposed sulfhydryl groups. Since the block of action on the receptors suggested a location of the disulfide bridges near the active site, it appeared possible to increase the strength of the reaction by introducing a group attracted by the anionic site, as was done in earlier studies with AcCh esterase and was also the reasoning for the use of TDF. The ethyl group of the maleimide was substituted by phenyltrimethylammonium—4-(*N*-maleimido)phenyltrimethylammonium iodide (MPTA). This compound appeared to be a potent affinity label when applied to the receptor reduced by DTT; the response of the electroplax to carbamylcholine is abolished by MPTA at 10^{-7} *M*. The tertiary analog of MPTA is not more potent than NEM. Adding the extra methyl group enhances the alkylation of the receptor about three hundredfold, which is about the same difference which was found between tertiary and quaternary analogs in other derivatives of receptor activators, as mentioned before.

Two other affinity labeling compounds for the AcCh receptor were designed by Silman: bromoacetylcholine bromide and [(p-nitrophenyl)-p-carboxyphenyl] trimethylammonium iodide. Both compounds are receptor activators; they depolarize the electroplax, but they form covalent bonds with the receptor only after exposure to DTT, i.e., presumably reacting with the nucleophilic sulfhydryl groups (Silman and Karlin, 1969; Silman, 1970).

Drastic changes of biological effects may result from the reduction of the disulfide bridges in the active site of the receptor by exposure to DTT. Hexamethonium, a bisquaternary reversible receptor inhibitor, becomes a receptor activator after reduction (Karlin and Winnick, 1968). The compound TDF, an irreversible inhibitor of the receptor at 10^{-4} M, becomes after reduction a potent reversible activator at 10^{-6} M (Podleski *et al.*, 1969). Thus apparently opposite biological actions, leading to excitation or inhibition, may be due to relatively small changes in the state of the receptor and to other factors in the membrane, and not, as it is widely assumed, to different excitatory or inhibitory transmitters. Such differences in the response cannot be explained on the basis of physiological effects, which are the result of a great variety of factors, but require an analysis of the molecular reactions involved. As an illustration for this view may be mentioned the well known dependence of the action of ATP on the concentration of Ca^{2+} ions in inducing either muscular contraction or relaxation.

2. CHARACTERIZATION OF THE RECEPTOR BY SNAKE VENOM TOXINS

The use of compounds prepared for affinity labeling for the receptor was not yet successful in experiments aimed at its isolation in its native state. The use of maleimide derivatives apparently still lacks the necessary specificity for permitting the isolation after the reduction of the receptor by dithiothreitol. Apparently SH groups belonging to other proteins present in the membrane may interfere. A promising new tool was recently introduced by Changeux *et al.* (1970) for the characterization and identification of the receptor: toxins of snake venom which react specifically with the receptor protein. One of them, α-bungarotoxin, a polypeptide of molecular weight of 8000 purified from the venom of the snake *Bungarus multicinctus*, which had been studied by Lee and his associates (Chang and Lee, 1963; Lee and Chang, 1966; Lee *et al.*, 1967). The toxin produces irreversible neuromuscular block; *d*-tubocurarine protects against the action of α-bungarotoxin. From these findings, Lee and Chang (1966) concluded that the toxin combines irreversibly with the cholinergic receptor at the motor end plate.

The toxin was applied by Changeux *et al.* (1970) to various preparations derived from the electric organ of *Electrophorus*, i.e., to the isolated electroplax and to isolated excitable membranes ("vesicles," see below) and to proteins obtained by fractionation of the electric tissue. The toxin (1 mg/ml) irreversibly blocks the electroplax response to carbamylcholine; *d*-tubocurarine protects against the effect of the toxin in agreement with the observations of Lee and his associates. The same effects were observed with isolated membranes of the electroplax. Finally, the authors tested α-bungarotoxin on the binding of radioactive decamethonium to a protein isolated from electric tissue and which presents in solution some properties characteristic of the cholinergic receptor protein. The toxin blocks, in equilibrium dialysis assay, the binding of decamethonium to the protein; it has no effect on AcCh esterase. Thus, the observations suggest specific binding to the physiological receptor of AcCh without affecting the catalytic sites of the enzyme. Still more recently, Changeux and his associates have been able to separate, in soluble extracts of electroplax membranes, the receptor protein from AcCh esterase by using the α-toxin of *Naja nigricollis* coupled to Sepharose. The toxin–Sepharose absorbs most of the receptor, but leaves essentially all the enzyme in the supernate (Meunier *et al.*, 1971).

The α-toxin of *Naja nigricollis* was also used by Changeux and his associates for testing the localization of the receptor in the electroplax. Antibodies to the toxin were prepared and coupled with fluorescent immunoglobulins (Bourgeois *et al.*, 1971). It was found that the toxin binds exclusively to the excitable membrane (Fig. 9). The distribution of the receptor (A) is not quite as uniform as that of the enzyme (B), shown on an equivalent segment of the excitable membrane. For the interpretation of this electron micrograph several factors must be considered in respect to the question of whether or not the receptor protein is as uniformly distributed in the synaptic and the conducting parts of the excitable membrane, as is AcCh esterase (see page 46). First, the total surface area of the synaptic junctions, in spite of their large numbers, forms less than 5% of the total surface area of the excitable membranes. The estimates of the relative surface areas were made many years ago before new and improved electron micrographs showed the very large number of the deep infoldings of the excitable membrane. The surface area of the synaptic junction may actually form a smaller fraction than 5% of the total. In any event, it would hardly account for the large part of the fluorescent segments of the total area of the membrane. Thus the distribution found contradicts the assumption that the receptor protein is

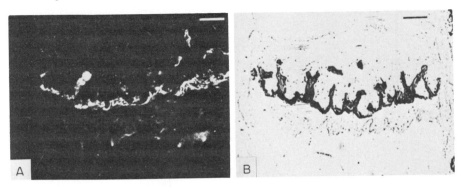

Fig. 9. Electron micrographs showing the localization of AcCh-receptor and esterase in the excitable membrane of eel electroplax. (A) Immunochemical localization of α-toxin of *Naja nigricollis* (reacting with the receptor) bound to a section of an electroplax. 5 μg/ml α-toxin, 0.2% formaldehyde, 100-fold dilution of rabbit and sheep serum. The length of the bar is 100 μm. (B) Localization of acetylcholinesterase on a section of electroplax by Koelle's reaction. Orientation of sections in A and B are identical. (From Bourgeois *et al.* 1971.)

localized only in the subsynaptic parts of the membrane. Second, a vast number of physicochemical data, accumulated over the years, indicate the presence as well as the functional activity of the receptor protein in both the conducting and synaptic parts of the excitable membrane of the electroplax (for a detailed discussion see Nachmansohn, 1971). Finally, the procedures demonstrating the presence of AcCh-esterase have been worked out and improved over two decades; in contrast, the method used for the localization of the AcCh-receptor protein is quite new and the authors indicate that they are working on its improvement and its adaptation to electron microscopy. Improved procedures may show a still more even distribution of the receptor protein. This is exactly what has happened with the localization of the enzyme when finally examination by electron microscopy fully agreed with the biochemical data. In view of all these factors it seems to the writer that the electron micrograph presented favors the assumption of the localization of the receptor parallel to that of the enzyme. Certainly, the limitation of the pharmacological effects of some cholinergic agents does not indicate that the receptor is present only in the subsynaptic membrane, as still maintained by some investigators. The contradiction between this conclusion, based on the external application of AcCh, and the biochemical, biophysical, and structural data, will be fully discussed in Section V. The presence of the AcCh-receptor protein in conducting fibers has recently been demonstrated in experiments of Denburg *et al.* (1972), who applied α-bungarotoxin and other tests for the AcCh-receptor protein to the axons of the walking leg of lobster, fibers which are rich in AcCh esterase and choline-*O*-acetyltransferase.

3. OXYGEN, SULFUR, AND SELENIUM ISOLOGS OF AcCH AND ITS CONGENERS

Small changes of structure and configuration of ligands reacting with proteins and enzymes may strongly modify the reaction and greatly affect biological activity. For a systematic study of these factors, Mautner and his associates prepared a great variety of sulfur and selenium isologs of AcCh, choline, benzoylcholine, and many of their congeners (Mautner and Günther, 1961; Günther and Mautner, 1964). The oxygen, sulfur, and selenium isologs are similar in size, but may greatly differ in electron distribution and configuration. The role of these two aspects of the ligands in their reaction with AcCh esterase and receptor have been analyzed in the last few years by Mautner and his associates; in the tests of the effects on the receptor, the electroplax preparation was generally used and in collaboration with the writer's laboratory.

Kinetic, spectroscopic, and dipole measurements of isologous esters performed by Mautner and his associates offer evidence that the electron distribution in the oxygen, sulfur, and selenium isologs is different (for a summary, see Mautner, 1967). Therefore, their ability to bind to the active sites of the two proteins and of inducing conformational changes could be different. Remarkably, strong differences have been found in the biological activity of the isologs. For instance, choline, even in 10^{-1} M quantities, is virtually unable to produce a depolarization. This is about 40,000-fold the concentration at which AcCh has a strong depolarizing action. The sulfur isolog, cholinethiol, has, on the other hand, only a slightly less depolarizing effect than AcCh itself; the difference is less than tenfold (Mautner *et al.*, 1966). Although the choline isologs increase in potency by the substitution of the oxygen atom by sulfur and selenium, the opposite is true for the esters. AcCh is more effective than acetylthiocholine and very much more than acetylselenocholine. It is interesting that the biological ester is by far the most active of the three isologs, whereas its hydrolysis product is the most inactive. On the other hand, the rates of hydrolysis by the enzyme are similar. The great contrast in potencies between the reactions of the isologs with receptor and those with esterase is another support for the difference between the active sites of the enzyme and the receptor.

The differences of conformation of the isologs were studied by X-ray analysis in order to distinguish between the effect of this factor on biological activities and the effect of electron distribution (Shefter and

Mautner, 1967, 1969). The results indicate that the structure of crystals of oxygen and sulfur isologs tend to differ, whereas those of sulfur and selenium are so similar as to make such molecules isosteric. AcCh, choline, and a series of related molecules show in the solid state that in general the N^+—C—C—O grouping is in the *gauche* formation (Canepa *et al.*, 1966; Shefter and Mautner, 1969); this is also the case in solution (Culvenor and Ham, 1966). In contrast, in acetylselenocholine crystals the nitrogen and selenium are trans to one another (Günther and Mautner, 1964; Shefter and Kennard, 1966). The trans conformation of the S—C—C—$\overset{+}{N}$ group in acetylthiocholine, and the corresponding group in acetylselenocholine are rather stable. Both compounds retain their trans conformation in solution, as was shown with nuclear magnetic resonance (Cushley and Mautner, 1970). In view of the almost identical conformation of sulfur and selenium isologs, the marked differences in biological activities, both with the receptor and with the enzyme, may be ascribed to differences in electron distribution. These and other observations support the assumption that both factors, conformation and electron distribution, play an essential role in biological activity.

The isologs offer an important possibility for testing similarities or differences of their reaction with the AcCh receptor protein in junctional and axonal membranes. If the active sites of the protein are similar, the potency of oxygen, sulfur, and selenium isologs applied to the two types of membranes should exhibit comparable differences, provided the compound can penetrate through the barriers surrounding axons and reach the receptor. When 2-dimethylaminoethylbenzoate and its sulfur and selenium isologs were tested (the ether oxygen was replaced by sulfur or selenium) on the electroplax membrane and on the squid giant axons, the differences in potency obtained with these three isologs on the two preparations were found to be remarkably similar (Rosenberg *et al.*, 1966). Similarly marked differences of potency were obtained in the reactions with the receptors of both preparations when the carbonyl oxygen was replaced by sulfur and selenium (Rosenberg and Mautner, 1967). The data offer a new kind of support for the presence and functional similarity of the AcCh receptor sites in axons and in junctions.

4. Photoregulation of Membrane Potential by Photochromic Substances

The strong effects of sunlight on growth and development of plants has long been known. One of the compounds known to function by photoregulation in many plant processes is the chromoprotein phytochrome that seems to act as an absorber and transducer of light energy (Hendricks and Borthwick, 1967). The activity of phytochrome has been suspected for some time to result in changes of membrane permeability. Recent observations of Jaffe (1970) suggest that AcCh regulates the phytochrome-mediated phenomena by effecting changes in ion fluxes across cell membranes of the plant which he tested.

The classic example of photoregulation by photochromic substances in animals is vision. It is based on the cis–trans isomerization of retinal (Wald, 1968). cis-Retinal reacts in the dark with the protein opsin to form rhodopsin. Light-induced isomerization of the cis-retinal to the all-trans configuration leads to nerve excitation. Absorption of only a few quanta of light leads to a neural response (Hecht et al., 1941). Indications begin to accumulate that photoregulation by photochromic substances may be a more widespread mechanism than has been suspected, not only in plants but also in animals.

Recently, Erlanger and his associates prepared a number of diazo compounds which exist as cis and trans isomers and are intraconvertible by means of light. The compounds share a p-phenylazophenyl group. Light causes a reversible shift in the cis-trans equilibrium about the nitrogen–nitrogen double bond of the compounds. The trans isomer predominates at 420 nm wavelengths in the light of a photoflood lamp; the cis predominates at 300 nm wavelengths in ultraviolet light. When these light-sensitive ligands were applied to chymotrypsin (Kaufman et al., 1968) or to AcCh esterase (Bieth et al., 1969), the potency of the two isomers as enzyme inhibitors was different. A photoregulation of the activity of the two enzymes was obtained on exposure of the enzyme system in presence of the azo compounds to different wavelengths.

In view of the effect of cis–trans isomerism in vision leading to neural stimulation and the effects obtained on AcCh esterase, it appeared of interest to test whether photoregulation of the potential across the excitable membrane of the electroplax could be obtained with ligands which are structural analogs of AcCh receptor activators and which isomerize by exposure to different wavelengths. Two diazo compounds were tested on the electroplax membrane: N-p-phenylazophenyl-N-phenylcarbamylcholine chloride (azo-CarCh) and p-phenylazophenyltrimethylammonium chloride (azo-PTA) (Deal et al., 1969). The compounds, derivatives of strong receptor activators, are strong receptor inhibitors. Both trans isomers, predominating at 420 nm, are markedly stronger inhibitors than the two cis forms, predominating at 320 nm. When the electroplax is depolarized by 20 μM carbamylcholine, the two compounds depolarize the membrane in low concentrations. Azo-CarCh in its trans form has

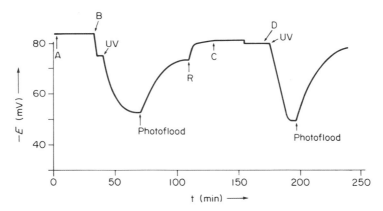

Fig. 10. Photoregulation of the potential across the innervated electroplax membrane of *Electrophorus*. Ultraviolet (Spectroline Model B-100) and photoflood lamps were turned on at the·times indicated. (A) Carbamylcholine (carch), 20 μM; (B) azo-carch, 1 μM; azo-carch, 3 μM; carch, 50 μM.

a strong repolarizing action in 1 μM concentration, whereas the cis form has a weak effect. The two isomers of azo-PTA have comparable effects, but at ten times higher concentrations. When the electroplax is exposed to carbamylcholine in the presence of 1 μM azo-CarCh under ultraviolet irradiation, a marked depolarization takes place. When a steady state is reached and the preparation is then exposed to a photoflood lamp, the membrane is repolarized by 20–30 mV (Fig. 10). Similar results were obtained with 20 μM azo-PTA.

The system may be considered as a model illustrating how one may link a cis–trans isomerization, the first step in the initiation of a visual impulse, with substantial changes in the potential difference across an excitable membrane. The changes of 20–30 mV in membrane potential are comparable to those occurring in the visual process, but it takes minutes to obtain the effects, whereas the biological response to light occurs in a millisecond. This difference in efficiency is not surprising. The active compounds in the retina are highly specialized; the light-sensitive molecules form part of the membrane and are almost certainly structurally organized; the potency of intracellularly active biological compounds is usually small when applied from the outside (see Section V). The interesting feature of these experiments is the demonstration of a neural stimulation induced by a change of conformation of a ligand of a kind similar to that occurring on the biological process of vision.

Among a number of new phenylazophenyl derivatives recently synthesized by Erlanger and Wassermann, two of the new compounds proved to be of particular interest in several respects. The one is 3,3′-bis [α-(tri-methylammonium)methyl]azobenzene dibromide (bis-Q) and the closely related 3-(α-bromoethyl)-3′-[α-(trimethylammonium)methyl]azobenzene bromide (QBr):

QBr   Bis-Q

trans-Bis-Q tested on the electroplax was found to be an extremely potent AcCh-receptor activator, one of the most potent ever found (Bartels *et al.*, 1971). When the concentration at half-maximal response (6–8 $\times$ 10^{-8} M) was used for comparison, this isomer was found to be 500 times more potent than carbamylcholine. No pure *cis* isomer was available, but the calculated activity for the *cis* isomer from the mixture indicates very low potency; it is possible that pure *cis* isomer may even lack activity. Increase of the concentration of *trans*-bis-Q to 2×10^{-6} M produces repolarization. D-tubocurarine blocks the depolarization by the low concentration of *trans*-bis-Q. The repolarization induced by 2×10^{-6} M *trans*-bis-Q takes place both in the presence and in the absence of *d*-tucoburarine, suggesting the possible existence of two binding sites of which only one competes with *d*-tubocurarine. The response to *trans*-bis-Q is inhibited by the reduction of the receptor sites with dithiothreitol.

Exposure of the electroplax to *trans*-QBr at a concentration of 2×10^{-7} M causes a slight depolarization of 5–10 mV. This action is readily reversed. *trans*-QBr inhibits the response to carbamylcholine. However, prior treatment of the cell with dithiothreitol results in irreversible inhibition of the effect of carbamylcholine: thus it is evident that *trans*-QBr is irreversibly attached to the reduced receptor, forming a covalent bond with a sulfhydryl group of the reduced receptor comparable to that with covalently attached *p*-(carboxyphenyl)trimethylammonium iodide reported by Silman (1970) and Silman and Karlin (1969).

The high potency of *trans*-bis-Q, depolarizing the membranes at exceedingly low concentrations (about 10^{-7} M) indicates a remarkably high degree of specificity, emphasized by the finding that the *cis* isomer has very little or no activity. Receptor activators, just as enzyme substrates, require greater specifications of molecular structure than receptor (or enzyme) inhibitors: factors such as electron distribution, configuration, complementarity, etc., apparently play a stronger role. Inhibitors may act by less specific interactions. It may be estimated from the now known approximate number of molecules of enzyme and receptor in the electroplax membrane, that only about 10^4 molecules of *trans*-bis-Q are required in the solution for the reaction with the receptor (see Bartels *et al.*, 1971). This remarkable degree of specificity may be useful in studies aimed at isolation and characterization of the receptor. The compound or appropriate derivatives may be used for the isolation of the receptor by affinity chromatography.

The covalent attachment of QBr to the receptor opens the possibility to find compounds in which photoregulation by photochromic substances covalently linked to the effector molecule may be produced in a more rapid and efficient way than by compounds acting from the outside.

E. Isolated Fragments of Excitable Membranes

Changeux *et al.* (1969) have prepared isolated fragments of excitable membranes which permit *in vitro* studies of a variety of properties of these membranes and of the proteins located there. The membranes are prepared in the following way. Homogenized suspensions of electric organs of *Electrophorus electricus* are sonicated and centrifuged in sucrose gradients. A particulate fraction rich in AcCh esterase contains membrane fragments free of cytoplasm. The membranes form close microsacs or vesicles. Most of them are derived from excitable membranes and are rich in AcCh esterase. Some of the microsacs, although only a minor fraction, are parts of the nonexcitable membranes; they contain ATPase. The two types differ not only by their chemical but also by their structural features as demonstrated by the electron micrographs shown in Figs. 11 and 12. The microsacs are incubated over night in a ^{22}Na containing medium. The suspension is then diluted into a nonradioactive medium. The ^{22}Na content of the microsacs as a function of time is then followed by rapid filtration on Millipore filters. The rate of ^{22}Na efflux is increased in the presence of AcCh-receptor activators (cholinergic "agonists") and blocked by receptor inhibitors such as *d*-tubocurarine or tetracaine. Thus the preparation permits to analyze the action of these compounds *in vitro* in a cell-free medium in a well-defined environment. The information obtained with these isolated membranes on the effects of excitation by cholinergic agents on ion fluxes appears to the writer to belong to one of the most pertinent advances in the biology of excitable membranes, especially in conjunction with parallel studies of the solubilized or isolated proteins and the isolated intact electroplax preparation.

Figure 13 shows the difference of the effects of carbamylcholine and *d*-tubocurarine on the rate of Na efflux when applied to microsacs formed by either excitable or nonexcitable membranes. Carbamylcholine accelerates the efflux of Na$^+$ ions from the excitable membranes, *d*-tubo-

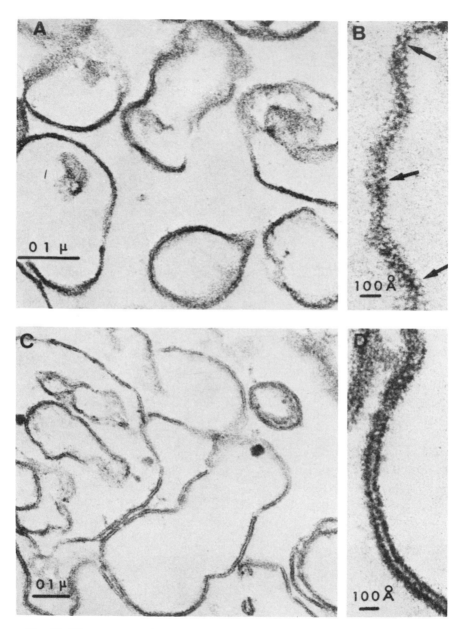

Fig. 11. Thin sections of microsacs formed by membrane fragment of electroplax. (A) Microsacs of AcCh esterase rich fraction. (B) High magnification showing that the membrane consists of a row of globular units (arrows). (C) Microsacs of ATPase rich fraction. Shape and size are more irregular, with membranes showing a triple layered structure (D). From Castaud *et al.* (1971).

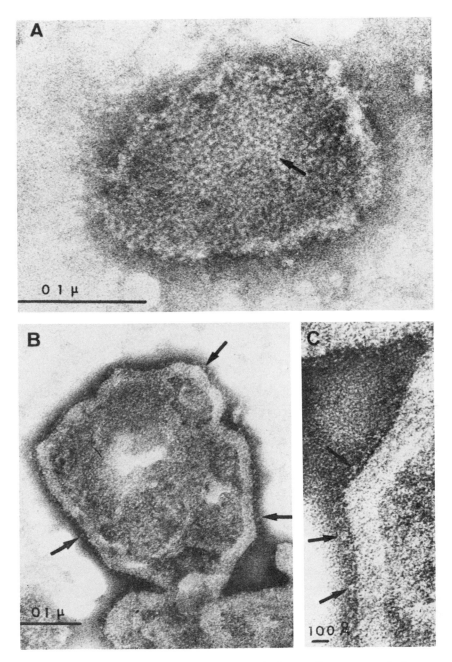

Fig. 12. Negative staining. (A) Microsacs of the AcCh esterase rich fraction negatively stained by sodium phosphotungstate. Membrane shows globular subunits. The arrow points to a regular array of subunits. (B), (C) Microsacs of ATPase-rich fraction. The membrane surface (C) is covered by globular knobs which are protruding at the edge of the membrane sheets (arrows).

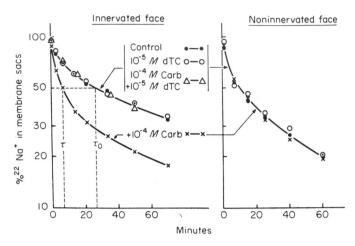

Fig. 13. Effect of carbamylcholine (Carb) and d-tubocurarine (d-TC) on the efflux of $^{22}Na^+$ from excitable microsacs. Left: excitable AcCh esterase-rich microsacs derived from innervated membranes. Right: nonexcitable ATPase-rich microsacs derived from noninnervated membranes. 27°C. τ and τ_0 are the times for half equilibration in the presence and in the absence of cholinergic agonist, respectively. (From Kasai and Changeux, 1971.)

curarine blocks this effect; both compounds have no effect on the non-innervated membranes (Kasai and Changeux, 1970, 1971).

An important result of these observations is the remarkable parallelism found with receptor activators and inhibitors when the effects on the membranes *in vitro* were compared with those obtained on steady state potentials of the intact electroplax (Fig. 14). The electrophysiological data are taken from Changeux and Podleski (1968). The same affinities are observed with the two activators used, carbamylcholine and decamethonium, for controlling the ion effluxes from the microsacs and the electric response in the intact cell. Even the sigmoid shape of the dose–response curve is similar in both systems, indicating cooperativity and allosteric effects. The Hill coefficients for AcCh (in the presence of eserine), carbamylcholine, decamethonium and phenyltrimethyl ammonium were found to be the same *in vivo* (intact cells) and *in vitro* (microsacs).

The effects of cholinergic agents on the rate of Na$^+$ ion efflux are readily reversible, as was shown by experiments in which carbamylcholine was first applied in 10^{-4} M concentration and then diluted to 7.5×10^{-6} M. The fast rate in presence of the high concentration returns to that of the control in which the low concentration was used. These observations thus contradict the assumption that the rate of Na efflux may be due to an irreversible disruption of the microsacs.

As was discussed before (see page 74), exposure of the electroplax to dithiothreitol (DTT) reduces S–S bridges in the neighborhood of the active site of the AcCh-receptor and modifies the response to chemical

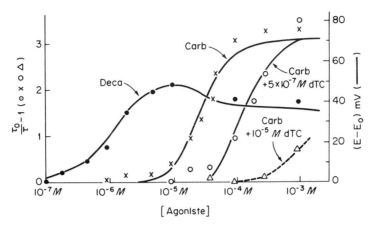

[Agoniste]

Fig. 14. Affinities to AcCh receptor tested on cellular level (isolated electroplax of *Electrophorus*) and compared to those on subcellular level (isolated membranes: "vesicles," "microsacs"). The dose-response curves on the electroplax of activators—carbamylcholine (carb), decamethonium (deca)—or inhibitors—*d*-tubocurarine (dTC)—have been taken from Changeux and Podleski (1968); they were evaluated as usual by electrical parameters (voltage changes are given on the right side). They were compared to the dose-response curves obtained by measuring the rate of ^{22}Na efflux from the "vesicles" (see text). τ is the period of time required for the 50% loss of the ^{22}Na of the vesicles in the presence of carbamylcholine, $\tau = \tau_0$ in its absence. (Taken from Kasai and Changeux, 1970.)

excitation, sometimes transforming receptor inhibitors into activators or vice versa. Similar effects were obtained with the isolated membrane fragments. After in vitro exposure of the microsacs to 10^{-13} *M* DTT for 15 minutes, the response to carbamylcholine is decreased, whereas that to hexamethonium, before exposure to DTT and inhibitor, is increased.

Among the many other observations the effects of pH and temperature appear of particular interest and may be briefly mentioned. The rate of Na efflux, $1/\tau_0$ min^{-1}, increased fourfold between pH 5.5 and 8.5, while the excitability, $\tau_0/\tau - 1$, remained constant. These data support the assumption that the sites for the recognition of the signal (the receptor protomer) differ from the ionophores which are responsible for the accelerated ion movements. Similarly, the rate of Na efflux increased with temperature in the presence as well as in the absence of decamethonium, but the relative response to 3×10^{-6} *M* of the compound did not change. The Q_{10} for the change of ion flux was about 1.8.

Another pertinent result of these studies with isolated membranes are the selective permeability changes effected by the action of carbamylcholine. The permeability to Na$^+$, K$^+$ and Ca^{2+} ions is increased, but not that to large ions, such as choline or tetraethylammonium, or uncharged permeants. The permeability to negatively charged ions is either slightly (Cl$^-$) or not at all (SO$_4^{2-}$) affected.

The most dramatic result, however, in the view of the writer, is the striking similarity of the various parameters of Na$^+$ and K$^+$ fluxes when the effects of chemical stimulation of the microsacs are compared to those of the electric response to squid giant axons, as summarized in Table II.

TABLE II
PROPERTIES OF MICROSACS AND SQUID GIANT AXONS[a]

Properties measured	Microsacs (10^{-4} M carb)	Squid giant axons (action potential)
Fluxes (moles/cm^2/sec)		
Φ Na	4×10^{-13} (effl.)	3.7×10^{-12} (infl.)
Φ K	2×10^{-12} (effl.)	8.0×10^{-11} (effl.)
Permeabilities (cm/sec)		
p Na	2×10^{-9}	3.3×10^{-9}
p K	1×10^{-8}	7.4×10^{-7}
Conductance (mho/cm^2)		
g Na	1×10^{-6}	3.3×10^{-6}
g K	9×10^{-6}	2.4×10^{-4}

[a] Comparison of Na$^+$ and K$^+$ ion fluxes, permeabilities and conductances obtained either by chemical stimulation of isolated membrane fragments with carbamylcholine (carb) or by electrical stimulation of squid giant axons.

The fluxes in moles/cm^2/sec, Φ_i and Φ_0, were calculated following the assumption of the independent principle of Hodgkin and Huxley (1952)

$$\Phi_i = \rho C_0 \frac{F}{RT} \frac{E}{e^{EF/RT} - 1}$$

$$\Phi_0 = \rho C_i \frac{F}{RT} \frac{E}{1 - e^{-EF/RT}}$$

where Φ_i and Φ_0 are the influx and outflux of the permeant ions, p the permeability, F the Faraday constant, R the gas constant, T the absolute temperature and E the given membrane potential. C_0 and C_i are the outer and inner concentrations of the ions. Since in the case of the microsacs E may be assumed to be 0, then

$$\Phi_0 = pC_i$$

Similarly, the conductances, mho/cm^2, were calculated from the relation between conductance and permeability according to Dodge and Frankenhaeuser (1959). Again assuming E and E_0 (the equilibrium potential for the considered ion) to be 0, the equation is

$$g = \rho \frac{F^2}{RT} C_i$$

The permeabilities p, cm/sec, are calculated from τ_0 obtained from the flux experiments by the relation $p = r/2\tau_0$ with r as the average radius of the microsacs (assumed to be 500 Å). Due to the uncertainties in the determination of r, a range of values is given for r (see Kasai and Changeux, 1971, p. 38).

Whereas the parameters of Na$^+$ and K$^+$ fluxes are similar when tested on isolated electroplax membranes and on squid giant axons, the values obtained with intact electroplax for K efflux are 2–3 orders of magnitude higher. This discrepancy obviously requires further investigations. It may not be real but due to a number of assumptions which require correction. One of the most difficult and problematic estimates is that of the surface area. Recent electron micrographs indicate an amount of folding of the excitable membranes very much greater than that which appeared in earlier electron microscope examinations. The surface area of the excitable membrane may be two orders of magnitude greater than estimated before. Since the actual area of the excitable membrane of the intact electroplax is an essential factor in the parameters measured, the results may be due to this single factor two orders of magnitude smaller than assumed in earlier studies. Other factors which formed the basis of these calculations may also require reevaluation. The sources of error in the calculation of the fluxes using squid giant axons or microsacs are more limited than those using intact electroplax.

The striking similarity of the various parameters Φ, g, and p of Na$^+$ and K$^+$ ions across the isolated excitable membranes of the microsacs stimulated chemically by carbamylcholine and across the membranes of the squid giant axon in response to electrical stimulation is one of the most dramatic supports in favor of the theory proposed by the writer more than two decades ago, i.e., that the action of AcCh on the proteins is responsible for the permeability changes in excitable membranes during electrical activity and that the function of the protein assembly activated by AcCh as the specific signal is similar in axonal and synaptic membranes. The presence of AcCh-receptor and esterase in both types of membranes and the evidence for their essential role in the electrical activity of axons makes the conclusion almost inescapable that the similarity of the effects of chemical and electrical stimulation on Na$^+$ and K$^+$ fluxes is due to the activation of a similar mechanism controlling the permeability changes. At junctions the excitable membrane is usually poorly protected or not at all, and AcCh applied externally is, therefore, capable of activating the protein assembly. In contrast, in axons the excitable membrane is almost always protected by an insulating cover. However, AcCh is released within the membrane by excitation. The similarities and differences between axonal conduction and synaptic transmission are fully discussed in Section V.

The preparation of isolated fragments of excitable membranes offers a new and valuable assay method for the analysis of the properties of excitable membranes. It permits to compare the behavior of the proteins associated with these membranes (i) in solution, (ii) in the subcellular structure, and (iii) in the intact cell. It provides the possibility of more detailed studies of the other important components of the membrane such as the phospholipids, the nature of the ionophores, the role of Ca^{2+} ions, of many other small molecules, etc. Some of the many still open problems have been discussed by Changeux and his associates and a few tentative interpretations have been given. One may disagree with some of them, and when more information begins to accumulate, modifications may become necessary. But the preparation offers a great opportunity and a challenge to the imagination of investigators interested in the many problems of excitable membranes on a molecular level.

V. Similarities and Differences between Axonal and Junctional Membranes

The fundamental notions applied to the problem of the role of AcCh in nerve activity were based, as described in this chapter, on bioenergetics and protein chemistry. This approach permitted the characterization and identification of the proteins associated with the function of AcCh. In addition, the results very soon indicated the necessity of a drastic modification of the original theory of the transmitter role of AcCh that was based on notions of classical electrophysiology and pharmacology. A quarter of a century has passed since the new concept began to emerge. It seems appropriate to reevaluate the original theory in the light of the vast amount of biochemical data accumulated. A test for the validity and strength of a concept is its ability to provide more satisfactory interpretations of the observations than the preceding theory.

A. *Reevaluation of the Basic Facts of the Transmitter Theory*

1. Effects of AcCh and Curare. Permeability Barriers Surrounding Conducting Membranes

Two facts have been emphasized from the beginning as chief support of the assumption that the mechanism of synaptic transmission differs fundamentally from that of conduction and that the role of AcCh is limited to the junction. The first one is the powerful pharmacological action of AcCh on junctions in contrast to its complete failure to affect axonal conduction. Similarly, more than a century ago, Claude Bernard observed that the activity of curare blocks only the transmission of the impulse from nerve to muscle, but does not affect conduction in nerve and muscle fibers. Curare is a receptor inhibitor with an affinity to the protein which is greater than that of AcCh and prevents, therefore, the depolarizing action of AcCh. The conducting membranes of axons are surrounded by Schwann cells. They form structural barriers usually impervious to certain quaternary nitrogen derivatives such as AcCh, curare, and related compounds. During the last twenty-five years, the presence of such barriers, preventing AcCh and curare from reaching the conducting membrane and reacting with the receptor, has been demonstrated in a great variety of experiments and with the use of many types of preparation. The barriers are usually either less effective or absent at the synaptic junctions. The conducting membrane of the muscle fiber is also surrounded by structural barriers, although no Schwann cell is present. The barriers explain the limitation of the action of AcCh applied from the outside (i.e., of its pharmacological action) to the membranes of the junction.

Curare has been applied by Claude Bernard to the whole frog sciatic nerve. As mentioned before, this nerve is formed by several thousand axons. The plasma membrane of myelinated fibers is surrounded by a heavy myelin sheath which may be several microns thick except at the nodes of Ranvier. The whole bundle is surrounded by a sheath of connective tissue impervious to many chemical compounds. As mentioned before, physostigmine affects the electrical activity of an isolated single fiber at concentrations four to five orders of magnitude lower than those acting on the whole nerve. The effective concentrations are similar to those acting on the neuromuscular junction (Dettbarn, 1960a). When curare (d-tubocurarine) is applied to single fibers, the electrical activity, recorded at the nodes of Ranvier, is rapidly and reversibly blocked (Dettbarn, 1960b). There is still a layer of Schwann cells at the nodes, although no myelin is present. This layer is still less pervious to quaternary than to tertiary nitrogen derivatives; neostigmine, a quaternary nitrogen derivative, affects electrical activity only in ten times higher concentrations than the tertiary physostigmine, although both compounds are nearly equally strong inhibitors in solution (Augustinsson and Nachmansohn, 1949b). Even the whole bundle becomes sensitive to AcCh, curare, and neostigmine after it has been desheathed and exposed to the detergent cetyltrimethylammonium (Walsh and Deal, 1959).

In unmyelinated axons the excitable membranes are usually also well protected against quaternary ammonium ions. The Schwann cell surrounding the squid giant axon is about 4000 Å thick and contains lipid. When these axons are exposed to AcCh, curare, and neostigmine in very high concentrations, no effects on electrical activity are observed (Bullock *et al.*, 1946). In contrast, the tertiary physostigmine and the tertiary analog of neostigmine block electrical activity reversibly, although the latter is a poorer enzyme inhibitor in solution than either the quaternary analog or physostigmine. When the axoplasm of the exposed axons was extruded, physostigmine was found to have penetrated, while neostigmine had not entered the interior. The presence of the barrier was also confirmed with ^{15}N-labeled AcCh (Rothenberg *et al.*, 1948). In contrast, ^{15}N-labeled trimethylamine readily penetrated to the inside. More recently, Hoskin and Rosenberg (1964) confirmed these findings with radioactively labeled AcCh, curare, and trimethylamine.

In addition to the experiments already mentioned, direct actions of AcCh and curare on the excitable membranes of axons have been demonstrated on several other preparations where the structural barriers are either inadequate to prevent the compounds from penetrating to the receptor and affecting the electrical parameters in the way predicted by theory, or where chemical treatment has reduced these barriers to a degree that the compounds are able to penetrate and become active. Rosenberg and his associates have shown that the electrical activity of squid giant axons is reversibly blocked by exposure to AcCh and curare when preceded by treatment with cottonmouth moccasin venom for a brief period, 20–30 min, in low concentrations, 20–30 μg/ml. The treatment itself had no effect on electrical activity (for summary, see Rosenberg, 1965, 1966). Examination with electron microscopy revealed marked structural alterations of the Schwann cell, but none of the plasma membrane (Martin and Rosenberg, 1968). Subsequent to the venom treatment, radioactively labeled AcCh and curare were found to have penetrated into the axoplasm, whereas they were absent in controls (Hoskin and Rosenberg, 1964). The active component in the snake venom was shown to be phospholipase A; its effect is fully accounted for by the formation of lysolecithin which by itself produces the same result as the treatment by the venom (Condrea and Rosenberg, 1968; Rosenberg and Condrea, 1968).

A direct action of AcCh, without any treatment, has been found on the axons of the walking leg of lobster (Dettbarn and Davis, 1963). The compound first prolongs markedly the descending phase and then blocks electrical activity and depolarizes the membrane. Under proper conditions, this effect is antagonized by atropine. Dettbarn (1963) demonstrated also an effect of curare on these axons. This preparation is tested in seawater, which has a high concentration of Ca^{2+} and Mg^{2+} ions. Since these ions are known to antagonize the action of curare, Dettbarn reduced the concentration of Ca^{2+} and omitted Mg^{2+} ions in the seawater. Under these conditions, he obtained rapid and reversible effects on electrical activity with curare. Apparently the Schwann cell in these axons does not offer the complete protection it does in other preparations. This is supported by the findings with electron microscopy (DeLorenzo *et al.*, 1968). When the axons are exposed to cottonmouth moccasin venom prior to the application of AcCh and its congeners, the potency of the compounds increases twenty to fifty times, indicating a still further reduction of the barriers.

Another preparation which reacts to AcCh after the removal of the sheath is the vagus nerve of rabbit (Armett and Ritchie, 1960), although the effect was considered to be a pharmacological curiosity (Ritchie, 1963).

On the other hand, even synaptic junctions frequently do not react to AcCh or curare, for instance, the neuromuscular junction of lobsters, although the concentrations of AcCh esterase in these junctions are extremely high. They do react, however, to lipid-soluble compounds such as physostigmine and DFP. Thus, the proteins there are present and functional, but apparently protected, as in axons, by barriers against quaternary nitrogen derivatives. Similar barriers were discovered in the nervous system of insects. The total absence of the action of AcCh on the insect nervous system was considered as evidence that it is not cholinergic. Therefore, the effect of organophosphate insecticides was attributed to a different factor than the inhibition of AcCh esterase. It took more than a decade of intensive investigations until this situation was clarified. The effect of AcCh was demonstrated after mechanical removal of the barriers and the close relationship between the toxic action of organophosphates and enzyme inhibition was established (for a summary, see O'Brien, 1960).

Another support for the presence of similar receptors in junctional and conducting parts of excitable membranes are the observations with oxygen, sulfur, and selenium isologs of AcCh or its congeners, as described in Section IV; the differences in potency between the isologs in their reaction with the receptor are strikingly similar in junctional and in conducting parts of the membranes of electroplax and squid giant axons, thus again supporting the assumption of similar receptors.

Recently, Karlin *et al.* (1970) made an estimate of the number of receptor molecules in a single electroplax and compared this number to that of molecules of AcCh esterase. The enzyme activity was determined in solution after the cell was kept in $1 M$ sodium chloride, a procedure which appears to solubilize most of the membrane-bound enzyme. In contrast, the estimate of receptors present per cell was based on experiments on the intact cell, using the usual pharmacological criteria. In essence, quaternary maleimide derivatives were applied to the intact reduced cell and the competition with hexamethonium, a bisquaternary compound, was used as basis for the calculation. While the calculations thus include all or most of the enzyme of an electroplax, the methods applied for estimating the receptor molecules permit in the best case the determination of the small fraction of the receptor present at junctions only.

One of the fundamental biochemical concepts, discussed by Krebs (1966), is the assumption that all cell constituents have a biological function, even if the function remains obscure for a long time. In view of all the data accumulated, it appears difficult to assume that the presence of the extraordinary concentration of AcCh esterase in the conducting parts of the electroplax membrane forming 95% of the total is not functional and has, in contrast to the synaptic parts of the membrane, no physiological significance, as the absence of receptor there would imply. The approach to the problem is an illustration of how difficult it is to reconcile views based on biochemical concepts with those derived from pharmacological ones.

2. Release of AcCh

The second observation, considered strong evidence for the theory of neurohumoral transmission, is the appearance of AcCh in the extracellular perfusion fluid of junctions after stimulation. However, it was stressed time and again by Dale and his associates (see Dale *et al.*, 1936) that no trace of AcCh appears in the absence of physostigmine, even after prolonged stimulation. This failure to detect any extracellular AcCh unless AcCh esterase, the highly effective inactivation mechanism, is blocked, suggests that the appearance in the outside fluid is an artifact attributable to incomplete hydrolysis. The observation of Otto Loewi (1921)—that stimulation of the heart vagus of frog in the absence of physostigmine leads to the appearance of a substance (*Vagusstoff*) in the perfusion fluid which mimics vagal stimulation when added to the perfusion fluid of another frog—is still frequently quoted in support of neurohumoral transmission. Loewi tried for many years in this country to repeat his original observation, but was unsuccessful. He attributed this failure to the use of a frog species (*Rana pipiens*) different from that used in Europe, where he has used *Rana esculenta.**

---

* However, at the International Congress of Biochemistry in Vienna in 1958, Professor F. Bruecke informed the writer when he visited his laboratory, that he and his collaborators had tried for two years to repeat Loewi's experiments with *Rana esculenta,* but were unsuccessful.

It has long been assumed that AcCh is released on stimulation exclusively from the nerve endings (Dale *et al.*, 1936). However, McIntyre and his associates have demonstrated in extensive investigations that AcCh appears in the perfusion fluid of muscle even after complete degeneration of the nerve endings, thus suggesting a release from the postsynaptic as well as the presynaptic membrane (McIntyre, 1959). In these studies an explanation is also offered for the negative results of the early experiments.

The detection of an efflux of AcCh from axonal membranes will be prevented by the structural barriers surrounding axons and preventing AcCh and curare from reaching the excitable membrane from the outside. However, in a preparation such as the axons of the walking leg of lobster, where the barriers are incomplete, an efflux may be expected and was indeed found when the axons were kept in physostigmine (Dettbarn and Rosenberg, 1966). Various ions were found to affect this efflux in a way similar to that described for junctions. It may be recalled that Lorente de Nó (1938) has vigorously objected to the assumption that AcCh metabolism is specific for the junction as early as 1938. He has demonstrated the appearance of AcCh on several sites outside synaptic junctions.

In summary, the release of AcCh may be demonstrated at sites outside the junction. It is an artifact, since no trace appears outside the cell, neither at junctions nor elsewhere, without inhibition of AcCh esterase, the physiological removal mechanism. Thus, the two fundamental facts on which the transmitter theory was originally based, the limitation of the pharmacological action to the junction, and the release of AcCh from the nerve terminal require a new interpretation in the light of the new data; they can no longer be considered evidence for the transmitter theory.

B. Theory Based on Biochemical Data

An alternative explanation accounts more satisfactorily for the events occurring during transmission and conduction on the basis of all new biochemical information accumulated. The new interpretation, moreover, integrates the observations on which the transmitter theory was based with the views of the many prominent neurobiologists who vigorously objected to the interpretation of a mechanism of transmission across junctions fundamentally different from conduction along nerve and muscle fibers (Erlanger, 1939; Lorente de Nó, 1938; Fulton, 1938; and many others).

1. Presence of AcCh Esterase and Receptor in Junctional Membranes

As repeatedly stated, it is now well established that both the receptor protein and the enzyme are present in the excitable membranes of nerve and muscle fibers. While light microscopy was unable to distinguish between pre- and postsynaptic membrane and to determine the precise localization in synaptic junctions, electron microscopy in combination with staining methods has well established that the enzyme is localized in both pre- and postsynaptic membranes. The first demonstration by Barrnett (1962) has been subsequently confirmed and greatly extended in many laboratories.

Masland and Wigton (1940) were the first to show that AcCh, curare, and neostigmine produce an effect on the terminal, i.e., presynaptic membrane as well as on the postsynaptic membrane. These observations have been confirmed, and during the last fifteen years largely extended in many laboratories (see Riker *et al.*, 1959; Werner and Kuperman, 1963). It appears that in many instances the terminal membrane is even more sensitive to these compounds than the postsynaptic membrane. Thus, the two proteins reacting during electrical activity with AcCh, receptor and esterase, are both present and functional in both pre- and postsynaptic membranes.

2. Discrepancy between Amounts of AcCh Pharmacologically Active and Externally Released

As discussed above, AcCh appears outside the membrane only in the presence of physostigmine. However, even in the presence of physostigmine, the amounts required to produce a response are surprisingly high. Such a sensitive preparation as the electroplax does not respond to 10^{-4} M AcCh in the absence of physostigmine; in its presence, 10^{-6} M AcCh is required. The amounts released per stimulus per terminal are probably at most of the order of 10^{-20} mole. This extraordinary discrepancy raises a serious difficulty in attributing to AcCh the role postulated in the original theory, namely that it is released from the terminal and acts as a transmitter of the impulse on the second cell, nerve, or muscle.

3. Role of AcCh at the Junction

On the basis of the biochemical data and in the light of all the available information, the following explanation appears to be the most satisfactory. AcCh acts within the terminal as within the conducting mem-

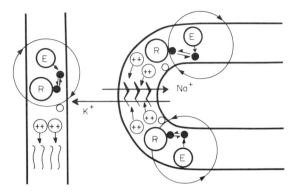

Fig. 15. Schematic presentation of the role of AcCh at junctions postulated to be essentially similar to that in the axon (see Fig. 3; symbols used are the same). As in the axonal membrane, AcCh is released by the current arriving at the terminal; it acts as the signal that initiates the reactions which amplify the signal and lead to increased permeability of ions in the terminal membrane. Potassium ions coming out of the nerve terminal act as transmitters of the impulse from cell to cell: many millions of them cross the nonconducting gap per 1000 molecules of AcCh released. However, these ion movements require the reactions within the terminal membrane initiated by AcCh. The flow of K^+ ions acting across the gap leads, directly or indirectly, to a release of AcCh in the postsynaptic membrane which then initiates the same series of reactions there.

brane as a signal which is released within the membrane, when the currents propagating the impulse along the axon reach the terminal (see Fig. 15). The same amplifier process takes place: the signal is recognized by the receptor in the terminal membrane and triggers the sequence of the reactions, resulting in the influx of Na^+ and efflux of K^+. Shortly after Cowan (1934) had demonstrated a strong efflux of K^+ ions from axons following stimulation, Feldberg and Vartiainen (1934) found an efflux of K^+ ions at synaptic junctions. In view of these findings, Eccles (1935) attributed the transmitter function across junctions to K^+ ions. However, at that time no chemical control was even envisaged. Today it is obvious that the movement of Na^+ and K^+ ions across the terminal membranes will require chemical control reactions, as in the axonal membranes. This control mechanism on both sides of the junction is triggered by AcCh. Many millions of K^+ ions, possibly as many as 40–50 millions, cross the gap for each one thousand molecules of AcCh released and effect directly or indirectly (e.g., H_3O^+, electric field, etc.) the release of AcCh in the postsynaptic membrane.

Half a century ago, Meyerhof developed the notion that cell reactions are chemically and energetically coupled. As we know today, most of them are structurally organized. It thus appears more plausible that the specific signal in one of the fastest cellular mechanisms known, our telephone system, is recognized by its target protein within the membrane where it is released, rather than by a target protein located in another cell. A structurally organized sequence of chemical reactions taking place within the membrane may be extremely efficient, fast, and precise; they will be energetically readily coupled with preceding and subsequent reactions. All these features we would indeed expect from chemical processes associated with bioelectricity.

The pharmacological action of AcCh, when applied to junctions, must be reinterpreted on the basis of the data described as mimicking the signal within the two membranes of the junction. Therefore, it acts not only on the postsynaptic membrane, but is able to produce antidromic impulses as is well known. However, as is the case in many actions of biologically active compounds applied externally to the cell, the effectiveness of AcCh applied from the outside is extremely small, since almost certainly only an infinitely small fraction will enter the membrane, which is rich in lipids, and will require in addition the presence of a strong inhibitor of the enzyme present in high concentrations.

4. FLOW OF CURRENT FROM NERVE TERMINALS

For many years, the failure to detect flow of current from nerve terminals seemed to exclude currents or ions as transmitting agents across junctions in contrast to axons. The apparent absence of flow of current was therefore considered strong support for the theory assuming chemical transmitters. The absence of current flow seemed particularly manifest in the case of the giant synapse of squid (Bullock and Hagiwara, 1957). In this preparation, parts of the pre- and postsynaptic activated axons are located side by side. It is possible to insert microelectrodes into the two axons so that the tips of the pipettes are approximately 2 mm apart. Thus, the current flow can be readily monitored. The failure to detect current flow appeared in this case particularly conspicuous. Biological systems are far too complex to permit theories to be built on the basis of negative experiments, even when performed with highly sensitive physical methods. An excellent illustration is provided by the early history of heat production measurements in nerve. Transmission by electrical currents was described by Furshpan and Potter (1959). Subsequently, flow of current from nerve terminals has been demonstrated by Hubbard (1963) and his findings were confirmed (Katz and Miledi, 1965). Recently, flow of current from both post- and presynaptic axons was found in the giant synapse of squid by Gage and Moore (1969). Thus, another objection to the assumption of a fundamentally similar mechanism of conduction and transmission has been removed.

5. Synaptic Vesicles

When it was discovered by electron microscopy that there are vesicles near the terminals (de Robertis and Bennett, 1955), it was suggested that these vesicles are a kind of little glands containing AcCh and releasing the transmitter during activity. Del Castillo and Katz (1957) proposed in essence the following mechanism: during activity the vesicles move to the nerve terminals and open there, each vesicle releasing packages or "quanta" of AcCh.

The difficulties of interpreting structural patterns seen with the electron microscope in terms of function has been discussed by Sjöstrand (1962). He vigorously objects to speculations and uncritical interpretations based exclusively on structural patterns. Some investigators stimulated by these patterns are carried by their imagination, he writes, to describe rather nice stories of functional sequences and events, without even any serious effort of proving or disproving their ideas. While it seems obvious that the vesicles, like any other cell structure, must have a functional significance, their role in synaptic transmission was first suggested exclusively on the basis of their presence, i.e., on the interpretation of a picture without any evidence of biochemical or biophysical data in support of the hypothesis. Vesicles have been found in other parts of the neuron; sometimes they are present at a high concentration on the postsynaptic site (Edwards *et al.*, 1958). Only in the last few years have serious efforts been made to identify the chemical content of the vesicles. However, the results, far from clarifying the situation, have multiplied the difficulties of linking the vesicles with synaptic transmission. The problem has been discussed in more detail elsewhere (Nachmansohn, 1969); therefore, only some essential features of the results will be mentioned here.

Presynaptic nerve terminals and their component organelles have been isolated by the use of mild procedures of homogenization and by sucrose density gradient centrifugation; they are referred to as synaptosomes (Gray and Whittaker, 1962; de Robertis *et al.*, 1962; for a recent summary, see Whittaker, 1968). The synaptosomes contain the vesicles, which may be isolated by further sucrose density gradient centrifugation. However, the results of the biochemical data offer more problems and difficulties than provide a satisfactory answer to the role of the vesicles in the function of AcCh. One serious difficulty, raised by Whittaker (1968), is that of the mechanism by which AcCh gets into the vesicles. There is no choline O-acetyltransferase in the vesicles. Labeled AcCh and choline penetrate into the cytoplasm of synaptosomes, but only negligible amounts enter into the vesicles, suggesting a permeability barrier between cytoplasm and vesicles, according to Marchbanks (1968,

1970). Similar results were obtained when radioactive compounds were injected intracerebrally and the vesicles were isolated. Another serious difficulty is the question of how AcCh released from the vesicles—by a mechanism still completely mysterious—would be able to reach the outside, i.e., the synaptic gap, where it is supposed to act as the transmitter to the second cell. The ester released from the vesicles should pass within microseconds through the cytoplasm and penetrate a membrane in which AcCh esterase is present in high concentration and which is rich in lipids and thus poorly pervious to quaternary ammonium ions. Whittaker makes the ad hoc assumption that there may be intercommunicating tubules, but they cannot be seen by electron microscope examination. Even the assumption of invisible channels would encounter innumerable difficulties. One could always add additional speculations to get around these difficulties. Finally, the observation that when the vesicles are isolated by density gradient centrifugation from the cytoplasm of the synaptosomes (about two thousand molecules of AcCh are found per vesicle) does not exclude the possibility that this fraction of AcCh comes from the cytoplasm in which most of the AcCh is present. Particularly in the light of the experiments with radioactive AcCh, it appears possible that AcCh, which is a strong cation, is absorbed during centrifugation to some negatively charged groups of the organelle which is a rather large particle compared to the size of AcCh, and is kept to the organelle by electrostatic and Van der Waals' forces.

In view of these and other difficulties, the role of the vesicles in the release of AcCh during activity appears quite obscure. The assumption that AcCh is released on excitation within the membrane from a bound form and acts on the target protein located at a distance of a few angstroms (i.e., the AcCh receptor, known to be present and functional in the terminal membrane) appears to the writer to be the more satisfactory view. Such an assumption is in line with the increasing evidence that membranes are formed by structurally well organized protein assemblies. The idea of a structural organization within the membrane seems all the more reasonable for processes associated with one of the fastest, most efficient, and most precise function of living cells, such as the permeability changes during electrical activity that propagate impulses at a speed of many meters per second.

C. Special Features of Junctions

The specific chemical forces underlying cellular mechanisms, such as motility, energy supply, vision, and genetic control, are remarkably similar throughout the animal kingdom. Modifications of some components of these systems in different species, in the course of evolution and phylogenetic development, are surprisingly small. It has become apparent that this basic similarity applies also to the specific chemical forces underlying bioelectricity, in particular to the specific proteins controlling the changes of ion permeability. The great diversity of bioelectrical phenomena may be explained by the nearly infinite variations of cellular structure and organization. The multiformity of shape, structure, and environment, especially at synaptic junctions, is bound to modify the effects of chemical reactions taking place in the membranes, presenting a very small fraction of the total mass of nerve or muscle fibers. As an illustration how markedly structural differences may affect electrical parameters is the great variation of conduction velocity in different types of axons, which varies from 0.1 to 100 m/sec, in spite of the fact that axons have a rather uniform and relatively simple cylindrical shape.

However, the most dramatic and most striking difference between junctions and fibers offer the pharmacodynamic action aspects. All or almost all cholinergic drugs act on synaptic junctions only. The effectiveness of the barriers protecting the excitable membranes of fibers has been documented and discussed before. The barriers even protect in the intact organism many synaptic junctions against many cholinergic drugs—for instance, most of the junctions located in the brain. One factor is the blood–brain barrier, which prevents quaternary or bisquaternary compounds from entering the brain. However, in view of the high concentration of lipids in the brain, this may be only a minor factor in the protection of the junctions located in the brain. Even tertiary nitrogen derivatives do not readily reach the excitable membrane in intact organs, as illustrated by the extraordinary discrepancy between the action of physostigmine on the whole sciatic nerve and on the nodes of Ranvier of an isolated single fiber discussed before. It came as a big surprise to many investigators believing in neurohumoral transmission that curare affects the muscle after denervation, not only at the junction but along the fiber. Rather wild speculations were proposed for explaining this phenomenon. Since it is well established that in the excitable membranes of muscle fibers both AcCh receptor and esterase are present and functional, and that tertiary ligands reacting with either one or the other of these two proteins affect electrical activity, the effect of curare after denervation is not as surprising as it seems. It suggests that the protecting barrier covering the sarcolemma becomes less impervious to curare after denervation, probably due to some structural modifications which may be very small.

It is of paramount and crucial importance not only for pharmacology but for medicine in general that cholinergic drug action be limited to the junctions. These pharmacodynamic aspects, however, do not provide an adequate basis for the interpretation of the biological function of AcCh. The gap between the views presented in this chapter and the theory of neurohumoral transmission is at present too large to be bridged. While many biochemists have now accepted the views presented here, it will probably take one or two more decades until biochemically well trained physiologists and pharmacologists will be accessible to the ideas based on bioenergetics and biochemistry.

VI. Bioenergetics of Nerve Impulse Conduction

A. *Heat Production*

Helmholtz first demonstrated heat production in muscular contraction in 1848, but failed to find any in stimulated frog nerve. The sensitivity of the instruments used was several orders of magnitude below that needed for these measurements. Several attempts by other investigators in the nineteenth century failed. Even early in this century, Hill (1912), using greatly improved thermoelectric instruments, was still unable to detect heat production in stimulated nerves. Bayliss (1915) wrote that "this result makes it impossible to suppose that any chemical process resulting in an irreversible loss of energy can be involved in the transmission of the nerve impulse." Hill, however, did not believe that nerve impulse conduction is a purely physical process and went on trying with still further improved instruments, until he finally, in 1925, succeeded in demonstrating heat production during nerve activity (for a summary of these developments, see Hill, 1964). Although the initial heat produced per stimulus appeared small, he felt that there was very strong evidence that chemical processes are involved in nerve impulse conduction and he gave to a lecture in Cambridge the challenging title, "Chemical Wave Transmission in Nerve" (Hill, 1932), with the intention of encouraging research on the chemistry of nerve activity. A few decades later, he and his associates, using much faster recording instruments than those available before, demonstrated a strong heat production and absorption, coinciding roughly with electrical activity, as mentioned before (Abbott *et al.*, 1958). While in the 1920s the heat per gram of nerve per impulse was estimated to be a few microcalories, the new values—which referred to grams of membrane per impulse in the light of the new information about biomembranes—were about 2 mcal. In this paper and at several later occasions, Hill stressed that there was hardly any alternative for explaining these results than the assumption that chemical processes must be associated with the permeability cycle.

In spite of this evidence of a strong heat production and absorption associated with the electrical activity, Hodgkin and his associates still consider the ion movements as a simple diffusion process, fully accounting for nerve impulse conduction. He and his associates have still repeatedly and explicitly rejected the assumption of any chemical reaction associated with the permeability cycle during electrical activity (see Baker *et al.*, 1962; Keynes and Aubert, 1964), although Hodgkin (1964) admits that the strong heat production is puzzling. The facts which are considered evidence against the assumption of a chemical reaction in electrical activity have been discussed elsewhere (Nachmansohn, 1964, 1966b) and do not need additional comments.

However, the interpretation of Howarth *et al.* (1968) of their recent observation that heat is produced and absorbed coinciding with the action potential is due to ion friction requires some comments. In their paper, the authors quote as to the possible source of the net heat, the "suggestion (Nachmansohn, 1959) that it is the heat of hydrolysis of acetylcholine broken down during the spike. However, since this is of the order of 3000 cal/mol (Nachmansohn, 1959) . . ." and the authors continue to make estimates on the basis of this figure. The figure of 3000 cal/mole given in the monograph quoted represents the ΔF of AcCh hydrolysis. The enthalpy, ΔH, which indicates the heat of hydrolysis, is not mentioned in the monograph. Neither on page 150 of the monograph quoted by the authors, nor in any publication of the writer, has the suggestion quoted by the authors been made. However, to the question of Keynes, "Would you expect (a) the release of AcCh, and (b) its action on the membrane permeability to be exothermic or endothermic reactions?" raised at the symposium on "Nerve as a Tissue" in 1964, published in 1966, the writer (Nachmansohn, 1966a, pp. 269–270) has given the following answer:

It is impossible to answer this question. What we definitely know is that there are a whole series of chemical reactions associated with the permeability cycle: the release of AcCh; the reaction with the receptor; the hydrolysis of AcCh; the release and neutralization of protons; ion mixing, possibly release and binding of Ca; rearrangement of polyelectrolytes and phospholipids, possibly with breaking and formation of hydrogen bonds and hydrophobic bonds. We have very little information about the many chemical reactions involved, and no notion of the associated enthalpies, not even in solution. The recent work of Katchalski has shown how extensively chemical reactions may be affected if they take place in a structure. In view of our ignorance of the reactions and enthalpies involved, your question cannot be answered. For the same reason it seems to me also im-

possible to draw any conclusions from heat measurements as to the underlying mechanism, or to predict whether one should expect heat production
or absorption just on the basis of one single reaction out of the many
involved.

In the light of these statements the above mentioned quotation by
Keynes and his associates appears quite surprising.*

It seems appropriate to elaborate this casual and improvised discussion
remark and to comment on the significance of thermodynamic parameters in the light of present information. For evaluating the heat produced
by chemical reactions, the following thermodynamic parameters are essential: the free energy change (ΔF), the heat or enthalpy change (ΔH),
and the entropy change (ΔS). In addition to having mixed up two
thermodynamic parameters, Howarth *et al.* (1968) consider in their
discussion of the strong heat produced and absorbed the entropy changes,
but do not specify the chemical processes responsible for them. This para-
meter reflects the change in the degree of randomness of the system
or the passage to differently ordered states; it has special significance
in macromolecular systems, since it has been recognized that the large
values of ΔS in such systems are mainly associated with conformational
changes. The value of ΔS may be as high as 40–50 cal/mole/deg, so that
$T \Delta S$ may amount to 15,000 cal/mole.

The thermodynamic parameters of many enzymes and protein systems
in solution have been evaluated. The complexity of these systems, even
in solution (i.e., in a steady state) frequently offers experimental and
theoretical difficulties which prevent precise separation of individual
parameters. As an illustration may be mentioned the analysis of these
parameters for the reaction of hemoglobin with oxygen, a protein of
which the tridimensional structure and the existence of four subunits
is known and in which conformational changes have been demonstrated
(Rossi-Fanelli *et al.*, 1964; Wyman, 1964).

B. Nonequilibrium Thermodynamics

If it is difficult to determine and to identify thermodynamic parameters
in well defined and isolated protein systems in solution, an analysis
of these parameters in intact structures appears at present impossible.
Classic thermodynamics is concerned primarily with equilibrium processes. Clausius, one of the founders of thermodynamics, was well aware
that the principles developed do not apply to systems which are not
in equilibrium, such as living cells. The Nernst equation and the Don-

---

* "Controversy may be distasteful; but misrepresentation, like libel, demands coun-
teraction." (Hill, 1964, p. 363.)

nan potentials have been successfully applied in the well known studies on the physical chemistry of permselective membranes by L. Michaelis, K. Sollner, T. Teorell, K. H. Meyer, J. F. Sievers, and many others. The synthetic membranes used consist of a polymeric matrix that carries fixed charges. The investigations laid the foundation for many important aspects of ion transport across membranes.

Although many of these studies were originally stimulated by biological interest, it is today recognized that many of the principles applied to the studies of synthetic membranes do not apply to biomembranes and that the latter require, therefore, a different treatment. As was pointed out at the beginning of this chapter, cell membranes are organelles in a highly active and dynamic state; they are the site of many proteins and enzymes and of many chemical reactions; they are not just passive barriers. Equilibrium is only reached at the death of the cell. Thus, the notions of classic thermodynamics are inadequate when applied to living cells or biomembranes. During the last few decades new principles and notions of nonequilibrium or irreversible thermodynamics have been developed by the work of Onsager (1931), Casimir (1945), Prigogine (1947), Katchalsky (1967), their associates, and many others. The central function of nonequilibrium thermodynamics is the rate of "irreversible" production of entropy, or the dissipation function which gives the dissipation of free energy per unit time and which is the more convenient function for the treatment of isothermal processes (see Katchalsky, 1967). Although some membrane problems have been approached in the light of the new notions, the available information is much too incomplete for a satisfactory analysis. However, even from the tentative discussions of some of the problems of biomembranes, it is apparent that many of the current concepts, still based on classic thermodynamics, will have to be revised in the light of the new notions of nonequilibrium thermodynamics.

Heat production and absorption measured in living cells must necessarily present overall values resulting from a great number of chemical reactions, many of them exergonic, many others endergonic. Any attempt to analyze these values must keep in mind the extraordinary complexity of chemical reactions in living cells or subcellular structures such as membranes, their chemical and energetic coupling, their dependence on structure and organization, the many control mechanisms, and the nonequilibrium character of these reactions. The interpretation of the strong heat produced and absorbed during electrical activity as being due simply to ion friction is hard to reconcile with the insight and information presently available about the biochemistry and biophysics of biomembranes in general and of excitable membranes in particular.

VII. Concluding Remarks—Concepts and Axioms in Science

In view of the two opposing theories on the role of AcCh in nerve activity, it appears appropriate to make a few comments on the decisive importance of concepts and basic notions for the progress of science—they determine the thinking, the methods of approach, the design of experiments, and the interpretation of the results by the investigator. However, as history has shown time and again, concepts once accepted prevail for a long time and it usually takes several decades until a new concept, no matter how sound and convincing, is accepted and the preceding one abandoned. A famous example is Stahl's phlogiston theory; it had so strongly penetrated the thinking of scientists of the eighteenth century, that leading scientists (Cavendish, Scheele, Fourcroy, and many others) refused to accept Lavoisier's view, who by using the balance explained the weight increase of metals on heating by oxidation. Priestley, the discoverer of oxygen, died in 1804 still believing in phlogiston. In his article, "Theoretical Concepts in Biological Sciences," Krebs (1966) quotes several interesting and illuminating examples of the rejection of new concepts, the hostility, vituperation, and vitriolic comments with which other colleagues met the new ideas. This response seems to be common and may reflect a natural human reaction. It reminds one of the story in the novel by Anatole France, *L'Ile des Pingouins*. A historian who has a new concept of writing history consults a member of the Academy. As advice he gets a strong warning to stay away from his plan: "If you have a new point of view, an original idea . . . you will surprise the reader. And the reader does not like to be surprised. He never seeks to find in a history anything but the foolishness he already knows. An original historian is the object of distrust, scorn, and universal distaste."

The objections to new concepts may not be as surprising as it may seem to those who think of science in terms of a human endeavor essentially based on objective data and experimental observations. When one realizes, however, to what extent basic notions and methods of approach affect our reasoning, conflicts and opposition to new ideas appear almost unavoidable. The biochemist or the molecular biologist, analyzing, for example, a problem on the cellular, subcellular, and molecular level, will use different criteria and notions from those used by a physiologist, who studies overall manifestations of complex preparations or organs characterized by a nearly infinite diversity of structure and organization. We have seen that even a subcellular structure, such as a cell membrane, is extremely complex and that biochemists have become increasingly aware during the last decade of the many big loopholes in our knowledge of chemical composition, ultrastructure, reaction mechanisms, etc. Nevertheless, it can hardly be questioned that many theoretical concepts of biology have been profoundly deepened and clarified by the biochemical approach. As Krebs writes in the article quoted, even in physical sciences there are certain basic concepts, although supported by a vast body of evidence, which cannot be rigidly proved or disproved, but are nevertheless real and essential; he quotes Max Planck (1922), "*Auch in der Physik gilt der Satz, dass man nicht selig werden kann ohne den Glauben.*" Biological "axioms" applied in the biochemical approach differ in many respects in a fundamental way from those of the physiological one.

In summary, it has been shown in this chapter how the notions of bioenergetics and of protein and enzyme chemistry applied to the problem of nerve conduction have provided new information about the properties of excitable membranes. For many decades the study of nerve impulse conduction was dominated by the analysis of electrical parameters and the underlying ion movements. In recent years it has become increasingly apparent to many biologists that the knowledge of the physical events, although necessary and important, is by itself not adequate for explaining the basic mechanism of bioelectricity; it must be supplemented by the knowledge of the chemical and molecular changes which take place in excitable membranes during electrical activity.

The evidence accumulated and summarized in this chapter strongly supports the two basic assumptions proposed by the writer more than two decades ago: first, the essential role of AcCh as a signal initiating the permeability changes in excitable membranes during electrical activity; second, the essential similarity of the function of AcCh in synaptic and conducting parts of the excitable membranes. These two fundamental concepts have found in the last few years increasing support and much new and striking evidence by the work in many laboratories.

It may be stressed, however, that no claim is made that the role of the two proteins or their properties and behavior are already elucidated or explain the chemical events in bioelectricity. The specific proteins involved form less than 10% of the volume of the excitable membrane. Even this specific protein assembly is far from being fully understood and is at present under intensive investigation in many laboratories, both as to its properties in solution and its behavior in the intact structure. In addition, it is almost certain that there are many other proteins in excitable membranes. We have at present no knowledge about their possible role. The composition and the mechanism of the elements referred to as ionophores, the precise role of phospholipids in the permeability changes, are still wide open questions. The role of Ca^{2+} ions widely, if not generally accepted to be essential in the excitatory processes is far from being resolved on a molecular level. Whether or not there are differences of molecular arrangements between synaptic and conducting parts of excitable membranes is unknown, but cannot be tested with the classical pharmacological methods. The enumeration of unsolved problems could be greatly extended. Fortunately, the increasing interest of biochemists in neurobiology and especially in the biochemical properties of excitable membranes is rapidly changing the situation and has, during the last few years, markedly contributed to the progress in the field. The imaginative and dynamic approach of Changeux and his associates, their development of a procedure that permits to use isolated fragments of membranes for the analysis of their molecular properties, has provided completely new and superb tools for the analysis of many of the open problems. The use of the mouse neuroblastoma introduced by Nirenberg and his associates is another example of a preparation bound to provide pertinent information; it permits the use of genetics in the studies of chemical aspects of nerve activity (Seeds *et al.*, 1970; Blume *et al.*, 1970). The use of highly refined physical methods in the analysis of molecular structure and properties, such as X-ray diffraction, nuclear magnetic resonance, optical rotatory dispersion and circular dichroism, and spin labeling, some of them so successfully applied, e.g., by Mautner and his associates in their investigations of sulfur and selenium isologs of AcCh in their reactions with the proteins, offer another example of the great new possibilities in this area. Only by the application of a variety of biochemical methods and different types of preparations can one expect to explain the physical events in chemical terms and to find satisfactory answers to the problem of the complex mechanism by which excitable membranes are capable of propagating nerve impulses.

ACKNOWLEDGMENTS

This work has been supported in part by grants from the U.S. Public Health Service Nos. NS-03304 and NS-07743, by the National Science Foundation, Grant No. NSF-GB-25362, and by the New York Heart Association, Inc.

REFERENCES

Abbott, B. C., Hill, A. V., and Howarth, J. V. (1958). *Proc. Roy. Soc., Ser. B* **148**, 149.

Aceves, I., and Machne, X. (1963). *J. Pharmacol. Exp. Ther.* **140**, 138.

Ariëns, E. J., ed. (1964). "Molecular Pharmacology," Vols. 1 and 2. Academic Press, New York.

Arnett, C. J., and Ritchie, J. M. (1960). *J. Physiol.* (*London*) **152**, 141.

Augustinsson, K.-B., and Heimburger, G. (1955). *Acta Chem. Scand.* **8**, 310.

Augustinsson, K.-B., and Nachmansohn, D. (1949a). *Science* 110, 98.
Augustinsson, K.-B., and Nachmansohn, D. (1949b). *J. Biol. Chem.* 179, 543.
Baker, P. F., Hodgkin, A. L., and Shaw, T. I. (1962). *J. Physiol. (London)* 164, 355.
Barrnett, R. J. (1962). *J. Cell Biol.* 12, 247.
Bartels, E. (1970). Personal communication.
Bartels, E. (1962). *Biochim. Biophys. Acta* 63, 365.
Bartels, E. (1965). *Biochim. Biophys. Acta* 109, 194.
Bartels, E. (1968). *Biochem. Pharmacol.* 17, 945.
Bartels, E., and Nachmansohn, D. (1965). *Biochem. Z.* 342, 359.
Bartels, E., and Nachmansohn, D. (1969). *Arch. Biochem. Biophys.* 133, 1.
Bartels, E., Wassermann, N. H., and Erlanger, B. F. (1971). *Proc. Nat. Acad. Sci. U.S.* 68, 1820.
Bayliss, W. M. (1915). "Principles of General Physiology." Longmans, Green, New York.
Benson, A. A. (1966). *J. Amer. Oil Chem. Soc.* 43, 265.
Benson, A. A. (1968). *In* "Membrane Models and the Formation of Biological Membranes" (L. Bolis and P. A. Pethica, eds.). North-Holland Publ. Co., Amsterdam, p. 190.
Berg, P. (1956). *J. Biol. Chem.* 222, 991 and 1015.
Berman, J. D., and Young, M. (1971). *Proc. Nat. Acad. Sci., U.S.* 68, 395.
Berman, R., Wilson, I. B., and Nachmansohn, D. (1953). *Biochim. Biophys. Acta* 12, 315.
Bieth, J., Vratsanos, S. M., Wasserman, N., and Erlanger, B. F. (1969). *Proc. Nat. Acad. Sci. U.S.* 64, 1103.
Blasie, J. K., and Worthington, C. R. (1969). *J. Mol. Biol.* 39, 417.
Blaustein, M. P., and Goldman, D. E. (1966). *J. Gen. Physiol.* 49, 1043.
Bloom, F. E., and Barrnett, R. J. (1966). *J. Cell Biol.* 29, 475.
Blume, A., Gilbert, F., Wilson, S., Farber, J., Rosenberg, R., and Nirenberg, M. W. (1970). *Proc. Nat. Acad. Sci. U.S.* 67, 786.
Bourgeois, J. P., Tsuji, S., Boquet, P., Pillot, J., Ryter, A., and Changeux, J.-P. (1971). *FEBS Lett.* 16, 92.
Brink, F. (1954). *Pharmacol. Rev.* 6, 243.
Brzin, M. (1966). *Proc. Nat. Acad. Sci. U.S.* 56, 1560.
Bullock, T. H., and Hagiwara, S. (1957). *J. Gen. Physiol.* 40, 565.
Bullock, T. H., Nachmansohn, D., and Rothenberg, M. A. (1946) *J. Neurophysiol.* 9, 9.
Bullock, T. H., Grundfest, H., Nachmansohn, D., and Rothenberg, M. A. (1947). *J. Neurophysiol.* 10, 11.
Canepa, F. G., Pauling, P., and Sörum, H. (1966). *Nature (London)* 210, 907.
Carr, C. W. (1953). *Arch. Biochem. Biophys.* 43, 147.
Casimir, H. B. G. (1945). *Rev. Mod. Phys.* 17, 343.
Chang, C. C., and Lee, C. Y. (1963). *Arch. Int. Pharmacodyn. Ther.* 144, 241.
Changeux, J.-P., and Podleski, T. R. (1968). *Proc. Nat. Acad. Sci. U.S.* 59, 944.
Changeux, J.-P., and Thiery, J. (1968). *BBA (Biochim. Biophys-Acta) Libr.* 11, 116.
Changeux, J.-P., Tung, Y., and Kittell, C. (1967). *Proc. Nat. Acad. Sci. U.S.* 57, 335.
Changeux, J.-P., Podleski, T. R., and Wofsy, L. (1968). *Proc. Nat. Acad. Sci. U.S.* 58, 2063.

Changeux, J.-P., Podleski, T. R., and Meunier, J.-C. (1969a). *J. Gen. Physiol.* **54**, 225S.

Changeux, J.-P., Gautron, J., Israel, M., and Podleski, T. R. (1969b). *C. R. Acad. Sci.* **269**, 1788.

Changeux, J.-P., Kasai, M., and Lee, C. Y. (1970). *Proc. Nat. Acad. Sci. U.S.* **67**, 1241.

Chu, S. H., and Mautner, H. G. (1966). *J. Org. Chem.* **31**, 308.

Clark, A. J. (1937). *In* "Handbuch der experimentellen Pharmakologie" (W. Heubner and J. Schueller, eds.), Vol. 4, p. 63. Springer-Verlag, Berlin and New York.

Cleland, W. W. (1964). *Biochemistry* **3**, 480.

Cole, K. S. (1965). *Physiol. Rev.* **45**, 340.

Condrea, E., and Rosenberg, P. (1968). *Biochim. Biophys. Acta* **150**, 271.

Cowan, S. L. (1934). *Proc. Roy. Soc., Ser. B* **115**, 216.

Cox, R. T., Coates, C. W., and Brown, M. V. (1946). *Ann. N.Y. Acad. Sci.* **47**, 487.

Cuatrecasas, P. (1970). *J. Biol. Chem.* **245**, 3059.

Cuatrecasas, P., Wilchek, M., and Anfinsen, C. B. (1968). *Proc. Nat. Acad. Sci. U.S.* **61**, 636.

Culvenor, C. C. J., and Ham, N. S. (1966). *Chem. Commun.* **15**, 537.

Cushley, R. J., and Mautner, H. G. (1970). *Tetrahedron* **26**, 2151.

Dale, H. H., Feldberg, W., and Vogt, M. (1936). *J. Physiol. (London)* **86**, 353.

Davis, F. A., and Nachmansohn, D. (1964). *Biochim. Biophys. Acta* **88**, 384.

Deal, W. J., Erlanger, B. F., and Nachmansohn, D. (1969). *Proc. Nat. Acad. Sci. U.S.* **64**, 1230.

Del Castillo, J., and Katz, B. (1957). *Colloq. Int. Cent. Nat. Rech. Sci.* p. 245.

DeLorenzo, A. J. D., Dettbarn, W.-D., and Brzin, M. (1968). *J. Ultrastruct. Res.* **24**, 367.

Denburg, J. L., Eldefrawi, M. E., and O'Brien, R. D. (1972). *Proc. Nat. Acad. Sci. U.S.* **69**, 177.

De Robertis, E. D. P., and Bennett, H. S. (1955). *J. Biophys. Biochem. Cytol.* **1**, 47.

De Robertis, E. D. P., de Iraldi, A. P., del Arnaiz, G. R., and Salganicoff, L. (1962). *J. Neurochem.* **9**, 23.

Dettbarn, W.-D. (1960a). *Biochim. Biophys. Acta* **41**, 377.

Dettbarn, W.-D. (1960b). *Nature (London)* **186**, 891.

Dettbarn, W.-D. (1963). *Life Sci.* **12**, 910.

Dettbarn, W.-D., and Davis, F. A. (1963). *Biochim. Biophys. Acta* **66**, 397.

Dettbarn, W.-D., and Rosenberg, P. (1966). *J. Gen Physiol.* **50**, 447.

Dodge, F. A., and Frankenhaeuser, B. (1959). *J. Physiol.* **148**, 188.

Eccles, J. C. (1935). *J. Physiol. (London)* **84**, 50P.

Edwards, G. A., Ruska, H., and de Harven, E. (1958). *J. Biophys. Biochem. Cytol.* **4**, 107.

Elbers, P. F. (1964). *Recent Progr. Surface Sci.*, **2**, 443.

Erlanger, J. (1939). *J. Neurophysiol.* **2**, 370.

Feld, E. A., Grundfest, H., Nachmansohn, D., and Rothenberg, M. A. (1948). *J. Neurophysiol.* **11**, 125.

Feldberg, W., and Vartiainen, A. (1934). *J. Physiol. (London)* **83**, 103.

Fulton, J. F. (1939). "Physiology of the Nervous System." Oxford Univ. Press, London and New York.

Furchgott, R. F. (1964). *Annu. Rev. Pharmacol.* **4**, 21.

Furshpan, E. J., and Potter, D. D. (1959). *J. Physiol.* (*London*) 145, 289.
Gage, P. W., and Moore, J. W. (1969). *Science* 166, 510.
Gray, E. G., and Whittaker, V. P. (1962). *J. Anat.* 96, 79.
Green, D. E., and MacLennan, D. H. (1969). *BioScience* 19, 213.
Green, D. E., and Perdue, J. F. (1966). *Proc. Nat. Acad. Sci. U.S.* 55, 1295.
Green, D. E., and Silman, I. (1967). *Annu. Rev. Plant Physiol.* 18, 147.
Günther, W. H. H., and Mautner, H. G. (1964). *J. Med. Chem.* 7, 229.
Hecht, S., Shlaer, S., and Pirenne, M. H. (1941). *J. Gen. Physiol.* 25, 819.
Hendricks, S. B., and Borthwick, H. A. (1967). *Proc. Nat. Acad. Sci. U.S.* 58, 2125.
Higman, H. B., and Bartels, E. (1961). *Biochim. Biophys. Acta* 54, 543.
Higman, H. B., and Bartels, E. (1962). *Biochim. Biophys. Acta* 57, 77.
Higman, H. B., Podleski, T. R., and Bartels, E. (1963). *Biochim. Biophys. Acta.* 75, 187.
Higman, H. B., Podleski, T. R., and Bartels, E. (1964). *Biochim. Biophys. Acta* 79, 138.
Hill, A. V. (1912). *J. Physiol.* (*London*) 43, 433.
Hill, A. V. (1932). "Chemical Wave Transmission in Nerve." Cambridge Univ. Press, London and New York.
Hill, A. V. (1960). *In* "Molecular Biology" (D. Nachmansohn, ed.), p. 153. Academic Press, New York.
Hill, A. V. (1964). "Trails and Trials in Physiology." Arnold, London.
Hinterbuchner, L. P., and Nachmansohn, D. (1960). *Biochim. Biophys. Acta* 44, 554.
Hobbiger, F. W. (1963). *In* "Handbuch der experimentellen Pharmakologie" (G. B. Koelle, ed.), Vol. 15, p. 921. Springer-Verlag, Berlin and New York.
Hodgkin, A. L. (1951). *Biol. Rev.* 26, 338.
Hodgkin, A. L. (1964). "The Conduction of the Nervous Impulse." Thomas, Springfield, Illinois.
Hodgkin, A. L., and Huxley, A. F. (1952). *J. Physiol.* 117, 500.
Hoskin, F. C. G., and Rosenberg, P. (1964). *J. Gen. Physiol.* 47, 1117.
Hoskin, F. C. G., Rosenberg, P., and Brzin, M. (1966). *Proc. Nat. Acad. Sci. U.S.* 55, 1231.
Howarth, J. V., Keynes, R. D., and Ritchie, J. M. (1968). *J. Physiol.* (*London*) 194, 745.
Hubbard, J. I., and Schmidt, R. F. (1963). *J. Physiol.* 166, 145.
Jaffe, M. J. (1970). *Plant Physiol.* 46, 768.
Kalckar, H. M. (1942). *Biol. Rev. Cambridge Phil. Soc.* 17, 28.
Kalckar, H. M. (1969). "Biological Phosphorylations. Development of Concepts." Prentice-Hall, Englewood Cliffs, New Jersey.
Kalderon, N., Silman, I., Blumberg, S., and Dudai, Y. (1970). *Biochim. Biophys. Acta* 207, 560.
Karlin, A. (1965). *J. Cell Biol.* 25, 159.
Karlin, A. (1967). *J. Theor. Biol.* 16, 306.
Karlin, A., and Bartels, E. (1966). *Biochim. Biophys. Acta* 126, 525.
Karlin, A., and Winnik, M. (1968). *Proc. Nat. Acad. Sci. U.S.* 60, 668.
Karlin, A., Prives, J., Deal, W., and Winnik, M. (1970). *Mol. Properties Drug Receptors, Ciba Found. Symp.* p. 247.
Kasai, M., and Changeux, J.-P. (1970). *C. R. Acad. Sci., Ser. D* 270, 1400.
Kasai, M., and Changeux, J.-P. (1971). *J. Mem. Biol.* 6, 73.

Katchalsky, A. (1964). In "Connective Tissue: Intercellular Macromolecules," p. 9. Little, Brown, Boston, Massachusetts.

Katchalsky, A. (1967). In "The Neurosciences" (G. C. Quarton, T. Melnechuk, and F. O. Schmitt, eds.), p. 326. Rockefeller Univ. Press, New York.

Katchalsky, A., Cooper, R. E., Upadhyay, J., and Wasserman, A. (1961). J. Chem. Soc., London p. 5198.

Katz, B., and Miledi, R. (1965). Proc. Roy. Soc. Ser. B 161, 453.

Kaufman, H., Vratsanos, S. M., and Erlanger, B. F. (1968). Science 162, 1487.

Kennedy, E. P. (1967). In "The Neurosciences" (G. C. Quarton, T. Melnechuk, and F. O. Schmitt, eds.), p. 271. Rockefeller Univ. Press, New York.

Kewitz, H., Wilson, I. B., and Nachmansohn, D. (1956). Arch. Biochem. Biophys. 64, 456.

Keynes, R. D., and Aubert, X. (1964). Nature (London) 203, 261.

Kirtley, M. E., and Koshland, D. E., Jr. (1967). J. Biol. Chem. 242, 4192.

Korey, S. R., de Braganza, B., and Nachmansohn, D. (1951). J. Biol. Chem. 189, 705.

Korn, E. D. (1966). Science 153, 1491.

Korn, E. D. (1969). Annu. Rev. Biochem. 38, 263.

Koshland, D. E., Jr. (1960). Advan. Enzymol. 22, 45.

Koshland, D. E., Jr., and Neet, K. E. (1968). Annu. Rev. Biochem. 37, 359.

Koshland, D. E., Jr., Strumeyer, D. H., and Ray, W. J., Jr. (1962). Brookhaven Symp. Biol. 15, 101.

Krebs, H. A. (1966). In "Current Aspects of Biochemical Energetics" (N. O. Kaplan and E. Kennedy, eds.), p. 83. Academic Press, New York.

Kremzner, L. T., and Wilson, I. B. (1963). J. Biol. Chem. 238, 1714.

Lawler, H. C. (1961). J. Biol. Chem. 236, 2296.

Lee, C. Y., and Chang, C. C. (1966). Mem. Inst. Butantan, Sao Paulo 33, 555.

Lee, C. Y., Tseng, L. F., and Chin, T. H. (1967). Nature (London) 215, 1177.

Lenard, J., and Singer, S. J. (1966). Proc. Nat. Acad. Sci. U.S. 56, 1828.

Leuzinger, W., and Baker, A. L. (1967). Proc. Nat. Acad. Sci. U.S. 57, 446.

Leuzinger, W., Baker, A. L., and Cauvin, E. (1968). Proc. Nat. Acad. Sci. U.S. 59, 620.

Leuzinger, W., Goldberg, M., and Cauvin, E. (1969). J. Mol. Biol. 40, 217.

Lipmann, F. (1940). J. Biol. Chem. 134, 463.

Lipmann, F. (1941). Advan. Enzymol. 1, 99.

Loewenstein, W. R. (1966). Ann. N.Y. Acad. Sci. 137, 403.

Loewenstein, W. R. (1967). J. Cell. Interface Sci. 25, 34.

Loewi, O. (1921). Pfluegers Arch. Gesamte Physiol. Menschen Tiere 189, 239.

Lorente de Nó, R. (1938). Amer. J. Physiol. 121, 331.

Lundin, S. J., and Hellström, B. (1968). Z. Zellforsch. Mikrosk. Anat. 85, 264.

McIntyre, A. R. (1959). In "Curare and Curare-like Agents" (D. Bovet, F. Bovet-Nitti, and G. B. Marini-Bettolo, eds.), p. 211. Elsevier, Amsterdam.

Marchbanks, R. M. (1968). Biochem. J. 106, 87; 110, 533.

Marchbanks, R. M. (1970). FEBS Symp. 21, 285.

Martin, R., and Rosenberg, P. (1968). J. Cell Biol. 36, 341.

Masland, R. L., and Wigton, R. S. (1940). J. Neurophysiol. 3, 269.

Mautner, H. G. (1967). Pharmacol. Rev. 19, 107.

Mautner, H. G., and Bartels, E. (1970). Proc. Nat. Acad. Sci. U.S. 67, 74.

Mautner, H. D., and Günther, W. H. H. (1961). J. Amer. Chem. Soc. 83, 3342.

Mautner, H. G., Bartels, E., and Webb, G. D. (1966). Biochem. Pharmacol. 15, 187.

Mazur, A. (1946). *J. Biol. Chem.* **164**, 271.

Membrane Proteins. (1969). Proc. Symp. spons. by New York Heart Association. Little, Brown, Boston, Massachusetts.

Meunier, J. C., Huchet, M., Boquet, P., and Changeux, J.-P. (1971). *C. R. Acad. Sci.* **272**, 117.

Meyerhof, O. (1937). *Ergeb. Physiol., Biol. Chem. Exp. Pharmacol.* **39**, 10.

Meyerhof, O. (1951). *In* "Phosphorus Metabolism" (W. D. McElroy and B. Glass, eds.), p. 3. Johns Hopkins Press, Baltimore, Maryland.

Meyerhof, O., and Suranyi, J. (1927). *Biochem. Z.* **191**, 106.

Millar, D. B., and Grafius, M. A. (1970). *FEBS Lett.* **12**, 61.

Monod, J., Changeux, J.-P., and Jacob, F. (1963). *J. Mol. Biol.* **6**, 306.

Monod, J., Wyman, J., and Changeux, J.-P. (1965). *J. Mol. Biol.* **12**, 88.

Nachmansohn, D. (1955). *Harvey Lec.* **49**, 57.

Nachmansohn, D. (1959). "Chemical and Molecular Basis of Nerve Activity." Academic Press, New York.

Nachmansohn, D. (1963a). *In* "Handbuch der experimentellen Pharmakologie" (G. B. Koelle, ed.), Vol. 15, p. 701. Springer-Verlag, Berlin and New York.

Nachmansohn, D. (1963b). *In* "Handbuch der experimentellen Pharmakologie" (G. B. Koelle, ed.), Vol. 15, p. 40. Springer-Verlag, Berlin and New York.

Nachmansohn, D. (1964). *New Perspect. Biol., Proc. Symp., 1963* p. 176.

Nachmansohn, D. (1966a). *In* "Nerve as a Tissue" (K. Rodahl, ed.), p. 269. McGraw-Hill, New York.

Nachmansohn, D. (1966b). *Ann. N.Y. Acad. Sci.* **137**, 877.

Nachmansohn, D. (1966c). *In* "Current Aspects of Biochemical Energetics" (N. O. Kaplan and E. Kennedy, eds.), p. 145. Academic Press, New York.

Nachmansohn, D. (1968). *Proc. Nat. Acad. Sci. U.S.* **61**, 1034.

Nachmansohn, D. (1969). *J. Gen. Physiol.* **54**, 187 S.

Nachmansohn, D. (1970). *Science* **168**, 1059.

Nachmansohn, D. (1971). *In* "Handbook of Sensory Physiology" (W. R. Loewenstein, ed.), Vol. 1, p. 18. Springer-Verlag, Berlin and New York.

Nachmansohn, D. (1972). Prefatory Chapter, *Ann. Rev. Biochem.* 1972, in press.

Nachmansohn, D., and Berman, M. (1946). *J. Biol. Chem.* **165**, 551.

Nachmansohn, D., and John, H. M. (1944). *Proc. Soc. Exp. Biol. Med.* **57**, 361.

Nachmansohn, D., and Lederer, E. (1939). *Bull. Soc. Chim. Biol.* **21**, 797.

Nachmansohn, D., and Machado, A. L. (1943). *J. Neurophysiol.* **6**, 397.

Nachmansohn, D., and Wilson, I. B. (1951). *Advan. Enzymol.* **12**, 259.

Nachmansohn, D., Cox, R. T., Coates, C. W., and Machado, A. L. (1943a). *J. Neurophysiol.* **6**, 383.

Nachmansohn, D., John, H. M., and Waelsch, H. (1943b). *J. Biol. Chem.* **150**, 485.

Namba, T., and Hiraki, K. (1958). *J. Amer. Med. Ass.* **166**, 1834.

Nernst, W. (1888). *Z. Physik. Chem.* **2**, 613.

Nernst, W. (1889). *Z. Physik. Chem.* **4**, 129.

O'Brien, R. D. (1960). "Toxic Phosphorus Esters." Academic Press, New York.

Onsager, L. (1931). *Phys. Rev.* **37**, 405; **38**, 2265.

Overton, E. (1902). *Arch. Gesamte Physiol. Menschen Tiere* **92**, 346.

Palade, G. E. (1970). Personal communication.

Palade, G. E. (1963). *In* "The Scientific Endeavor: Centennial Celebration of the National Academy Sciences," p. 179. Rockefeller Univ. Press, New York.

Planck, M. (1890). *Ann. Physik. Chem.* **39**, 161 and **40**, 561.

Planck, M. (1922). "Gesammelte Reden und Aufsaetze." Hirzel, Stuttgart.

Podleski, T. R. (1966). Ph.D. Thesis, Columbia University, New York.
Podleski, T. R. (1967). *Proc. Nat. Acad. Sci. U.S.* **58**, 268.
Podleski, T. R. (1969). *Biochem. Pharmacol.* **18**, 211.
Podleski, T. R., and Nachmansohn, D. (1966). *Proc. Nat. Acad. Sci. U.S.* **56**, 1034.
Podleski, T. R., Meunier, J. C., and Changeux, J.-P. (1969). *Proc. Nat. Acad. Sci. U.S.* **63**, 1239.
Prigogine, I. (1947). "Etude thermodynamique des phenomenes irreversibles." Dunod, Paris.
Prince, A. K. (1967). *Proc. Nat. Acad. Sci. U.S.* **57**, 1117.
Racker, E. (1965). "Mechanisms in Bioenergetics." Academic Press, New York.
Racker, E. (1967). *Fed. Proc., Fed. Amer. Soc. Exp. Biol.* **26**, 1335.
Racker, E. (1969). *J. Gen. Physiol.* **54**, 38s.
Riker, W. F. (1953). *Pharmacol. Rev.* **5**, 1.
Riker, W. F., Jr., Werner, G., Roberts, J., and Kuperman, A. S. (1959). *Ann. N.Y. Acad. Sci.* **81**, 328.
Ritchie, J. M. (1963). *Biochem. Pharmacol.* **12**, (S), 3.
Robertson, J. D. (1960a). *In* "Molecular Biology" (D. Nachmansohn, ed.), p. 87. Academic Press, New York.
Robertson, J. D. (1960b). *In* "Progress in Biophysics" (B. Katz and J. A. V. Butler, eds.), p. 10, 343. Pergamon, Oxford.
Rosenberg, P. (1965). *Toxicon* **3**, 125.
Rosenberg, P. (1966). *Mem. Inst. Butantan, Sao Paulo* **33**, 477.
Rosenberg, P., and Condrea, E. (1968). *Biochem. Pharmacol.* **17**, 2033.
Rosenberg, P., and Dettbarn, W.-D. (1967). *Toxicon* **4**, 296.
Rosenberg, P., and Mautner, H. G. (1967). *Science* **155**, 1569.
Rosenberg, P., Mautner, H. G., and Nachmansohn, D. (1966). *Proc. Nat. Acad. Sci. U.S.* **55**, 835.
Rosenberry, T. L., and Bernhard, S. A. (1971). *Biochemistry* **10**, 4114.
Rosenberry, T. L., Chang, H. W., and Chen, Y. T. (1972). *J. Biol. Chem.* **247**, 1555.
Rossi-Fanelli, A., Antonini, E., and Caputo, A. (1964). *Advan. Protein Chem.* **19**, 73.
Rothenberg, M. A., and Nachmansohn, D. (1947). *J. Biol. Chem.* **168**, 223.
Rothenberg, M. A., Sprinson, D. B., and Nachmansohn, D. (1948). *J. Neurophysiol.* **11**, 111.
Rothfield, L., and Finkelstein, A. (1968). *Annu. Rev. Biochem.* **37**, 463.
Schoffeniels, E. (1957). *Biochim. Biophys. Acta* **26**, 585.
Schoffeniels, E. (1959). Thèse d'agrégation, Univ. de Liège, Liège.
Schoffeniels, E., and Nachmansohn, D. (1957). *Biochim. Biophys. Acta* **26**, 1.
Seeds, N. W., Gilman, A. G., Amano, T., and Nirenberg, M. W. (1970). *Proc. Nat. Acad. Sci. U.S.* **66**, 160.
Shefter, E., and Kennard, O. (1966). *Science* **153**, 1389.
Shefter, E., and Mautner, H. G. (1967). *J. Amer. Chem. Soc.* **89**, 1249.
Shefter, E., and Mautner, H. G. (1969). *Proc. Nat. Acad. Sci. U.S.* **63**, 1253.
Silman, I. (1970). *FEBS Symp.* **21**, 337.
Silman, I., and Karlin, A. (1969). *Science* **164**, 1420.
Sjöstrand, F. S. (1962). *Symp. Int. Soc. Cell Biol.* **1**, 47.
Sjöstrand, F. S. (1963). *J. Ultrastruct. Res.* **9**, 561.
Sjöstrand, F. S., and Barajas, L. (1968). *J. Ultrastruct. Res.* **25**, 121.

Sjöstrand, F. S., and Barajas, L. (1970). *J. Ultrastruct. Res.* **32**, 293.
Stadtman, E. R. (1952). *J. Biol. Chem.* **196**, 527 and 535.
Stadtman, E. R., Novelli, G. D., and Lipmann, F. (1951). *J. Biol. Chem.* **191**, 365.
Tasaki, I. (1968). "Nerve Excitation." Thomas, Springfield, Illinois.
Vanderkooi, G., and Green, D. E. (1971). *BioScience* **21**, 409.
Van Holde, K. E., and Baldwin, R. L. (1958). *J. Phys. Chem.* **62**, 734.
Wald, G. (1968). *Science* **162**, 230.
Walsh, R. R., and Deal, S. E. (1959). *Amer. J. Physiol.* **197**, 547.
Webb, G. D. (1965). *Biochim. Biophys. Acta* **102**, 172.
Werner, G., and Kuperman, A. S. (1963). In "Handbuch der experimentellen Pharmakologie" (G. B. Koelle, ed.), Vol. 15, p. 570. Springer-Verlag, Berlin and New York.
Wescoe, W. C., and Riker, W. F., Jr. (1951). *Ann. N.Y. Acad. Sci.* **54**, 438.
Whittaker, V. P. (1965). *Progr. Biophys. Mol. Biol.* **15**, 39.
Whittaker, V. P. (1968). *Proc. Nat. Acad. Sci. U.S.* **60**, 1081.
Wilson, I. B. (1952). *J. Biol. Chem.* **197**, 215.
Wilson, I. B. (1960). In "The Enzymes" (P. D. Boyer, H. Lardy, and K. Myrbäck, eds.), 2nd rev. ed., Vol. 4, p. 501. Academic Press, New York.
Wilson, I. B., and Cabib, E. (1956). *J. Amer. Chem. Soc.* **78**, 202.
Wilson, I. B., and Ginsburg, S. (1955). *Biochim. Biophys. Acta* **18**, 168.
Wilson, I. B., and Harrison, M. A. (1961). *J. Biol. Chem.* **236**, 2292.
Wilson, I. B., and Quan, C. (1958). *Arch. Biochem. Biophys.* **73**, 131.
Wofsy, L., Metzger, H., and Singer, S. J. (1962). *Biochemistry* **1**, 1031.
Wyman, J., Jr. (1964). *Advan. Protein Chem.* **19**, 233.
Yphantis, D. A. (1964). *Biochemistry* **3**, 297.

Addendum

During this period since this chapter has been revised and written in its present form, significant advances have been reported in the exploration of membranes in general and of the properties and function of the proteins in excitable membranes. Among the most actively pursued problems has been the isolation and characterization of the AcCh-receptor by many investigators applying a variety of techniques. Much progress has been achieved, although many essential features of this protein remain still under discussion. Unfortunately, the incorporation of the new information into the page proofs is impossible. However, within the space available, three developments may be mentioned since they are pertinent to the basic concept of the role of AcCh in excitable membranes.

1. E. Neumann and A. Katchalsky (*Proc. Nat. Acad. Sci. USA*, **69**, 933, 1972) have found electric impulses to be capable of inducing conformational changes of biopolymers and proposed a mechanism for the structural transitions induced. The electric fields applied were about 20

kV/cm. These observations may have an important bearing on the effects of electrical stimulation of excitable membranes. The field strength applied is equivalent to 20 mV for a membrane of 100 Å thickness. It is known that, in some axons, a drop of 15 to 20 mV in the resting potential is the threshold which may initiate an action potential. Thus, the first step in nerve excitation may be a conformational change of the storage protein leading to the release of AcCh.

2. On the initiative of the late Aharon Katchalsky, an attempt has been made at an integral interpretation of nerve excitability, incorporating electrophysiological, biophysical, and biochemical events during electrical activity (Neumann, E., Nachmansohn, D., and Katchalsky, A. *Proc. Nat. Acad. Sci. USA,* in press). A quantitative interpretation requires the application of nonequilibrium thermodynamics; this is at present impossible due to the anisotropy of biomembranes and our ignorance of the molecular structure. However, the information available allows one to suggest at least a qualitative picture of the molecular events in the membrane associated with electrical activity. The integral model proposed assumes that the gateway for ions is surrounded by a certain number of basic excitation units (BEU) formed by subunits in which the three proteins processing AcCh (storage, receptor, enzyme) are interlocked. The gateway becomes permeable to ions by the cooperative subunits of the BEU. In order to initiate an action potential a certain critical number of BEU is assumed to be activated within a certain time interval; a smaller number of activated subunits may lead to a subthreshold graded response and a still smaller number may remain ineffective. The model is an extension of the chemical theory and offers an interpretation of the threshold for the generation of an action potential, of the problem of all-or-none versus graded responses to stimulation and of the stimulus effect itself.

3. Recently, cells have been grown in tissue culture with properties similar to those of neurons. The cells were developed by fusion of mouse neuroblastoma with rat glia cells; they have processes which resemble nerve fibers and contain the proteins associated with AcCh. The fibers produce action potentials by electrical stimulation or by electrophoretic application of AcCh. (Dr. Bernd Hamprecht, personal communication.)

3

SOME ASPECTS OF THE BIOPHYSICS OF MUSCLE

KENNETH L. ZIERLER

I. Introduction

The following events lead to and accompany contraction and subsequent relaxation of skeletal muscle in response to a nerve impulse. Acetylcholine is released from terminals of motor nerves and diffuses across clefts to motor end plates, which are specialized receptors on muscle fibers. The combination of acetylcholine and receptor leads to an unknown change in structure of the membrane of the end plate as a result of which the permeability of the membrane of the end plate to Na^+ and K^+ is increased. It is the increase in permeability to Na^+ that is responsible for depolarization of the end plate. The altered electric field set up by depolarization is effective over only a small volume in the vicinity of the end plate. When this end plate potential, as the electrical potential difference across the depolarized end plate is called, is sufficiently large, muscle membrane in the immediate vicinity of the end plate is altered.

The alteration of the membrane is manifested first by increased permeability to Na^+. As a result there is net movement of Na^+ from interstitial fluid across sarcolemma. The electrical potential difference across sarcolemma, which at rest is negative inside with respect to the outside of the fiber, reverses its sign at the point of increased Na^+ permeability. This local transient change in electrical potential difference across sarcolemma is an action potential. The electric field of the action potential causes a structural change in neighboring sarcolemma, increasing permeability to Na^+ and producing an action potential at this point.

In this way, the action potential is self-propagating over the surface of a muscle fiber. During this propagation, the action potential arrives at the opening of the transverse tubular system, the T system, the walls of which can be considered continuations of sarcolemma. It is surmised that depolarization is spread, either by electrotonus or by a propagated action potential, along the T system, penetrating the whole depth of the muscle fiber.

During the course of this propagation along the T system, the spreading depolarization arrives at interfaces between the T system and the terminal cisternae of sarcoplasmic reticulum. It is not entirely clear whether or not this interface is a tight junction, but it is probably quite permeable to Na^+ and K^+. There is indirect evidence that the concentration of Na^+ and K^+ in sarcoplasmic reticulum is similar to that of interstitial fluid and not to that of sarcoplasm, and there is evidence that there is a more substantial portion of conductance into sarcoplasm than can be accounted for by sarcolemma plus T system. It is therefore at least conceivable that the action potential may continue its propagation from the T system along the walls of sarcoplasmic reticulum. If so, there would be an electric field moving along every myofibril.

There may be many effects of the moving electric field. At least one of them is that Ca^{2+} is released from some place in the lumen or from the wall of sarcoplasmic reticulum. At least one such site, and perhaps the main, if not the only, site is the terminal cisterna.

From the wall of sarcoplasmic reticulum, Ca^{2+} diffuses to contractile proteins that are the myofilaments. There is thereby initiated a reaction that causes a change in the relation between the two types of myofilaments, thick and thin. This change in relation between thick and thin filaments causes tension to be exerted along the long axis of the myofibers or causes the muscle to shorten, depending on external conditions, that is, on whether or not the muscle ends are fixed so that it can not shorten. Whether the muscle shortens or exerts tension but does not shorten, the process is called a muscle contraction, and the fundamental events are the same.

If a muscle does shorten and in so doing moves a load, it has done work, sometimes called external work. If there is no external shortening no external work has been done because work is defined as a force exerted over a distance. However, whenever a muscle contracts, whether or not it does external work, at least some contractile elements shorten and, in so doing, stretch certain material in series with them. Such extensible material, whatever its composition, is a series elastic element. Stretch of series elastic elements by contractile elements is internal work.

Unless the muscle is stimulated at a sufficiently high frequency, the membrane repolarizes, the contracted contractile element reverts to its resting state, Ca^{2+} is recaptured from sarcoplasm by sarcotubules, and the Na^+ excess is pumped out.

Work, external or internal, is a form of energy expenditure. In living tissue, energy expenditure comes ultimately from ordered chemical reactions. A series of such reactions occurs during contraction and another series during immediate restitution of noncontraction or resting state. Over some longer period, a third series of ordered chemical reactions must occur to restore fully the chemical state of muscle to that which prevailed just before contraction occurred. This process is called recovery.

With these series of ordered chemical reactions, heat is generated. Because for many years it has been possible to measure the time course of evolution of heat by muscle with resolution not generally possible for the case of individual chemical reactions in living tissues, much emphasis has been placed on correlations among the time course of heat production and mechanical events and changes in anatomic relationships. Measurement of heat, or of heat plus work, gives a measure of the total energy expenditure. The observed relations between heat and work, heat and contractile force, and so on, can give reliable information at the phenomenological level that places constraints on any molecular model of muscle contraction. In the following sections we shall consider in more detail from a biophysical viewpoint certain aspects of the series of events surrounding muscle contraction.

II. Excitation

A. *Electrotonus*

Muscle, like nerve, is excitable. It conducts an electric current. It is capable of generating an electrical potential difference across barriers separating sarcoplasm from extracellular fluid. Under suitable circumstances, electrical properties of these barriers are altered, and associated with these alterations there are generated electrical currents and potential differences propagated along, and possibly through, muscle fibers.

Thus, there are two forms of spread of an electrical potential difference over the surface of a muscle fiber. The spread that occurs as though the fiber is passive (that is, the fiber does not alter in such a way as to generate an action potential) is called electrotonus. In contrast, the propagated wave of altered electrical potential difference in response to a suitable stimulus resulting in the change in barrier properties is the action potential. If it were not for those properties of muscle that permit it to display electrotonus, there could not be a propagated action potential. Therefore, we will first consider electrotonus, then consider how the properties that permit electrotonus can initiate a propagated disturbance, and finally consider the phenomena underlying the propagated action potential.

In 1879 Hermann first proposed a model for electrotonus in nerve which is still accepted with scant change. The model is an electrical cable with a leaky insulating sheath immersed in a conducting solution. It has therefore been likened to a submarine cable and is hence known as the cable model. Because the core of the cable is conducting, the theory is also known as the core conductor theory.

The external solution and the cable core are considered to be purely resistive conductors. Specifically, it is assumed that the total internal and the total external longitudinal (or axial) currents flowing across any plane perpendicular to the axis of the cable (or fiber) are proportional to the longitudinal voltage gradient at the internal and external surface, respectively.

The cable sheath or muscle membrane (to be specified later) is made of a parallel unit that can carry current between the external solution and the core. There are many possible models of such membrane units, but essentially each unit can be represented by a capacitance shunted by a resistance. That the membrane includes capacitative elements is shown by the fact that when an applied stimulus is removed, the electrical potential difference across the membrane does not suddenly collapse to its steady value but approaches it more or less with a single time constant which can be equated to the product of a resistance and a capacitance.

There are, then, three regions—intracellular denoted by subscript i; membrane, by m; and extracellular, by e. A general solution for membranes of arbitrary shape has been developed by Hellérstein (1968), from whom the following treatment is largely drawn.

Figure 1 illustrates the model for Hellerstein's equations. The core or intracellular phase can contain three current sources: (1) an applied current density J_a, (2) currents $-g_i \, \partial V_a / \partial x$ due to applied potentials V_a, (3) currents due to potentials generated by phenomena in the core of intracellular phase $-g_i \, \partial V_i / \partial x$, where the constant uniform conductivity of the core is g_i and x is in the direction through the membrane, perpendicular to tangents to the membrane surface.

As a result of the fact that there is only negligible reactance in the intracellular phase, charges cannot accumulate in the intracellular phase.

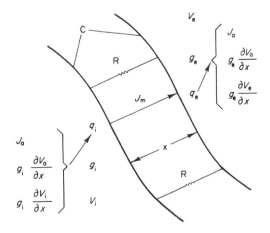

Fig. 1. Core conductor model. Adapted from Hellerstein (1968).

But because there is membrane capacitance C, expressed as capacitance per unit area, a surface charge density, q_i, accumulates on the inner membrane surface. The value of q_i is depleted by the transmembrane current density J_m, expressed as charge/(area $\times$ time), which in turn, tends to accumulate a charge density q_e on the outer membrane surface. The value of q_e is depleted by three current sinks—(a) the applied current density J_a, (b) currents $-g_i \, \partial V_a/\partial x$, due to applied potentials, V_a, and (c) currents into the external medium $-g_e \, \partial V_e/\partial x$, where g_e is the uniform conductivity of the external medium.

From this description we immediately state that the change in charge density on the inner surface during the time interval dt is

$$\frac{\partial q_i}{\partial t} = J_a - g_i \left(\frac{\partial V_a}{\partial x} + \frac{\partial V_i}{\partial x} \right) - J_m \tag{1}$$

and on the outer surface is

$$\frac{\partial q_e}{\partial t} = J_m - J_a + g_e \left(\frac{\partial V_a}{\partial x} + \frac{\partial V_e}{\partial x} \right) \tag{2}$$

But the membrane current density is simply

$$J_m = - \frac{V_e - V_i}{R} \tag{3}$$

where R is the distributed resistance through the membrane, given as resistance times unit area, and the difference between membrane charges is

$$q_e - q_i = 2C(V_e - V_i) \tag{4}$$

where C is the capacitance per unit area of membrane.

From Gauss's law, integration over the Gaussian surface containing inner and outer membrane surfaces yields, since J_a and V_a are not generated by the membrane charge density,

$$\frac{\partial V_e}{\partial x} - \frac{\partial V_i}{\partial x} = - \frac{q_e + q_i}{\epsilon_0} \tag{5}$$

where ϵ_0 is the permitivity, taken to be that of free space on the assumption that the core and the external solution have no reactance, no capacitative effects.

Differentiation of Eq. (5) with respect to time and substitution for $\partial q_i / \partial t$ and $\partial q_e / dt$ from Eqs. (1) and (2) yield

$$\epsilon_0 \frac{\partial}{\partial t} \left(\frac{\partial V_e}{\partial x} - \frac{\partial V_i}{\partial x} \right) = -q_e \left(\frac{\partial V_e}{\partial x} + \frac{\partial V_a}{\partial x} \right) + g_i \left(\frac{\partial V_i}{\partial x} + \frac{\partial V_a}{\partial x} \right)$$

g_e and g_i are of the same order and ϵ_0/g, the relaxation time, is of the order of 10^{-11} sec. Therefore, over times of the order of milliseconds, which are the smallest with which we are concerned, the left hand side of the above equation is negligible and we have the approximation

$$g_e \left(\frac{\partial V_e}{\partial x} + \frac{\partial V_a}{\partial x} \right) - g_i \left(\frac{\partial V_i}{\partial x} + \frac{\partial V_a}{\partial x} \right) \tag{6}$$

Subtraction of Eq. (1) from (2) with substitution of Eqs. (3), (4), and (6) gives the pair of boundary equations

$$-g_i \frac{\partial V_i}{\partial x} - g_i \frac{\partial V_a}{\partial x} + J_a = C \frac{\partial}{\partial t} (V_i - V_e) + \frac{V_i - V_e}{R} \tag{7}$$

$$-g_e \frac{\partial V_e}{\partial x} - g_e \frac{\partial V_a}{\partial x} + J_a = C \frac{\partial}{\partial t} (V_i - V_e) + \frac{V_i - V_e}{R} \tag{8}$$

The left hand side of each equation contains the three terms of the driving current and the right hand side gives the response of a parallel *RC* circuit to that current.

Since we assumed no reactance and conductivity approximately that of salt solutions, it can be shown that

$$\begin{aligned} \nabla^2 V_i &= 0 \\ \nabla^2 V_e &= 0 \end{aligned} \tag{9}$$

Equations (7), (8), and (9) are Hellerstein's final form of the equations for electrotonic potentials for membranes of arbitrary shape.

With the assumption that longitudinal currents are proportional to longitudinal voltage gradients at the membrane surfaces

$$\begin{aligned} I_i &= -\frac{1}{r_i} \left(\frac{\partial V_i}{\partial z} + \frac{\partial V_a}{\partial z} \right) \\ I_e &= -\frac{1}{r_e} \left(\frac{\partial V_e}{\partial z} + \frac{\partial V_a}{\partial z} \right) \end{aligned} \tag{10}$$

where the z direction is axial, r equals longitudinal resistance per unit length, and the I equals total longitudinal current, we can reduce the general equations to the cable equations for a cylinder.

If a is the radius of the cylinder and its wall thickness is negligible compared to its radius (a condition that will need to be relaxed later), then by definition,

$$\begin{aligned} I_i &= \int_0^a 2\pi r J_z \, dr \\ I_e &= \int_a^\infty 2\pi r J_z \, dr \end{aligned} \tag{11}$$

If we express the current density J_z in terms of the potential gradient and differentiate with respect to z, Eqs. (11) become

$$\frac{\partial I_i}{\partial z} = -2\pi g_i \int_0^a x \frac{\partial^2 (V_i + V_a)}{\partial z^2}\, dx$$

$$\frac{\partial I_e}{\partial z} = -2\pi g_e \int_a^\infty x \frac{\partial^2 (V_e + V_a)}{\partial z^2}\, dx$$

(12)

where the x direction is now radial.

From Laplace's equation for a cylindrically symmetrical potential

$$\frac{\partial}{\partial x}\left(x \frac{\partial V}{\partial x} \right) = -x \frac{\partial^2 V}{\partial z^2}$$

so that Eqs. (12) become

$$\frac{\partial I_i}{\partial z} = 2\pi g_i \left[x \frac{\partial (V_i + V_a)}{\partial x} \right]_0^a$$

(13a)

$$\frac{\partial I_e}{\partial z} = 2\pi g_e \left[x \frac{\partial (V_e + V_a)}{\partial x} \right]_a^\infty$$

(13b)

Equation (13a) has zero as its lower limit since $\partial (V_i + V_a)/\partial x$ is not unbounded at $x = 0$. Equation (13b) must have zero as its upper limit if the source is spatially finite, for if the source is spatially finite it must eventually appear to be a point source as x grows large. Therefore $\partial (V_e + V_a)/\partial x$ decreases at least as greatly as $1/x^2$. Therefore,

$$\frac{\partial I_i}{\partial z} = 2\pi a g_i \left[\frac{\partial (V_i + V_a)}{\partial x} \right]_a$$

(14a)

$$\frac{\partial I_e}{\partial z} = -2\pi a g_e \left[\frac{\partial (V_e + V_a)}{\partial x} \right]_a$$

(14b)

Since Eqs. (10) were defined at the membrane surfaces, that is at $x = a$, differentiation of Eqs. (10) with respect to z, substitution from Eqs. (14), and addition give

$$\frac{\partial^2 (V_e - V_i)}{\partial z^2} = 2\pi a \left[r_e g_e \frac{\partial (V_e + V_a)}{\partial x} + r_i g_i \frac{\partial (V_i + V_a)}{\partial x} \right] \qquad x = a \quad (15)$$

Substitution of Eq. (6) into (15) gives

$$\frac{\partial^2 (V_e - V_i)}{\partial z^2} = 2\pi a (r_e + r_i) g_e \frac{\partial (V_e + V_a)}{\partial x} \qquad x = a \quad (16)$$

We then solve Eq. (16) explicitly for $g_e\, \partial (V_e + V_a)/\partial x$ and substitute this value into the boundary Eq. (8) to get

$$\frac{\partial^2 \psi}{\partial z^2} - \frac{1}{\lambda^2}(\psi - R J_a) = \frac{RC}{\lambda^2}\frac{\partial \psi}{\partial t} \qquad x = a \quad (17a)$$

where we define

$$\lambda^2 = \frac{R/(2\pi a)}{r_e + r_i}$$

and

$$\psi = -(V_e - V_i)$$

Equation (17) is a cable equation for electrotonic spread along a thin-walled cylinder; $-(V_e - V_i)$ is the transmembrane potential; $R/(2\pi a)$ is membrane resistance of a unit length of cylinder; J_a is an applied current. If there is no applied source, then Eq. (17a) becomes the usual cable equation,

$$\frac{\partial^2 \psi}{\partial z^2} - \frac{1}{\lambda^2} \psi = \frac{RC}{\lambda^2} \frac{\partial \psi}{\partial t} \tag{17b}$$

We are interested in the answer to the following question which will lead to a useful form of the cable equation in a steady state. Let the transmembrane potential ψ be the same ψ_r along the entire length of a cylindrical fiber in a steady state with the fiber at rest. (By rest we mean that a nerve fiber is not conducting or receiving local depolarizing or hyperpolarizing currents over the length under consideration, or, for the case of a muscle fiber, that the fiber is not contracting or receiving local currents.) Now over some length of fiber surface from $z = 0$ to $z = -s$ a steady change in ψ is produced, either by application of a continuous steady pulse J_a from a source outside the tissue or by some alteration in properties of the tissue capable of altering V_e or V_i. The question we ask is, what is the shape of the curve describing the disturbance of transmembrane potential, $\Delta\psi = \psi - \psi_r$, as a function of the distance z from the zero point at which $\Delta\psi(0) = \psi(0) - \psi_r$, the steady state disturbance over the range $z = 0$ to $-s$?

Since we are dealing with the steady state, $\partial\psi/\partial t$ of Eqs. (17) is zero, and equation (17a) becomes

$$\frac{\partial^2 \psi}{\partial z^2} = \frac{1}{\lambda^2} (\psi - RJ_a)$$

The factor $(\psi - RJ_a)$ is now equivalent to the disturbance $\Delta\psi(z)$ because we measure only one transmembrane potential at z, $\psi(z)$, which is made to differ from ψ_r either because J_a was applied at $z = 0$ (note that $J_a = 0$ for $z > 0$) or because ψ_r was changed to $\psi(0)$ at $z = 0$ by some biological process.

We note that since ψ_r is a constant,

$$\frac{\partial^2 \Delta\psi(z)}{\partial z^2} = \frac{\partial^2 [\psi(z) - \psi_r]}{\partial z^2} = \frac{\partial^2 \psi}{\partial z^2}$$

so that Eq. (17a) can be written for these conditions

$$\frac{\partial^2 \Delta\psi}{\partial z^2} = \frac{1}{\lambda^2} \Delta\psi$$

The solution is

$$\Delta\psi(z) = \Delta\psi(0)\, e^{-z/\lambda} \tag{8a1}$$

as can be verified by differentiation. Equation (18a) is the form in which the cable equation is usually written for a steady state disturbance. It states simply that in this model the disturbance in transmembrane potential $\Delta\psi = \psi - \psi_r$ decays monoexponentially along the length z of the cylinder surface with the length (or space) constant λ, the axial distance from the zero point at which the disturbance falls to $1/e$ or 37% of its value at the zero point.

An equivalent form of Eq. (18a) which makes explicit the change with distance of the transmembrane potential itself is obtained by substitution for $\Delta\psi$, to obtain

$$\psi(z) = \psi(0)\, e^{-z/\lambda} + \psi_r(1 - e^{-z/\lambda}) \tag{18b}$$

There are two important things to notice about λ: (a) from the definition of λ^2, as the membrane gets leakier R decreases and λ gets smaller; (b) λ is a function of the radius of the cylinder. The value r_i is a function of the radius of the cylinder and may be written $r_i = R_i/(\pi a^2)$. Since r_i is usually very much larger than r_e, the definition of λ^2 then becomes approximately

$$\lambda^2 = Ra/2R_i \tag{19}$$

from which we see that the larger the radius of the fiber the greater the space constant λ.

For example, membrane resistance R for a frog sartorius muscle 100 μ diameter may be 4000 Ω cm^2. If R_i is taken as the resistance of physiological salt solution, about 50 Ω cm, then the space constant λ is about 4.4 mm. If the fiber had a diameter of 30 μ, then λ would be only 2.4 mm. But if the diameter is only 500 Å, as in the transverse tubular system, and if cable theory still holds, then for the same R and R_i, λ is only 100 μ.

Actually, for frog sartorius muscle at 5°C in isotonic Ringer solution, membrane resistance of fibers averaging 56 μ was 3700 Ω cm^2, core resistance R_i was 169 Ω cm, and λ was 1.8 mm (Adrian *et al.*, 1970).

One of the conclusions we reach from such calculations is that a potential applied at or originating at the myoneural junction can not spread unattenuated over the length of a 1 or 2 cm muscle fiber purely by electrotonus. There must be some other method by which action potentials are propagated.

B. Initiation of an Action Potential

In 1937, Rushton published his classic paper entitled, "Initiation of the Propagated Disturbance." Although, as Rushton wrote, it had been supposed for some time that the mechanism of propagation is stimulation of the inactive region just in front by the advancing action potential, it had not been demonstrated how this occurred and it was only within the year of his analysis that the mechanism was demonstrated at the electrical level.

Blair and Erlanger (1936) and Hodgkin (1937) both observed that a nerve pulse induced by electrical stimulation could be blocked so that it did not propagate beyond the blocked region (whether the block was induced by anodal polarization, by application of cold, or by pressure), and that there was produced downstream from the block an altered potential that persisted long enough so that a second action potential upstream from the block could add enough to the altered potential downstream to reinitiate propagation for an action potential. Clearly, the persistence of the altered potential is a reflection of the RC characteristics of the membrane and distribution of the altered potential beyond the blocked region is related to the spread of electrotonus discussed previously.

Such experiments illustrate that the intensity of the propagated action potential is normally many times greater than necessary to excite the neighboring portion of nerve. Hodgkin (1937) estimated it to be perhaps ten times greater. But this does not tell us all we need to know about the initiation of an action potential, say at the cathode of a stimulating current.

Rushton (1937) attacked this problem on the basis of the core conductor model. He considered first the case of stimulation of a nerve by a bipolar electrode with the cathode at $z = 0$ and the anode at $-z_a$. Since all currents flowing through the core away from the electrode except for the region between the poles must return through extracellular fluid, $I_i = -I_e$ in the region beyond the electrodes. Rushton assumed that a charge accumulated in the membrane as a result of the stimulus. In the notation we used in the previous section, this assumption is that a net charge, $|q_e - q_i|$, accumulates on the surfaces of the membrane. When this reaches a critical value Δq_c, an abrupt change occurs in the properties of the membrane and an action potential appears. Thus the problem reduces to the demonstration that $|q_e - q_i|$ grows with time at some point z beyond the stimulating electrode so that it can reach the critical value Δq_c.

From Eq. (4) and (18), we see at once that if the action potential has extended to a distance z_2 from the cathode, so that the net charge at z_2 is the critical value Δq_c, then for some plane at the distance z greater than z_2 the electrotonic spread gives

$$|q_e - q_i| = \Delta q_c \, e^{-(z-z_2)/\lambda} \qquad z > z_2 \tag{20}$$

Consider that the active region, i.e., the region over which $|q_e - q_i| = \Delta q_c$, extends from $-z_1$ to $+z_2$ on either side of the cathode at $z = 0$ and we enquire into the conditions that permit continued accumulation of net charge on the membrane at some $z > z_2$ after the external stimulus is withdrawn. Therefore, what we are asking is that at z, $d|q_e - q_i|/dt$ must at least be greater than zero after the stimulus is withdrawn.

On the basis of a development of cable theory related closely to the model that led to Eq. (17), Rushton showed that if $\partial|q_e - q_i|/\partial t$ is positive, then the length of the active region, $z_1 + z_2$, must exceed a certain constant:

$$z_2 + z_1 > -\lambda \ln\left(1 - \frac{2R\,\Delta q_c}{\alpha\psi_r}\right) \tag{21}$$

where α is the time constant of the equivalent RC component for longitudinal currents through the membrane and ψ_r is the resting transmembrane potential.

Rushton called the length $z_2 + z_1$ the liminal length, and the constant $1 - (2R\,\Delta q_c/\alpha\psi_r)$ the propagation constant. This constant is independent of the nature of the stimulus. Rushton's liminal length is not only necessary for propagation, but, as he showed, it is also sufficient, for it guarantees an initial positive velocity of propagation.

There are several places in which Rushton's concept of liminal length is important. The first place concerns initiation of an action potential in the region just beyond the motor end plate. In response to normal stimulus through the motor nerve, an electrotonic potential spreads beyond the end plate. This potential never becomes large compared to the action potential, but when it reaches a certain size there occurs an abrupt change. Instead of decaying electrotonically, the potential increases rapidly to the magnitude of the peak of an action potential, which then propagates. According to Rushton, it is not directly the peak size of the electrotonic end plate potential that is critical, but it is the surface outside the end plate that has been charged to a critical level that leads to the abrupt generation of an action potential. We see from Eq. (21) that the liminal length necessary to contain this charge is a function of the length constant for electrotonic spread λ, the *RC* characteristics of the membrane, the critical charge, and the transmembrane potential difference ψ_r prevailing before the end plate potential was generated. While these are explicitly independent of the height of the end plate potential, the region over which the spreading electrotonic potential is greater than some critical value necessary to charge the membrane to Δq_c is not really independent of the height of the end plate potential, as we see by inspection of Eq. (18).

While Rushton's concept of a critical net charge on the membrane is one explanation for initiation of the abrupt change leading to an action potential, it is not the only possible explanation. For example, the electric field through the membrane may itself be a powerful orienting force at the level of molecular structure. A transmembrane potential of 100 mV, across a membrane of 100 Å is equivalent to a linear gradient of 10^5 V across 1 cm, which could be an effective electrophoretic voltage gradient. A change in this gradient might be sufficient to restructure some critical component of the membrane, altering membrane resistance and membrane time constants.

Although we do not know how the rising and spreading end plate potential produces the abrupt change that leads to an action potential, Rushton's concept of a liminal length still seems valid, and as we shall see, we do know descriptively the changes in electrical properties that permit generation of an action potential.

A second place in which Rushton's liminal length concept may be useful is in consideration of the spread of voltage changes beyond the sarcolemma during a propagated action potential. As we shall see, there is good reason to hold that during the spread of action potentials over the sarcolemma there is depolarization along the transverse or T tubular system, and it is difficult to see how one can fail to consider spread along sarcoplasmic reticulum. The question is, is this spread through the T system and possibly through sarcoplasmic reticulum only by electrotonus or is there a propagated action potential?

Adrian *et al* (1969) addressed themselves to the question of radial spread of contraction in single fibers of frog semitendinosus muscle. On the basis of cable theory, and on the basis of an assumed geometry of a network of T tubules over the cross section of the fiber, they estimated a space constant λ_T for electrotonic spread through the T network. They concluded that for a fiber of 50 μ diameter, λ_T was 60 μ and that this was just barely adequate to allow an action potential at the surface to spread by electrotonus along the T system to activate myofibrils at the center. There are, however, a number of assumptions necessary to make such a calculation, the validity of which is uncertain. For example, it is not clear that the usual solution to the cable equation [Eq. (17)] applies when membrane thickness is appreciable compared to the radius of the cylinder, as it is for the T tubules. It is also not clear that spatial summation of the potential, which is neglected in the development of the cable equation on the assumption that the field about one cable has no influence on that about its neighbor, does not apply to the network of cylindrical T tubules in which individual tubules may lie so close together that summation of electrotonic potentials may occur, and a critical level may be exceeded more safely than Adrian's calculations imply. On the other hand, it is conceivable that such summation may lead to spread beyond the liminal length and an action potential may be propagated.

C. The Resting Membrane Potential and Its Relation to the Distribution of Ions

1. THE RESTING MEMBRANE POTENTIAL AND THE GOLDMAN–HODGKIN–KATZ EQUATION

Until it was possible to impale a muscle fiber with a micropipette and measure the electrical potential difference across some barrier between the inside and outside of the fiber, the exact magnitude of the potential of resting muscle was not known. However, it had been known for years that when a muscle or nerve was injured so that some local area of the membrane became so leaky that this area mirrored the intracellular potential, the injured area was negative relative to the bathing solution. Even though it is now known that the magnitude of this injury potential is less than the true resting potential, the observation did demonstrate that nerve and muscle cells are polarized, negative on the inside. It was also known that potassium concentration was very much greater and sodium and chloride very much less inside the cell than outside. It was an inspired hypothesis that Julius Bernstein proposed in 1902, and while it is no longer tenable in all its details, owing to newer information, it undoubtedly focused attention on the quantitative relation between ion distribution and electrical potential that has made the modern view possible. Indeed, this may have been one of the most important contributions to neurophysiology of this century, and it was made about 50 years before it was possible to prove that its main features are correct.

Bernstein proposed, rather simply, in a paper entitled "Investigations on Thermodynamics of Bioelectric Currents," that the Nernst theory for the potential difference between two solutions of the same electrolyte at different concentrations could be applied quantitatively to explain the mechanism by which an electrical potential difference is generated across membranes of excitable cells. Bernstein knew from the work of J. Katz in 1896 that potassium was by far the major intracellular cation in muscle and that muscle chloride content was low and might be zero. Because there was no way at the time to measure intracellular water content, it was not possible to determine intracellular ion concentrations precisely, yet Bernstein was inspired to propose that the muscle membrane might be permeable to potassium, impermeable to other cations and to anions, and that the electrical potential difference across the membrane would then be given quantitatively by the Nernst equation

$$-E = -E_K = \frac{RT}{F} \ln \frac{(K^+)_i}{(K^+)_o} \tag{22}$$

where E is the transmembrane electrical potential difference, representing the same quantity $\psi = V_i - V_e$ used in our earlier equations, R is the gas constant, T the absolute temperature, F the faraday, ln the natural logarithm, $(K^+)_i$ and $(K^+)_o$ the potassium activity inside and outside the cell, respectively. Activity, (K^+), is defined by the product of an activity coefficient α and concentration (K^+). Since it was assumed and is now known experimentally, though with some uncertainty, that the activity coefficients inside and outside are about the same, the activity ratio is often written simply as a concentration ratio in the Nernst equation. E_K is the equilibrium potential for potassium; that is, it is the electrical potential difference at equilibrium between two solutions separated by a membrane through which K^+ and no other ion can diffuse when the concentration ratio at equilibrium is $[K^+]_i/[K^+]_o$.

Bernstein tested his proposal by showing that the observed injury potential for frog sartorius muscle varied approximately linearly with temperature. In early experiments, increases in $[K^+]$ in the bathing solution decreased the absolute value of the injury potential (i.e., made it less negative). This response is called depolarization. Although it was in the correct direction, the depolarization was not exactly that predicted by the Nernst equation, but the experiments were handicapped by technical problems, including inaccurate measurements of true transmembrane potential differences and true ion concentrations.

In 1941, Boyle and Conway wrote a monumental paper on the distribution of ions in frog muscle and showed that muscle was permeable to chloride. Seven years later, when it was possible to use a radioisotope

of sodium to trace sodium movement, Levi and Ussing (1948) obtained
the first strong evidence by this method that muscle was also permeable
to sodium.

Another major technical advance occurred on another front. Hodgkin
and Huxley (1939), using a silver–silver chloride electrode, and Curtis
and Cole (1942), using a glass micropipette filled with concentrated potassium chloride solution which served to conduct to the outside, impaled
the giant axon of the squid and, with the circuit completed through
an extracellular electrode, obtained the first reliable measurements of
the electrical potential difference across the axon membrane. Graham
and Gerard (1946) and Ling and Gerard (1949) applied Curtis and
Cole's technique to measure transmembrane electrical potential differences in frog muscle. Hereafter we will refer to the transmembrane
electrical potential difference across muscle at rest as the resting membrane potential, denoted either by ψ_r or by E_r. E_r was found close
to, but probably slightly less than, the theoretical value predicted for
the potassium equilibrium potential E_K by the Nernst equation [Eq.
(22)], despite the fact that the membrane was now known to be permeable to both sodium and chloride as well as to potassium. Furthermore,
variations in external $(K^+)_o$ in the bathing solution produced close to
the predicted variation in E_r over a wide range of $(K^+)_o$ from normal
to many times normal values, though not for less than normal values.
Thus, more than 45 years after Bernstein's proposal, experiment seemed
to support his quantitative predictions, even though one of his assumptions, impermeability to Na^+ and to Cl^-, was not tenable. Clearly Bernstein's hypothesis needed modification.

An important step toward a solution was made by Goldman (1943),
who proposed a treatment of the relation between transmembrane potentials, concentrations of various ions on either side of the membrane,
ion mobilities, and other properties, as we shall see, which was made
solvable in a simple, useful form by the assumption that electrical potential gradient through the membrane was constant. The equation has become known, therefore, as the constant field equation.

In 1949, Hodgkin and Katz applied the equation, with some simplifica-
tion in the development, to analysis of data from studies of the squid
giant axon. The equation, which became known as the Goldman–Hodgkin–Katz equation, has now been applied to analysis of data concerning
bioelectric phenomena from a number of cell types, including muscle.
The Goldman–Hodgkin–Katz equation has made it possible not only to
understand resting membrane potentials, but also action potentials, and
it has become one of the most productive tools in neurophysiology.
For this reason we derive it below, using the constant field assumption,

although it has been possible to derive the same equation without the constant field assumption.

Let J_K, J_{Na}, and J_{Cl} be the contributions to the inward current density J across the membrane made by K^+, Na^+, and Cl^-, respectively. Assume that ions move in the membrane under the influence of both chemical potential and electrical potential in a manner essentially similar to that in free solution. Then the current densities through the membrane due to K^+, Na^+, and Cl^- through a plane parallel to the surface of the membrane at a distance x from the outer boundary at $x = 0$ are

$$-J_K = RTu_K \frac{dC_K}{dx} + C_K u_K F \frac{dV}{dx}$$

$$-J_{Na} = RTu_{Na} \frac{dC_{Na}}{dx} + C_{Na} u_{Na} F \frac{dV}{dx} \qquad (23)$$

$$-J_{Cl} = -RTv_{Cl} \frac{dC_{Cl}}{dx} + C_{Cl} v_{Cl} F \frac{dV}{dx}$$

where u_K, u_{Na} and v_{Cl} are mobilities of ions and C_K, C_{Na}, and C_{Cl} are activities of ions in the membrane at x, and V is the electrical potential at x.

We now make three assumptions: (1) that the membrane is homogeneous so that the mobilities are the same throughout the membrane, (2) that in the steady state J_K, J_{Na}, and J_{Cl} are constant through the membrane, and (3) that dV/dx is constant through the membrane. This last assumption is the assumption of a constant field. If dV/dx is constant, then the gradient in potential must be simply the difference between the potential outside and the potential inside, divided by the thickness of the membrane α, or

$$\frac{dV}{dx} = \frac{V_i - V_e}{\alpha} = \frac{\psi}{\alpha}$$

It is this substitution that simplifies Eqs. (23) to an ordinary first order linear differential equation to permit direct integration through the thickness of the membrane from $x = 0$ to $x = \alpha$. For example, for the case of K^+, rearrangement of Eq. (23) with the substitution of ψ/α, gives

$$\frac{dC_K(x)}{dx} + \frac{\psi F}{RT\alpha} C_K(x) + \frac{J_K}{RTu_K} = 0$$

the solution of which gives

$$J_K = \frac{u_K F \psi}{\alpha} \frac{C_K(0) - C_K(\alpha)e^{\psi F/RT}}{e^{\psi F/RT} - 1}$$

It is now assumed that the activity $C_K(0)$ at the outer edge within the membrane is directly proportional to the activity of potassium in the external fluid $(K^+)_o$, and that the activity $C_K(\alpha)$ at the inner edge within the membrane is directly proportional to the activity of potassium in the internal fluid $(K^+)_i$, and that the two constants of proportionality are identical because the membrane is assumed to be homogeneous. Therefore,

$$C_K(0) = \beta_K(K^+)_o$$

$$C_K(\alpha) = \beta_K(K^+)_i$$

where β_K is the constant of proportionality and in essence is a partition coefficient between the membrane and aqueous solution. This assumption permits us to substitute in the equation for J_K to obtain

$$J_K = P_K \frac{F^2\psi}{RT} \frac{(K^+)_o - (K^+)_i e^{\psi F/RT}}{e^{\psi F/RT} - 1} \tag{24a}$$

where we introduce the symbol

$$P_K = u_K \beta_K RT/(\alpha F) \tag{25}$$

where P_K is called the permeability coefficient for potassium through the particular membrane. It has dimensions of velocity (centimeters per second), and at a given temperature and membrane thickness, it can be changed only by changing mobility of K^+ or by altering the membrane in such a way that the partition coefficient β_K is altered.

There will be an equation of the form of Eq. (24a) for J_{Na} and one for J_{Cl}, with appropriate attention to sign,

$$J_{Na} = P_{Na} \frac{F^2\psi}{RT} \frac{(Na^+)_o - (Na^+)_i e^{\psi F/RT}}{e^{\psi F/RT} - 1} \tag{24b}$$

$$J_{Cl} = P_{Cl} \frac{F^2\psi}{RT} \frac{(Cl^-)_i - (Cl^-)_o e^{\psi F/RT}}{e^{\psi F/RT} - 1} \tag{24c}$$

Total current density through the membrane is $J = J_K + J_{Na} + J_{Cl}$. Addition of Eqs. (24) gives

$$J = \frac{F^2\psi}{RT} \frac{1}{e^{\psi F/RT} - 1} \{[P_K(K^+)_o + P_{Na}(Na^+)_o + P_{Cl}(Cl^-)_i]$$
$$- [P_K(K^+)_i + P_{Na}(Na^+)_i + P_{Cl}(Cl^-)_o]e^{\psi F/RT}\} \tag{26}$$

Equation (26) gives the relation between the membrane current J, the membrane potential ψ, and ion permeabilities and activities.

Now we note in Eq. (26) that if $J = 0$, that is, if there is no net current, then we are left with only the portion of Eq. (26) in brackets. With rearrangement, the logarithm yields

$$\psi = -\frac{RT}{F} \ln \frac{P_K(K^+)_i + P_{Na}(Na^+)_i + P_{Cl}(Cl^-)_o}{P_K(K^+)_o + P_{Na}(Na^+)_o + P_{Cl}(Cl^-)_i} \qquad (27)$$

Equation (27) is the Hodgkin and Katz formulation of the Goldman equation. The justification for setting $J = 0$ and thus obtaining Eq. (27) is that in the steady state, the sum of the current densities due to cation movement must equal the sum of the current densities due to anion movement, since cation and anion movement must be coupled electrostatically, and, since the charge is opposite, there is no net current density. Thus, in the steady state, $J_K + J_{Na} = -J_{Cl}$. This means that the Goldman–Hodgkin–Katz equation applies only to steady states.

Goldman (1943) assumed a constant field, which made it possible to derive the relatively simple expression for ψ given in Eq. (27). He pointed out that if the membrane contained a large number of dipolar ions near their isoelectric point, these ions could minimize distortions of the field, particularly at low currents, and so tend to maintain a nearly constant field. We have noted that it has been possible to derive Eq. (27) by other means without assuming a constant field.

Only the three monovalent ions, K^+, Na^+, and Cl^-, are considered in the equation for the membrane potential because they make by far the largest contribution. Three things are required for an ion to contribute to the potential: (a) it must have a concentration or activity gradient, (b) its concentration at least on one side of the membrane must be large enough to permit it to make a significant contribution to the current density, and (c) its permeability coefficient P must be sufficiently large to permit it to make a significant contribution to the current density. Hydrogen ion, for example, meets two of these requirements. There is a concentration gradient for hydrogen ion, and the membrane is quite permeable to it. But at a concentration of 10^{-7} M compared to 0.1 M for potassium, sodium, and chlorine, there is simply not enough hydrogen ion to make an impression on ψ. Calcium ion has a gradient and a large enough concentration, but its ability to penetrate the normal membrane is so small that it doesn't contribute significantly to ψ.

If an ion moves from one region to another only under the influence of its chemical potential gradient and the electrical potential gradient, eventually, in a closed system, a state will be reached in which the chemical potential gradient tends to drive the ion in a direction opposite that of the electrical potential gradient, and the magnitudes of the ion movement in the two directions become constant and equal. If that state is reached, the ion makes no contribution to net current density, even though many ions of that species may be flowing across the membrane.

There is good reason, from the work of Hodgkin and Horowicz (1959b) to think that this is true for Cl^- in frog muscle. In that case J_{Cl} in Eq. (24c) is zero and Eq. (24c) yields

$$(Cl^-)_i - (Cl^-)_o e^{\psi F/RT} = 0$$

or

$$\psi = -\frac{RT}{F} \ln \frac{(Cl^-)_o}{(Cl^-)_i}$$

But from the Nernst equation, Eq. (22), this is the definition of the chloride equilibrium potential. Therefore,

$$\psi = E_{Cl} \qquad \text{if } J_{Cl} = 0 \tag{28}$$

With this possibility we return to Eqs. (26) and (27) and note that if $Cl^-)_i = (Cl^-)_o e^{\psi F/RT}$, then the Goldman-Hodgkin–Katz equation degenerates to

$$\psi = -\frac{RT}{F} \ln \frac{P_K(K^+)_i + P_{Na}(Na^+)_i}{P_K(K^+)_o + P_{Na}(Na^+)_o} \qquad \text{if } J_{Cl} = 0 \tag{29}$$

in which case, $J_{Na} = -J_K$. We can use Eq. (28) to determine the relative permeability coefficient for Na^+ compared to K^+,

$$b_{Na} = P_{Na}/P_K$$

Division of numerator and denominator of the logarithmic term in Eq. (29) by P_K, with rearrangement and elevation to powers of e, yields

$$b_{Na} = \frac{(K^+)_i - (K^+)_o e^{-\psi F/RT}}{(Na^+)_o e^{-\psi F/RT} - (Na^+)_i} \qquad J_{Cl} = 0 \tag{30}$$

The permeability coefficients are not constants but are functions of ψ or of concentrations. For this reason it has been difficult to measure them accurately by means of Eq. (27). To do this, one needs to hold the external concentrations of two ions constant, varying only that of the third ion, hope that no changes occur in internal concentrations, and observe the effect on ψ. If the values of P are constant, and if one does this experiment separately for each ion, one can calculate P_{Na}, P_K, and P_{Cl}. However, if $J_{Cl} = 0$ and $E_{Cl} = \psi$, one can not calculate P_{Cl} this way, because no chloride terms appear in Eq. (29). And if the values of P vary with ψ, one cannot make the calculation. In the early papers it was assumed that values of P were nearly constant over small ranges of ψ.

Despite difficulties in exact measurements, it is clear that for resting muscle, b_{Na} from Eq. (30) is small, between 0.01 and 0.05, which means that P_K is twenty to one hundred times P_{Na}. P_{Cl} is probably greater than P_K, perhaps twice as great.

There have been numerous tests of the Goldman–Hodgkin–Katz equation and a very large body of data has now accumulated that is consistent with it (Eq. 27). For the case of frog sartorius muscle at rest, Richard Adrian (1956) has collected quantitative data based on measurements of intracellular and extracellular concentrations (but see later for discussion of true sarcoplasmic concentrations) and on membrane potentials. Hodgkin and Horowicz (1959b) calculated permeability coefficients for frog sartorius at 20°C: $P_K = 2 \times 10^{-6}$ cm/sec, $P_{Na} = 2 \times 10^{-8}$, $P_{Cl} = 4 \times 10^{-6}$. These figures may be slightly in error due to uncertainties in intracellular concentrations.

For a mammalian muscle (extensor digitorum longus of the rat) at about 25°C, we calculate from our own data on the basis of flux experiments (to be described in the section on ion fluxes) that $P_K = 10^{-7}$ cm/sec and $P_{Na} = 7 \times 10^{-9}$ cm/sec. Notice that this mammalian muscle is very much less permeable to K^+, by an order of magnitude, than is frog muscle, and the ratio of P_{Na}/P_K is 0.07, about seven times greater than in frog muscle. It is this relatively greater permeability to Na^+ in mammalian muscle that accounts for the fact that the absolute value of E_r in mammalian muscle is substantially less than the absolute value of E_K, whereas in frog muscle E_r is only a few millivolts less than E_K.

2. MEMBRANE RESISTANCE AND CAPACITANCE
—ROLES OF SARCOLEMMA AND T SYSTEM

It had been known for some time that the capacitance of muscle membranes was much greater than that of nerve, which is about 1 $\mu F/cm^2$. For various muscle preparations, the capacitance was calculated at anywhere from 2.5 to about 40 $\mu F/cm^2$. The highest values were obtained in crustacean muscle, usually crab or crayfish, and in that case it was suggested that estimates of membrane (sarcolemma) surface area were far too small because the surface membrane folded upon itself to produce clefts, but this left unexplained the high apparent capacitance of frog muscle, which had no such infoldings.

With electron microscopy, it was discovered that in skeletal muscle from all species the sarcolemma appeared to continue into a tubular network that penetrated into the center of the fiber, mainly at a plane of a cross section of the fiber, with either one or two of these networks per sarcomere, depending on the species. This network, the transverse tubular or T system, has a plasma membrane of about the same thickness as that of the bilamellar sarcolemma plasma membrane, and it does also apparently have a basement-membrane, an amorphous, Schiff-positive layer continuous with a layer of similar appearance lying outside the plasma membrane. The total diameter of the T tubules is perhaps 500 Å, so that membrane thickness is about the same as the radius of the lumen. The T system probably forms some sort of annular structure around each myofibril so that the whole T lattice probably has a more or less honeycomb appearance.

There are numerous openings of the T system to the outside, hence to interstitial or bathing fluid. Ferritin and albumin-^{131}I, added to bathing solutions, have been identified inside T tubules, but not in sarcoplasmic reticulum, which is a term we shall restrict to the longitudinal system, including terminal cisternae.

Porter and Palade (1957) suggested that the function of the T system was to propagate the action potential into the center of the fiber so that contraction of core myofibrils would occur more rapidly. Huxley and Taylor (1958) showed that, whether an action potential was propagated through the T system or spreading depolarization was by electrotonus, an argument into which they did not enter but to which we shall return, subthreshold depolarizations applied to the surface of frog muscle fibers did not produce a twitch unless they were applied at the Z line, which is the plane of the T tubules in the frog. Small depolarizations caused visible twitches only of surface myofibrils. Larger ones caused all myofibrils to twitch.

It was only a short step from this to suspect that the extra capacitance of muscle membrane might lie in the T tubules. Falk and Fatt (1964) measured membrane capacitance and time constants in response to high frequency currents under voltage clamp conditions and concluded that the extra capacitance could lie in what they called the sarcoplasmic reticulum, in which they included the T system. Their data for frog sartorius fitted a model in which a surface membrane (sarcolemma?) resistance R_m of 3100 Ω cm^2 shunting a surface membrane capacitance C_m of 2.6 μF/cm^2 is in series with another path (T system?) with resistance R_T of 330 Ω cm^2 and capacitance C_T of 4.1 μF/cm^2, all referred to unit area of fiber surface. In fact, the area of the T tubules has been estimated at three to seven times the area of the fiber surface, so that the capacitance and resistance values are not quite correct with respect to the surface of the T system. The point is that about 60% of the membrane capacitance of frog sartorius muscle probably lies in some component other than the surface membrane, presumably at least the membrane of the T system. Notice that the estimated resistances differ by a factor of ten. For this reason, the time constant of the RC element in what we shall tentatively call the T system is only about 1 msec compared to about 20 msec for the total membrane time constant.

More direct demonstration of the capacitative contribution of the T system was made possible by the discovery that soaking a muscle in solutions made hypertonic with glycerol and then returning it to isotonic Ringer's solution ruptured T tubules, at least in superficial fibers (Howell, 1969). Such fibers might have normal resting potentials (but many were at least slightly depolarized) and they propagated action potentials. But they did not twitch in response to electrical stimuli, a discovery reported earlier by Fujino *et al.* (1961) before the morphological effects of the glycerol treatment were understood.

More complete studies of the electrical effects associated with glycerol treatment were reported by Gage and Eisenberg (1969a,b) and Eisenberg and Gage (1969). It is not known how the glycerol treatment produces its effect, but it has been speculated that during soaking in the hypertonic solution (glycerol plus Ringer) the glycerol makes the contents of the T tubules hyperosmotic with respect to sarcoplasm. When the muscles are then placed in isotonic Ringer the osmotic effect moves sarcoplasmic water into the T system, tending to concentrate sarcoplasmic contents to some extent, and also moves the bathing solution into the T system, increasing the volume of the T system until rupture occurs. It is quite likely that this treatment does produce other effects It probably makes the surface membrane leakier. Gage and Eisenberg selected for further study only fibers that maintained fairly normal resting potentials. In such fibers, with the data analyzed on the basis of the cable core conductor equation, membrane resistance R_m was 3700 Ω cm^2, essentially normal, but membrane capacitance C_m was only 2.2 μF/cm^2, compared to 6.1 μF/cm^2 in fibers from muscles exposed to glycerol but not returned to isotonic Ringer.

The evidence, then, is strong that the membranes of the T system, continuous with (=in series with) the fiber surface membranes, contribute to the electrical properties of the muscle fiber. The quantitative nature of this contribution depends on the validity of a number of assumptions. One of these is the question of whether or not the core conductor equations used in the determinations are really appropriate to a system in which membrane thickness is of the same order as lumen radius. Another question is whether or not the membrane electrical circuitry continues from the T system through sarcoplasmic reticulum including terminal cisternae and longitudinal reticulum.

It had been suggested that the contact between T tubule and terminal

cisternae is a tight junction. The question of morphological fact has
been debated back and forth. At the time of this writing, the most
recent study (Franzini-Armstrong, 1970) failed to uncover any structures
resembling tight junctions in frog sartorius. The flattened surfaces of
the terminal cisternae were separated from the T tubules by 120–140
Å. However, at periodic intervals of 300 Å the terminal cisternae membrane forms small projections, the tips of which join the T system through
some amorphous material which covers only 30% or less of the T tubule
surface. Thus, although we shall later refer to other sorts of experiments
in which electrical coupling between T tubules and sarcoplasmic reticulum seems to be required, the morphological evidence is that the junction
does not seem to be well designed for direct conductive or capacitative
coupling because of its spacing. However, the longitudinal component
of sarcoplasmic reticulum does broaden tremendously from its narrow
lumen to form the large surfaces of the terminal cisternae that do approximate the T tubules, even though the gap may be 120 Å instead of
only 40 Å as one might prefer for easy conduction. We shall see later
that there are studies of Na^+ flux that are best interpreted at present
by a model in which the junction between the T tubule and the terminal
cisternae is freely permeable to Na^+. Furthermore, as we shall see, there
are reasons for suspecting that Ca^{2+} is released from terminal cisternae
into sarcoplasm in response to depolarization of the T tubule, which
requires a signal of some sort from T tubule to sarcoplasmic reticulum.
While the junction between T and terminal cisternae may not look like
a tight junction and may exhibit wider separation than one wishes for
capacitative coupling, it is even less likely that the signal is by chemical
transmitter, since there is clearly none of the familiar machinery (e.g.,
mitochondria) for synthesis of a chemical transmitter in the lumen or
wall of the T tubules.

Atlhough I do not agree that electrical coupling between T tubule
and terminal cisternae can be eliminated, there is another possible mechanism by which information might be transmitted. Franzini-Armstrong
describes amorphous connections, at intervals of about 300 Å, in the
gap between T tubule and terminal cisternae. If these are protein or
complex glycoproteins, it may be that their tertiary structure is altered
by the electrical field. If these connections have two stable configurations
and if the switch from configuration 1 to configuration 2 is determined
by some critical electrical field density (a shift from 10^5 V/cm to 5×10^4
V/cm, for example), then, if the connecting proteins can transmit a
structural change to the protein wall of terminal cisternae, one might
alter permeability (= conductance) of walls of terminal cisternae by
a critical depolarization across the wall of the T tubule.

3. ROLE OF INDIVIDUAL IONS AND OF INDIVIDUAL SURFACES
   IN MEMBRANE CONDUCTANCE

In our development of the Goldman–Hodgkin–Katz equation, we began with the statement that the contribution of an ion to the current density through the membrane was determined by the electrical potential and by its chemical potential gradient. We can state this for the kth ion species by

$$J_k = g_k(\psi - E_k) \tag{31}$$

where g_k is the conductance of the membrane to the kth ion and E_k is the Nernst equilibrium potential for the kth ion.

Summation over all ion species gives total current density through the membrane

$$J = \Sigma J_k = \psi \Sigma g_k - \Sigma (g_k E_k)$$

whence

$$\psi = RJ + \Sigma (\bar{g}_k E_k) \tag{32}$$

where $R = 1/\Sigma g_k$ is total resistance through the membrane and $\bar{g}_k = g_k/\Sigma g_k$ is partial conductance with respect to the kth ion.

As for the case of the Goldman–Hodgkin–Katz equation, we assume that at rest there is a steady state in which $J = 0$, so that Eq. (32) degenerates to

$$\psi = \Sigma (\bar{g}_k E_k) \tag{33}$$

The circuit described by Eq. (33) is simply one of parallel elements, each element being a voltage source and a resistance.

For the case of muscle membrane, k is K^+, Na^+, and Cl^-. Hutter and Noble (1960) found that when the external solution was made free of Cl^-, by substitution of anions to which the membrane was impermeable, the membrane conductance of frog sartorius muscles was reduced by 68%; that is, about two-thirds of the membrane conductance is to Cl^-.

In 1959, Hodgkin and Horowicz (1959b) reported that when frog muscle was bathed in Cl^--free solutions, variations in $(K^+)_o$ altered ψ in the manner predicted by the Nernst equation, as though $\psi = E_K$. However, when $(Cl^-)_o$ was changed at constant $(K^+)_o$, although there was an initial shift in ψ toward the new value of E_{Cl}, after some time (10–60 minutes), the membrane potential drifted back to its original value, suggesting that Cl^- had redistributed itself in such a way as to return E_{Cl} to the previous value of ψ.

This led Hodgkin and Horowicz (1960) to explore effects of sudden changes in $(Cl^-)_o$ or $(K^+)_o$ on membrane potential of single fibers dissected from frog semitendinosus muscle. Change in $(Cl^-)_o$, either up or down, produced an appropriate change in membrane potential ψ, which was 50% complete in less than 0.3 sec. When $(K^+)_o$ was increased, the expected depolarization also occurred rapidly, about the same as for the case of changes in $(Cl^-)_o$, but when $(K^+)_o$ was decreased from the high level, the expected repolarization took ten times as long to reach half completion. These observations suggested to Hodgkin and Horowicz that K^+ was retained temporarily in some special region (T tubules?), and since the asymmetric response did not occur with Cl^-, they suggested that Cl^- conductance lay mainly through some readily accessible surface, say sarcolemma, but K^+ conductance lay mainly through a less accessible surface, say T tubules. They could not explain the asymmetry to K^+.

Adrian and Freygang (1962) took up the question by studies of conductance in frog sartorius. Their technique depended on insertion of three electrodes into a single fiber in the bundle of fibers. Through one they passed a constant current and through the other two, placed near the end of the fiber, they recorded electrotonic potentials. On the basis of cable theory, they calculated membrane conductance. By alterations of ionic composition in the bathing solution, they were able to produce changes in membrane potential, and these changes were in the direction predicted by the Goldman–Hodgkin–Katz equation, Eq. (27).

Adrian and Freygang and, more recently, Adrian, *et al.* (1970), confirmed that conductance varies with membrane potential, a well known phenomenon in both nerve and muscle. The slope of a plot of current as a function of membrane potential gives $dJ/d\psi$ which is defined as g_m, the membrane conductance. With hyperpolarization (increasingly more negative values of transmembrane potential) g_m gets increasingly smaller, and over this range the membrane current is negative (that is, the current is moving from outside to inside). Current is zero at the value of the resting potential, and current becomes more positive as the membrane is depolarized. At the same time, g_m increases progressively. The increase in g_m is gradual at first, but becomes abrupt in the vicinity of -40 mV, which is about the value of the critical depolarization that leads to an action potential. This sort of change in conductance with the change in current is called rectification; large conductance for outward current, small conductance for inward current.

Under certain conditions, particularly when the bathing solution is rich in K^+, the change in conductance is in the opposite direction; small conductance for outward current, large conductance for inward current. This is called anomalous rectification.

It was in an effort to explain anomalous rectification that Adrian and Freygang undertook their studies. Although they failed in this effort and anomalous rectification is still unexplained, the data they obtained compelled attention to the possible role of internal tubular systems, T tubules and sarcoplasmic reticulum, in electrical phenomena. Their data were consistent with the interpretation that the surface membrane was permeable to both Cl^- and K^+, but that the T tubule was permeable only to K^+.

Stronger evidence for this view came from experiments by Eisenberg and Gage (1969) on muscles in which T tubules were ruptured by glycerol treatment. On the assumption that sodium conductance g_{Na} could be neglected in resting frog muscle (an assumption that is plausible for frog muscle but not for mammalian muscle), for fibers about 50 μ in diameter at approximately normal ψ, with experimental design such that T tubules were present or absent and chloride conductance g_{Cl} was present or absent, they found that about 80–90% of the conductance through the surface membrane was to Cl^-, and that K^+ conductance g_K through the nonsurface membrane, considered by them to be T tubules, was about twice as great as through the surface membrane. However, if this nonsurface membrane is that of the T tubules, then, since the total surface area of T tubules is about four times that of sarcolemma, the specific conductance per square centimeter of surface is only half as great through T tubular walls as through sarcolemma. Some real values, approximated from Eisenberg and Gage, are:

g_{Cl}	sarcolemma	220 μmho/cm^2 of sarcolemma
g_K	sarcolemma	28 μmho/cm^2 of sarcolemma
g_{Cl}	T tubule	0 μmho/cm^2 of T tubule surface
g_K	T tubule	14 μmho/cm^2 of T tubule surface

To obtain relative contributions to total conductance, multiply g_K (T tubule) by 4, to get $g_K = 28 + (4 \times 14) = 84$, and g_{Cl} is 220, or total membrane conductance g_m is 300 μmho/cm^2 of fiber surface, of which 70% is to Cl^- and 30% to K^+.

D. The Action Potential

A propagating action potential is preceded by a wave of currents that depolarize the membrane to a critical level. In 1939, for giant axon of the squid, Cole and Curtis concluded that when an action potential passed a localized region, the membrane resistance decreased abruptly from 1000 Ω cm^2 to 25 Ω cm^2 with no change in membrane capacitance. The decrease began immediately with the rising phase of the propagated action potential. At that point, the membrane current changed from outward to inward. That is, the initial depolarization, akin to what would be electrotonus if it did not become great enough to lead to an action potential, was associated with an outward current. Over this range, the current–voltage relations are such that there is only a small decrease in membrane resistance. But suddenly, membrane resistance nearly collapses and simultaneously current shifts from outward to inward.

The action potential is not simply a discharge of the resting potential to ground, but there is an overshoot with reversal of potential from −90 mV at rest to, say, +30 mV at the peak of the action potential (Fig. 2).

How are this increase in conductance, reversal of current, and reversal of potential to be explained? In 1902, Overton noticed that frog muscle was inexcitable in solutions in which $(Na^+)_o$ was less than 10% of that in Ringer. Hodgkin and Katz (1949) proposed that since the sodium equilibrium potential E_{Na} is normally positive, +30 mV or greater, the reversal of current and the overshoot of the action potential to +30 mV could be explained if the effect of critical depolarization were to increase permeability of the membrane to Na^+ specifically. The idea has been tested in both nerve and muscle, and it has been amply demonstrated that the height of the action potential is proportional to E_{Na}. That Na^+ actually does move from the bath into the axon during the action potential was shown by Keynes and Lewis (1951), who demon-

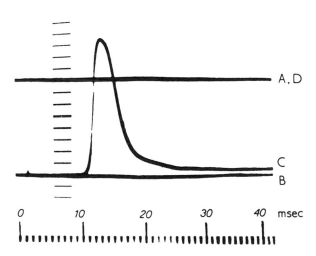

Fig. 2. Resting and action potential of a muscle fiber recorded at 6°C. External diameter of electrode tip, 0.5 μ; resistance of microelectrode, 57 MΩ. Records A and D were obtained with the microelectrode outside the fiber at the beginning and end of the experiment; B and C were obtained with it inside in the resting and stimulated conditions. From Nastuk and Hodgkin (1950).

strated an increase in axoplasmic sodium concentration after a series of action potentials amounting to 3.5×10^{-12} moles per square centimeter of membrane per impulse. The amount of Na^+ that must be transferred to discharge the membrane and recharge it with opposite sign is given as the product of the voltage change, which is the height of the action potential and the capacity of the membrane. Since muscle membranes have several times the capacity of the axon, the quantity of Na^+ transferred per impulse must be somewhat greater, and measurements verify that it is (Hodgkin and Horowicz, 1959a).

Since the action potential is transient (time from onset to peak is less than a millisecond and time from peak to 50% return to resting potential is also less than a millisecond), there must be some other source, presumably, of current flow in a direction opposite that of Na^+ inflow current, counteracting and overcoming the Na^+ current. This source, in theory, could be either outward K^+ current or inward Cl^- current or both. The evidence is that it is in fact due to increased outward K^+ current, a result of increased permeability to K^+.

Normally, before the rapidly falling action potential in muscle has quite returned to the resting value, there is a small depolarization by a few millivolts followed by a slow repolarization back to the resting membrane potential over a period of a few milliseconds. This wave is the afterpotential, or, since we will deal with an even later potential, the early afterpotential. The late afterpotential is unique for skeletal muscle. It does not occur in nerve.

The spike of the action potential, the decay of the spike, and the early afterpotential can all be explained on the basis of alterations in conductance to specific ions through the membrane, with knowledge of the equilibrium potentials of the ions.

The sequence of events is this. At rest P_{Na}/P_K is about 0.01 for frog sartorius, greater for rat extensor digitorum longus. At the level of critical depolarization at which the action potential arises, there is an abrupt increase in membrane conductance to hundreds of times its resting value, from about 0.3 mmho/cm^2 at rest to about 700 or more mmho/cm^2 during the early phase of the action potential (Adrian *et al.*, 1970). The increase is due to an increase in both sodium conductance g_{Na} and potassium conductance g_K, but the time courses of the changes in conductance are quite different. The value of g_{Na} increases from nearly zero at rest to about 60 mmho/cm^2 during the peak of the action potential, at which time g_K has risen to only 10 mmho/cm^2. So P_{Na}/P_K shifts to about 6 at this time.

With the tremendous early increase in g_{Na}, there is a large inward sodium current J_{Na}, tending to drive the membrane potential ψ toward the positive equilibrium potential for sodium E_{Na}. Adrian *et al.* (1970)

interpret their data as showing that the peak of the action potential is actually more positive than expected, which they attribute to artifacts of the voltage clamp method, but which I suspect is more likely due to a gross underestimate of sarcoplasmic sodium concentration and hence to underestimate of E_{Na}, which I will discuss in the section on ion fluxes and distributions. On the basis of our estimate of E_{Na}, we suspect that the rising outward K^+ current, together with decreasing inward Na^+ current prevents the peak of the action potential from ever reaching E_{Na}.

During the falling phase of the action potential, g_{Na} is decreasing, and at the time at which the early afterpotential is reached, P_{Na}/P_K is reduced to perhaps 0.03 in frog sartorius (Adrian *et al.*, 1970). Potassium conductance and total conductance decrease toward normal resting values as the membrane is repolarized. This explains why the membrane potential approaches the initial resting potential but it does not explain why the early depolarizing afterpotential is at a level less than the resting potential.

The explanation offered by Adrian *et al.* (1970) is that during the action potential the large net outward K^+ current represents K^+ movement out of sarcoplasm into some space in which it accumulates temporarily, and thereby raises the local $(K^+)_o$ so that over this region E_K is less negative than expected from E_K calculated on the basis of K^+ concentration in the bathing solution. From the equilibrium potential for the late current in voltage clamp experiments Adrian *et al.* (1970) calculated that the extra K^+ accumulated in a space equal to one-sixth to one-third of fiber volume, an order of magnitude greater than the volume of the T system. We shall see in the section on ion fluxes and distribution that this space is likely to be sarcoplasmic reticulum.

Gage and Eisenberg (1969b) have excellent evidence that the afterpotential is due to currents flowing across some membranes not on the surface of the fiber. When they destroy the T tubules of frog sartorius by glycerol treatment, there is no afterpotential component to the action potential.

III. Sodium and Potassium Fluxes and Distribution in Muscle

A. *Distribution of Water in Muscle*

We have already seen that it has been known for about three-quarters of a century that muscle fibers are rich in potassium compared to their environment, and poor in sodium and chloride.

Knowledge of the exact concentrations had to await some means of measuring the volume of sarcoplasmic water. This was accomplished, it was thought, by using a substance distributed only in extracellular fluid, such as inulin. After the muscle was soaked for a sufficiently long period in a solution containing inulin in concentration $[In]_o$ so that the inulin equilibrated to concentration $[In]_o$ in all fluid accessible to it, inulin content In of the tissue was measured. Since the inulin content consisted only of extracellular inulin at concentration $[In]_o$ in an unknown volume V_o, extracellular volume, or, more precisely, the volume of distribution of inulin V_{In} can be calculated:

$$V_o = V_{In} = In/[In]_o \tag{34}$$

If total water V_T is measured by drying the tissue, then intracellular water can be calculated as

$$V_i = V_T - V_o \tag{35}$$

Sodium or potassium content, Na_T or K_T, is the sum of extracellular sodium Na_o or potassium K_o, and intracellular sodium Na_i or K_i, respectively. Since extracellular concentrations are known, intracellular contents and concentrations are determined from

$$K_i = K_T - K_o = K_T - V_o[K]_o \tag{36}$$

$$[K]_i = \frac{K_T - V_o[K]_o}{V_i} \tag{37}$$

There is, of course, a pair of similar equations for sodium. Since $[K]_o$ is small compared to $[K]_i$, errors in estimate of V_o have relatively little effect on the estimate of $[K]_i$, but since $[Na]_o \gg [Na]_i$, errors in estimate of V_o have a great effect on estimate of $[Na]_i$.

There are a number of reasons for wanting to measure $[Na]_i$ and $[K]_i$. In the context of our previous discussion, an important reason is to be able to calculate the equilibrium potentials E_{Na} and E_K. It is therefore important to consider the validity of the marker of extracellular space, inulin. What space is it measuring? From the fact that other large molecules, specifically ferritin and radioiodinated albumin, penetrate interstitial space and the T tubule but not sarcoplasmic reticulum, demonstrated by electron microscopy and autoradiography, we assume that inulin is distributed in interstitial fluid and in T tubules but not in sarcoplasmic reticulum or sarcoplasm.

The sucrose space $Su/[Su]_o$ in muscle exceeds the inulin space.

Duggan and Martonosi (1970) have found that sucrose enters preparations of fragmented sarcoplasmic reticulum but inulin does not.

We must therefore think of at least three water spaces in muscle, as Harris (1963) pointed out: (a) V_o, interstitial water space plus volume of water in T tubular lumen, (b) V_{SR}, volume of water in lumen of sarcoplasmic reticulum, and (c) V_i, volume of sarcoplasmic water. There have been several estimates of the size of these spaces, and they are in pleasing agreement. V_o is measured as inulin space and $(V_o + V_{SR})$ can be estimated from the following.

First, the quantity of rapidly moving ^{24}Na or of sodium rapidly exchanging for external potassium represents a space (quantity of ion/extracellular concentration) equal to 22% of wet weight in freshly dissected muscles (Carey and Conway, 1954), 30% of wet weight in muscles bathed in Ringer's solution (Levi and Ussing, 1948), and 25% of wet weight in muscles bathed in Ringer–bicarbonate solution (Harris, 1965; Rogus and Zierler, 1970). We obtain the same value for the space represented by sodium rapidly exchanging for external lithium. The spaces so calculated, each on the assumption that the concentration of the ion in SR is the same as in the bathing solution, give a value of V_{SR} in agreement with Peachey's (1965) estimate from inspection by electron microscopy.

Second, if one is now willing to assume that sodium concentration in SR is the same as in extracellular fluid, $[Na]_{SR} = [Na]_o$, then on the basis of the two-component muscle model (described in more detail in Section III,E), from studies of Na^* efflux, V_{SR} and V_i can be estimated. Support for the likelihood that these estimates of volume do represent separate spaces comes from experiments in which rat extensor digitorum longus muscles were soaked in solutions of varying tonicity (Rogus and Zierler, 1970). The changes in estimated V_i and V_{SR} were in opposite directions. V_i shrunk in Ringer's made hypertonic by added sodium chloride, and V_{SR} increased. V_i increased and V_{SR} decreased in hypotonic Ringer's (with reduced sodium chloride). The volume changes in V_i are proportional to the reciprocal of the osmolality of the bathing solution, as predicted for an osmometer. The volume changes in V_{SR} are predictable by a model in which the osmolality of V_{SR} equilibrates rapidly with that of V_o, and in which water moves into V_{SR} from V_i when the solutions in V_o and V_{SR} are hypertonic, and in the reverse direction when they are hypotonic. V_{SR} decreases linearly with the reciprocal of the osmolality of the bathing solution. Changes in V_{SR} and V_i estimated by this method agree with those measured by electron microscopy of frog sartorius muscles exposed to anisotonic solutions (Birks and Davey, 1969). In isotonic Ringer's, our estimates are that in extensor digitorum longus muscles of young male rats, V_o is about 13% of muscle wet weight, V_{SR} is about 12%, and V_i is about 55%.

B. Concentrations of Sodium and Potassium

The classic equations for measuring concentration of sarcoplasmic sodium and potassium, Eq. (36) and (37), fail because we now have additional unknowns, the quantity or concentration of sodium or of potassium in SR. Although we do not know the ion concentrations in SR, we have already implied that there are several lines of evidence that suggest they are similar to those in extracellular fluid. Specifically, when we assumed $[Na]_{SR} = [Na]_o$ we calculated from Na^* efflux experiments a volume V_{SR}, which agrees with Peachey's estimate of V_{SR} by electron microscopy and which changes when muscles are soaked in anisotonic solutions in quantitative agreement with electron microscopic estimates by Birks and Davey.

These latter studies are quite important. We interpret the volume changes in anisotonic solutions to mean that extracellular water exchanges with sarcoplasmic water by two routes: sarcolemma and membranes of SR. If SR contained only a low concentration of sodium and if the junction between T tubule and terminal cisterna were only poorly permeable to sodium, then SR would tend to lose water to T tubules and thence to extracellular space when the muscle is soaked in hypertonic solution, whereas it in fact swells. This can only mean that water enters SR from sarcoplasm more rapidly than it leaves SR into T tubules. The quantitative aspects of the relation between V_{SR} and external osmolality lead to the conclusion that probably $[Na]_{SR} = [Na]_o$. We suspect that it is also true that $[K]_{SR} = [K]_o$. We suspect that this is true because if the T tubule–terminal cisternae junctions are freely permeable to Na^+ and Li^+, they are likely to be freely permeable to K^+ also. Moreover, recall that Adrian et al. (1970) estimated that the equilibrium potential for potassium during the phase of delayed K^+ current, the phase of the afterpotential of the action potential, could be accounted for if the K^+ lost from muscle during activation had accumulated in a volume representing one-sixth to one-third that of muscle in which the K^+ concentration behaved as though it were extracellular with regard to its effect on transmembrane potential. In light of what we have been saying, this volume seems likely to be SR.

With these assumptions, that $[Na]_{SR} = [Na]_o$ and $[K]_{SR} = [K]_o$, we tabulate contents, concentrations, and equilibrium potentials for rat extensor digitorium longus muscle at 25°C (Table I). Notice that E_{Na}

TABLE I

WATER, SODIUM, AND POTASSIUM DISTRIBUTION IN RAT EXTENSOR
DIGITORUM LONGUS MUSCLE[a]

Substance	Magnitude
Total water	0.80 ml/gm wet weight
Extracellular water	0.13 ml/gm wet weight
Volume of SR	0.126 ml/gm wet weight
Sarcoplasmic volume	0.54 ml/gm wet weight
Sodium content, SR	20 mEq/kg wet weight
Sodium content, sarcoplasm	0.75 mEq/kg wet weight
[Na], sarcoplasm	1.4 mEq/liter sarcoplasmic water
E_{Na}	118–126 mV
Potassium content, SR	0.6 mEq/kg wet weight
Potassium content, sarcoplasm	98 mEq/kg wet weight
[K], sarcoplasm	180 mEq/liter sarcoplasmic water
E_K	94 mV

[a] Temperature 25°C. Calculations based on assumption that sodium concentration in SR equals that in extracellular water and potassium concentrations are similarly equal. Data from Rogus and Zierler (Unpublished)

is considerably larger than previous estimates. Notice also that more than 95% of muscle fiber sodium that formerly was estimated to be in fiber water is actually not in sarcoplasm but in SR, if the principles behind our calculations are correct.

C. *Definition of Active and Passive Fluxes and the Ratio of Passive Fluxes*

By flux we refer to a unidirectional movement in dimensions of quantity per time, often quantity per time per unit surface area. The quantity is usually moles. Flux from the outside of a cell to the inside is influx. Flux from inside to outside is efflux. Flux may be active or passive. If the force driving the fluxes is only the electrochemical potential difference, the fluxes are said to be passive. If there is an additional force coupled to some metabolic process, the added component of the flux is said to be active.

When we speak of flux we are interested in unidirectional flux—influx or efflux. In previous sections, when we considered contributions of individual ions to net inward current density, we could speak only of the difference between inward and outward current because we had no way of measuring either current separately. Similarly, the standard equations of thermodynamics define only net flux, the difference between influx and efflux, and do not define either flux separately. There is, of course, a quantitative relation between net current density and net flux defined by Faraday's constant.

If the contribution of the kth species of univalent cation to the net current density due to an electrochemical potential difference with respect to that ion is J_k, then the net passive flux of that ion is

$$M_i - M_o = J_k/F \tag{38}$$

where i refers to influx and o to efflux, both passive.

From Eq. (24), which we developed in preparation for the Goldman–Hodgkin–Katz equation, and Eq. (38), it follows that

$$M_i - M_o = P_k \frac{F\psi}{RT} \frac{(C_k^+)_o - (C_k^+)_i e^{\psi F/RT}}{e^{\psi F/RT} - 1} \tag{39}$$

where $(C_k^+)_o$ and $(C_k^+)_i$ are the concentrations of the cation outside and inside the cell, respectively.

Now consider an isotope of the kth species, which we identify by an asterisk superscript. We make the assumption that mobilities, solubilities and permeabilities of the two isotopes are the same, so that $P_k^* = P_k$. Then

$$M_i^* - M_o^* = P_k \frac{F\psi}{RT} \frac{(C_k^+)_o^* - (C_k^+)_i^* e^{\psi F/RT}}{e^{\psi F/RT} - 1} \tag{40}$$

The ratio of the net fluxes of the two isotopes is obtained by dividing Eq. (39) by Eq. (40),

$$\frac{M_i - M_o}{M_i^* - M_o^*} = \frac{(C_k^+)_o - (C_k^+)_i e^{\psi F/RT}}{(C_k^+)_o^* - (C_k^+)_i^* e^{\psi F/RT}}$$

Now perform an experiment in which at zero time the external solution contains none of the original species and the internal solution contains none of its isotope, or

$$(C_k^+)_o = (C_k^+)_i^* = 0 \qquad \text{at} \quad t = 0$$

Then at zero time $M_i = M_o^* = 0$, and at zero time the ratio of net fluxes becomes the ratio of efflux of the original species to influx of its isotope, or

$$\frac{M_o}{M_i^*} = \frac{(C_k^+)_i e^{\psi F/RT}}{(C_k^+)_o^*} \qquad \text{at} \quad t = 0$$

But if the relation given above for unidirectional flux is true for the case of two isotopes, it must also be true for unidirectional fluxes of a single isotope, including the endogenous species. Furthermore, in this case, since zero time is arbitrary, it will be true at any time. Therefore,

$$\frac{M_o}{M_i} = \frac{(C_k^+)_i}{(C_k^+)_o} e^{\psi F/RT} \tag{41}$$

Equation (41) is sometimes known as the Ussing equation (1949), although Ussing attributes its ancestry to Behn (1897), and Teorell developed it independently (1949).

We recall the Nernst equation, Eq. (22), for the equilibrium potential of a monovalent cation,

$$-E_k = \frac{RT}{F} \ln \frac{(C_k^+)_i}{(C_k^+)_o}$$

Combination of Eqs. (22) and (41) gives

$$\frac{M_o}{M_i} = e^{-(E_k - \psi)F/RT} \tag{42}$$

The absolute magnitude of the expression $E_k - \psi$ is called the driving potential. When the observed membrane potential ψ equals the equilibrium potential E_k, from Eq. (42), passive efflux of the kth species M_o equals passive influx M_i. The ratio M_o/M_i, given by either Eq. (41) or (42) is determined from experimentally measurable quantities and has been a valuable result for the following reason.

If a tissue is in a steady state with respect to an ion, then the internal concentration of the ion $(C_k)_i$ must be constant. We designate efflux of the ion, whether active or passive by ϕ_{ko} and influx by ϕ_{ki}. Then the change in the amount of the ion inside the cells during the time interval between t and $t + dt$ is

$$V_i \, d(C_k)_i = S(\phi_{ki} - \phi_{ko}) \, dt \tag{43}$$

where V_i is volume in which the ion appears at concentration $(C_k)_i$ and S is the surface area of the barriers across which the fluxes occur.

In the steady state, since $d(C_k)_i/dt = 0$, $\phi_{ki} = \phi_{ko}$. If we measure $E_k - \psi$ and find that the ratio of passive fluxes M_o/M_i is not unity, then it follows that at least some of the flux cannot be passive, must be active, and either $\phi_{ki} > M_{ki}$ or $\phi_{ko} > M_{ko}$ or both. The magnitude of active flux, that is, the active component of the total flux, is therefore given by

$$A_i = \phi_i - M_i$$
$$A_o = \phi_o - M_o \tag{44}$$

for influx and efflux, respectively.

From our data from rat extensor digitorum longus muscle, the ratio of passive fluxes for potassium is $(M_o/M_i)_K = 2.96$ and for sodium $(M_o/M_i)_{Na} \approx 1/1800$. This result assures us that in rat muscle nearly all the sodium efflux must be driven by some force linked to a metabolic process and that about two-thirds of potassium influx is also so linked.

D. The Sodium Pump, Sodium–Potassium-Activated Ouabain-Inhibited ATPase, and Electrogenesis

In the previous section we determined that there must be a force coupled with some metabolic process that accounts for virtually all sodium efflux from muscle. This process is called a sodium pump, and through the efforts of many people, including particularly Skou who also reviewed the subject (1965), it has been well demonstrated that there is in many tissues a particular ATPase associated with some membrane fractions, activated by Mg^{2+} and by certain concentrations of both Na^+ and K^+, which must both be present, and inhibited by cardiac glycosides such as ouabain. Variations produced experimentally in activity of this specific ATPase in assay systems correlate well with variations in sodium efflux or inversely with variations in sodium content of tissues, so that it is generally accepted that either this specific ATPase or some process linked closely with it is an important, if not the only, sodium pump.

For several years after the existence of such a membrane-bound ATPase had been well documented in other tissues, there was some uncertainty and disagreement about its existence in skeletal muscle. However, through the work of Samaha and Gergely (1965) and Rogus *et al.* (1969), there is now no doubt that such a Mg^{2+}–Na^+–K^+-activated, ouabain-inhibited ATPase exists in skeletal muscle.

There is, however, some doubt as to its localization. Although this ATPase is often called a membrane ATPase, it has not been localized clearly, and often is found distributed among various centrifugal fractions from tissue homogenates. In rat skeletal muscle, however, it is found only in a fraction rich in fragmented sarcoplasmic reticulum (Rogus *et al.*, 1969), and although this does not eliminate the possibility that the ATPase may lie in sarcolemma fragments, it is at least consistent with the possibility that the sodium pump in skeletal muscle may be associated with the wall of SR, and that the major efflux path of sarcoplasmic sodium may be across the wall of SR, through the lumen of SR, across the junction with T tubules, and through the lumen of T tubules to the outside.

It has been argued for some time that sodium pumping required a considerable fraction of the cell's energy expenditure and of the resting energy expenditure of skeletal muscle. Several workers calculated the cost of sodium pumping and, depending on their assumptions, estimated that anywhere from a few percent to more than all the oxygen consumption of the cell was spent in maintaining the sodium pump. Recently, however, a simple experiment by Ismail-Beigi and Edelman (1970) seems to have given a reasonable result. Comparison of oxygen consumption by several sorts of mammalian tissue slices before and after exposure to large concentrations of ouabain reveal that apparent total block of the sodium pump reduces oxygen consumption by about 30%. It has not been ruled out that the reduced oxygen consumption was not due to the altered internal environment, secondary to inhibition of the pump, rather then to subtraction of the oxygen requirement for regeneration of ATP spent by the pump.

We saw also in the previous section that part of potassium influx could not be accounted for by passive flux. Put this together with evidence that sodium accumulates in muscle if there is no external potassium and that the specific membrane ATPase system that is inhibited by ouabain requires K^+, and it becomes plausible that active K^+ influx is coupled with active Na^+ efflux, a proposal first made by Hodgkin (1958) for muscle and nerve.

Hodgkin proposed a 1:1 coupling; that is, for each equivalent of Na^+ pumped out, one equivalent of K^+ is pumped in. Such a pump would, of course, have no effect on net current density across the membrane. However, evidence began to accumulate, beginning with observations by Kernan (1962) and including particularly contributions by Mullins and Noda (1963), Mullins and Awad (1965), and Cross *et al.* (1965), that the coupling is not necessarily 1:1. It has been impossible experimentally to arrive at an exact ratio, but Mullins and Awad conclude that in frog muscle the ratio is at least 4:1, Na^+ efflux exceeding active K^+ influx. Such a pump contributes to the inward current density; that is, a coupled ion pump with a ratio other than 1:1 is not electroneutral but is electrogenic.

There have been a number of nice demonstrations that the coupled cation pump in muscle is electrogenic, including one by Frumento (1965), but the experiments I find most intriguing are those of Kerkut and Thomas (1965) on snail nerve cells. Kerkut and Thomas stimulated the sodium pump by injecting Na^+ into cells of the abdominal ganglion of the common snail. In 10 min the membrane potential increased in absolute value by 30 mV. This hyperpolarization could not be produced by injection of K^+, and was inhibited by ouabain or by reduction in

$[K^+]_o$. It is the great size of the hyperpolarization that sets these observations apart from earlier ones on less responsive preparations and leaves no doubt that an effect occurred. The size of the effect rules out the possibility that hyperpolarization was due to increased $[K^+]_i$ produced by the pump, with a corresponding rise in membrane potential in accordance with the Goldman–Hodgkin–Katz equation, because $[K^+]_i$ could not have increased to the necessary extent in these experiments. Furthermore, hyperpolarizations have been recorded in excess of the potassium equilibrium potential. If such calculations are not in error, this must mean that the $Na^+ : K^+$ coupled pump is contributing to the membrane potential under these conditions.

While these observations demonstrate that the pump can be electrogenic under special circumstances, they do not demonstrate that it is electrogenic in resting muscle. In our studies of Na^+ and K^+ flux in rat extensor digitorum longus muscle at rest, we find Na^+ influx, which must equal active Na^+ efflux, 2.7×10^{-12} moles sec^{-1} cm^{-2}, and the active component of K^+ influx is 2.73×10^{-12} moles sec^{-1} cm^{-2}. Therefore, in this mammalian muscle at rest, if the active K^+ influx and active Na^+ efflux are due entirely to the same pump, the coupling ratio is 1:1 and the pump is not electrogenic.

E. Sodium and Potassium Fluxes: A Two-Component Model

In order to measure unidirectional flux of an ion, one must use an identifiable and measurable isotope of the endogenous ion. Many studies of ion flux are carried out in the following way. The tissue is bathed in a solution containing an appropriate radioisotope. At various time intervals, the tissue is removed from the bath, sometimes washed briefly or blotted, and its radioactivity measured. We can refer to the time period covered by such experiments as the loading period. At the end of the loading period, which is usually intended to be sufficiently long to permit specific activity of the fluxant to be the same everywhere, the tissue is removed from contact with the loading solution and there begins some sort of measurement of net efflux into solutions containing, ideally, zero radioisotope concentration. This measurement consists either of measuring the radioisotope content of effluent from the tissue as a function of time, or of measuring the quantity of radioisotope remaining in the tissue as a function of time, by some sort of external monitoring device. We refer to the time period covered by this phase of the experiment as the washout period. The washout period need not be preceded by a period of interrupted loading during which repeated measurements of tissue tracer content are made, but often there is a single long loading period with minimal disturbance of the tissue.

We shall see that although, with a few assumptions, we may be able to measure flux of the tracer, we cannot translate this into flux of the endogenous substance without more information. Furthermore, we are often interested not so much in the absolute flux as in the properties of membranes that permit the flux in response to known or assumed driving forces. It will turn out that interpretation of all tracer experiments of this sort has depended on a model, which we will look into later. For the time being, let us see what we can do with a relatively model-free analysis.

The tracer content of the tissue N_T^* includes extracellular tracer N_o^* and intracellular tracer N_i^*. We define tracer fluxes as $\phi_{o,B}^*$ from bath into extracellular fluid, $\phi_{B,o}^*$ from extracellular fluid to bath, ϕ_i^* from extracellular fluid to intracellular fluid, ϕ_o^* from intra- to extracellular fluid, all in units of quantity per unit of time per square centimeter of arbitrary membrane surface S. Then

$$\frac{dN_o^*}{dt} = S(\phi_{o,B}^* + \phi_o^* - \phi_{B,o}^* - \phi_i^*) \tag{45}$$

$$\frac{dN_i^*}{dt} = S(\phi_i^* - \phi_o^*) \tag{46}$$

$$\frac{dN_T^*}{dt} = S(\phi_{o,B}^* - \phi_{B,o}^*) \tag{47}$$

What is determined experimentally, except for the case of single fiber preparations, is N_T^*. Its derivative is in explicit terms only of uninteresting fluxes, $\phi_{o,B}^*$ and $\phi_{B,o}^*$. However, it is plausible that tracer in extracellular fluid might exchange far more readily with that in the bath than it does with that inside the cell. During the early loading period, N_i^* can be taken to be so small that ϕ_o^* can be neglected. If, during the time at which ϕ_o^* can be neglected, flux into extracellular fluid from the bath becomes matched by the sum of fluxes out of extracellular fluid, then

$$\phi_{o,B}^* = \phi_{B,o}^* + \phi_i^*$$

and

$$\frac{dN_o^*}{dt} = 0$$

and

$$\frac{dN_T^*}{dt} = S\phi_i^* = \frac{dN_i^*}{dt} \qquad \text{for} \quad t_1 < t < t_2$$

during loading, where t_1 is time at which $dN_o^*/dt = 0$ and t_2 is such that for $t > t_2$, ϕ_o^* is not negligible.

A similar problem exists during the washout period. Again, except for the case of single fiber preparations, what is determined is N_T^*, not N_i^*, and Eq. (47) applies. However, if the tracer concentration in the bath is maintained close to zero, then it can be expected that N_o^* will fall relatively rapidly toward zero so that ϕ_i^* may become negligible. In that case, $\phi_{o,B}^*$ is virtually zero and $\phi_{B,o}^*$ is close to ϕ_o^*, so that

$$\frac{dN_T^*}{dt} = -S\phi_o^* = \frac{dN_i^*}{dt} \qquad \text{for} \quad t_1' < t' < t_2'$$

during washout, where t_1' is time at which N_o^* becomes sufficiently close to zero and t_2' is such that for $t' > t_2'$, ϕ_i^* is not negligible compared to ϕ_o^*, $t' = 0$ is time at which washout begins.

It is the existence of these uncertainties that has made it desirable to study flux into and from a single muscle fiber. Hodgkin and Horowicz (1959a) have performed such studies on frog muscle. While these studies avoid the difficulties of dealing with distributions of tracer through interstitial fluid and in theory lead to cleancut determination of ϕ_i^* and ϕ_o^*, they introduce a different sort of problem. Particularly for the case of sodium, the quantity of radioactivity that can be put into the frog muscle fiber is so small that, during washout, residual radioactivity falls to background noise level too soon to get all the information desired from the curves. There is no question that the derivative of the quantity of tracer remaining in the single fiber $N_i^*(t)$ as a function of time is the tracer efflux, $-S\phi_o^*$, but this does not give one an understanding of mechanism, which requires us to relate flux to driving force, and this in turn requires accurate presentation of the whole washout curve down to $N_i^* = 0$.

To circumvent this latter difficulty, investigators have fit available portions of the experimental tracer curves to various models. Hodgkin and Horowicz (1959a) found that their data from Na* flux from a single fiber fitted a single exponential of the form $N_i^*(t) = N_i^*(0)e^{-kt}$, but in general they report measurements for a time period not longer than the estimated time constant, that is, $N_i^*(t)$ at the end of the experiment was generally still about 40% or more of the initial value. This is generally not long enough to justify confidence that the curve is really described by a single exponential.

There have been many studies of K⁺ and Na⁺ flux into and out of muscle by many able people, too numerous to cite. For an excellent review, see Caldwell (1968). In general, in whole muscle preparations, washout of K* is described more or less by a single exponential, but washout of Na* is not. For the case of K*, the same problem applied to whole muscle as I indicated applied to washout of Na* from single fibers. The apparent time constant for K* efflux is so large, 4 hrs or longer, that it has not been practicable to continue the experiment sufficiently long to determine the shape of the entire washout curve.

In the case of Na* efflux, there is no doubt that the curve of $N_T^*(t)$ during washout from whole muscle, whether from frog (Keynes and Steinhardt, 1968; Horowicz *et al.*, 1970) or rat (Rogus and Zierler, unpublished), is not described by a single exponential. Questions are: (a) Is efflux from every fiber described by a single exponential; is there a family of time constants for, say, the family of fiber diameters so that the overall curve is a weighted sum of exponentials due to differences among fibers? (b) Is there really no exponential washout from any fiber or collection of fibers, but is the observed curve an expression of diffusion through some very complicated architecture (see below for general statement)? (c) Is the washout of Na* described by a sum of *n* exponential terms, and are there in reality *n* pools of sodium in muscle?

No absolute answers to these questions exist, but an affirmative answer to question (c) has much to recommend it. With regard to the two other questions I point out the following. There is indeed a population of fiber diameters, but these are approximately normally distributed and heavily dominated by a small range of diameters (Carey and Conway, 1954; Creese, 1968). If efflux from each fiber is described by a single time constant and if the time constant is proportional to fiber diameter, then it can be shown that for the real distribution of fiber diameters, washout will appear to be described by a single time constant (Zierler, 1966). However, the most important arguments against departure from a single exponential being due to distributed properties among muscle fibers are found in details of our study of Na* movement from rat extensor digitorum longus muscle under a variety of conditions in which shifts in intercepts without shifts in time constants can be accounted for only on the basis of altered fiber properties and not on the basis of selective responses from some fibers and not from others. This effectively rules out the possibility that Na+ movement from an individual fiber should be describable by a single exponential over the entire course of Na* washout. As we shall see, one must follow the washout curve until 99% or more of the tracer is washed out in order to see what is going on, and those experiments that found single exponentials over a period of one time constant simply could not detect the part of the story that I will describe below.

But first we must consider question (b), that the curvature is due simply to the fact that we should use diffusion equations. Diffusion equations cannot be written precisely without specifying geometry and boundary conditions, and these are not really known exactly for muscle. However, we do know that among alternative possible forms that such diffusion equations might take is a series of error–function complements, or a series of trigonometric functions, or a series of Bessel functions. It is conceivable that such equations might be approximated by an infinite series of exponentials weighted heavily by the first term or two, and these terms having no implications whatsoever as to the existence of separate sodium pools. However, we think that these separate pools do exist, and the reason is that we can change pool size and volumes of distribution and obtain appropriate changes in the exponential terms of the washout curve.

The main observations on Na* washout from rat extensor digitorum longus muscle are that in normal glucose–Ringer solution there is a very rapid initial washout of Na* with a time constant of about 2 min. Although about half the total sodium of this muscle is in the extracellular (inulin) space, it is possible for the contribution of washout from extracellular space to be missed after the first 2 min of washout. We deal now with the observed curve for washout time greater than 2 min. It is an experimental fact that from that time on the radioactivity remaining in muscle is described by

$$N_i^*(t)/N_i^*(0) = A_1 e^{-r_1 t} + A_2 e^{-r_2 t} \qquad (48)$$

where the As and rs are constants. With loading and washout in glucose–Ringer solution, the rate constant of the fast component r_2 was about 5.5 hr^{-1}, r_1 was about 1.2 hr^{-1}, A_1 was 0.03 and A_2 0.97.

A model that is described by these results is one given by Keynes and Steinhardt (1968) and extended by us. Consider that there are two pools of fiber sodium; one is in sarcoplasm, the other in sarcoplasmic reticulum, SR. Consider that sodium can move back and forth between sarcoplasm and the lumen of SR by crossing the walls of SR. Consider that sodium can also leave sarcoplasm across sarcolemma and that it can also leave SR by crossing the junction between terminal cisternae and T tubules. Thus, sodium has two routes between sarcoplasm and interstitial fluid. One is directly across sarcolemma. The other is across SR walls into SR lumen, across SR–T tubule junction, into T tubule and thence into interstitial fluid.

Improbable as it may seem that any relatively long narrow tube such as SR, with a length of about 2 μ and a diameter over the longitudinal component of a few hundred angstroms, could be really homogeneous with respect to chemical potential, if diffusion through lumen of SR is rapid relative to movement across the walls of SR, either into sarcoplasm or into T tubules, then the model may be acceptable.

If we accept the model tentatively, then we write the set of differential equations describing the distribution of sodium, calling sarcoplasm component 1 and SR component 2. For washout of tracer into zero external radioactivity

$$\frac{dN_1^*}{dt} = \lambda_{12}N_2^* - (\lambda_{01} + \lambda_{21})N_1^*$$

$$\frac{dN_2^*}{dt} = \lambda_{21}N_1^* - (\lambda_{02} + \lambda_{12})N_2^* \tag{49}$$

$$\frac{dN_i^*}{dt} = \frac{dN_1^*}{dt} + \frac{dN_2^*}{dt}$$

where the coefficients λ_{ij} are read "into i from j." The solution of Eq. (49) is of the form of Eq. (48); $\lambda_{ij} = k_{ij}/V_j$, where V_j is either V_1 or V_2, volume of sarcoplasm or of SR. The observed rate constants r_1 and r_2 are defined by

$$r_1 + r_2 = \lambda_1 + \lambda_2$$
$$r_1r_2 = \lambda_1\lambda_2 - \lambda_{12}\lambda_{21}$$

where $\lambda_1 = \lambda_{01} + \lambda_{21}$ and $\lambda_2 = \lambda_{02} + \lambda_{12}$, and the observed intercepts A_1 and A_2 are defined by

$$(r_2 - r_1)N_i^*(0)A_1 = (r_2 - \lambda_{01})N_1^*(0) + (r_2 - \lambda_{02})N_2^*(0)$$
$$A_1 + A_2 = 1$$

If the muscle was loaded for a sufficiently long time so that the specific activity of sodium in the kth component was the same as that in every other component, then

$$N_i^*(0)/N_i = N_1^*(0)/N_1 = N_2^*(0)/N_2$$

and, if $N_i^*(0)$ and N_i are measured and if $N_1^*(0)$ and $N_2^*(0)$ can be calculated, then N_1 and N_2, the steady state quantities of sodium in sarcoplasm and SR, respectively, can be calculated.

Furthermore, if one makes the reasonable assumption, discussed earlier, that the concentration of sodium in SR, $[Na]_2$, is the same as that in interstitial fluid and in the bathing solution, $[Na]_o$, then the volume of sarcoplasmic reticulum, $V_2 = V_{SR}$, can be calculated. Since total fiber water V_i is calculated as the difference between total water, obtained by drying, and inulin space, sarcoplasmic water volume, $V_1 = V_i - V_2$, can be calculated. With calculation of sodium content N_1 and volume V_1 of sarcoplasm, sodium concentration in sarcoplasm $[Na]_1$ is calculated.

There are far more unknowns than there are equations, so that there are no unique solutions for these volumes, contents, and concentrations, not to mention the rate constants across all the membranes. Nevertheless, by methods of inequalities, Rogus and Zierler have solved the system for a number of the unknowns, including all volumes, contents, and concentrations, and the rate constants λ_1, λ_{02}, and λ_{12}. We have not yet been able to make the important segregation of λ_1 and λ_{01} and λ_{21}, that is, separate rates from sarcoplasm into interstitial fluid λ_{01} or from sarcoplasm into SR λ_{21}. The importance of this segregation is that if one of these rate constants is determined mainly by the sodium pump, segregation of the rate constants could localize the sodium pump. Similarly, such segregation could help localize action of various agents, such as peptide hormones, that act on membranes. Do they act on sarcolemma or on SR? The main results of these calculations are as follows:

a. The volume of SR, V_2, under standard conditions lies between a formal lower limit of $12.6 + 1.1\%$ of wet weight and·a formal upper limit of $12.7 + 1.1\%$. Thus, the upper and lower limit differ by only 1% and are well within experimental error of one another. This remarkable result means that SR volume is determined as reliably as if there had been a unique formal solution. The value, which corresponds to 14.5% of fiber volume, agrees well with Peachey's estimate (1965) of SR volume from electron microscopy. Further evidence that the method really estimates SR volume lies in experiments with muscles soaked in various hypo- and hypertonic solutions. When Na* efflux is measured, and V_1 and V_2 are calculated, V_1, sarcoplasmic volume, increases and decreases precisely as predicted for an osmometer; V_2 decreases and increases in the opposite direction, but as predicted for a structure with walls more permeable to water than to sodium, with respect to the interface with sarcoplasm, but which rapidly equilibrates its salt and water concentration with the extracellular phase.

b. At least 96 and probably 97% of all sodium measured as muscle fiber sodium Na_i is not in sarcoplasm but is in the lumen of SR.

c. Under standard conditions, the concentration of sodium in sarcoplasm is between 1.1 and 1.7 mEq per liter sarcoplasmic water. This is many times less than previous estimates.

d. The low concentration of sodium in sarcoplasm leads to a revised estimate of the ratio between sarcoplasmic and external sodium concentrations. This ratio is between 96 and 132. Therefore, the equilibrium potential for sodium E_{Na} is no less than $+118$ and may be as great as $+126$ mV. This means that the peak of the action potential falls far short of E_{Na} and probably explains the observations of Adrian *et al.* (1970) that the action potential peak was greater than expected on the basis of calculations that equated all measured fiber sodium with sarcoplasmic sodium.

e. Efflux of sodium from sarcoplasm by both routes (across sarcolemma and into SR) is at a rate of only 0.6–1.0% of the efflux of sodium from SR, and nearly all the latter, by a factor of at least 1000, is from SR to T tubule. Nearly all the movement of fiber sodium to the outside is by way of SR to T tubule and not by way of sarcoplasm across sarcolemma.

f. The decrease in sodium flux occurring when other ions, for example lithium, are substituted for sodium in the bathing solution is accounted for quantitatively by substitution of lithium for sodium in SR, and gives no support to the notion of sodium–sodium exchange diffusion which has been postulated to explain results of such experiments.

We have not discussed potassium flux. As noted, studies of potassium flux were carried out before the two-component model was formulated and have not been reexamined in that context. And it will be difficult to do so because with an apparent time constant of 4 or 5 hrs, it would require about 24 hrs to follow the decay down to 1% of initial value, and a good deal longer if a second, slower, exponential term appeared. It has not been possible to continue experiments on mammalian muscle for so long a period, because they cannot maintain a steady state *in vitro* for 24 hrs. However, if the two-component model is correct and if $[K]_{SR} = [K]_o$, then there is about 150 times as much potassium in sarcoplasm as in sarcoplasmic reticulum, and the chances are that what we are seeing in K^* efflux experiments is largely K^* efflux from sarcoplasm, because it can be shown that the larger intercept of the two-component model is associated with the component with the larger ion content.

If the muscle is loaded to uniform specific activity before washout of K^* begins, then

$$\frac{[K^*]_i(0)}{[K]_i} = \frac{\phi_o^*(0)}{\phi_o}$$

What is observed experimentally, assuming extracellular K^* is negligible, is a function of fiber K^* content, $K_i^*(t)$, and the derivative of this observed residue function,

$$\frac{dK_i^*(t)}{dt} = -S\phi_o^*(t)$$

Therefore, with no further specifications about the model, the true derivative of the K_i^* residue at zero time measures ϕ_o^*, from which ϕ_o can be calculated. Thus, although we cannot yet dissect K^* flux experiments to give an estimate of relative contributions of sarcolemma and SR wall, we can measure ϕ_o. In rat extensor digitorum longus muscle at 25°C, potassium efflux is 4×10^{-12} moles sec^{-1} cm^{-2}, which leads to an estimate of P_K, the Goldman–Hodgkin–Katz permeability coefficient to potassium, of 10^{-7} cm sec^{-1}.

IV. Mechanics

A. *Excitation–Contraction Coupling*

This important subject, currently in the forefront of investigations in muscle physiology, is considered in detail elsewhere in this book. For the sake of continuity of our story, I outline the main features. Calcium, released from sarcoplasmic reticulum by some process initiated by adequate depolarization of the T tubule wall, sets in motion a series of events that leads to shortening and exertion of tension.

Studies of excitation–contraction coupling have not yet been quantitative, and many questions remain. It is proposed that calcium is stored in resting muscle in the terminal cisternae of SR, which make extensive surface approximations (though apparently not tight junctions) with T tubules. It is not yet known whether walls of T tubules are depolarized by action potentials propagated from sarcolemma or by electrotonus, but in any case, adequate depolarization of T tubules can lead to a decrease in calcium content of terminal cisternae and an increase in calcium content of myofibrils (Winegrad, 1968). It seems plausible to assume that somehow depolarization has started migration of calcium from terminal cisternae to myofibrils. It is generally supposed that the migration is simply diffusion, that depolarization releases Ca^{2+} bound to some component of terminal cisternae, that Ca^{2+} ions diffuse and are sequestered by powerful calcium-binding proteins in myofibrils, the most effective of which is troponin in the thin filaments (Ebashi and Endo, 1968).

After contraction has ceased, radioautography shows a slow decrease in myofibrillar Ca^{2+} and an increase in Ca^{2+} in the walls of or in the lumen of the longitudinal component of SR (Winegrad, 1968). It is argued from this observation that the Ca^{2+} pump is in the walls of the L system, the longitudinal component of SR, and that Ca^{2+} diffuses from the lumen of the L system to its site of accumulation at rest in terminal cisternae.

Most of this is as yet only speculative and many steps need to be examined. The notion that Ca^{2+} simply diffuses from terminal cisternae to myofibrils to initiate contraction, and simply diffuses from myofibrils to L system to end it may be difficult to accept, since diffusion is not spatially reliable and is slow. It is difficult to imagine the abrupt onset of active state, with almost negligible temporal dispersion, if the process depends on diffusion of Ca^{2+} to sites distributed over a broad family of distances from terminal cisternae. And we need to explain the skewness of the twitch and of what we will call the active state. Why does it take so much longer to turn it off than to turn it on?

B. The Twitch—Series and Parallel Elastic Elements—The Active State

Let us first describe muscle contraction at the gross level of observation. A whole muscle, with tendons attached, is placed in an apparatus with its long axis along the line of gravity. We select a muscle with parallel fibers, such as the sartorius of the frog, so that the force vector of individual fibers is parallel to that of the whole muscle. The tendons are secured so that the muscle is extended to some known length. Stimulating electrodes or plates are positioned about the muscle.

There are two general conditions under which the experiment may be performed. If the muscle is permitted to contract against some fixed resistance, so that the distance between the tendons decreases, the muscle is shortening against a fixed load; this is *isotonic* contraction. If the ends of the muscle are fixed in a stiff apparatus so that the muscle can not change its overall length, but it does exert a force when stimulated as it pulls against the fixed position, this is *isometric* contraction.

If there is a single adequate stimulus, normal muscle responds with a single contraction. The curve of shortening against time in an isotonic contraction looks very much like the curve of tension (or force) against time in an isometric contraction. In both cases there is delay following the stimulus, an upslope that is quite slow compared to that of the action potential, a rounded peak, and an even slower decay to baseline (Fig. 3).

If the muscle is stimulated repetitively at sufficiently small frequency, it responds with one twitch per stimulus. As stimulus frequency increases, the upstroke of one twitch adds to the tail of the preceding twitch, and the mechanical response proceeds with time almost as though it were the integral of the twitch. Eventually, the response to repeated

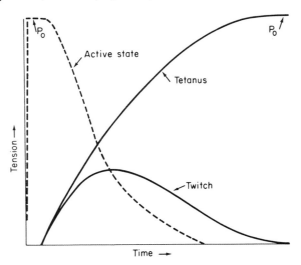

Fig. 3. Relation between tension and time for twitch, tetanus, and active state. P_0 is the maximum tension of which the fiber is capable. Stimulus begins at zero time. More recent estimates are that peak tension of active state in response to an impulsive stimulus is only about 0.9 of peak tetanic tension. From Zierler (1968).

stimuli appears to rise smoothly to a plateau which can be maintained for a finite period under proper conditions until the stimulus ceases when, after a delay, it falls to baseline along a curve somewhat as though the transient upstroke had been rotated 180° about the time axis. This smooth curve is a *tetanus,* either isometric or isotonic. The frequency of stimulation that just produces it is to a certain extent an artifact of the lumped time constant of the recording system, but that approximate frequency is the tetanizing frequency and is a characteristic of muscle types.

The force exerted during the plateau of an isometric tetanus considerably exceeds the peak force of an isometric twitch of the same muscle at the same length. A curve can be constructed relating the plateau force of an isometric tetanus to muscle length. This is the classic length–tension diagram (Fig. 4). If a muscle is not stimulated but is stretched passively, it resists stretch; it exerts tension. In skeletal muscles, with parallel fibers, there is no detectable passive resistance to stretch until the muscle reaches a length approximately equal to that it is thought to have in the body during rest, hence called rest length. This happens to be the length at which maximum active tension can be exerted, and this has often been used as a practical criterion for rest length. The conventional notation is that muscle length is denoted by L, rest length by L_0, the force or tension exerted by muscle by P, and

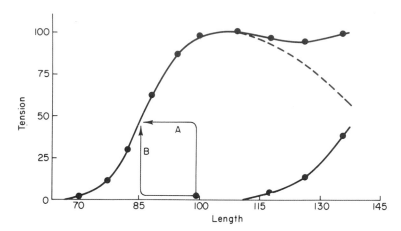

Fig. 4. Relation between length and tension. Rest length L_0 and maximum tension P_0 are set at 100. The monotonic increasing curve at lower right represents effect of passive stretch and is the contribution of parallel elastic components. The upper solid curve is the observed relation for the system as a whole. The broken line represents behavior of contractile elements obtained by subtraction of passive stretch curve from overall curve. Paths A and B indicate that whether muscle contracts isotonically, A, so that it shortens from, say, L_0 to 0.85 L_0, at a load of, say, 0.5 P_0, or isometrically from length 0.85 L_0 it reaches the same final state of tension, 0.5 P_0, and length. From Zierler (1968).

the maximum force the muscle can exert by P_0. If the muscle is stretched passively to lengths greater than L_0, tension increases monotonically, concave upwards, generally rising steeply at 1.3 L_0 and greater.

When the forces exerted in response to passive stretch are subtracted from the observed length–tension curve of stimulated muscle, an active length–tension curve is obtained. It is clear that the passive length–tension curve can be attributed to the response of elastic elements in parallel with the contractile machinery. These are called, therefore, parallel elastic elements. The remaining active length tension curve is zero for lengths less than 0.6–0.7 L_0, rises to a peak at L_0 and falls to zero at 1.3–1.6 L_0.

If the muscle is stimulated isotonically at L_0 it can lift a load equal to P_0, and in so doing it will not shorten. But if the load is less than P_0, the muscle will shorten from L_0 to the corresponding length L, defined previously by the L–P curve for isometric tetanus.

Several lines of evidence indicate that at least part of the delay between stimulus and overall mechanical response is due to the fact that contractile elements must first stretch some previously slack elastic elements in series with them. The evidence is based on experiments in which muscles are loaded with a certain weight at the onset of contraction, with later automatic substitution of a different load, and on quick stretch experiments in which, at various times before peak tension is reached, the muscle is transiently stretched to some higher tension and then released. On release, the tension is greater than it would have been had there been no stretch superimposed on the natural response to the electrical stimulus, but it is never greater than maximum tetanic tension. There are, then, *series elastic elements*, and the necessity for stretching them obscures our ability to determine the time course of contraction of the contractile components.

When a muscle shortens and, in so doing, lifts a weight it does work, sometimes called external work W_e, defined as exertion of a force over a distance.

$$W_e = P \, dx = -P \, \Delta L$$

where ΔL is the shortening. Thus, it is only under isotonic conditions (as opposed to isometric) that the muscle can perform external work, measured by a change in its overall length. Since the muscle cannot shorten at load P_0, it does no external work at maximum load. And, as we have seen, there is a muscle length, say 0.6–$0.7 \ L_0$, at which the muscle can exert no tension, so it also does no work. Therefore, if we plot external work against force from zero to P_0, we obtain a curve that rises from zero to a maximum, from which it falls back to zero. Each point on the curve can be obtained from the length–tension diagram as the area of the rectangle of height P and width $(L_0 - L)$ for $L \leq L_0$.

This does not mean that muscles contracted under isometric conditions do no work. We know they consume more oxygen and evolve more heat than at rest. Their mechanical work consists of stretching the series elastic elements, and we can define this as internal work. It does not move bones about a joint, which is the result of shortening of skeletal muscle. However, it is necessary not only because the mechanical connections are such that the muscle must do internal work before it can shorten and do external work, but also because it must stretch the series elastic elements if it is to stiffen the muscle in order to fix position at a joint, which is what an isometric contraction does.

We imagine that, as a result of some (chemical) process initiated by excitation–contraction coupling, there is a change in the contractile machinery, which we identify anatomically with the thick and thin filaments, such that its stress–strain properties are altered from those of a relatively compliant system at rest to those of a stiff system in contraction. This new state of stress-strain properties is called the *active state*. An active state tension curve is a plot of force versus time in response to a single impulsive stimulus. It is the force exerted by the contractile machinery at the designated time, stripped in imagination of all attachments that obscure the time course.

There are several methods for inferring the time course of active state tension. They lead to the same general shape. There is a rapid, almost but not quite a step, rise in tension almost to the maximum tetanic tension at the fixed length. This tension is maintained for some period, usually tens of milliseconds, depending on temperature, on the characteristics of the muscle (fast or slow), and to some extent on length or tension, and on a number of environmental factors (ions, drugs). Usually, when the term "duration of active state" appears in the literature, it is intended to mean duration of the plateau of active state tension. Active state decays slowly. The time at which it falls to half maximum value is several times greater than the duration of the plateau. One of the unanswered questions is why the active state curve is so skewed. The active state tension curve explains a number of phenomena.

a. The plateau of active state tension explains fusion and tetanus. The fusion frequency is about the frequency that gives time intervals approximately equal to the duration of the plateau of active state tension.

b. It explains the brief maintenance of tetanic tension after electrical stimulus ceases, for the delay before tetanic tensions falls equals the duration of the active state plateau.

c. Together with knowledge that there are series elastic components, the twitch is explained. While the contractile elements are pulling out the slack in series elastic components, the active state is already at plateau. Then, when the muscle is shortening, active state plateau ceases and active state tension declines. The peak of twitch tension is reached during the decline of the active state, and from that time on, until the muscle returns to rest, the overall twitch tension exceeds that of the active state, due to the contribution of the series elastic components. The muscle in twitch never gets to exert the tension it does in tetanus because that latter tension is the active state plateau tension, and that plateau does not last long enough to be reached by the slowly responding twitch.

C. Velocity of Contraction—Force–Velocity and Length–Velocity Curves

We have already seen that when a muscle contracts it takes time to reach peak tension if it contracts isometrically, or to shorten as much as it is going to if it contracts isotonically. If we impose a series of isotonic loads on a muscle and observe the speed with which the muscle shortens at each load, we notice that the heavier the load the more slowly the muscle shortens. At any given load, velocity of shortening is not constant throughout shortening. In this section we will look into the relationships between force, length, and speed of shortening.

A family of force–velocity curves, each at a different initial length, is shown in Fig. 5. A number of formal expressions have been proposed to fit such curves, some purely empirical, and some on theoretic grounds. Fenn and Marsh (1935) were the first to propose an exponential equation for P as a function of velocity (initial velocity) v, but they did so empirically. Polissar (1952) and Ramsey (1955) proposed different exponential equations on different theoretic grounds. Ramsey's model is that velocity is proportional to the distance remaining to be shortened.

The best known formulation is that of A. V. Hill (1938), which fits a hyperbole to the curve with the arbitrary choice of two constants. Incidentally, all the proposed equations have a minimum of two arbitrary constants. Hill's equation is often given as

$$(P + a)(v + b) = \text{constant} = (P_0 + a)b \tag{50}$$

where a is an arbitrary constant with dimensions of force P, and b

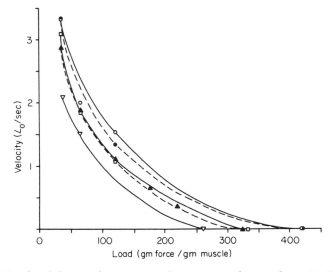

Fig. 5. Family of force–velocity curves from rat gracilis muscle at 16.4°C. Muscle weight, 83 mg; L_0, 2.9 cm. Note that velocity is greatest at L_0 at any load. Open circles, L_0; open squares, 0.86 L_0; open triangles, 0.79 L_0; filled circles, L_0; and filled triangles, 1.21 L_0. From Zierler (1968).

is an arbitrary constant with dimensions of velocity v. If we define v_0 as the velocity at $P = 0$, then v_0 is the maximum velocity and

$$b = av_0/P_0 \tag{51}$$

With Eqs. (50) and (51), we can express dimensionless velocity v/v_0 as a function of dimensionless force P/P_0,

$$\frac{v}{v_0} = \frac{a}{P_0} \frac{1 - (P/P_0)}{(a/P_0) + (P/P_0)}. \tag{52}$$

Abbott and Wilkie (1953) extended Hill's observations to lengths less than L_0 and found the same constants, a and b, applied to a family of curves each with its own $P_0(L)$, the maximum tension the muscle can exert at a given length, or

$$(P + a)(v + b) = [P_0(L) + a]b \tag{53}$$

Bahler *et al.* (1968) extended the observations to initial lengths greater than as well as less than L_0 and found that each initial length had its own maximum velocity $v_0(L)$; that is, that Eq. (53) probably holds over all lengths where, from Eq. (51),

$$a/b = P_0/v_0 = P_0(L)/v_0(L) \tag{54}$$

We will see later that it is quite important on theoretic grounds to decide whether or not v_0 is constant, independent of length, or varies as described by Eqs. (53) and (54). The fact is that experiments are never done at true zero load. There must be some load to prevent slack in the muscle. Therefore, published estimates of $v_0(L)$ are based on extrapolations. The reader will have to study published curves and decide for himself whether or not it is reasonable to extrapolate to the same v_0 or to a family of different $v_0(L)$ for a family of L, as done in Fig. 5. It is my opinion that extrapolations to a common v_0 are not justified and that the curves appear to be headed toward a family of $v_0(L)$.

Carlson (1957) was the first to display phase plane diagrams, plots of velocity against length, for frog skeletal muscle. Sonnenblick (1965) reported such diagrams for cat papillary muscle, and Bahler *et al.* (1968) carried out a systematic study for rat gracilis anticus muscle. A family of such trajectories is generated by allowing the muscle to shorten against different isotonic loads from the same initial length, or by allowing the muscle to shorten against the same load at different initial lengths (Fig. 6).

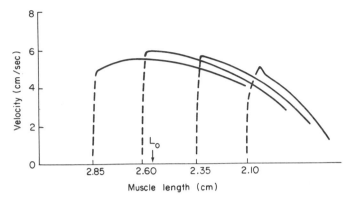

Fig. 6. Family of length–velocity curves from rat gracilis muscle, showing effects of different initial lengths at some load, in this case 6.94 gm or 328 gm per square centimeter of muscle cross section. L_0 2.55 cm. From Zierler (1968).

When the muscle shortens from an initial length of L_0 or less, maximum velocity of that trajectory is reached almost at once with gradually increasing deceleration until shortening is completed. When the muscle shortens from an initial length greater than L_0, its initial velocity is less than maximum. Velocity increases to a maximum at L_0, from which it decreases along a curve shaped like that of a muscle that began to shorten with the same load at L_0. In fact, all the curves at the same load from different initial lengths almost lock in on an apparently common parent curve, as though instantaneous velocity at any absolute length were fixed independent of initial length.

It is clear that the main factors in determining velocity of shortening at any instant are the absolute load and the absolute length at that instant. There are secondary factors because the velocity at a given length and load is less if the muscle began to shorten at a longer length, and the longer it takes a muscle to reach that length after it begins to shorten, the smaller the velocity at that length. By varying frequency of stimulation, it can be shown that the modulating factor is not time but the number of stimuli received by the muscle before it reaches that length. This suggests that the chemical history of the muscle is important, that the muscle exhausts some energy-yielding metabolic system, *in vitro* at least, not replenished rapidly enough to maintain velocity of shortening, and this in turn suggests that velocity of shortening is energy dependent.

We can therefore consider velocity of shortening (Bahler *et al.*, 1968) as

$$dL/dt = f(L,P,S)$$

where S is frequency of stimulation. It was found empirically that the expression could be written

$$dL/dt = f_1(L,P)f_2(S)$$

and that for small values of t and S, $f_2(S)$ was approximately unity, but that $f_2(S)$ decreased for larger S and t. An empirical linear expression for $f_2(S,t)$ was found. With it,

$$v = \frac{dL/dt}{f_2(S,t)} = f_1(L,P) \tag{55}$$

where v is the ideal velocity of shortening as a function of length and tension, that is, the velocity of shortening if there were no deterioration of the energy-supplying systems that are vulnerable to the history of stimulation (number of stimuli delivered or the product of frequency of stimulation and the duration of stimulation St).

We saw from the length–tension diagram that length and tension are interdependent to the extent that P determines the final L reached in isotonic shortening from some arbitrary initial length. But since we can fix initial conditions arbitrarily, v remains a function of both L and P. We can combine the length–tension, length–velocity, and force–velocity curves to obtain a sculptured surface that gives all the information (except for the stimulus frequency and duration dependent modulation) at this level of observation.

D. A Gross Phenomenological Description of Contraction

We have just seen that velocity of shortening is dependent on the total number of stimuli previously administered, suggesting energy dependence. There is a lot of evidence that the force of contraction is energy dependent. For example, as a muscle preparation deteriorates and is less able to exert maximum tension, the tension it exerts is matched linearly by decreasing initial heat production (Fales and Zierler, 1969). These observations suggest that the same energy source may be spent for both speed of shortening and for exerting tension, and the force–velocity curve suggests that there may be a trade-off. If energy is spent in speed, it can not be spent in force. These considerations lead us to a model of muscle contraction at the phenomenological level; that is, at the level of observation of force, shortening, and velocity without any effort to explain the mechanisms by which the muscle contracts.

We state simply (Bahler *et al.*, 1967) that as a result of a stimulus the contractile component of muscle is activated to become a force generator. The generated force either appears as tension measured across the whole muscle or is dissipated internally against a hypothetical internal load, and this latter dissipation is a function of the velocity with which the contractile element shortens. Formally,

$$P_g(L,t) = P + B(L,v) \tag{56}$$

where P_g is the force generated by the contractile component, L is the length of the contractile component (not of the whole muscle), t is time, P is the tension observed across the whole muscle, and B is the force dissipated by the internal load, a function of length L and velocity v, with which the contractile component shortens. Notice that L and v are overall observed values corrected for effects of series elastic elements.

We know from the length–tension curve that force is a function of length. We also know that we can measure the stress–strain properties of the series elastic elements and obtain a force K as a function of stretch ΔL. Since force flows through all series elements, the force K across series elastic elements is also exactly the force P across the whole muscle, or

$$P = K(\Delta L) \tag{57}$$

Under isometric conditions the contractile components stretch series elastic elements by a length ΔL and themselves shorten by the same ΔL so that there is no change in overall muscle length. It was shown experimentally that the length of the contractile component varies by less than 6% during an isometric twitch, so that we can make the approximation that during an isometric twitch the length of the contractile component is constant. Therefore under isometric conditions the force generated is, from Eqs. (56) and (57)

$$[P_g(t)]_{\text{isometric}} = [B(v)]_{\text{isometric}} + K(\Delta L) \tag{58}$$

where $[P_g(t)]_{\text{isometric}}$ is the active state curve. All the information necessary to solve Eq. (58) for the active state force is obtained from the isometric twitch, the force–velocity curve for the muscle, and the tension–extension curve for the series elastic elements.

The value of Eq. (58) lies in that it gives an experimentally determinable function of the contractile components independent of any hypothesis about chemical bases or ultramicroscopic mechanics, and it therefore provides quantitative information that all molecular hypotheses of muscle contraction must be required to explain.

Before we leave this section, it is worth emphasizing a few fundamentals concerning force. Mechanical force is transmitted through a series system, like volume flow through a hydrodynamic system or current through a series of resistors in an electrical system. The force must be the same across all elements of the series. The implication of this is that serial addition of identical sarcomeres lengthens a fiber and permits it to shorten to a greater extent, but it can not increase the maximum force that a fiber can generate. It can only affect the time course with which that force is attained across the whole length of the fiber. On the other hand, an increase in the number of contractile elements in parallel, assuming no adverse interaction between parallel units, must result in an increase in maximum force. Since the cross-sectional area of a muscle is in some measure a function of the number of fibers, and since the cross-sectional area of a fiber is in some measure a function of the number of fibrils and of myofilaments within fibrils, it is not surprising that maximum tension P_0, which varies from muscle to muscle, is remarkably constant when expressed per unit cross-sectional area of muscle, about 2 or 3 kg of force per square centimeter.

E. Mechanical Properties of Single Fibers and Single Sarcomeres

In 1940, Ramsey and Street succeeded in dissecting viable muscle fibers from frog semitendinosus in such a way as to be able to obtain isometric length–tension curves. The properties were remarkably like that of the whole muscle. There was a parallel elastic component revealed by passive stretch. The active length–tension curve had the same general shape as that from whole muscle, and it is quite evident that the length–tension curve of whole muscle can be generated as a sum of length–tension curves of parallel fibers.

The question naturally arises as to the length–tension curve of a single sarcomere. I have already noted that in a mechanical system composed of elements in series, like sarcomeres in a fiber, the force across the whole system is simply the force across one of the elements. If the elements are force generators, the force is determined by the element that generates the largest force. It had been observed (Happel, 1926; Fischer, 1926) that muscles do not shorten all at once in a twitch, but that the central portions shorten first, tending initially to stretch both ends. The proposed explanation is that the central sarcomeres, with the whole fiber at rest length, tend to be at sarcomere rest length, where $P = P_0$, but distal sarcomeres tend to be at shorter lengths, where $P < P_0$.

If the force exerted by a sarcomere, or, more properly, by a half-sarcomere between the Z and M lines, is proportional to the number of effective cross bridges between thick and thin filaments, and there is some evidence that it may be, then we would expect the length–tension curve for a sarcomere to behave as follows. When the muscle is stretched so that thick and thin filaments do not overlap, no cross bridging is possible and no tension is exerted. As progressive overlap between thick and thin filaments occurs there is a linear increase in the number of cross bridges and there is expected a linear increase in tension until overlap is maximum. This occurs when the thin filaments meet at the midpoint of the sarcomere. This point must be called L_0 if it is defined as that length at which maximum tension P_0 is exerted. But we know that tension decreases as muscles shorten from L_0. If this is also true in individual sarcomeres, then cross bridges must be broken as the sarcomere shortens, and this can occur if one myofilament interferes with another or if shortening is associated with widening at the Z line (for which there is electron microscopic evidence) to such an extent that the distance between thin and thick filaments is increased sufficiently to reduce the probability that bridges can be formed. X-ray diffraction studies confirm that myofilaments separate laterally as shortening occurs (for review, see Elliott *et al.*, 1970). Whatever the mechanism of decreasing the number of cross bridges with shortening, there is expected a decrease in tension with shortening. There is a finite limit to shortening of skeletal muscle; it cannot shorten to less than the length of a thick filament, about 1.5 μ, or about 70% of L_0. At that length there should still be a substantial number of cross bridges; tension should not be zero.

Length–tension curves of single sarcomeres have been observed by Gordon *et al.* (1966) and in general they resemble the curve described above on theoretical grounds. They differ in that tension at the shortest length, which was in fact about 0.7 L_0, fell to as little as 10–20% of P_0. It is not yet obvious from electron microscopy that as few as 10–20% of cross bridges remain at length 0.7 L_0, but of course one can not determine by microscopy whether or not bridges are in fact contributing to tension.

Edman and Grieve (1966) have obtained force–velocity and active state curves for a single sarcomere at rest length, 2.1 μ, in a single fiber of frog semitendinosus muscle at 4–5°C. The force–velocity curve was fitted satisfactorily to Hill's Eq. (50). The active state curve looked remarkably like those obtained for whole muscle: a rapid rise in less than 10 msec, a plateau maintained for about 50 msec, and a fall in tension much less rapid than the rise, but possibly somewhat more rapid than calibrated for whole muscle.

V. Heat

A. Resting Heat Production

Muscles produce heat even when they are at rest. Indeed, this is the major source of body heat in the basal state, although it is only that the total bulk of muscle is so large that makes it so. It has been quite difficult to measure resting heat production. I am aware of only one report in the literature. Hill and Howarth (1957) state that "in a large number of experiments [on frog sartorius] the resting heat in oxygen was measured at or near 20°C after long soaking in normal Ringer. The mean value was 10×10^4 ergs gm^{-1} min^{-1}." This value, about 2.4×10^{-3} cal gm^{-1} min^{-1}, corresponds to the heat expected in association with oxygen consumption of 0.48 mm^3 gm^{-1} min^{-1}, in agreement with a number of observations of oxygen uptake by frog sartorius muscles. I pointed out (Section III) that a substantial portion of resting oxygen consumption, and therefore, we now presume, of resting heat production, is in association with the operation of the sodium pump.

B. Heat Production Associated with Muscle Contraction—Isometric

Most of what we know about heat we learned from A. V. Hill and his colleagues. Hill (1965) has collected this extensive work and commented on it in a book that all students of the subject must read. When a muscle contracts isometrically, heat is liberated in excess of that produced at rest. When a muscle is stimulated so that it shortens against an external load, it produces even more heat. If the experiments are carried out in an atmosphere of pure nitrogen with muscles poisoned by iodoacetic acid so that both aerobic metabolism and glycolysis are denied them, the extra heat (above resting heat) appears as a unimodal function of time which is called the *intial heat*. If the experiments are carried out in an adequate oxygen environment with no iodoacetic acid, additional heat is produced, usually somewhat, but not entirely, later than the initial heat. This is called *recovery heat* because it is dependent on metabolic processes, aerobic or glycolytic, that can be associated with restoration of the basal state of muscle composition, probably in particular with regeneration of special phosphate bonds as in ATP and phosphocreatine. In mammalian muscles, which are largely lipid metabolizing, most of the recovery process is probably due to oxidation of fatty acids (Zierler *et al.*, 1968).

Our further attention in this section will center on initial heat. The total initial heat produced in response to an isometric twitch is called activation heat. In an isometric tetanus, the total initial heat is called maintenance heat, and it is generally regarded as a summation of activation heats produced by iterative activation of muscle with each stimulus. Initial heat under isometric conditions has been thought of at least tacitly as related to the active state tension curve. Activation heat flow from muscle outlasts the active state tension curve, probably mainly due to the heat capacity of muscle. The specific heat of a tissue, usually about 0.8, is determined largely by its water content. Dissipation of heat to the sensing instrument from the interior of the muscle is relatively slow, so that the time course of heat dissipation is longer than that of heat generation. This makes it difficult to make precise comparisons with nice time resolution between mechanical and thermal events, but the recording of thermal events can still be sufficiently rapid and quantitative to lead to certain conclusions.

At rest length L_0, initial heat of an isometric twitch at 20°C is about 3.5 mg cal gm^{-1} per twitch. The quantity of initial heat in an isometric contraction depends on the length of the muscle. But of course, the tension exerted is also a function of muscle length. A. V. Hill noticed that what he called the isometric heat coefficient PL/Q, tension $\times$ length/heat produced, was nearly constant, or activation heat is proportional to the product of length and tension over a limited range or variation in muscle length. This constant relation is maintained as a muscle fatigues. As it loses the ability to develop maximum tension at a given length, heat production decreases almost exactly proportionately. These observations suggest that the same chemical processes that are responsible for tension development are responsible for heat production.

As is discussed in detail elsewhere in this book, it has been assumed that the immediate chemical source of energy driving the mechanical process in muscle is the free enthalpy of splitting of adenosine triphosphate (ATP) to ADP and inorganic phosphate. Because ATP content of muscle does not change after a small number of twitches, it has been assumed that the ATP split during contraction is regenerated rapidly by transphosphorylation of ADP from phosphorylcreatine (PC) with creatine as the product. Accordingly, the net reaction is simply

$$PC \rightarrow C + P_i$$

Carlson *et al.* (1967) measured heat and PC splitting in frog sartorius under isometric conditions. These are obviously very difficult and demanding experiments and there is accordingly some uncertainty. The

178   *Kenneth L. Zierler*

interesting thing is that on the assumption that all the initial heat was produced by splitting PC an *in vivo* enthalpy of hydrolysis of PC was calculated. It was 10.6 kcal mole^{-1}, which is close to the figure given from *in vitro* data and which supports the hypothesis that this reaction is the major source, and perhaps the only source, of initial heat.

C. Heat and Isotonic Contraction

"Whenever a muscle shortens upon stimulation and does work . . . an extra amount of energy is mobilized," wrote Fenn in 1923. The phenomenon he described is now known as the Fenn effect, although he credited Heidenhain with the original observation in 1864. A. V. Hill and his colleagues did an enormous amount of work quantitating Fenn's observation. The extra amount of energy, meaning heat that is in addition to that occurring with isometric contraction, is called shortening heat, illustrated in Fig. 7. For a long time it was thought to be proportional only to the distance shortened, so that if ΔL is the distance shortened, then $a \, \Delta L$ is shortening heat, where a is a constant. Initial

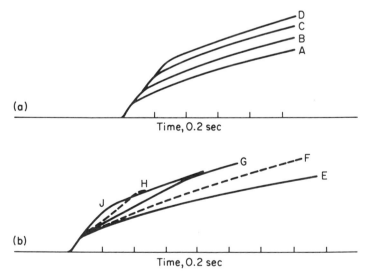

Fig. 7. Cumulative heat production during isotonic shortening. Tetanus at 0°C; L_0, 3.25 cm; muscle weight, 85 mg. (a) Shortening different distances under constant load, 1.9 gm, from A, isometric (no shortening), to D, 9.6 mm shortening. (b) shortening constant distance under different loads from E, isometric; F, 31.9 gm; to J, 1.9 gm. From Hill (1938).

heat in an isotonic contraction, then, can be written as

$$Q = A + a \, \Delta L$$

where A is the activation heat, or maintenance heat for a tetanus. Later, Hill reported that a is not constant but increases with increasing load. Most recently, Woledge (1968) has restudied the problem in rectus femoris muscle of the tortoise and found that a is a constant.

It is, in a sense, arbitrary to divide initial heat of isotonic contractions into maintenance (or activation) and shortening heats since they are not separable experimentally. Furthermore, since these terms were coined before the role of series elastic elements was as well known, we now realize that the contractile components do shorten even in an apparent isometric contraction. If there is such a thing as shortening heat, it must exist also under isometric conditions.

The chemical reaction that supplies the energy for mechanical responses in isometric contraction probably serves the same role when there is shortening. Evidence for this is, for example, Wilkie's (1968) observation that the same *in vivo* molar enthalpy change of -11 kcal mole^{-1} is calculated under isometric and isotonic conditions from the observed heat production, work produced, and decrease in PC content. This should be no surprise. We have known from the force–velocity curves that muscle can be considered to trade force for speed of shortening; it spends its energy on one or the other, and which it does depends on external arrangements such as load and stretch.

One of the fascinating observations, which may or may not have been coincidence, is that for frog sartorius the force constant a in the force–velocity equation

$$(P + a)(v + b) = (P_0 + a)b$$

has about the same value as the shortening heat constant a in the isotonic initial heat equation. Woledge (1968) has found that this is not universally true. In tortoise rectus femoris, the shortening heat constant is always less than the force constant, on the average about half as much, and both are much smaller than for frog muscle ($0.04 \, P_0$ for tortoise, $0.25 \, P_0$ for frog).

The most important implication of the fact that shortening heat per distance shortened is approximately constant, independent of the tension and therefore of the velocity of shortening, is that this eliminates the possibility of treating muscle as a pure viscoelastic body. Heat evolved in association with displacement of a mass through a viscous body is proportional to velocity of displacement. There is no stored energy in muscle, analogous to that stored in a stretched spring. The energy of contraction must be generated by chemical reactions on demand.

There are many more problems concerning muscle heat. For example, if Ca^{2+} is pumped actively back into the lumen of the longitudinal SR after relaxation of the contractile components, should we be able to detect heat associated in time with this reaction? A scholarly and stimulating inquiry into the whole question of muscle heat and its genesis has been done by Mommaerts (1969).

VI. Unanswered Questions

Many aspects of muscle biophysics are dealt with by others elsewhere in these volumes. In particular there are a number of hypotheses about mechanisms of contraction at the molecular level. Their resolution clearly is a major remaining problem. But there are others.

These questions have surely occurred to many students of muscle physiology, biochemistry, and biophysics. There are two reasons why they have not been answered yet. In some cases the mathematical models that form the basis for such studies rest on assumptions that might be tenuous or on special cases, such as those that lead to the core conductor equations or the equations for ion fluxes, so that further advances may require more general mathematical solutions. But mainly, in nearly every case I cite below, the questions are unanswered because their solution seems to require techniques or equipment not yet available or fresh approaches.

To begin at the beginning, despite excellent, thoughtful, careful, and imaginative contributions, we still do not know how depolarization moves from sarcolemma radially through the T system. Is it by electrotonus or is there a propagated action potential? What is the message from T system to terminal cisternae? Is it electrical or chemical? And if the latter, which seems highly unlikely, what is its nature? How does this message evoke release of Ca^{2+} from terminal cisternae? Is it really true that Ca^{2+} then moves to myofibrils only by random thermal motion, a process that seems slow and chancey? Exactly what does the Ca^{2+} do? What really is mechanochemical coupling?

How is force transmitted through a muscle fiber? Through a sarcomere?

What is the signal for relaxation? Does some reaction just run its course in the absence of further stimulus to contraction? If so, does this mean that each stimulus detonates a quantum of reactive mix? If so, what is the mix and where are the quanta? Does Ca^{2+} simply diffuse back to SR from myofibrils? Where is the Ca^{2+} pump and when does it act?

Where is the Na^+ pump; specifically, where is the Na^+–K^+-activated, ouabain-inhibited ATPase? Is there more than one Na^+ pump? How are ions distributed in muscle? What governs their flux? What is the role of ordered and highly ordered water in diffusion processes inside muscle, and does ice-like water exclude certain solutes?

What arc the properties of muscle membranes (and which elements of the membranes are involved) that permit specific changes in ion conductance? That permit nonspecific changes? Various substances modify these properties. How do they do it?

We think we know a great deal about muscle, but we still have so much to learn.

REFERENCES

Abbott, B. C., and Wilkie, D. R. (1953). *J. Physiol. (London)* **120**, 214.

Adrian, R. H. (1956). *J. Physiol. (London)* **133**, 631.

Adrian, R. H., and Freygang, W. H. (1962). *J. Physiol. (London)* **163**, 61.

Adrian, R. H., Constantin, L. L., and Peachey, L. D. (1969). *J. Physiol. (London)* **204**, 231.

Adrian, R. H., Chandler, W. K., and Hodgkin, A. L. (1970). *J. Physiol. (London)* **208**, 607.

Bahler, A. S., Fales, J. T., and Zierler, K. L. (1967). *J. Gen. Physiol.* **50**, 2239.

Bahler, A. S., Fales, J. T., and Zierler, K. L. (1968). *J Gen. Physiol.* **51**, 369.

Behn, U. (1897). *Ann. Phys. Chem. [3]* **62**, 54.

Bernstein, J. (1902). *Arch. Gesamte Physiol. Menschen Tiere* **92**, 521.

Birks, R. I., and Davey, D. F. (1969). *J. Physiol. (London)* **202**, 171.

Blair, E. A., and Erlanger, J. (1936). *Amer. J. Physiol.* **117**, 355.

Boyle, P. J., and Conway, E. J. (1941). *J. Physiol. (London)* **100**, 1.

Caldwell, P. C. (1968). *Physiol. Rev.* **48**, 1.

Carey, M. J., and Conway, E. J. (1954). *J. Physiol. (London)* **125**, 232.

Carlson, F. D. (1957). *In* "Tissue Elasticity" (J. W. Remington, ed.), pp. 55–72. Amer. Physiol. Soc., Washington, D.C.

Carlson, F. D., Hardy, D., and Wilkie, D. R. (1967). *J. Physiol. (London)* **189**, 209.

Cole, K. S., and Curtis, H. J. (1939). *J. Gen. Physiol.* **22**, 649.

Greese, R. (1968). *J. Physiol. (London)* **197**, 255.

Cross, S. B., Keynes, R. D., and Rybová, R. (1965). *J. Physiol. (London)* **181**, 865.

Curtis, H. J., and Cole, K. S. (1942). *J. Cell. Comp. Physiol.* **19**, 135.

Duggan, P. F., and Martonosi, A. (1970). *J. Gen. Physiol.* **56**, 147.

Ebashi, S., and Endo, M. (1968). *Progr. Biophys. Mol. Biol.* **18**, 123.

Edman, K. A. P., and Grieve, D. W. (1966). *J. Physiol. (London)* **184**, 21P.

Eisenberg, R. S., and Gage, P. W. (1969). *J. Gen. Physiol.* **53**, 279.

Elliot, G. F., Rome, E. M., and Spencer, M. (1970). *Nature (London)* **226**, 417.

Fales, J. T., and Zierler, K. L. (1969). Amer. J. Physiol. 216, 70.
Falk, G., and Fatt, P. (1964). Proc. Roy. Soc., Ser. B 160, 69.
Fenn, W. O. (1923). J. Physiol. (London) 58, 175.
Fenn, W. O., and Marsh, B. S. (1935). J. Physiol. (London) 85, 277.
Fischer, E. (1926). Pfluegers Arch. Gesamte Physiol. Menschen Tiere 213, 352.
Franzini-Armstrong, C. (1970). J. Cell. Biol. 47, 488.
Frumento, A. S. (1965). Science 147, 1442.
Fujino, M., Yamaguchi, T., and Suzuki, K. (1961). Nature (London) 192, 1159.
Gage, P. W., and Eisenberg, R. S. (1969a). J. Gen. Physiol. 53, 265.
Gage, P. W., and Eisenberg, R. S. (1969b). J. Gen. Physiol. 53, 298.
Goldman, D. E. (1943). J. Gen. Physiol. 27, 37.
Gordon, A. M., Huxley, A. F., and Julian, F. J. (1966). J. Physiol. (London) 184, 170.
Graham, J., and Gerard, R. W. (1946). J. Cell. Comp. Physiol. 28, 99.
Happel, P. (1926). Pfluegers Arch. Gesamte Physiol. Menschen Tiere 213, 336.
Harris, E. J. (1963). J. Physiol. (London) 166, 87.
Harris, E. J. (1965). J. Physiol. (London) 177, 355.
Hellerstein, D. (1968). Biophys. J. 8, 358.
Hermann, L. (1879). In "Handbuch der Physiologie" (L. Hermann, ed.), Vol. II, pp. 1–196. Vogel, Leipzig.
Hill, A. V. (1938). Proc. Roy. Soc., Ser. B 126, 136.
Hill, A. V. (1965). "Trails and Trials in Physiology." Williams & Wilkins, Baltimore, Maryland.
Hill, A. V., and Howarth, J. V. (1957). Proc. Roy. Soc., Ser. B 147, 21.
Hodgkin, A. L. (1937). J. Physiol. (London) 90, 183 and 211.
Hodgkin, A. L. (1958). Proc. Roy. Soc., Ser. B 148, 1.
Hodgkin, A. L., and Horowicz, P. (1959a). J. Physiol. (London) 145, 405.
Hodgkin, A. L., and Horowicz, P. (1959b). J. Physiol. (London) 148, 127.
Hodgkin, A. L., and Horowicz, P. (1960). J. Physiol. (London) 153, 370.
Hodgkin, A. L., and Huxley, A. F. (1939). Nature (London) 144, 710.
Hodgkin, A. L., and Katz, B. (1949). J. Physiol. (London) 108, 37.
Horowicz, P., Taylor, J. W., and Waggoner, D. M. (1970). J. Gen. Physiol. 55, 401.
Howell, J. N. (1969). J. Physiol. (London) 201, 515.
Hutter, O. F., and Noble, D. (1960). J. Physiol. (London) 151, 89.
Huxley, A. F., and Taylor, R. E. (1958). J. Physiol. (London) 144, 426.
Ismail-Beigi, F., and Edelman, I. S. (1970). Proc. Nat. Acad. Sci. U.S. 67, 1071.
Kerkut, G. A., and Thomas, R. C. (1965). Comp. Biochem. Physiol. 14, 167.
Kernan, R. P. (1962). Nature (London) 193, 986.
Keynes, R. D., and Lewis, P. R. (1951). J. Physiol. (London) 114, 151.
Keynes, R. D., and Steinhardt, R. A. (1968). J. Physiol. (London) 198, 581.
Levi, H., and Ussing, H. H. (1948). Acta Physiol. Scand. 16, 232.
Ling, G., and Gerard, R. W. (1949). J. Cell. Comp. Physiol. 34, 383.
Mommaerts, W. F. H. M. (1969). Physiol. Rev. 49, 427.
Mullins, L. J., and Awad, M. Z. (1965). J. Gen. Physiol. 48, 761.
Mullins, L. J., and Noda, K. (1963). J. Gen. Physiol. 47, 117.
Nastuk, W. L., and Hodgkin, A. L. (1950). J. Cell. Comp. Physiol. 35, 39.
Overton, E. (1902). Arch. Gesamte Physiol. Menschen Tiere 92, 346.
Peachey, L. D. (1965). J. Cell Biol. 25, 209.
Polissar, M. J. (1952). Amer. J. Physiol. 168, 766.

Porter, K. R., and Palade, G. E. (1957). *J. Biophys. Biochem. Cytol.* 3, 269.

Ramsey, R. W. (1955). *Amer. J. Physiol.* 181, 688.

Ramsey, R. W., and Street, S. F. (1940). *J. Cell. Comp. Physiol.* 15, 11.

Rogus, E., and Zierler, K. L. (1970). *Fed. Proc., Fed. Amer. Soc. Exp. Biol.* 29, 455 (abstr.).

Rogus, E., Price, T., and Zierler, K. L. (1969). *J. Gen. Physiol.* 54, 188.

Rushton, W. A. H. (1937). *Proc. Roy. Soc., Ser. B* 124, 210.

Samaha, F. J., and Gergely, J. (1965). *Arch. Biochem. Biophys.* 109, 76.

Skou, J. C. (1965). *Physiol. Rev.* 45, 596.

Sonnenblick, E. H. (1965). *Circ. Res.* 16, 441.

Teorell, T. (1949). *Arch. Sci. Physiol.* 3, 205.

Ussing, H. H. (1950). *Acta Physiol. Scand.* 19, 43.

Wilkie, D. R. (1968). *J. Physiol. (London)* 195, 157.

Winegrad, S. (1968). *J. Gen. Physiol.* 51, 65.

Woledge, R. C. (1968). *J. Physiol. (London)* 197, 685.

Zierler, K. L. (1966). *J. Gen. Physiol.* 49, 423.

Zierler, K. L. (1968). *In* "Medical Physiology" (V. B. Mountcastle, ed.), Vol. II, pp. 1128–1171. Mosby, St. Louis, Missouri.

Zierler, K. L., Maseri, A., Klassen, G., Rabinowitz, D., and Burgess, J. (1968). *Trans. Ass. Amer. Physicians* 81, 266.

4

ENERGY NEED, DELIVERY, AND UTILIZATION IN MUSCULAR EXERCISE

SUNE ROSELL and BENGT SALTIN

I. Introduction

A great number of independent factors may limit man's ability to perform physical work. The traditional view is focused on the importance of the central circulation and the amount of oxygen that can be transported. This concept is based on the assumption that the skeletal muscle has an ability to use more oxygen than can be delivered by the cardiovascular system. The availability of essential substrates for the energy release in the muscle in prolonged exercise or in very intense work is an often overlooked factor. Before going into details about metabolic adjustments during work under normal conditions, some aspects will be given on energy demand in exercise and different ways to define the magnitude of the work.

A. Absolute and Relative Work Loads

At the onset of exercise, oxygen uptake increases gradually, and within a couple of minutes, it reaches a new level (steady state), which is proportional to the performed work. This is the case as long as the need for energy does not surpass the amount which can be released aerobically, which is defined by the subject's maximal oxygen uptake. The work load can thus be expressed in absolute terms, i.e., oxygen uptake in liters per minute or in bicycle exercise in kilometers per minute (100 kpm/min = 16.35 watts)—or in relative terms as the observed oxygen uptake during the exercise, which may be expressed in percent of the individual's maximal oxygen uptake (see Fig. 1).

Provided a certain mechanical efficiency (bicycle work ≈ 22%) is assumed, it is also possible to express in relative terms work loads above the subject's aerobic power. The heart rate response during work is also an indication of the relative work load. However, the variation in maximal heart rate between subjects is great (Reeves and Sheffield, 1971), and moreover, it never reaches a steady state during prolonged exercise (Hartley et al., 1970). Thus the heart rate is an inferior index.

When subjects with different maximal oxygen uptake perform exercise to exhaustion at the same relative work load, approximately the same work time is attained (Saltin, 1971). The degree of training seems to be of minor importance, even though well trained subjects may perform slightly better. This is especially true at work loads demanding more than 90% of maximal oxygen uptake. Above this work rate, the subject's anaerobic power plays a great relative role in the magnitude of the energy release.

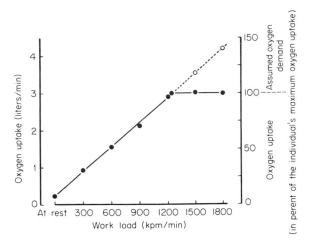

Fig. 1. Oxygen uptake for one individual at rest and during different work loads. The present data correspond with earlier published data in this field. The solid circles denote determined values (mechanical efficiency 22.5%) and the open circles denote assumed demand of oxygen at work loads which for this particular subject are above his maximal oxygen uptake (>3.0 liters/min) and covered by anaerobic processes. These latter work loads are called supramaximal work loads, and the work load just enough to load the oxygen uptake maximally is called maximal work load. As can be noted from the ordinate on the right, the work load can be expressed in relative terms, setting the maximal oxygen uptake to 100%.

Generally speaking, the time a person can perform at a certain work load is inversely related to the severity of the exercise (Fig. 2). Thus, when a bicycle work load is performed under optimal conditions demanding 100% or more of the individual's maximal oxygen uptake, maximal work time may vary from seconds up to 10 min. At a work rate of 90%, the work time that can be attained is about 30 min. Work time is further increased and approximates 90 min at 85% of the maximal oxygen uptake. Below a work load of 60% of maximal oxygen uptake, the exercise may go on for many hours (>5 hr), and usually there is no easily definable point of exhaustion (I. Åstrand, 1960).

The need for energy in the aforementioned examples varies markedly and is related to the individual's aerobic and anaerobic power. In healthy adults, maximal oxygen uptake varies in relation to sex, age, and training status between 2.5 and 6.5 liters/min (P-O. Åstrand and Christensen, 1964; Saltin and Åstrand, 1967).

The maximal amount of energy derived from anaerobic sources may be approximately 15–25 kcal (Hermansen, 1969). In brief maximal bicycle exercise, 15–30 kcal/min are released, but since the work can only be sustained for 2–3 min, the total energy output is 30–90 kcal. At lighter

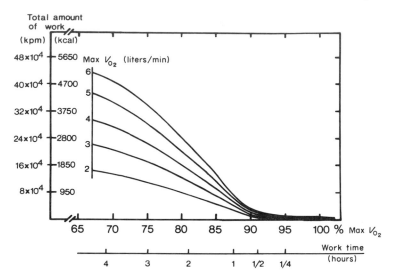

Fig. 2. Approximate amount of work (kpm or kcal) that can be performed by well motivated subjects with different maximal oxygen uptake when they exercise to exhaustion at work loads demanding 100% or less of their maximal capacity. The scale at the bottom of the figure indicates the approximate time a subject can stand a work load of different relative intensities. (From data reported by Saltin, 1971).

work loads which can be maintained for hours, a combustion rate of 500–1000 kcal/hour has been reported. (Hedman, 1957; P-O. Åstrand *et al.*, 1963; Hermansen *et al.*, 1967). In more extreme situations, such as long distance running, skiing, bicycling, and lumber jacking 5000–10,000 kcal a day is likely to be combusted.

In many studies in humans and animals, only one or a few muscle groups are involved in the exercise. The ideal situation would then be to relate the work performance by the muscle or group of muscles to their maximal capacity. This is in many situations difficult to do. One method sometimes used is to relate the force in each contraction to the maximal isometric force in the actual muscle group (Rohmert, 1960). It is impossible to make a direct comparison between this relative scale and the one described above.

II. Extramuscular Lipid Stores

A. *Liver*

According to Rowell *et al.* (1965; Rowell, 1971), the splanchnic arteriovenous concentration differences for phospholipids and triglycerides varies randomly with time during heavy exercise in man, and they could not detect any net production of phospholipids and triglycerides by the liver. They concluded that no more than 5% of the total energy expenditure could have been derived from the liver in the form of these two lipid fractions.

B. Adipose Tissue

The average man has about 15 kg of adipose tissue, which by far is the largest energy source. As much as 90% of the weight of adipose tissue may be comprised of triglycerides. As a consequence, adipose tissue may yield close to the theoretical 9.4 cal/gm of pure lipids. There is considerable evidence that the elevated concentration of free fatty acids (FFA) in the plasma pool during exercise is due to an increased mobilization from the white adipose tissue.

The work of Dole (1956) and R. S. Gordon and Cherkes (1956) has shown that the mobilization of FFA is due to lipolysis, i.e., hydrolysis of triglycerides stored in the adipocytes (see Rizack, 1965). The nucleotide adenosine 3′,5′-monophosphate (cyclic AMP) is now generally believed to function as an intracellular mediator in the action of lipolytic agents via an adenyl cyclase system in the cell membrane (see Butcher, 1970). Lipolysis leads to an outflow of FFA and glycerol from adipose tissue. Of these two products of lipolysis, glycerol is generally used as a measure of the rate of lipolysis, since it is only utilized in white adipose tissue to only a minor extent or not at all (Steinberg and Vaughan, 1965). The reason for this is that adipose tissue has very limited capacity to phosphorylate glycerol (Margolis and Vaughan, 1962; Robinson and Newsholme, 1967). FFA, on the other hand, is generally reesterified to a certain extent, as indicated by the finding that often less than 3 moles of FFA per mole of glycerol are released *in vitro* (Steinberg and Vaughan, 1965) or appear in the venous outflow from adipose tissue *in vivo* (Fredholm and Rosell, 1968).

1. CATECHOLAMINES AS LIPOLYTIC AGENTS

The adrenergic neurohumoral system is considered to have a key position as a physiological regulator of the enhanced lipolysis in adipose tissue during exercise (see Havel, 1965; Fredholm, 1970). One reason for this is that catecholamines are potent lipolytic agents in several species, including dog and man (Rudman *et al.*, 1965). Further, the blood level of norepinephrine increases during physical activity (Vendsalu,

1960), especially at work loads requiring about 75% or more of maximal oxygen uptake (Häggendahl *et al.*, 1970). The level of circulating norepinephrine may predominantly reflect the activity in the sympathetic nerves, since diffusion and overflow from the nerve terminals to the blood appears to be one of the principal inactivating mechanisms for the adrenergic transmitter, especially during exercise (Hertting and Axelrod, 1961; Folkow *et al.*, 1967). Changes in the plasma concentration of epinephrine, on the other hand, may reflect the activity in the adrenal glands only.

It has also been shown that in human volunteers pretreated with propranolol (4.2–5.3 mg per kilogram of body weight)—an adrenergic β-receptor blocking agent—the elevation of plasma FFA during and after treadmill exercise was approximately 50% lower than corresponding values for the controls (Muir *et al.*, 1964). This action of propranolol was presumably due to a reduced mobilization of FFA, since this is characterized as a β-receptor effect in man (Pilkington *et al.*, 1962).

A change in the physical activity from rest to exercise elicits a rapid elevation of the glycerol concentration in plasma as well as an increased rate of FFA turnover. This prompt reaction has also been taken as evidence for the view that the sympathetic nervous system is responsible for fat mobilization, at least initially, during exercise (Havel *et al.*, 1964).

2. CIRCULATING CATECHOLAMINES VERSUS SYMPATHETIC INNERVATION

Although catecholamines are potent lipolytic agents, recent reports indicate that in dogs, blood-borne norepinephrine or epinephrine are not as effective in producing lipolysis as the activity in sympathetic nerves to adipose tissue (Ballard *et al.*, 1971; Rosell, 1966).

To increase the lipolytic rate in subcutaneous adipose tissue, the concentrations of added norepinephrine or epinephrine in the arterial blood had to exceed 0.01 μg/ml, and such levels have been reported for norepinephrine only at supramaximal work loads in man and during stressful conditions, including bleeding, in dogs. On the other hand, both the omental and the subcutaneous adipose tissue in the inguinal region in dogs react with increased lipid mobilization, even during stimulation of the adrenergic sympathetic nerves with a frequency of 1 per second. The maximal release rate is obtained with a frequency of about 3 per second (Rosell, 1966). These frequencies should be compared with the calculated frequency in sympathetic nerves to the vascular bed of skeletal muscle during resting conditions (1–2 per second) and during maximal reflex activation (6–10 per second) (Folkow, 1952), and it may be reasonable to believe that the sympathetic nerves to the adipose

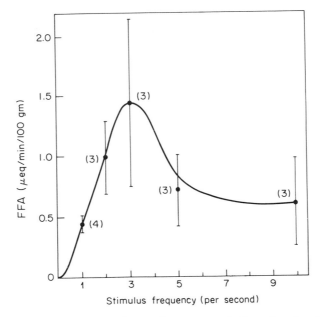

Fig. 3. Relation between stimulation frequency and the rate of net release of FFA during the stimulation period. The figures in parentheses indicate the number of stimulations. The vertical lines represent the standard error of the mean. (From Rosell, 1966).

tissue have the same frequency characteristics. The data indicate that the sympathetic innervation of adipose tissue plays a much greater role in FFA mobilization than circulating catecholamines from, for example, the adrenal medulla. Not even at very high plasma concentrations may circulating norepinephrine and epinephrine contribute significantly to FFA release, since these values are only found at or above maximal work intensities, where carbohydrates may be the only combusted fuel and the entry of FFA to the plasma pool may be counteracted by high blood lactate concentrations (see below).

3. Regional Differences in the Sympathetic Control

Although there is experimental evidence for the view that the sympathetic nerves are the more important of the two links of the adrenergic neurohumoral system, they do not seem to regulate the release of FFA from all lipid stores in the body. Instead, there seem to be regional differences which may be great enough to be of physiological importance. Thus, the canine mesentery does not respond to sympathetic nerve stimulation with enhanced lipolysis, in contrast to the situation in canine subcutaneous and omental adipose tissue (Ballard and Rosell, 1969, 1971; Rosell, 1966). Aronowsky et al. (1963) suggested that the function of adipose tissue in some locations may be to support or insulate rather than to furnish energy by means of mobilization of FFA. They came to this conclusion after having found that orbital fat and paw fat of the cat did not respond with increased lipolysis upon addition of norepinephrine in vitro. Whether this is also the sole function of mesenteric adipose tissue remains to be evaluated, since there are a multitude of factors, in addition to the adrenergic neurohumoral system, that have lipolytic actions as will be discussed below.

4. CENTRAL NERVOUS CONTROL

Most of the information concerning the influence of sympathetic nervous activity on lipid mobilization stems from studies in which these nerves have been activated peripherally. Such studies do not provide information concerning the central nervous control of lipid metabolism (i.e., the location of the integrative areas), the afferent nervous influence, and the efferent pattern of activity. Therefore, there is only limited information concerning the central nervous control of lipolysis.

In contrast, the central control of the cardiovascular system has been worked out in detail. Cardiovascular response patterns which resemble those found during physical activity can be evoked from corticohypothalamic integration centers. It would of course be of interest to know to what extent the sympathetic nervous outflow to the adipose tissue is an integrative part of such response patterns.

One visceromotor adjustment that can be evoked from the corticohypothalamic integration areas is the "defense reaction," which includes increased cardiac output, augmented muscle blood flow due to activation of cholinergic sympathetic vasodilator nerves, and concomitant vasoconstriction in most other vascular regions, including the kidney, skin, and gastrointestinal tract. In addition, there is an increased release of catecholamines from the adrenals (see Uvnäs, 1960). The defense reaction is supposed to be elicited in situations involving emotional distress, such as anticipatory adjustments to flight or fight (Eliasson et al., 1951; Abrahams et al., 1964; Bolme and Novotny, 1969; Lisander, 1970). The concomitant metabolic adjustments have been studied to only a limited extent. However, electrical stimulation of the defense areas in the hypothalamus in anaesthetized dogs does not seem to elevate the plasma concentration of FFA or glycerol (Orö et al., 1965). Similarly, the blood flow in subcutaneous adipose tissue does not change significantly during hypothalamic defense area stimulation (Ngai et al., 1966). These findings indicate that activation of the sympathetic nerves to the adipose tissue does not constitute an essential link in the defense reaction. This may be surprising in view of the opinion that the visceromotor adjustments are supposed to occur in anticipation of strenous exercise including flight or fight. However, such situations certainly require work of maximal intensity, and therefore carbohydrates, rather than fat, may be the most important fuel. This and other aspects of the metabolic link of the visceromotor adjustments in connection with the defense reaction deserve further study.

Orö *et al.* (1965) found an elevated concentration of plasma FFA following electrical stimulation in diencephalic areas closely related to H_1 and H_2 fields of Forel from which cardiovascular responses similar to those seen during exercise on a treadmill can be elicited (Rushmer *et al.*, 1961). The elevation of plasma FFA concentrations was eliminated after treatment with ganglionic blocking agents, but not by adrenalectomy, indicating that the effect was due to activation of sympathetic nerves to the adipose tissues. These experiments indicate that there are discrete diencephalic areas from which the mobilization of lipids in adipose tissue is regulated via the sympathetic nervous outflow. It is not known, however, how incoming nervous activity may modify this outflow, and experiments designed to reflexively change the sympathetic activity to adipose tissue have not been successful so far. For example, activation of the baroreceptor reflex by occlusion of the common carotids does not cause any marked changes in plasma FFA concentration (Orö and Fröberg, 1964) or in the release of FFA from subcutaneous adipose tissue in the dog. Moreover, the vascular resistance in the subcutaneous adipose tissue is hardly affected by occlusion of the carotids (Ngai *et al.*, 1966). These negative results may indicate that the adipose tissue, the subcutaneous tissue at least, does not participate in the reflex adjustments on the peripheral resistance elicited via the carotid baroreceptors. This reflex is intimately connected to the blood pressure homeostasis, and there is the interesting possibility that the adipose tissue may be linked to "metabolic" reflexes of some hitherto unknown sort.

5. Local Modification of Lipolysis

There are local modifications of the sympathetic lipomobilizing effects. Thus, Issekutz and Miller (1962) and Issekutz *et al.* (1965) suggested that lactate may serve as an inhibitor of FFA mobilization and recent experiments have demonstrated that sodium L($+$)-lactate at blood concentrations about 10 mM inhibits the release of FFA caused by electric stimulation of the sympathetic nerves to the subcutaneous adipose tissue.

On the other hand, the release of glycerol was not significantly affected by sodium L(+)-lactate infusion (Fredholm, 1971). The decreased release of FFA concomitantly with an unchanged glycerol release suggested an enhanced reesterification of fatty acids in the adipose tissue (Fredholm, 1970). These findings suggest that not only the rate of hydrolysis of triglycerides, but also the rate of reesterification may be of importance in the regulation of the outflow of FFA from adipose tissue. Since the latter effect was shown to operate at plasma levels of lactate which occur during heavy muscular exercise, it is likely that lactate may function as a physiological brake on the lipid inflow to the plasma pool in situations where carbohydrates may be the predominant source of fuel. Thus, lactate may prevent the working skeletal muscle from being flooded with fatty acids that they have little use for (Fredholm, 1970).

6. LIPOLYTIC HORMONES

In addition to the adrenergic neurohumoral system, there are also fast acting hormones from the pituitary gland (ACTH and TSH) and from the pancreas (glucagon). These have been shown to have a lipolytic action *in vitro*, at least in some species (Rudman, 1963). However, norepinephrine and epinephrine are probably the only hormones that may be of physiological significance in situations like exercise (see Goodman, 1968; Scow and Chernick, 1970).

Growth hormone and glucocorticoids may induce lipolysis after a latency of more than 1 hr (Engel *et al.*, 1959; Zahnd *et al.*, 1960; Pearson *et al.*, 1960; Fain *et al.*, 1965) and are therefore called slow acting hormones (Scow and Chernick, 1970). Probably, the mechanism of action is different from the fast acting hormones, since their lipolytic action is counteracted by inhibitors of RNA synthesis (actinomycin) and of protein synthesis (puromycin). These findings indicate that the slow acting lipolytic hormones act via an enhanced RNA and protein synthesis rather than via the cyclic AMP (cAMP) system (Fain *et al.*, 1965). Gollnick *et al.* (1971) found in rats that mobilization of FFA as a consequence of exercise was diminished by hypophysectomy as well as by blockade of the adrenergic β-receptors with propranolol (6 mg/kg intraperitoneally 30 min prior to the start of exercise). Further, hypophysectomy combined with blockade of adrenergic β-receptors almost completely abolished the exercise-induced elevation in adipose and plasma FFA concentration. Although these data do not indicate which pituitary factor or factors may be of importance for the lipid mobilization during exercise, they do suggest that the pituitary is of importance during strenuous exercise. Hunter *et al.* (1965a,b) noted a marked increase in growth hormone in plasma in subjects during a 12.8 km walk at 6.4 km/hr on a trademill. Exercise at this rate expends energy approximately five times faster than at rest. The elevated plasma level of growth hormone was associated with a fall in respiratory exchange ratio (R) and a gradually rising plasma FFA concentration. These results indicate that growth hormone may have a role in maintaining the release of FFA during exercise, since the action may last for several hours (Scow and Chernick, 1970).

A fall in the blood glucose concentration has been suggested to be a stimulus to growth hormone secretion (Roth *et al.*, 1963a,b). This may be of physiological importance, since even decreases as small as 10 mg per 100 ml blood following insulin have been shown to significantly elevate the plasma level of the growth hormone (Luft *et al.*, 1966; Greenwood and Landon, 1966).

The other slow acting lipomobilizing principle, glucocorticoids, potentiates the lipolysis produced by growth hormone in fat cells isolated from adipose tissue of fasting normal rats (Fain *et al.*, 1965) and may therefore contribute to the lipolytic rate during exercise.

7. Insulin as a Regulator of Lipolysis

Insulin may be another factor of importance in the mobilization of FFA during exercise. The plasma concentration of immunoreactive insulin (IRI) decreases during prolonged work of 20% of maximal oxygen uptake or more (Cochran *et al.*, 1966; Rasio *et al.*, 1966; Hunter and Sukhar, 1968). At work loads of 50–70% of maximal oxygen uptake, the decrease in the plasma concentration averaged between 50 and 60% of the preexercise value (Pruett, 1970a,b, 1971). Insulin has an antilipolytic action, and the possibility thus exists that this action is diminished during prolonged work, thus favoring elevated rates of entry of FFA from the adipose tissue.

The findings that plasma insulin levels are not increased (Schatch, 1967) and are generally decreased during severe exercise might be due to the action of epinephrine, which inhibits insulin release from the pancreas. Porte *et al.* (1966) have shown that the expected rise in the plasma concentration of IRI in response to glucose, for example, is inhibited by the infusion of epinephrine at a rate which produces plasma levels similar to those found during stressful conditions. Such a mechanism may also be consistent with the finding of Pruett (1970b) that during exercise to exhaustion at 70% of maximal oxygen uptake, the IRI concentration decreased, even though the glucose levels increased.

There is thus the possibility that catecholamines may increase glycogenolysis and inhibit the release of insulin during exhaustive work. Such an adjustment would save carbohydrates for the central nervous system.

8. SUMMARY

Adipose tissue is the largest source of energy. At present the quantitative importance of the different factors that may promote inflow of FFA from adipose tissue during exercise and the interplay between them is unclear. However, it seems that the sympathetic nervous outflow plays a predominant role for lipolysis in adipose tissue from certain locations. This system is suited for rapid elevations of the lipolytic rate. Blood lactate may serve as an inhibitor of FFA entry into the plasma pool during exercise of high intensity when carbohydrate rather than fat is required as fuel. In addition growth hormone may act as a lipolytic agent, especially during work of long duration and falling plasma glucose levels. The glucocorticoids seem to potentiate the lipolytic effect of growth hormone and may therefore be of importance. The plasma insulin level tends to decrease at work intensities above 20% of maximal oxygen uptake. This might promote entry of FFA from adipose tissue. Epinephrine blocks the insulin secretion from the pancreas, but it is debated whether or not the plasma levels of epinephrine observed during exercise are high enough to exert such an effect, and if so, to what an extent.

III. Extramuscular Carbohydrate Stores

A. Blood and Liver

Eight to 10 gm of glucose is available in the blood and in interstitial fluid. However, the main extramuscular carbohydrate stores, in addition to those built into different membranes and cell structures and therefore not immediately available for energy metabolism, are found in the liver. In different animals there are wide variations in the normal glycogen content of the liver, which may be due in part to species differences but other factors including diet and the degree of physical activity also affect liver glycogen content. In man, after a mixed diet and no exercise during the preceding 24 hr, the liver glycogen content seems to vary between 40 and 50 mg/kg (Hultman and H:son Nilsson, 1971). With a liver weight of 1.4–1.8 kg, the total amount of glycogen stored in the liver would then be 55–90 gm. One day of starvation or food intake without carbohydrates is sufficient to produce almost complete depletion of liver glycogen stores. By contrast, a carbohydrate-enriched diet for 24 hours may double the normal glycogen content (Hultman and H:son Nilsson, 1971).

Another physiologically interesting way to estimate the magnitude of available hepatic carbohydrates is to determine the release of glucose from the splanchnic area during exercise. Here glucose may be derived from glycogenolysis in the liver as well as from gluconeogenesis. Results are available from several studies (Rowell *et al.*, 1965, 1966, 1968; Hultman, 1967a; Wahren *et al.*, 1971) which show that there is gradual enhancement of hepatic glucose release with time during exercise. Moreover, the greater the relative work load, the greater the production of glucose. During the first hour of a prolonged exercise period, 15–25 gm of glucose can be supplied by the liver; a maximum rate of 30–35 gm in one 1 hr period has been reported (Hultman, 1967a). These values are in very good agreement with the 24 gm/hr reported by Hultman and H:son Nilsson (1971). They based their calculations of liver glucose production on direct determinations of liver glycogen content after 1 hr of exercise. Rowell (1971) estimated that at the most, one-third of the glucose produced during exercise comes from gluconeogenesis. The main precursors in glucose synthesis are lactate, pyruvate, and glycerol, all of which are taken up by the liver during exercise.

Recent data in human volunteers indicate that alanine is released in significant amounts from skeletal muscle (Felig *et al.*, 1970). This is interesting, since alanine is extracted from the blood by the liver and is the principal gluconeogenic amino acid (Mallette *et al.*, 1969). During exercise on a bicycle ergometer resulting in a three- to ninefold increase in oxygen consumption, Felig and Wahren (1971) observed that alanine was the only amino acid released in significant amounts from the leg. The arterial concentration of alanine also rose. Felig and Wahren (1971) suggested that alanine may be synthesized from pyruvate by transamination. Thus, the carbon skeleton of alanine may be derived from the breakdown of glucose in the working skeletal muscle. Alanine may therefore be of importance as a substrate for hepatic glucose production and, according to Felig and Wahren (1971), covering up to 10% of total hepatic release of glucose. Even though the alanine–glucose cycle has yet to be established and its quantitative role during exercise may be modest, it opens up an interesting possibility. In starvation or with a noncarbohydrate diet for an extended period of time, the conversion of alanine to glucose in the liver may be an important factor in keeping blood glucose at a physiological level.

B. Release of Carbohydrates from the Liver

The regulation of hepatic glucose release during exercise is not well understood. One triggering mechanism is reduced blood glucose concentration, which may enhance glucose production through increased adrenergic activity or by an increase in the amount of circulating catecholamines. A slight fall in blood glucose is often seen at the onset of work (Christensen, 1931), but no gradual, pronounced, further reduction is usually observed, especially not during very heavy exercise (Hermansen et al., 1970).

Circulating norepinephrine augments hepatic glucose production (Costin et al., 1971). The question, however, is whether the very minor increase observed during brief exercise at low work intensities is sufficient to produce this regulatory action. Norepinephrine concentrations in arterial plasma are not significantly increased ($>0.8-1$ µg/liter) until work loads exceed 75% of maximal oxygen uptake (Häggendahl et al., 1970). No data are available for plasma norepinephrine concentrations during prolonged exercise.

During exercise, an accelerated increase in hepatic glucose production appears to coincide with near depletion of muscle glycogen. A feedback mechanism via some stimulus in the working muscle has been discussed and may seem tempting, but it is also very premature.

Several authors report that hypoxia may influence hepatic glucose release. Thus, data by Rowell et al. (1968), for example, indicate that liver glucose production during heavy exercise in hot surroundings is inversely related to the oxygen content of the hepatic vein. A release rate as high as 1.2 gm/min of glucose has been observed during short periods of exercise when the oxygen content in the vein was less than 1 ml per 100 ml. In these cases with very high glucose release, there is also a release of lactate. It can then be assumed that the entire production of liver glucose in these circumstances comes from glucogenolysis, as gluconeogenesis is dependent upon an adequate oxygen supply. In this context, it can be pointed out that although the oxygen consumption of the liver is unchanged during exercise, the splanchnic blood flow is inversely related to the relative work stress imposed on the subject. It only amounts to 10–20% of its resting value (1.5 liters) at very heavy work. During prolonged exercise, there is a further gradual fall in blood flow to the liver from this level.

Pyruvate, lactate, and glycerol are taken up by the liver, and they can all be precursors of glucose synthesis in the liver. Alanine has also been said to play a role. An exact quantification of the magnitude of gluconeogenesis at different exercise levels can not be made, but 10–15 gm/hr appears to be an upper limit.

The release of glucose from the liver is influenced by blood sugar concentration, hypoxia, and hormones such as glucagon, insulin, epinephrine, and norepinephrine. The exact interplay among these different factors, so important in the regulation of the blood glucose concentrations within narrow limits, is still unclear.

C. Summary

During exercise in man under normal conditions, 55–90 gm of glycogen may be available in the liver. In addition to the 10 gm circulating in extramuscular compartments, there is the amount of glucose that can be synthesized in the liver during exercise.

IV. Amino Acids as a Source of Energy

A man of normal weight has approximately 10–11 kg of protein. It is unlikely that any of this is stored solely as an energy source, as are lipids in the adipose tissue and carbohydrates in muscle and liver. Proteins have other functions, but this does not exclude them as an energy source. One argument put forward as early as 1866 (Pettenkofer and Voit, 1866) against the hypothesis that they are used as energy source is that nitrogen excretion is not elevated during exercise, and it seems to be generally agreed that protein do not play any significant role in energy provision during exercise (see also Margaria and Foa, 1939). However, as pointed out above, the alanine–glucose cycle may be of some importance to gluconeogenesis (Felig and Wahren, 1971).

V. Uptake of Substrate from Blood

A. Uptake of Plasma FFA in Skeletal Muscle

The uptake of FFA from blood into skeletal muscle seems to be regulated by factors outside the muscle, rather than by some uptake mechanism within the muscle (Hagenfeldt and Wahren, 1968). This opinion is based on the finding that there is a linear relationship between the inflow of FFA (plasma flow $\times$ arterial FFA concentration) and the uptake into skeletal muscle. Consequently, skeletal muscle blood flow and arterial FFA concentration are factors of importance in FFA uptake. There are many studies both in experimental animals and man, showing that muscle contraction is one of the most potent vasodilating procedures in skeletal muscle, and further, that nervous, hormonal and local factors tend to distribute the increased cardiac output during exercise to skeletal muscles, thus promoting inflow of FFA (see P-O. Åstrand and Rodahl, 1970).

Changes in the arterial concentration of FFA during exercise has been studied by Carlsson and Pernow (1961), who found that during the first 10–15 min of exercise there was a decrease in the plasma FFA concentration which gradually turned into an increase during prolonged work. At the termination of the exercise, this increase was accentuated. A similar pattern was found in untrained dogs during exercise, whereas in trained dogs there was no initial decrease (Paul, 1971). This general picture of the changes in plasma FFA concentration is to a great extent influenced by the type, intensity, and duration of work, as well as by the physical condition of the subject. Havel et al. (1967) proposed that the initial reduction in FFA levels were due to circulatory readjustments and a descrepancy between release of FFA from adipose tissue and its uptake in skeletal muscle. As has already been discussed (Section II,B,5), there is also the possibility of a delayed release of FFA due to accumulation of lactate, which is most pronounced at the onset of work. This seems to be the case in the untrained dogs in which there was an appreciable increase in the blood lactate concentration (Paul, 1971).

In human volunteers, there is a simultaneous uptake and release of FFA in skeletal muscles both at rest (Rabinowitz and Zierler, 1962) and during forearm work (Hagenfeldt and Wahren, 1968). Hagenfeldt and Wahren found that the release averaged about 50% of the simultaneous uptake of FFA and is therefore of quantitative importance. A similar release of FFA has also been observed in femoral venous blood from exercising men (Havel et al., 1967). Determination of the rate of fractional uptake of FFA in working muscle may thus fail to yield any information about the extent to which plasma FFA is used as a fuel in the working muscle.

It has been found that in exercising man and dog (Havel et al., 1963; Issekutz et al., 1964) and in the working human forearm (Hagenfeldt and Wahren, 1968), the oxidation of FFA derived from plasma could not account for all of the total fat oxidation, as calculated from R measurements.

B. Turnover of Plasma FFA

To get a quantitative measurement of the combustion of plasma FFA in the working intact man or experimental animal, the inflow of FFA from the plasma pool, as well as the fraction of carbon dioxide derived from plasma FFA, must be determined. From such studies it may be possible to relate the role of the different substrates, including FFA, to the work load in relation to the maximal capacity of the subject. Havel *et al.* (1963) found that when well trained subjects walked on a treadmill at a rate of approximately 6 km/hr, resulting in an average heart rate of 100 beats/min, and with lactate concentration no higher than 1.6 mmoles/liter, the turnover of FFA averaged 28 mmoles/min, and 57–87% of the plasma FFA entering the tissues was oxidized. This accounted for 41–49% of the total energy output during exercise. This was also true for subjects who had been fasted overnight.

Paul (1970, 1971) presented data concerning the FFA metabolism in trained dogs running on a treadmill at different speeds. All the work was of the aerobic type, as judged by the almost unchanged blood lactate concentrations. The oxygen uptake amounted to approximately 40 ml/kg per minute, which was presumably less than 50% of aerobic power of the dogs. Plasma concentration, uptake and turnover of FFA were all increased with increasing work loads. Between 54 and 87% of the total oxygen consumption was used to oxidize plasma FFA. It is notable that at the highest work load, the relative amount of oxygen used for FFA oxidation diminished. Both the work loads and the degree of FFA utilization in this study were similar to those obtained by Havel *et al.* (1963). Thus, there is experimental evidence for the view that plasma FFA may account for 80% or more of the total substrate consumption during work of low to moderate intensity, i.e., work loads where no lactate is produced. However, there are no experimental data available to indicate the part played by plasma FFA during work at higher intensities in man or intact animals. Experiments on isolated canine gracilis muscle did not reveal any significant net uptake of FFA during muscular contractions (Karlsson *et al.*, 1971). In these experiments, the work performed was calculated to be equivalent to about 70% of maximal capacity. Appreciable amounts of lactate were found in the muscle as well as in the venous blood.

C. Uptake of Other Substrates

In the resting human forearm, Hagenfeldt and Wahren (1968) found a net uptake of approximately 50% of the arterial β-hydroxybutyrate and acetoacetate. During a 60 min period of forearm exercise, the arteriovenous difference diminished continuously and acetoacetate was actually produced by the muscle during the latter part of the exercise. Even if the fractional uptake of these keton bodies is high, due to their low arterial concentrations, their contribution as an energy source for the muscle is negligible (Hagenfeldt and Wahren, 1968).

As mentioned above, Rowell et al. (1965) concluded that the release of triglycerides from the liver could at most account for 5% of energy expenditure during exercise. This opinion is supported by the finding of Havel et al. (1967) that plasma triglycerides supplied less than 10% of the fatty acids burned during leg exercise.

The concentration of plasma triglycerides are known to decrease as a result of regular physical training (Holloszy et al., 1964; Björntorp, 1971; Grimby et al., 1971), which is presumable due to increased utilization in skeletal muscle. This may be of importance to the regulation of plasma lipid concentration, but does not seem to contribute very much to the energy supply for the exercising muscle.

D. Uptake of Blood Glucose

As early as the 1920s, changes in blood glucose levels occurring during exercise were determined (for references, see Christensen, 1931). Usually, a gradual slight decrease has been observed at the beginning of exercise. Only under special circumstances—including starvation, fat and protein diets, and heavy exercise for several hours—have blood glucose concentrations of less than 50 mg per 100 ml blood been obtained (Fig. 4). In some of these experiments, central nervous system effects have been observed, sometimes severe enough to cause exhaustion (Christensen, 1931; Christensen and Hansen, 1939). At the termination of the work period, the blood glucose level is restored to or above the resting level, except under the conditions mentioned above.

Plasma glucose is taken up in skeletal muscle during exercise but seems to be of minor quantitative importance under most conditions. Paul (1971) found that the metabolism of plasma glucose did not account for more than 7% of the total oxygen consumption in untrained dogs. The oxygen uptake increased sevenfold to 57 ml/kg per minute,

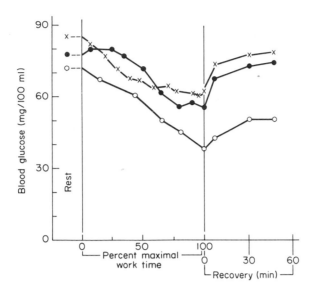

Fig. 4. Mean blood glucose concentration during exercise to exhaustion at 70% of maximal oxygen uptake. The work was performed after a period on different diets (● = mixed diet with an average work time of 126 min, ○ = fat and protein diet and 59 min, × = carbohydrate diet and 189 min.) At the end of the exercise preceded by a fat and protein diet, the subjects had severe headaches, were disoriented, and were unable to count the number of persons in the room. (Saltin and Heimansen, 1967.)

and there was a concomitant elevation of the lactate concentration to 40 mg per 100 ml. Similarly, in trained dogs which showed no elevation of the blood lactate concentration at a work load requiring an oxygen consumption of 48 ml/kg per minute, plasma glucose utilization did not represent more than 9% of the total oxygen uptake. The diminished plasma level of insulin during work (Pruett, 1971) may be a contributing factor to the limited importance of glucose as a fuel for the skeletal muscles during exercise, and thus the glucose will be saved for the central nervous system. In man, Wahren *et al.* (1971) have demonstrated a significant glucose uptake in the exercising forearm, which gradually increased with time. On the average, this could cover approximately 25% of the calculated carbohydrate combustion if completely oxidized. The question arises whether such a relatively high carbohydrate combustion occurs only when small muscle groups are activated. Under these circumstances, the insulin level may not decrease.

It has been suggested that muscular contraction releases an insulin-like humoral factor which enhances glucose uptake in skeletal muscle (Goldstein, 1965). It has not, however, been possible to isolate and identify such a factor chemically. Support for the importance of plasma glucose as a substrate during exercise has been provided by Chapler and Stainsby (1968). They found a marked increase in glucose uptake by the canine gastrocnemius plantaris muscle *in situ*. Following electric stimulation of the motor nerves with frequencies of 5 per second, the glucose combustion provided about 80% of the total substrate consumption in the muscle. These results are in contrast to those from similar experiments on the canine gracilis muscle by Costin *et al.* (1971) and Karlsson *et al.* (1971) in which blood glucose, at most, accounted for only 12% of the total energy expenditure.

VI. Use of Energy Stores in Skeletal Muscles

A. Muscle Glycogen Stores

Weiss (1871) demonstrated 100 years ago that muscle glycogen is utilized during work. It was long thought that the glycogen stored in the muscle was used almost exclusively for anaerobic glycolysis and that the amount utilized for oxidation was negligible. This is not the case, and the major part of the glycogen broken down in the muscle ends up in lactate only in the initial phase of exercise and during extremely intensive work (Hermansen *et al.*, 1967; Hultman, 1967a; Karlsson, 1971).

The amount of glycogen stored in the muscles of the body only varies from approximately 9 to 16 gm/kg wet muscles in normal circumstances (Hultman, 1967a). This small variation may be due to the fact that the percentage distribution of different types of fibers is rather constant in different human muscles (Edström and Nyström, 1969). The lowest values for glycogen content are usually in arm and shoulder muscles, and the higher values are in the lower extremities. The results obtained for muscle glycogen using needle biopsy samples taken from humans agree fairly well with results using other muscle sampling techniques and with what has been observed in mixed muscles of other species.

In any estimate of the total amount of glycogen stored in skeletal muscle, muscle mass must be known. If the muscle mass amounts to 40% of body weight, a total of 300–400 gm of glycogen may be found in the muscles. Christensen and Hansen (1939) and Hedman (1957) measured oxygen uptake and calculated RQ to estimate the magnitude of carbohydrate stores in the body. They arrived at approximately the same value (350–400 gm) for subjects who performed prolonged heavy exercise to exhaustion. Their figure also included glucose taken from extramuscular sources, but did not include any glycogen breakdown resulting in lactate production.

As mentioned above, liver glycogen is very much influenced by diet and starvation. Man's muscle glycogen is not significantly affected by a couple of days of starvation (Hultman, 1967a), but an increased carbohydrate intake results in a muscle glycogen content which is greater than after a mixed diet (Bergström et al., 1967). The storage of glycogen is further enhanced if prolonged heavy physical exercise precedes the intake of a high-carbohydrate diet. As much as 50–60 gm/kg wet muscles have been reported in humans (Karlsson and Saltin, 1971). If this high value is representative of all skeletal muscle in the body, glycogen stores of more than 1000 gm can be built up. Bergström et al. (1967) have also demonstrated that after a special regimen with exercise and a carbohydrate-enriched diet resulting in a mean muscle glycogen content of 40 gm/kg wet muscle, at least 800 gm of carbohydrates are oxidized during the exercise.

In recent years, a great deal of attention has been devoted to the factors that enhance storage of muscle glycogen. The key enzyme in the regulation of glycogen synthesis in the muscle cell is glycogen synthetase. This enzyme exists in two forms, a glucose 6-phosphate (G-6-P)-dependent form (D-form) and a form which is G-6-P independent (I-form). The two forms are interconvertible by mechanisms involving phosphorylation of the I-form and dephosphorylation of the D-form (Larner et al., 1969). Hormones such as insulin and even testosterone and estrogen stimulate glycogen synthesis in the muscle cell by increasing the percentage of the enzyme's I-form (Adolfsson and Ahrén, 1971). It has also been demonstrated that the activity of the I-form is inversely related to the glycogen content of the muscle cell (Hultman et al., 1971). However, this relationship only holds true up to normal levels for muscle glycogen content.

The D-form of the enzyme is activated when G-6-P concentration in the muscle cell is high (Piras and Staneloni, 1969). This only occurs during heavy exercise (Karlsson et al., 1970). In this situation, ATP concentration of the cell is also reduced slightly, which further enhances the activity of the D-form.

Under most physiological conditions activation of the I-form of the enzyme glycogen synthestase seems to be the most important factor in the glycogen synthesis occurring in the skeletal muscle. The D-form of the enzyme is normally only activated during heavy exercise. However, none of the possibilities described above can provide a satisfactory explanation for the enhanced glycogen synthesis observed 1–2 days after vigorous exercise when high-carbohydrate food is eaten (Bergström *et al.*, 1967). In this situation a *de novo* synthesis of enzyme protein may be brought into play, although this has yet to be proven.

1. Muscle Glycogen Depletion during Work

The reintroduction of a needle biopsy technique (Bergström, 1962) to take samples from human muscles, originally described by Duchenne (see Natrass, 1968) provides an opportunity for direct study of the utilization of muscle glycogen. In most of the studies published to date, a bicycle ergometer has been used for the exercise, and the lateral portion of the quadriceps muscles has been the target of biopsies. Muscle blood flow (Grimby *et al.*, 1967) and temperature measurements (Saltin *et al.*, 1968) have indicated that this particular muscle group is active during bicycle exercise in a manner which seems to have a linear relationship to the work performed.

The rate of muscle glycogen depletion is also related to the work rate (Saltin and Karlsson, 1971). In the first hour of exercise, a work load which demands 25–30% of the individual's maximal oxygen uptake, only 10% of the muscle's normal glycogen content is depleted (Fig. 5).

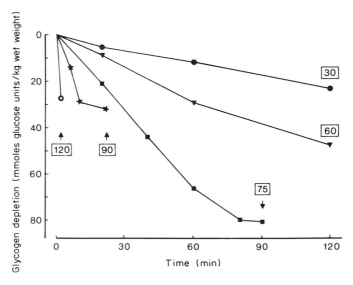

Fig. 5. Glycogen depletion in the quadriceps muscle during bicycle exercise of different intensities. Arrows indicate points of exhaustion, the numbers in the rectangles indicate percent of maximum V_{O_2}. (Saltin and Karlsson, 1971.)

A significant further decrease in muscle glycogen content cannot be observed at this work intensity with continued work. At intensities up to 50–60% of maximal oxygen uptake, there is only a slightly higher rate of muscle glycogen utilization. Above this work level, a significantly higher rate of glycogen depletion takes place in the muscle, and at 75% of maximal oxygen uptake, the muscle glycogen content is reduced by 50% within 45 min and is depleted after a further 45 min period of exercise. Subjects are exhausted in this situation. Even faster rates of glycogen depletion have been observed during maximal exercise leading to exhaustion within minutes. One-fourth of the glycogen stores in the muscles are then consumed within 2–4 minutes.

If the aforementioned values are expressed per minute of exercise, only 0.2–0.3 glucose unit per kilogram of muscle per minute is utilized at 20% of maximal oxygen uptake. The corresponding values are 0.5, 1.4, and 3.4 glucose units per kilogram per minute at 50, 75, and 100% of maximal oxygen uptake, respectively. When the maximal dynamic work load is far more than can be accommodated by aerobic processes (determined by the subject's maximal oxygen uptake), a muscle glycogen breakdown of up to 10–15 glucose units per kilogram of muscle per minute has been observed. However, these values are less than 30% of those reported in maximal isometric exercise (Bergström *et al.*, 1971) and in tetanized isolated muscle preparations (Piras and Staneloni, 1969).

The rate of muscle glycogen depletion at a given work load is only altered to a minor extent by high or low initial muscle glycogen levels, provided an adaptation period with special diets does not precede the exercise (Bergström *et al.*, 1967). This is part of the explanation for the finding that the time subjects can perform a work load demanding 65–80% of their maximal oxygen uptake is linearly related to the initial value for the muscle glycogen content in the thigh muscles (Fig. 6). However, a higher work rate cannot be maintained for extended periods just by increasing the muscle glycogen stores (Karlsson and Saltin, 1971). The pace seems to be set primarily by the individual's maximal oxygen uptake and his technique (mechanical efficiency) in the type of work in question.

Experiments have been performed in order to study factors involved in the regulation of glycogen utilized during exercise. Hultman (1967b) tried glucose infusion, but this did not significantly alter the rate of muscle glycogen depletion if exercise was heavy. In lighter work, a longer work time was observed, and there was a tendency during the latter part of the work period for glycogen utilization to decline. This was the case both when the rate of glucose infusion was adjusted

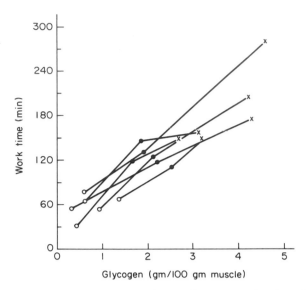

Fig. 6. Glycogen content obtained in the quadriceps muscle after different diets eaten during 3 days and the work time to exhaustion at a work load demanding ≈70% of the subject's maximal oxygen uptake. All diets provided 2800 kcal; ● = mixed diet; ○ = protein and fat diet, × = carbohydrate diet. (Saltin and Hermansen, 1967.)

so as to maintain blood glucose levels within a physiological range and when the rate was more than doubled.

Inhibition of the mobilization of FFA from adipose tissue by nicotinic acid during exercise leads to an increase in the utilization of muscle glycogen but no clear-cut effect on the subject's physical work capacity could be seen (Bergström *et al.*, 1969). Thus it was possible to modify to a certain extent the rate of muscle glycogen depletion during exercise by supplying the exercising muscle with an increased or reduced amount of extramuscular substrate. The changes were, however, not enough to significantly affect work performance.

2. The "Contractile Factor" versus Sympathetic Control Regulating Glycogenolysis

Glycogen is hydrolized to glucose-1-PO_4 by the enzyme phosphorylase. This enzyme exists in two forms—phosphorylase *b*, which is normally inactive except in the presence of elevated levels of 5-AMP, and phosphorylase *a*, which is active in the absence of 5-AMP.

Phosphorylase *b* can be converted to phosphorylase *a* by dimerization via the adenylcyclase system and associated enzymes. This process requires an energy input from ATP. One unit of phosphorylase *a* can be cleaved into two units of phosphorylase *b* by the enzyme phosphorylase phosphatase. An alternative method also exists for the conversion of the enzyme from the *b* to the *a* form. The enzyme phosphorylase kinase can be activated by an increased calcium level without a change in the activity of the protein kinase (Drummond *et al.*, 1969; Brostrom, *et al.*, 1971). The increase in calcium is of the same magnitude as that produced by muscular contraction. This mode of activation along with the increased activity of phosphorylase *b* due to excess 5-AMP is referred to as the contractile factor for stimulation of glycogenolysis.

The activation of glycogenolysis during exercise seems also to be regulated by the activity of the sympathetic nervous system. Experimental evidence shows that the catecholamines are potent activators of glycogenolysis in *in vitro* systems (Drummond *et al.*, 1968) and that the activity of the sympathetic nervous system increases during exercise as judged by elevated levels of catecholamines in the blood (Häggendahl *et al.*, 1970). In experiments on rats (Gollnick, 1972) an almost normal utilization of glycogen during exercise was observed in spite of blockade of the β-receptors with propranolol or destruction of the terminal nerve endings of the sympathetic nerves with 6-hydroxydopamine. Moreover in humans β-receptor blockade did not markedly influence the muscle glycogen breakdown during exercise (Harris *et al.*, 1971). This probably is an indication of the importance of the contractile factor for the activation of skeletal muscle phosphorylase. In this regard it has been shown that a nearly complete conversion of phosphorylase *b* to phosphorylase *a* can occur in the electrically stimulated muscle in as little as 3 seconds (Danforth *et al.*, 1962). This occurs without the formation of cyclic-AMP or conversion of the protein kinase to the active form. A direct activation of phosphorylase *b* without conversion to phosphorylase *a* also occurs when the ATP/AMP ratio declines. The reduction in ATP that occurs during heavy exercise as reported by Karlsson and Saltin (1970) could stimulate glycogenolysis. The direct stimulation of phosphorylase *b* has been demonstrated during muscular contraction in the phosphorylase-deficient mouse (Danforth and Lyon, 1964) and may be important in the isolated working rat heart (Øye, 1967). However, this mechanism for increasing glycogenolysis would seem to be relatively unimportant during exercise because the release of calcium can produce such a rapid and complete conversion of phosphorylase *b* to phosphorylase *a* (Drummond *et al.*, 1969; Brostrom *et al.*, 1971). On this basis it would appear that the contractile factor could easily account for the breakdown of glycogen during exercise.

B. Intramuscular Lipid Stores

1. MAGNITUDE

In the skeletal muscle, cell lipids are stored. In most animal species, half or more of the stored lipids consists of compounds (phospholipids, lecithin) which constitutes an important part of the cell structures. As an energy source, triglyceride could be of significance, but skeletal muscle does not contain large amounts of this substrate. On an average, 20 mg per gram of wet muscle of triglyceride may be available in the skeletal muscle (Masoro *et al.*, 1966; Fröberg *et al.*, 1971), which means a total of 80 gm at least in the skeletal muscle of the body.

2. USE OF INTRAMUSCULAR LIPIDS

Direct measurements of changes in the local lipid stores during exercise have given contradictionary results. Masoro *et al.* (1966) could not find any changes in the local stores of fat (triglyceride, phospholipids) which could account for the discrepancy between FFA uptake and estimated fat oxidation reported by other authors. He studied monkeys and stimulated (1–3 per second) the muscles for 5 hours. With human subjects, Hultman (1967a) and Morgan *et al.* (1971) were also unable to demonstrate a significant reduction in lipid stores of the muscle after prolonged exercise. Moreover, Hultman (1967a) pointed out the large variation in triglyceride content in muscle specimens obtained with the needle technique. Recently, however, Fröberg *et al.* (1971) did observe a decrease in the triglyceride content in healthy volunteers during exercise to exhaustion on a bicycle. The mean work time was 99 min. Muscle specimens were obtained from the lateral vastus of the femoral muscle by needle biopsies. The concentration of triglycerides decreased from 10.4 to 7.8 μmoles per gram of wet muscle on the average. This accounted for a contribution of 75% of total FFA metabolism during work. There is no information in this study on whether or not the muscle triglyceride depletion mainly takes place in the red fibers, as is the case in rats (Fröberg *et al.*, 1971). To the above discussions should be added that fat cells in close anatomical relation to the muscle may be able to supply FFA to muscle simply by diffusion (Havel *et al.*,

1963). This possibility, as well as the findings of Fröberg *et al.* (1971), deserve further examination.

C. Effect of Training

Subjects who are well trained and have a high aerobic power (expressed in milliliters per kilogram per minute) have a lower R, not only at the same absolute, but also at the same relative work loads as compared to subjects with low maximal oxygen uptakes (Christensen and Hansen, 1939). The best-trained subjects also exhibit a muscle glycogen depletion which is substantially smaller at the same absolute work load (Hermansen *et al.*, 1967). At the same relative work load, only minor differences are observed between trained and untrained subjects in the calculated amount of carbohydrates used. This is explained by the fact that the subjects with the highest maximal oxygen uptake perform at a higher absolute work load and oxygen uptake. This results in the intriguing situation that a fixed amount of carbohydrates is apparently combusted at a given relative work load regardless of training status (different maximal oxygen uptake in milliliters per kilogram per minute). Muscle glycogen depletion is also very similar at the same relative work load in trained and untrained subjects (Hermansen *et al.*, 1967). The major difference is a slightly higher rate of depletion in muscle glycogen due to a high lactate production during the first minutes of exercise in the untrained subjects at work loads demanding more than 50% of their maximal oxygen uptake.

The results described above are primarily based on cross-sectional studies of trained and untrained subjects. The same general type of response is obtained in subjects studied before and after several months of physical conditioning (Saltin and Karlsson, 1971). It may then be concluded that physical training, besides improving the subject's maximal oxygen uptake, also affects the muscle metabolism by increasing capacity to combust fat. A saving effect of the muscle glycogen stores although of quantitatively less importance, is also accomplished by a less marked lactate production at the beginning of submaximal exercise.

In recent studies, an increased mitochondrial volume has been observed as an effect of training (Kiessling *et al.*, 1971). Mitochondrial enzymes such as succinate dehydrogenase are also increased (Björntrop *et al.*, 1970). It is most likely that these changes occur mainly in red and intermediate forms of muscle fibers. In the trained state, the metabolic pattern also resembles that of these fibers. Even though the red and intermediate forms of muscle fibers may have a larger oxidative capacity after training, it has been questioned whether there are only quantitative differences in the red fibers that explain the metabolic pattern observed after training. In animals, an increased number of red or highly oxidative fibers has also been observed (Barnard *et al.*, 1970). However, no evidence of such changes is available in humans.

D. Summary

Large amounts of energy in the form of glycogen and lipids are stored in the skeletal muscles. Ample evidence is available which shows that muscle glycogen is utilized during exercise at all work intensities. At very low work loads, 0.2 mmoles of glucose per kilogram of muscle per minute are used. At work loads demanding 50–60% of maximal oxygen uptake, there is a fivefold increase in this amount, and at maximal bicycle work and maximal isometric contraction, 10–15 and 30–40 mmoles glucose units per kilogram of muscle per minute is utilized, respectively. The maximal rate of glucose oxidation amounts to 1.5–2.0 mmoles per minute per kilogram of wet muscle.

In earlier studies no reduction in the lipid store of the muscle was observed, but the most recent studies indicate that triglycerides may be consumed at a rate corresponding to half of the fat oxidation in the muscle. The phospholipids are not affected by exercise. The regulation of the utilization of intramuscular energy store will be discussed later.

VII. Substrate Consumption at Different Work Intensities and Duration

A. Structure, Recruitment, and Function of Muscle Fibers

As is evident from the foregoing discussion, hardly any published studies have taken into account all the available energy stores and determined their relative importance during work of different intensities and duration. Before an attempt is made to summarize the known data on the energy consumption of the skeletal muscles during different degrees of exercise, some other factors possibly involved in the metabolic adjustments that take place during work will be discussed. One of these factors is the relationship between structure and function of the muscle fiber and how different fibers of the muscle are recruited at different phases and intensities of the work.

Migrating species of different animals have red muscles rich in mitochondria, capillaries, and fat (Pette, 1971). Also, glycogen is stored and can be oxidized in the red fibers. However, fat is the only substrate stored in large quantities in migrating species before undertaking long journeys. Glycogen is stored with water (2–4 gm water per gram glycogen) (Puckett and Wiley, 1932) in liver and skeletal muscle, making it less suitable than fat as an energy store (Olsson and Saltin, 1970). Muscles in animals predominantly used for rapid escape, protection, or catching food are mainly of the white type structurally, which means a low content of myoglobin, mitochondria, and fat but rich in phosphagen (ATP + CP) and glycogen (Pette, 1971). Studies of fish muscles *in vitro* reveal a lower oxygen consumption and capacity to oxidize fat in white as compared to red fibers (Bilinski, 1963; M. S. Gordon, 1968).

Skeletal muscle is mixed (Edström and Nyström, 1969; Gollnick *et al.*, 1972), but there is considerable evidence that fibers within the same motor unit is structurally homogenous (Romanul and van der Meulen, 1967; Edström and Kugelberg, 1968). As demonstrated by Henneman *et al.*, 1965; Henneman and Olsson, 1965), it seems quite clear that the different sizes of the motoneurons in the spine have a functional significance. The motoneuron to type A muscle fibers, which in a histochemical classification are mainly white fibers (Stein and Padykula, 1962), have high thresholds and discharge phasically (Stubbs and Blanchaer, 1965; Kugelberg and Edström, 1968). Types B and C muscle fibers, with small motoneurons and lower thresholds, consist of two histochemically distinct types of red fibers. The C fibers are especially rich in fat. The intensity and duration of the neurogenic stimulus is then the determining factor for which motor units are activated. This may also mean that the general pattern for the metabolic response is settled.

In this connection, it may be worth mentioning that the term "submaximal work" can only be used when speaking of the muscle as a whole, since a single fiber contracts according to an "all or none" law. If the muscle fiber is stimulated, the splitting of ATP is under nearly all circumstances enough to activate it maximally. What one sees in submaximal exercise is that not all fibers are activated; there is a mixed population of fibers in the muscle, some are at rest and some are rapidly contracting (Davies, 1971).

The data presented above are mainly obtained in animals, and to what extent they can be applied to exercising man is an open question. This is especially true as it has been shown that although human muscles are mixed in regard to muscle fiber composition, only two distinctly different type of fibers exist (Gollnick *et al.*, 1972). One of these possesses high and the other low, myosin ATPase activity at alkaline pH (Guth and Yellin, 1971). According to Barnard et al (1970) a high myosin ATP-ase activity occurs in muscle fibers with fast twitch (FT) characteristics and a low activity in muscle fibers with slow twitch (ST) characteristics. As indicated by alfaglycerophosphate dehydrogenase activity, FT fibers always have a higher glycolytic capacity than ST fibers. In both fiber types an equal amount of glycogen can be found. The oxidative capacity, as apparent from DPNH-diaphorase activity, is usually lower in the FT fibers as compared to the ST fibers. However, a continuum of oxidative capacities exists in both fiber types depending at least partly on the training status of the muscles. This means that the oxidative capacity in FT fibers of endurance athletes may be higher than in ST fibers of untrained men.

If the results obtained on animals are combined with our present knowledge of human muscle fiber composition the following partly hypothetical model for the recruitment pattern of human skeletal motor units can be achieved. At very low work intensities with rhythmic monotonous exercise, i.e., pedaling a bicycle ergometer or walking on a treadmill, small motoneurons innervating ST fibers are activated. At increasing work loads, larger motor cells are also included, leading to the contraction of FT muscle fibers. In very sudden and nonrhythmic contraction a transient activation of FT fibers may occur also at low work rates.

At the onset of submaximal exercise, lactate formation is probable when the oxygen deficit becomes larger than 1.2 liters (Karlsson, 1971). This occurs in untrained men at work loads demanding 50–60% and in trained men at 60–70% of their maximal oxygen uptake (Saltin and Karlsson, 1971). It is usually accepted that the lactate production takes place in the glycogen-rich white fibers. In a state of relative hypoxia, as for example during the first minute of light exercise, nothing speaks against the possibility of a lactate formation in red fibers in which glycogen is also stored. In fact, lactate may also be formed at the start of the work at work loads lower than 50% of maximal oxygen uptake, but in such a case the amount is small and all of it may be oxidized within the muscle itself so that no significant accumulation of lactate in the muscle occurs (Knuttgen and Saltin, 1972) and blood lactate is not increased. In more prolonged periods of work, there is a continuous lactate production only at work loads above 70% of maximal oxygen uptake (Karlsson, 1971). This may be an indication of a more permanent recruitment of ST fibers.

The glycogen in muscle fibers may be depleted in prolonged exercise. However, these fibers can continue to contract, and FFA is thereafter the main substrate. Under these circumstances, a gradual decrease in R should be observed during exercise, which has also been long known to occur in man (Krogh and Lindhard, 1920; Christensen, 1932). At heavy work loads (maximal oxygen uptake >75%), very little or no reduction in R is observed during the exercise (Saltin and Hermansen, 1967).

We do not know to what extent the recruitment of motor units may circulate among similar types of fibers in the muscle during submaximal exercise (Grimby and Hannertz, 1968; Buchtal and Schmalbruch, 1970). If it occurs, it may then be part of the explanation as to why light work intensities can be sustained for very long periods of time. In prolonged heavy work, when the subjects are exhausted, they can always continue the exercise if work load is reduced (Pernow and Saltin, 1971). The explanation for this is still unclear. One explanation may be that a rather large number of glycogen dependent fibers (type A or FT) must be stimulated in order to develop the force that is needed in each contraction at very high work loads. When the glycogen is depleted in these particular fibers they are not able to contract any further, but exercise that can be performed with ST muscle fibers may be continued.

B. Metabolic Factors Limiting Exercise (Fig. 7)

1. Work Loads Demanding Less than 60% of Maximal Oxygen Uptake

R values indicate that 50–80% of the oxidized substrate comes from fat. This is supported by the findings on the turnover rate of plasma FFA as well as by the observed low rate of muscle glycogen depletion. No more than about 60% of the FFA oxidized is taken up from the blood. At very low work rates, the uptake of glucose from the blood may cover more than half of the carbohydrate oxidation, but even above 25% of maximal oxygen uptake, glycogen depletion provides most of glucose that is utilized. If the exercise continues for more than 1 hour, the importance of the extramuscular stores of substrates, especially plasma FFA, is enhanced and at least three-fourths of the energy may come from these sources. If the subjects are exhausted with central nervous system symptoms after many hours of work, a low blood glucose concentration is a common finding.

2. Work Loads Demanding between 60% and 90% of Maximal Oxygen Uptake

R data indicate that 50–80% of the calories that are combusted are derived from carbohydrates. This means that 2–4 gm of glycogen may

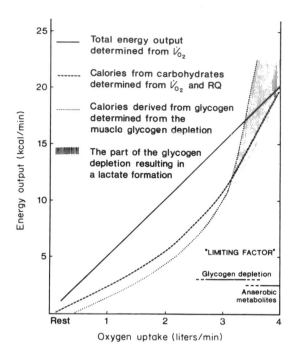

Fig. 7. Energy output and relative contribution of different substrates used during the first 10–20 min of work at different work intensities up to maximal exercise for a subject with a maximal oxygen uptake of 4 liters/min.

be utilized per minute. At least 80% of this is derived from the depletion of local muscle glycogen stores. After 1 hour of exercise, and especially during the latter part of the exercise, a greater contribution of carbohydrate from extramuscular sources may take place. The relative contribution of plasma FFA and FFA from local stores in the muscle to fat oxidation is not certain. It should be noticed that there is a certain delay in the FFA mobilization at these work intensities and that lactate is produced both at onset of work and more continuously during the exercise period. The lactate formed does not appear to be wasted, since it can be used both in the liver for the synthesis glucose and as a substrate in heart and skeletal muscle (Jorfeldt, 1970). Maximal work time at these work intensities varies from 30 to 300 minutes, and the point of exhaustion coincide with depleted glycogen stores if work time exceeds 1 hour.

3. Work Loads Demanding More than 90% of Maximal Oxygen Uptake

Some minor amounts of fat may be oxidized, but at these work intensities, carbohydrates are almost exclusively used as substrates. Since there is no, or a very minor, uptake of glucose from the blood, almost all of the energy expenditure is derived from the glycogen locally stored in the muscle. At exhaustion, which usually occurs within 30 min of exercise, the muscle glycogen stores are not depleted. In this situation, however, the accumulation of anaerobic metabolites may be a limiting factor (Karlsson, 1971).

REFERENCES

Abrahams, V. C., Hilton, S. M., and Zbrozyna, (1964). *J. Physiol. (London)* **171**, 189.

Adolfsson, S., and Ahrén, K. (1971). *In* "Muscle Metabolism during Exercise" (B. Pernow and B. Saltin, eds.), pp. 257–272. Plenum Press, New York.

Aronovsky, E., Levari, R., Kornbleuth, W., and Wertheimer, E. (1963), **2**, 259.

Åstrand, I. (1960). *Acta Physiol. Scand.* **419**, Suppl. 169.

Åstrand, P-O., and Christensen, E. H. (1964). *In* "Oxygen in the Animal Organism" (F. Dickens, E. Neil, and W. F. Widdes, eds.), pp. 295–301. Pergamon, Oxford.

Åstrand, P-O., and Rodahl, K. (1970). *In* "Textbook of Work Physiology," McGraw-Hill, New York.

Åstrand, P-O., Hallbäck, I., Hedman, R., and Saltin, B. (1963). *J. Appl. Physiol.* **18**, 619.

Ballard, K., and Rosell, S. (1969). *Acta Physiol. Scand.* **77**, 442.

Ballard, K., and Rosell, S. (1971). *Circ. Res.* **28**, 389.

Ballard, K., Cobb, C. A., and Rosell, S. (1971). *Acta Physiol. scand.* **81**, 246.

Barnard, R. J., Edgeton, V. R., and Peter, J. B. (1970). *J. Appl. Physiol.* **28**, 762.

Basu, A., Passmore, R., and Strong, J. A. (1960). *Quart. J. Exp. Physiol.* **45**, 312.

Bergström, J. (1962). *Scand. J. Clin. Lab. Invest.* **14**, Suppl. 68.

Bergström, J., Hermansen, L., Hultman, E., and Saltin, B. (1967). *Acta Physiol. Scand.* **71**, 140.

Bergström, J., Hultman, E., Jorfeldt, L., Pernow, B., and Wahren, J. (1969). *J. Appl. Physiol.* **26**, 170.

Bergström, J., Harris, R. C., Hultman, E., and Nordesjö, L-O. (1971). *In* "Muscle Metabolism During Exercise" (B. Pernow and B. Saltin, eds.), pp. 341–356. Plenum Press, New York.

Bilinski, E. (1963). *Can. J. Biochem. Physiol.* **41**, 107.

Björntorp, P. (1971). *In* "Muscle Metabolism During Exercise" (B. Pernow and B. Saltin, eds.), pp. 493–504. Plenum Press, New York.

Björntorp, P., Fahlén, M., Holm, J., Scherstén, T., and Szostak, V. (1970). Scand. J. Clin. Lab. Invest. **26**, 145.

Bolme, P., and Novotny, J. (1969). *Acta Physiol. Scand.* **77**, 58.

Brostrom, C. D., Hunkeler, F. L., and Krebs, E. G. (1971). *J. Biol. Chem.* **246**, 1961.

Buchtal, F., and Schmalbruch, H. (1970). *Acta Physiol. Scand.* **79**, 435.

Butcher, R. W. (1970). *In* "Adipose Tissue. Regulation and Metabolic Functions," (B. Jeanrenaud, D. Hepp, eds.) p. 5. Thieme, Stuttgart.

Carlsson, L. A., and Pernow, B. (1961). *J. Lab. Clin. Med.* **58**, 673.

Chapler, C. K., and Stainsby, W. N. (1968). *Amer. J. Physiol.* **215**, 995.

Christensen, E. H. (1931). *Arbeitsphysiologie* **4**, 128.

Christensen, E. H. (1932). *Arbeitsphysiologie* **5**, 463.

Christensen, E. H., and Hansen, O. (1939). *Skand. Arch. Physiol.* **81**, 137.

Cochran, B., Jr., Marbach, E. P., Poucher, R., Steinberg, T., and Gwinnp, G. (1966). *Diabetes* **15**, 838.

Costin, J. C., Saltin, B., Skinner, N. S., and Vastagh, G. (1971). *Acta Physiol. Scand.* **81**, 124.

Danforth, W. D., Helmreich, E., and Cori, C. F. (1962). *Proc. Nat. Acad. Sci.* **48**, 1191.

Danforth, W. H., and Lyon, J. B. (1964). *J. Biol. Chem.* **239**, 4047.

Davies, R. E. (1971). *In* "Muscle Metabolism During Exercise" (B. Pernow and B. Saltin, eds.), p. 551. Plenum Press, New York.

Dole, V. P. (1956). *J. Clin. Invest.* **35**, 150.

Drummond, G. I., Harwood, J. P., and Powell, C. A. (1969). *J. Biol. Chem.* **244**, 4235.

Edström, L., and Kugelberg, E. (1968). *J. Neurol., Neurosurg. Psychiat.* **31**, 424.

Edström, L., and Nyström, B. (1969). *Acta Neurol. Scand.* **45**, 257.

Eliasson, S., Folkow, B., Lindgren, P., and Uvnäs, B. (1951). *Acta Physiol. Scand.* **23**, 333.

Engel, H. R., Bergenstal, D. M., Nixon, W. E., and Patten, J. M. (1959). *Proc. Soc. Exp. Biol. Med.* **100**, 699.

Fain, J. N., Kovacev, V. P., and Scow, R. O. (1965). *J. Biol. Chem.* **240**, 3522.

Felig, P., and Wahren, J. (1971). *In* "Muscle Metabolism During Exercise" (B. Pernow and B. Saltin, eds.), pp. 205–214. Plenum Press, New York.

Felig, P., Pozefsky, T., Marliss, E., and Cahill, G. F., Jr. (1970). *Science* **167**, 1003.

Folkow, B. (1952). *Acta Physiol. Scand.* **25**, 49.

Folkow, B., Häggendahl, J., and Lisander, B. (1967). *Acta Physiol. Scand., Suppl.* **307**.

Fredholm, B. B. (1970). *Acta Physiol. Scand., Suppl.* **354**, 1.

Fredholm, B. B. (1971). *Acta Physiol. Scand.* **81**, 110.

Fredholm, B. B., and Rosell, S. (1968). *J. Pharmacol. Exp. Ther.* **159**, 1.

Fröberg, S. O., Carlsson, L. A., and Ekelund, L-G. (1971). *In* "Muscle Metabolism During Exercise" (B. Pernow and B. Saltin, eds.), pp. 307–314. Plenum Press, New York.

Goldstein, M. S. (1965). (E. S. Leibel and G. A. Wrenshall, eds.), *In* "On the Nature and Treatment of Diabetes" Int. Congr. Ser. No. 84, p. 308, Excerpta Med. Found., Amsterdam.

Gollnick, P. D. (1973). *In* "Limiting Factors of Physical Performance" (J. Keul, ed.). Georg Thieme, Stuttgart.

Gollnick, P. D., Amstrong, R. B., Piehl, K., Saltin, B., and Saubert, C. W. IV (1972). *J. Appl. Physiol.* **33**, 312.

Goodman, H. M. (1968). *Ann. N.Y. Acad. Sci.* **148**, 419.

Gordon, M. S. (1968). *Science* **159**, 87.

Gordon, R. S., Jr., and Cherkes, A. (1956). *J. Clin. Invest.* **35**, 206.

Greenwood, F. C., and Landon, J. (1966). *Nature (London)* **210**, 540.

Grimby, G., and Hannertz, J. (1968). *J. Neurol., Neurosurg. Psychiat.* **31**, 565.

Grimby, G., Häggendahl, E., and Saltin, B. (1967). *J. Appl. Physiol.* **22**, 305.

Grimby, G., Wilhelmsen, L., Björntorp, P., Saltin, B., and Tibblin, G. (1971). *In* "Muscle Metabolism During Exercise" (B. Pernow and B. Saltin, eds.), pp. 469–481. Plenum Press, New York.

Guth, L., and Yellin, H. (1971). *Exp. Neurol.* **31**, 277.

Hagenfeldt, L., and Wahren, J. (1968). *Scand. J. Clin. Lab. Invest.* **21**, 314.

Häggendahl, J., Hartley, L. H., and Saltin, B. (1970). *Scand. J. Clin. Lab Invest.* **26**, 337.

Hartley, L. H., Pernow, B., Häggendahl, J., Lacour, J., De Lattre, J., and Saltin, B. (1970). *J. Appl. Physiol.* **29**, 818.

Havel, R. J. (1965). *In* "Handbook of Physiology" (Amer. Physiol. Soc., J. Field, ed.), Sect. 5, p. 575. Williams & Wilkins, Baltimore, Maryland.

Havel, R. J., Naimark, A., and Borchgewink, C. F. (1963). *J. Clin. Invest.* **42**, 1054.

Havel, R. J., Carlson, L. A., Ekelund, L-G., and Holmgren, A. (1964). *J. Appl. Physiol.* **19**, 613.

Havel, R. J., Pernow, B., and Jones, N. L. (1967). *J. Appl. Physiol.* **23**, 90.

Hedman, R. (1957). *Acta Physiol. Scand.* **40**, 305.

Henneman, E., and Olsson, C. B. (1965). *J. Neurophysiol.* **28**, 581.

Henneman, E., Somjen, G., and Carpentier, D. O. (1965). *J. Neurophysiol.* **28**, 560.

Hermansen, L. (1969). *Med. Sci. Sports* **1**, 32.

Hermansen, L., Hultman, E., and Saltin, B. (1967). *Acta Physiol. Scand.* **75**, 129.

Hermansen, L., Pruett, E. D., Osenes, J. B., and Giere, F. A. (1970). *J. Appl. Physiol.* **29**, 13.

Hertting, G., and Axelrod, J. (1961). *Nature* (*London*) **189**, 66.

Holloszy, J. O., Skinner, J. S., Toro, G., and Cureton, T. K. (1964). *Amer. J. Cardiol.* **14**, 753.

Hultman, E. (1967a). *Scand. J. Clin. Lab. Invest.* **19**, Suppl. 94.

Hultman, E. (1967b). *Circ. Res.* **20**, Suppl. 1.

Hultman, E., and H:son Nilsson, L. (1971). *In* "Muscle Metabolism During Exercise" (B. Pernow and B. Saltin, eds.), pp. 143–152. Plenum Press, New York.

Hultman, E., Bergström, J., and Roch-Norlund, A. E. (1971). *In* "Muscle Metabolism During Exercise" (B. Pernow and B. Saltin, eds.), pp. 273–288. Plenum Press. New York.

Hunter, W. M., and Sukhar, M. Y. (1968). *J. Physiol.* (*London*) **196**, 110P.

Hunter, W. M., Fonseka, C. C., and Passmore, R. (1965a). *Quart. J. Exp. Physiol.* **50**, 406.

Hunter, W. M., Fonseka, C. C., and Passmore, R. (1965b). *Science* **150**, 1051.

Issekutz, B., Jr., and Miller, H. J. (1962). *Proc. Soc. Exp. Biol. Med.* **110**, 237.

Issekutz, B., Jr., Miller, H. J., Paul, P., and Rodahl, K. (1964). *Amer. J. Physiol.* **207**, 583.

Issekutz, B., Jr., Miller, H., Paul, P., and Rodahl, K. (1965). *Amer. J. Physiol.* **209**, 1137.

Jorfeldt, L. (1970). *Acta Physiol. Scand.* **78**, Suppl. 338.

Karlsson, J. (1971). *Acta Physiol. Scand., Suppl.* **358**.

Karlsson, J., and Saltin, B. (1971). *J. Appl. Physiol.* (in press).

Karlsson, J., Diamant, B., and Saltin, B. (1970). *Scand. J. Clin. Lab Invest.* **26**, 385.

Karlsson, J., Rosell, S., and Saltin, B. (1972). *Pflügers Arch.* **331**, 57.

Kiessling, K. H., Piehl, K., and Lundqvist, C. G. (1971). *In* "Muscle Metabolism During Exercise" (B. Pernow and B. Saltin, eds.), pp. 97–102. Plenum Press. New York.

Knuttgen, H., and Saltin, B. (1972). *J. Appl. Phyisol.* (in press).

Krogh, A., and Lindhard, J. (1920). *Biochem. J.* 14, 290.

Kugelberg, E., and Edström, L. (1968). *J. Neurol., Neurosurg. Psychiat.* 31, 405.

Larner, J., Villar-Palasi, C., Goldberg, N. D., Bishop, J. S., Huiging, F., Wenger, J. I., Sasko, H., and Brown, W. E. (1969). *In* "Progress in Endocrinology" (C. Gurl, ed.), pp. 135–147. Excerpta Med. Found., Amsterdam.

Lisander, B. (1970). *Acta Physiol. Scand., Suppl.* 351.

Luft, R., Cerasi, E., Madison, L. L., von Euler, U. S., Della Casa, L., and Roovete, A. (1966). *Lancet* 2, 254.

Margaria, R., and Foa, P. (1939). *Arbeitsphysiologie* 10, 553.

Margolis, S., and Vaughan, M. (1962). *J. Biol. Chem.* 237, 44.

Masoro, E. J., Rowell, L. B., McDonald, R. M., and Steiert, B. (1966). *J. Biol. Chem.* 241, 2626.

Morgan, T. E., Cobb, L. A., Short, F. A., Ross, R., and Gunn, D. R. (1971). *In* "Muscle Metabolism During Exercise" (B. Pernow and B. Saltin, eds.), pp. 87–96. Plenum Press, New York.

Muir, G. G., Chamberlain, D. A., and Pedoe, D. T. (1964). *Lancet* 2, 930.

Natrass, F. J. (1968). *In* "Research in Muscular Dystrophy," p. 11. Pitman, London.

Ngai, S. H., Rosell, S., and Wallenberg, L. (1966). *Acta Physiol. Scand.* 68, 397.

Olsson, K-E., and Saltin, B. (1970). *Acta Physiol. Scand.* 80, 11.

Orö, L., and Fröberg, S. (1964). *Acta Med. Scand.* 176, 65.

Orö, L., Wallenberg, L. R., and Bolme, P. (1965). *Acta Med. Scand.* 178, 697.

Øye, I. (1967). *Acta Physiol. Scand.* 70, 229.

Paul, P. (1970). *J. Appl. Physiol.* 28, 127.

Paul, P. (1971). *In* "Muscle Metabolism During Exercise," (B. Pernow and B. Saltin eds.) pp 225–247. Plenum Press, New York.

Pearson, O. H., Dominiquez, J. M., Greenberg, E., Pazianos, E., and Pette, D. (1971). *In* "Muscle Metabolism During Exercise" (B. Pernow and B. Saltin, eds.), pp. 33–50. Plenum Press, New York.

Ray, B. S. (1960). *Trans. Ass. Amer. Physicians* 73, 217.

v. Pettenkofer, and Voit, C. (1866). *Z. Biologie,* II Band. pp. 459–593. München. Oldenburg.

Pernow, B., and Saltin, B. (1971). *J. Appl. Physiol.* 31, 416.

Pilkington, T. R. E., Lowe, R. D., Robinson, B. F., and Titterington, E. (1962). *Lancet* 22, 216.

Piras, R., and Staneloni, R. (1968). *Biochemistry* 8, 2153.

Porte, D., Graber, A., Kuzuya, T., and Williams, R. H. (1966). *J. clin. Invest.* 45, 228.

Pruett, E. D. R. (1970a). *J. Appl. Physiol.* 28, 199.

Pruett, E. D. R. (1970b). *J. Appl. Physiol.* 29, 155.

Pruett, E. D. R. (1971). *In* "Muscle Metabolism during Exercise" (B. Pernow and B. Saltin, eds.), pp. 165–176. Plenum Press, New York.

Purkett, H. L., and Wiley, F. H. (1932). *J. biol. Chem.* 96, 367.

Rabinovitz, D., and Zierler, K. L. (1962). *J. clin. Invest.* 41, 2191.

Rasio, E., Mallaise, W., Franckson, J. R., and Conrad, V. (1966). *Arch. Intern. Pharmacodyn.* 160, 485.

Reeves, T. J., and Sheffield, L. T. (1971). *In* "Physical Fitness and Coronary

Heart Disease (Andrèe-Larsen and Malmberg, eds.), pp. 209–216. Munksgaard, Copenhagen.

Rizack, M. (1965). In "Handbook of Physiology" (Amer. Physiol. Soc., J. Field, ed.), Sect. 5, p. 5. Williams & Wilkins, Baltimore, Maryland.

Robinson, J., and Newsholme, E. A. (1967). Biochem. J. 104, 2C.

Rohmert, W. (1960). Int. Z. Angew. Physiol. Einschl. Arbeitsphysiol. 18, 123.

Romanul, F. C. A., and van der Meulen, J. P. (1967). Arch. Neurol. (Chicago) 17, 387.

Rosell, S. (1966). Acta Physiol. Scand. 67, 343.

Roth, J., Glick, S. M., Yalow, R. S., and Berson, S. A. (1963a). Science 140, 987.

Roth, J., Glick, S. M., Berson, S. A., and Yalow, R. S. (1963b). Metab., Clin. Exp. 12, 577.

Rowell, L. B. (1971). In "Muscle Metabolism During Exercise" (B. Pernow and B. Saltin, eds.), pp. 127–142. Plenum Press, New York.

Rowell, L. B., Masoro, E. J., and Spencer, M. J. (1965). J. Appl. Physiol. 20, 1032.

Rowell, L. B., Kranig, K. K., III, Evans, T. O., Kennedy, J. W., Blackman, J. R., and Kusami, F. (1966). J. Appl. Physiol. 21, 1773.

Rowell, L. B., Brengelmann, G. L., Blackman, J. R., Twiss, R. P., and Kusumi, F. (1968). J. Appl. Physiol. 24, 475.

Rudman, D. (1963). J. Lipid Res. 4, 119.

Rudman, D., Di Girolamo, M., and Garcia, L. A. (1965). In "Handbook of Physiology" (Amer. Physiol. Soc., J. Field, ed.), Sect. 5, pp. 533–539. Williams & Wilkins, Baltimore, Maryland.

Rushmer, R. F., Franklin, D. L., Van Citters, R. L., and Smith, O. A. (1961). Circ. Res. 9, 675.

Saltin, B. (1971). Scand. Rehab. Med. 3, 39.

Saltin, B., and Åstrand, P-O. (1967). J. Appl. Physiol. 23, 353.

Saltin, B., and Hermansen, L. (1967). In "Nutrition and Physical Activity" (G. Blix, ed.), pp. 32–46. Almqvist & Wiksell, Stockholm.

Saltin, B., and Karlsson, J. (1971). In "Muscle Metabolism During Exercise" (B. Pernow and B. Saltin, eds.), pp. 395–400. Plenum Press, New York.

Saltin, B., Gagge, A. P., and Stolwijk, J. A. J. (1968). J. Appl. Physiol. 25, 679.

Schatch, D. D. (1967). J. Lab. Clin. Med. 69, 256.

Scow, R. O., and Chernick, S. S. (1970). Compr. Biochem. 18, 19–49.

Stein, J. M., and Padykula, H. A. (1962). Amer. J. Anat. 110, 103.

Steinberg, D., and Vaughan, M. (1965). In "Handbook of Physiology" (Amer. Physiol. Soc., J. Field, ed.), Sect. 5, pp. 335–347. Williams & Wilkins, Baltimore, Maryland.

Stubbs, S., and Blanchaer, M. C. (1965). Can. J. Biochem. 43, 463.

Uvnäs, B. (1960). In "Handbook of Physiology" (Amer. Physiol. Soc., J. Field, ed.), Sect. 1, pp. 1131–1162. Williams & Wilkins, Baltimore, Maryland.

Vendsalu, A. (1960). Acta Physiol. Scand. 49, Suppl. 173.

Wahren, J., Ahlborg, G., Felig, P., and Jorfeldt, L. (1971). In "Muscle Metabolism During Exercise" (B. Pernow and B. Saltin, eds.), pp. 189–204. Plenum Press, New York.

Weiss, S. (1871). Sitzungsber. Akad. Wiss. Wien 64, 1.

Zahnd, G. R., Steinke, J., and Renold, A. E. (1960). Proc. Soc. Exp. Biol. Med. 105, 455.

THE CONTROL OF MUSCULAR ACTIVITY BY THE CENTRAL NERVOUS SYSTEM

MARION HINES

I. Introduction

Reluctance to credit the central nervous system with control of skeletal musculature marked the attitude of the public and of physicians for centuries. Severe head injuries were not correlated with ensuing change in the use of skeletal muscle. For example, during the siege of the Castle of Birr, the Earl of Kildare received a bullet wound which produced partial paralysis and difficulty with speech. He was called to England by the ministers of Henry VIII to defend himself against an accusation of working with the Catholics in Ireland. In contrast to his previous fluent defense before Wolsey, his speech was slow and hesitating. This hesitation was interpreted as guilt rather than attributed to aphasia. Later, even Charcot met with skepticism when he presented his correlations of changes in the use of skeletal muscle with lesions of the central nervous system. The Royal Society denied Marshall Hall the privilege of publication of his later studies upon the central nervous system of small reptiles and amphibians, because they were reluctant to accept the presence of the central nervous system as indispensable for a motor reponse to a sensory stimulus.

On the other hand, Magendie's experiment met with no such antagonism. That the dorsal roots "were destined for sensation" and that the ventral roots carried motor impulses easily became a part of the body of neurological fact. It was many years (1860) before Brongeest showed that the hind legs of the frog became limp and toneless after the dorsal roots of the lumbar plexus were severed homolaterally. It remained for Sherrington, however, to demonstrate conclusively that the nerve fibers in the dorsal roots, which maintained tone in skeletal muscle, were distributed with the motor nerves which innervated skeletal muscle.

The spastic state which accompanied hemiplegia in man was not recognized as such until the 1870s. The facets of this state include brisk and irradiating tendon reflexes, increase in resistance to passive movement of the clasp-knife type, and clonus. Spastic hemiplegia in man, most frequently the result of hemorrhage into the internal capsule, left at death only one lesion in the spinal cord. This lesion was confined to a loss of nerve fibers in the dorsal half of the lateral funiculus and in the ventral funiculus, the sites of the pyramidal tract. Thus it was that the spasticity and the paralysis were assigned to the loss of a single system.

Although Hughlings Jackson did not allocate this dual change in skeletal muscle following such severe injuries to particular fiber tracts within the central nervous system, he did (1932) consider that "the process of dissolution is not only a 'taking off' of the higher but at the same time a 'letting go' of the lower." Thus not only were certain normal functions lost, but also other functions were released, due not to irritation by pathological processes, but rather to loss of control over lower levels of the central nervous system by its higher ones.

Therefore, in spastic hemiplegia in man, the functions lost would be the skilled movements of the muscles of the extremities contralateral to the lesion and those which were "let go" (the phenomena of release, Walshe, 1929), the hypertonus, the briskness of the tendon reflexes, the exaggeration of reflexes of labyrinthian origin, associated movements, and the positive Babinski. In spite of Hughlings Jackson's suggestion, these two divergent results continued to be interpreted as the sequelae to injury of or loss of a single corticifugal system.

The single corticifugal system was described as stemming from large pyramidal cells (Holmes and May, 1909) in the V layer of Brodmann's area 4 (1903) and as penetrating the spinal cord, thus forming the tractus corticospinalis. The demonstration that this corticifugal system was interrupted in paralysis of cortical or of capsular origin required years of correlation of clinical and pathological findings (Charcot and Pitres, 1895). The modification of that correlation has been equally difficult. Other origins for this system, such as the postcentral gyrus (Vogts and Sachs, see Foerster, 1923) and the pyramidal cells in the III layer of area 4 (Spielmeyer, 1906) were long ignored, just as were other corticofugal systems such as the corticopontile (Dejerine, 1901), the corticothalamic (Dejerine, 1901; Vogt, see Foerster, 1923), the corticorubral, and the corticonigral (von Monakow, 1909, 1910). Indeed, the motor cortex became and remained limited for many years to the precentral gyrus in spite of Ferrier's 1876 record of motor activity obtained by electrical stimulation of the parietal, temporal and occipital lobes.

If the newer implications of the control of tone had been accepted by the majority of clinical neurologists, Bucy's impassioned appeal (1957) for discard of the allocation of spastic paralysis to "an upper neuron lesion" would have been unnecessary. Neurologists are after all as conservative as the tissue they study. In this study, therefore, the phasic and tonic activity of the control of skeletal muscle will be related to the fiber tracts and their cells of origin, not only within the central nervous system, but in the peripheral as well.

II. Normal Use of Skeletal Muscle by the Primate

The ability to control the contractions of skeletal muscle that characterizes any particular animal depends upon many diverse nerve fibers within the peripheral and central nervous systems. This control is of course initially dependent upon both the sensory and motor innervation of these muscles. The sensory innervation of skeletal muscles in the mammal is found upon specialized muscles, the intrafusal fibers of muscle spindles, upon the tendons, and about the tendinous attachments of the extrafusal fibers. The motor innervation of muscle fibers attached to bones in mammals is generally simple; a single terminal is found upon each muscle fiber. These terminals and the muscle fibers innervated by them are grouped as units. Such a unit is made up of the total number of muscle fibers innervated by the terminal branchings of one motor cell body within the central nervous system. No one knows whether this unit of muscle fibers is contained within one single muscle bundle or distributed among several anatomically discrete bundles. The control of the use of skeletal muscle is mediated, therefore, by way of single terminals upon individual muscle fibers acting as a group.

The control of the use of this musculature presents many facets. Some of these facets are generalized and characterize a genus; others are more specialized and distinguish species and subspecies. The domestic cat uses its musculature very similarly to that of other members of the feline family. That use, although comparable, is different from that made by dogs. Great variation delineates the various members of the primate family. Consequently, the results of studies of the cat or the dog cannot be transferred to the rabbit or to the monkey. And certainly, although suggestive, the analysis of the control of skeletal musculature characteristic of the macaque cannot be transferred to man.

The primate shares with all other forms the maintenance of posture and necessity for progression. The primate is differentiated from other mammals by the ability to use the upper extremity for the manipulation of objects; and man from other primates by his bipedal posture and bipedal progression. In man, the upright position is maintained not by an excess of tone in the extensors but rather by a nice balance between contraction of extensor and flexor sheets in the lower extremity—with a certain emphasis on the isotonic contraction of the anatomical flexors. There is fixation at both the pectoral and pelvic girdles, together with contraction of the M. rectus abdominalis. Bipedal progression is in one sense quadrupedal, for as one leg is protracted, the opposite arm is also protracted. Subsequent retraction of the lower extremity is accompanied by retraction of the contralateral upper extremity. Again, the girdles are fixed so that they do not twist on the trunk. The young monkey (3 months to 1 year of age) frequently uses bipedal progression. He fixes the pectoral girdle either by elevation of his arms or by the maintenance of simultaneous protraction. In these positions, the upper extremity fails to mimic quadrupedal progression. After a year of age, the macaque rarely progresses except in the usual quadrupedal manner.

Two modes of organization distinguish the use of skeletal muscle, namely, mass organization and discrete organization (Tower, 1940). In the former, the innervation begins in the proximal muscles and flows down the limb to involve successively more distally lying muscles (Foerster, 1936a). This mode of innervation in the intact primate is preceded by some degree of fixation of the muscles of the girdles and of those which hold the extremity to the girdle. Discrete organization is distinguished by innervation of distally lying muscles, which, as the movement progresses, may either include more proximally lying muscles in active contraction or be accompanied by fixation of more proximal muscle groups, including those of the girdle to which the extremity is attached.

The discrete use of skeletal muscle, directed toward an object desired by the primate, is the complex resultant of interdependent activity of many muscle groups employed in definite and distinct ways, capable of partial analysis. Discrete control of skeletal muscle requires ease of initiation of the contraction, a grading of contraction (either simultaneous or consecutive), the ability to stop this contraction at any phase of its execution, and the ability to start again in the same phase or to redirect the aim of the whole movement. An enduring substratum of posture is a prerequisite of such activity, for posture must be maintained in an easy and natural way to free hand and upper extremity for the examination of objects or the manipulation of tools, and to anticipate by adjustment of the muscles of the trunk and girdles the next stage of directed movements. The variants in cooperation which make skilled movements successful are fixation of proximal musculature and increment and decrement of tone in muscles of the trunk, in those muscles which attach the extremities to the girdles as well as in those which lie proximal to the actively contracting muscle groups. In the attainment of skilled muscular activity, these facets of use of skeletal muscle of the trunk or girdles are as important as the phasic activity of the muscles of the digits of the hand.

Furthermore, skilled movements require participation of both upper extremities, the one which leads and the other which cooperates. The activity of the muscles of the leading extremity is initiated distally, that of the cooperating extremity, proximally. The manner of use of the muscles within each of the two extremities is dissimilar. In Beevor's (1903) analysis, movement by the leading extremity is initiated by prime movers, aided in their contraction (1) by the synergists or cooperating muscle groups and (2) by the antagonists (which in this type of activity are more frequently slightly contracted than completely relaxed), as well as (3) by the fixation of more proximally lying muscle groups. On the other hand, the contraction of the musculature of the cooperating extremity may move down the arm involving in its passage either extensor of flexor sheets of muscle. These synergistic activities may be stopped, reversed, or fixed in any stage of their activity, receiving either an increment or decrement of tone.

How does the central nervous system manage to accomplish this beautiful use of skeletal muscle in the primate? Because all experimental data derived from man is dependent upon chance vascular accidents or upon the intervention of neural surgery, it is necessary to utilize such primates as the macaque (the apes are too expensive) to analyze the relation of different fiber tracts to the use of skeletal musculature by the primate. Nonetheless, when the results of comparable lesions are considered, those which follow experimental ablations of the central nervous system of the *Macaca mulatta* suggest a similar relationship of results to site of injury as in man. Besides the method of ablation, which allocates to the part removed the loss apparent in the resulting change in the use of the primate's muscles, there are two other methods. Surfaces of the cortex cerebri or cerebelli can be stimulated under anesthesia in the macaque, without anesthesia in man. Deep structures in animals can be stimulated and the position of the buried electrode subsequently determined. Electrical recordings may be made from the surface of the cortices or from deeper structures. Our understanding of the functional contribution of different parts of the central nervous system to the control of the use of skeletal muscle in animals is dependent upon our interpretation of the results of these methods of study.

III. Segmental Organization

Segmental organization of the central nervous system is more theoretical than real; this is so because organization of the spinal cord into levels or segments is superimposed upon the medulla spinalis by the enveloping bony covering of the vertebra. No single segment or level is able to function alone. Skeletal muscles are innervated both afferently and efferently by spinal roots from several levels. Within the central gray matter, motor cells are grouped as nuclei which innervate particular muscles and sensory cells are grouped as nuclei related to different peripheral nerve fibers carrying different varieties of sensory impulses. Other groups of cells, designated as adjustor nuclei, occur and are characteristic of several segments of the spinal cord. Each dorsal root contains nerve fibers which innervate sensory endings in the skin and subcutaneous connective tissue as well as those found in striated muscles;

in some dorsal roots (thoracic and upper lumbar), visceral afferents are added. Each ventral root contains large efferent nerve fibers (α fibers) innervating skeletal muscles and small nerve fibers (γ fibers) terminating upon the intrafusal fibers of muscle spindles. To these may be added in some ventral roots the visceral efferents of the autonomic nervous system.

The control of the use of skeletal muscles of the body, as opposed to those of the head, is exercised through arrangements of neurons found within the central gray matter of the spinal cord. These arrangements of neurons, which form one of the fundamental physiological mechanisms, are designated by the term reciprocal innervation.

Sherrington's reciprocal innervation has been given a new form by the results of intracellular recording of postsynaptic potentials by Eccles and his collaborators (Eccles, 1957). A single nerve fiber from sensory endings on muscle spindles divides within the central gray matter, sending a lateral branch to synapse with the motor cells innervating the M. quadriceps femoris and a medial branch to synapse with cells which Eccles believes to lie in Ramóny Cajal's (1909) intermediate nucleus. The axons of the cells of this nucleus which lie in this and in more caudal levels synapse with motor cells which innervate the M. biceps semimembranosis. The monosynaptic pathway is interpreted as excitatory, the disynaptic as inhibitory. The central distribution of the nerve fibers from Golgi endings on tendons and on the insertions of extrafusal muscle fibers differ from that of the sensory ending of the muscle spindle by the intercalation of a neuron in the former pathway.

These two proprioceptive endings have recently exchanged their previously allotted functions. The Golgi endings present a high threshold to stretch and their discharge increases during active contraction (Hunt and Kuffler, 1951b). On the other hand, the spindle afferents have a low threshold to stretch and their discharge rate decreases or ceases during active contraction. The sensory endings upon the intrafusal fibers are activated by contraction, elicited by stimulation of their polar or motor endings via small nerve fibers which leave the ventral root with the larger nerve fibers destined for the extrafusal muscle fibers (cat; Hunt and Kuffler, 1951a). This arrangement allows the activation of the sensory ending of the spindle to be controlled centrally as well as independently of the isotonic contraction of extrafusal muscle fibers. The "small nerve reflex" can be regulated not only by stretch upon the muscle itself (Hunt, 1952), but also by stimulation of touch, pressure and pain receptors in the skin and subcutaneous connective tissue (Hunt, 1951).

The interaction between these two proprioceptive organs is able to produce some regulation between antagonistic muscles of the same level. For, as the tension rises within a muscle in the process of contraction, the activity of the tendon endings increases and the spindle activity decreases. The motor neurons which by chance have not been activated become less excitable. The monosynaptic response from the spindle is inhibited and the antagonist response is facilitated. The opposing muscle is gradually stetched; the spindle afferents then excite the motor neurons to that muscle, the synergists are facilitated and the antagonists are inhibited.

These studies have implicated (Eccles, 1957) the intermediate nucleus of Cajal as the inhibitory nucleus in single reciprocal innervation and suggest that the maintenance of tone, assigned for years to nerve impulses entering the spinal cord via the dorsal root, is secondary to the excitation of the cellbodies (as yet unidentified) of the small nerve fibers innervating the polar ending on intrafusal muscle fibers.

Ever since Brongeest found that the skeletal muscles of the hind legs of a frog lost their tone after the dorsal roots of the lumbosacral plexus were cut and Sherrington's follow-up with deafferentation of the muscles of the upper extremity of a monkey, tone in a particular skeletal muscle has been considered to be maintained at the segmental level by the dorsal roots which supply it. If, however, the dorsal roots of L_3 to L_5, which supply the sensory innervation to the M. quadriceps femoris of the macaque are severed, the palpable loss of tone will not be as great as if the dorsal roots of L_3 and L_7 are severed. Indeed, surgical division of these two roots is followed by a great decrement of tone in the whole extremity on the side of the lesion. The dorsal root of L_6 contains more of these facilitatory nerve fibers than does that of L_7 (the total number of nerve fibers is also much greater). Bilateral division augment this condition to such an extent that the extremities hang loosely when the monkey sits in the examining chair. On the other hand, division of the dorsal roots of S_1 and S_2 increases tone in all the muscles of the lower extremity on the operated side, particularly in those of the extensors of the knee, the adductors of the thigh, the ventroflexors and the invertors of the foot. The tendon reflexes of the M. quadriceps femoris, of the hamstrings, and the ventroflexors of the ankle, as well as of the invertors of the foot, are brisk. The inhibition and facilitation evoked by the removal of the activity of these dorsal roots are of tone, not of movement. As phenomena, they are confined to the lower extremity and exaggerated in both of these extremities whenever these particular roots are severed bilaterally (Hines and Knowlton, 1952).

This inhibition has no relation to the inhibition of reciprocal innervation, as studied by Eccles and his collaborators (Eccles, 1957), for that inhibition is restricted to that of the homolateral opposing flexors. The site in the spinal cord of the inhibitory nuclei which change the polarization of motor cells in so many nuclei of the ventral horn is as unknown as that of its counterpart, the site of facilitatory nuclei. They may not exist. The intermediate nucleus selected by Eccles as an inhibitory nucleus is probably not the one concerned, because the increment in tone is found in the flexors as well as in the extensors of the knee. There is, in the macaque, at all levels of the spinal cord in which Clark's nucleus is not found, a discrete group of large nerve cells in the dorsomedial part of the central gray matter.

IV. Longitudinal Organization

A. *Contribution of Sensory Systems*

Each sensory system in the central nervous system contributes to the use of skeletal muscle. Skilled movements are directed by the eyes and by the ears as well as by proprioceptive systems. General cutaneous sensibility plays an appreciable role in some motor adjustments, whereas it is difficult to discover the role of the chemical senses of taste and smell. Indeed, sight can direct motor performance rather adequately.

The neuron pathways which ascend for short distances within the spinal cord and form with short descending systems the fasciculus proprius are able to maintain even in spinal man multiple reflex activity (Kuhn, 1950). Although tone waxes and wanes in the extensors of the lower extremity of these injured men, the upright posture of standing can in some individuals be assumed with help but not maintained. The adjustments between flexors and extensors of the lower extremities and the fixation of the trunk and girdles necessary for maintenance of standing are lacking.

The ascending systems within the spinal cord fall naturally into two groups—exteroceptive and proprioceptive. The exteroceptive group include the spinothalamic tracts and the spinotectal tracts carrying general cutaneous sensations to the thalamus and midbrain, respectively. The four proprioceptive systems are the two spinocerebellar tracts (dorsal and ventral) and the fasciculi gracilis et cuneatus. A transient diminution of tone in the ipsilateral muscles below the lesion follows section of the dorsal spinocerebellar tract (Ferraro and Barrara, 1935). No report has been made of the results of a similar lesion in the corresponding ventral tract. The ataxia of syphilis has been assigned to degeneration of the dorsal columns. Similarly, Foerster (1936a) described ataxia and the loss of the sense of position and passive movement on the side below a stab wound which severed the dorsal funiculus.

The nerve fibers of the spinocerebellar tract, activated by impulses from muscle spindles, and entering the dorsal root via group Ia and group II fibers (Laporte *et al.,* 1956), synapse on the giant nerve cells of Clark's Column (Szentágothai and Albert, 1955) forming large and small terminals. The area of contact of each large terminal extends for several hundred square microns; that of smaller ones is not larger than the ordinary boutons terminaux. Group Ia and group II nerve fibers furnish the giant terminals. Some of the smaller synaptic terminals may be axons of nerve cells which lie in the dorsal horn, others may be collaterals of spinocerebellar fibers. The same tract fiber can be activated by afferent volleys given different muscle nerves. Consequently, the synaptic relay could possess a function of coordination (Laporte *et al.,* 1956). No study has been reported which can correlate nerve endings in the periphery with degenerating boutons on the scattered cells of Stilling. Since the ventral spinocerebellar tract (Foerster and Gagel, 1932) receives additions at all levels of the spinal cord, it may play a significant role in control of skeletal muscles of the extremities.

The dorsal root fibers which enter the dorsal columns to form the fasciculus gracilis and the fasciculus cuneatus belong to the large group of nerve fibers within the dorsal root. Peripherally, these nerve fibers innervate muscle spindles and Golgi–Mazzoni tendon terminals. Single impulses experimentally evoked in the nerve fibers of the dorsal column of the fasciculus cuneatus have a remarkable excitatory effect on the neurons in the nucleus cuneatus, with a synaptic delay of only 0.6 msec (Therman, 1941). Anatomically, the axons of these neurons and those of the nucleus gracilis have two destinations, the cortex of the cerebellum and the cortex of the parietal lobe via the medial lemniscus and the nucleus ventralis posterior lateralis. In the cat, axons of cells which lie within these two nuclei ascend among the descending fibers of the pyramids and terminate in the motor and sensory cortices without a synapse in any thalamic nucleus (Brodal and Wahlberg, 1952), a spino-cortical tract.

The proprioceptive organ of the head is the labyrinth. The labyrinth presents two types of endings: (1) the crista, capped by the cupola, found in the ampullae of the semicircular canals, and (2) the macula, covered by an otolith, located within the utriculus and the sacculus. The function of the sacculus has resisted all attempts at analysis. The cristae are sensitive to increments in angular acceleration; there is one for each plane in space. The motor adjustments to these displacements are compensatory. These compensatory movements involve the extrinsic muscles of the eyes and the muscles of the trunk. The muscles of the trunk respond, particularly when the crista of the horizontal canal is stimulated. The eye muscles are related via the fasciculus longitudinalis medialis to each of the three canals. The lateral and medial recti deviate the eyes opposite to the direction of the rotation in the horizontal plane; the dorsal and ventral recti, to that of the anterior canal; the superior and inferior oblique, to that of the posterior canal. The utriculus is stimulated maximally only when the otolith hangs from the macula and minimally when it presses upon the macula (Magnus and de Kleijn; see Magnus, 1924). This is the organ which effects tone in skeletal muscle. In the position of maximal stimulation, after the dorsal roots of C_1 to C_3 have been cut, the decerebrate animal shows maximal extensor tone (Magnus, 1922).

Although the exteroceptive sensory systems may direct the use of skeletal muscle in skilled performance, they seem to have little to do with the maintenance of tone or the interpretation of position in space. Rather, the evidence presented suggests that the proprioceptive systems are those which aid in the inhibition or in the facilitation of tone in skeletal muscle. Also, at the spinal level, the proprioceptive afferent by its central connections facilitates contraction of extensors and inhibits that of the opposing flexors and the reverse (Granit, 1956). The remarkable excitatory action, recently discovered by electrical stimulation and recording of single nerve fibers in the dorsal spinocerebellar tract and in the fasciculus cuneatus, suggests that stimulation of only a few muscle afferents are necessary to produce this result.

B. Contributions by Descending Systems

The descending systems which control the use of skeletal muscle in the primate fall naturally into two groups—pyramidal and extrapyramidal. These systems were named for their relative anatomical positions in the medulla oblongata. They are identified as lying side by side within the internal capsule, the pyramidal group medial to a large group of the extrapyramidal. The former group occupies the middle third of the basis pedunculi (man, Bucy *et al.*, 1964; monkey, Bucy *et al.*, 1966). The extrapyramidal systems are found within the tegmentum of the brain stem whence they pass caudalward into the reticular formation of the medulla oblongata. More caudally, they lie in the ventrolateral and ventromedial division of the spinal cord (cf. fig. 1. Goldberger, 1969).

The pyramids of the primate (macaque or man) contain two descending systems: (1) the old corticospinal tract and (2) the new corticospinal tract. The former stems from the cells of Betz in layer V, from pyramidal cells situated within layer III (Spielmeyer, 1906) and in layer V (Wohlfahrt, 1932) within the whole of area 4 and from area 6, which contributes degenerated nerve fibers to the pyramids after injury of that region in man (Minckler *et al.*, 1944). In the macaque, however, neither Levin (1936) nor Hines (1943) found any degenerated myelin in the pyramids after the removal of the whole of area 6.

The new corticospinal tract stems from the parietal lobe. Its cells of origin were demonstrated by retrograde chromatolysis—subsequent to lesion in the corticospinal tract at C_4 in the macaque (Levin and Bradford, 1937)—to pyramidal cells of layer III of area I as well as the pyramidal cells of layers III and V of the remaining parietal lobe. Peele (1942) (also macaque, Marchi method) found degenerating myelin in the ipsilateral pyramid and in the corticospinal tract contralateral to lesions in the postcentral gyrus and to those in area 5 as low as the lumbar levels. Degenerated myelin following lesions of area 7 extended ipsilaterally only to the cervical cord. For the sake of clarity the old corticospinal tract may be called the frontospinal tract and the new the parietospinal tract (Wagley, 1945).

Although the axis cylinders which stem from nerve cell bodies in the parietal lobe pass caudal with old corticospinal tract, they do not share the terminal nuclei which characterize the "older" system.

The nucleus proprius of the dorsal horn of the spinal cord receives terminals from cell bodies which are found in layer V of the postcentral gyrus of the parietal lobe. The nucleus gracilis and nucleus cuneatus as well as the nucleus sensibilis of the V cranial nerve entertain endings which stem from nerve cell bodies found in areas 5 and 7 of the parietal lobe (macaque) (Kuypers, 1960; Liu and Chambers, 1964; Lawrence and Kuypers, 1968). These nuclei in the spinal cord and medulla oblongata return nerve fibers to that particular cortical space which was the origin of the terminals which each nucleus had received as a part of the parietospinal tract, thus forming a feedback system.

Besides these two corticospinal systems, Wahlberg and Brodal (1953) find degenerating axis cylinders among the sound nerve fibers of the pyramids of the ventral and lateral corticospinal tract in the spinal cord of the cat after removal either of the occipital lobe or of the temporal lobe.

The extrapyramidal systems fall naturally into two categories—(1) those which terminate in the dorsal sensory nuclei of the thalamus and (2) those which end in the nuclei of the brain stem. The thalamic

nuclei which receive such corticifugal systems are those of sensory projection, those of association, and those which are recruiting. The first
two systems form looped circuits with the afferent fibers to the cortex,
the last via the nuclei of association, reverberating circuits which produce asynchronous cortical rhythms (Starzl and Whitlock, 1952). Before
the intensive use of electrical devices for stimulation and recording of
nervous activity, the function of the efferents from the sensory cortices
to the thalamic nuclei were interpreted by Head as inhibitory.

Besides these corticothalamic systems which originate in layer VI
(Winkler, 1929), there are others. Many corticifugal systems share common destinations. The superior colliculus receives axons from the frontal
lobe (area 4) (Mettler, 1935), the parietal lobe (area 7) (Peele, 1942),
and the occipital lobe (areas 18 and 19) (Mettler, 1935; Lemmen, 1951),
(area 22) (Lemmen, 1951). The inferior colliculus, moreover, restricts
its cortical axons to. the temporal lobe, whereas the pretectal area receives fibers from the parietal lobe (areas 5 and 7) (Peele, 1942). The
nucleus ruber entertains fibers from the frontal lobe areas 4, 6, and
4s (Mettler, 1948a)—this author found none from the anterior division
of 4 (Hines, 1943), from the postcentral gyrus (Mettler, 1935; Lemmen,
1951), and from area 19 of the occipital lobe (Lemmen, 1951). The
subthalamus receives efferent fibers from the area frontalis agranularis
(Levin, 1936; Verhaart and Kennard, 1940; Hines, 1943). The substantia
nigra has a broad spectrum of corticifugal systems originating from the
area frontalis agranularis and area 8 of the frontal lobe (Levin, 1936;
Verhaart and Kennard, 1940; Hines, 1943), from the superior gyrus of
the temporal lobe (area 22) (Mettler, 1936; Bucy and Klüver, 1940),
from areas 3 and 5 of the parietal lobe (Peele, 1942), and perhaps from
area 19 of the occipital lobe (Lemmen, 1951).

C. Core of the Brain Stem

At the present time, the control of skeletal muscle cannot be discussed
without a critical appraisal of the function of the tegmentum and reticular formation of the medulla oblongata. This great core of the brain
stem which defied analysis for so many years has begun to yield some
factual material to the inquisitive deep recording and stimulating electrodes. The interruption of the continuity of these fibers and their cell
bodies, either in the region of the hypothalamus (Ranson, 1939) or
cephalad to the level of the superior colliculus (Peterson *et al.*, 1949),
produced a monkey distinguished by loss of motor initiation, a masklike
face, and a loss of the desire to eat. The long ascending sensory systems

were intact and the corticospinal tracts were uninjured. Besides this "loss of the will to move," these animals slept (Bremer, 1935).

The reticular system contains both ascending and descending systems. The descending systems stem from the corpus straitum, the hypothalamus, the midbrain tegmentum, and even from the cerebral cortex. The ascending systems within this complex region arise from reticular nuclei, from the spinal cord, or are collaterals of ascending sensory systems. Not only do collaterals from each of the four lemniscus systems enter the reticular formation, a recent rediscovery of Winkler's (1929) findings, but also axons from the nucleus fasciculus solitarius (Allen, 1923), from the posterior accessory optic tract—fibers not medulated either in man (Marburg, 1903) or in the monkey (Gillilan, 1941)—from the ascending visceral sensory systems, as well as from the vestibular system (Gerebtzoff, 1940) enter this region. Thus, the reticular system receives impulses from each kind of sensory system found in the mammalian body. Nonetheless, some organization is emerging. The Scheibels (Scheibel et al., 1955) find that convergence of afferent impulses on units of the reticular formation, although widespread, is not unlimited. Indeed, convergence patterns on individual units vary greatly.

The function of ascending systems in the reticular formation may not be restricted to keeping the cerebral cortex awake by way of recruiting thalamic nuclei (Starzl and Whitlock, 1952). There are direct spinoreticular systems which can be followed in the degenerated state as far cephalward as the level of the abducens nucleus (Morin et al., 1951). Some of these fibers synapse in the lateral reticular nucleus of the medulla and terminate medially and laterally in the midgrain tegmentum. The peripheral fibers innervate joints and tendons. The nucleus of origin is anatomically unknown (cat; Morin, 1953).

Throughout the reticular formation, many anatomically discrete groups of nerve cells exist. At the level of the nuclei of each cranial nerve in man, the reticular formation presents specialized nuclei, some of which are medially placed, others laterally. Brodal (1956) finds that the medial two-thirds of the reticular formation contains the cells of origin of the long ascending and of the long descending tracts of this region. Into this region many corticoreticular fibers terminate (from motor cortex and area 24, cat, Rossi and Brodal, 1955); and from this region some of the reticulospinal fibers originate, coinciding with the inhibitory area (Magoun and Rhines, 1946; Niemer and Magoun, 1947). The site of the cells of origin of these reticulospinal systems found rostrally in the pons and midbrain is less extensive than that of the facilitatory region of Magoun and his associates. In other words, the inhibitory impulses reaching the spinal cord arise in the medial part of the posterior division of the reticular formation (Magoun and Rhines, 1947; Ward, 1947), whereas the facilitatory impulses stem from cells located more rostrally, both laterally and medially in the tegmentum. This facilitatory system receives contributions from the basal part of the diencephalon, from the corpus striatum (globus pallidus in particular) (Mettler, 1941, 1943; 1944), and perhaps from the nuclei of the thalamus. Furthermore, it is augmented by fibers from the midbrain and the rostral part of the medulla oblongata. The reticulospinal tracts carrying facilitation lie throughout the lateral funiculus, concentrated near its lateral border (Niemer and Magoun, 1947). The reticulospinal tracts carrying inhibitory impulses are found in the ventral funiculus and in the ventral part of the lateral funiculus (Niemer and Magoun, 1947; Wagley, 1945). Unilateral lesions in these regions of the spinal cord (Wagley, 1945) show bilateral chromatolysis in the nerve cells of the nucleus reticularis lateralis in the midbrain. Other origins of these inhibitors are found in the caudal reticular formation, for Bodian (1946) found many large degenerated cells in this region during the spastic stage of experimental poliomyelitis in the monkey. The inhibitory fibers may terminate on adjustor neurons in the intermediate area of the spinal cord gray (cat; van Harreveld and Marmost, 1939). Electrophysiological analysis of the results which follow or accompany the activation of descending facilitatory systems (i.e., the vestibulospinal and reticulospinal, cat) led Lloyd (1941) to conclude that these fibers terminate directly upon the ventral horn cells and upon neighboring nerve cells. The indirect pathway implicated nearby neurons in such a way that the facilitation of the ventral horn cell was augmented and its duration prolonged.

Within the generalized reticular-tegmental region, three specialized nuclei have developed: the nucleus ruber, the substantia nigra, and the subthalamus. The nucleus ruber projects its axons cephalward to the globus pallidus, to the arcuate and ventrolateral nuclei of the thalamus, and to the subthalamus, and caudalward to the spinal cord (monkey; Carpenter, 1956). Injury to this nucleus alone is followed by a great reluctance to move, without any change in the use of skeletal musculature. The hypokinesis which followed brain stem lesions reported above must have included division of the corticorubral tracts. Certainly, the lesion in the spinal cord, which resulted in an unwillingness to use the muscles on the side of the lesion, severed the rubrospinal tract as well as reticulospinal pathways (Wagley, 1945). Unilateral lesion of the corresponding brachium conjunctivum plus either unilateral injury to the nucleus ruber (Orioli and Mettler, 1957) or a lesion in the ipsilateral substantia nigra (Carrea and Mettler, 1955) adds only hypokinesis to the preexisting results. The substantia nigra receives a wealth of corticifugal fibers, reserves a special site for their reception, according to cortical origin, and has not as yet divulged its function even to electrolytic lesions (Ranson, see Cannon *et al.,* 1944). Stereotaxic bilateral injury to the subthalamus in the macaque (Carpenter *et al.,* 1950) is followed by bilateral choreoid hyperkinesis, which can be eliminated or decreased by an 8% reduction of the nervous tissue within the globus pallidus (bilaterally).

The cerebellum, related on the afferent side to the exteroceptors and proprioceptors of the spinal cord, to nuclei in the reticular formation, to the tectum of the midbrain, and to the motor cortex, the parietal lobe, and the occipital and temporal lobes, projects upon the reticular formation via pathways which originate in one or more of the three cerebellar nuclei—the medial, or fastigial; the intermediate, or globosus and emboliformis; and the lateral, or dentatus. The bulk of the projection from the nucleus dentatus as the crossed ascending limb passes into the reticular formation, transverses the nucleus ruber (the uncrossed ascending limb bypasses this nucleus) to end in the ventrolateral nucleus of the thalamus and in the globus pallidus. The descending limb has three components: (1) to three discrete nuclei in the tegmentum of the pons, (2) entering the medial longitudinal faciculus, and (3) to the cervical spinal cord (Carrea and Mettler, 1954). Cerebellar ataxia was found to result from injury either to the dorsal or intermediate components of the brachium conjunctivum destined for the large celled nucleus ruber or to the descending limb of the superior brachium. Cerebellar tremor followed lesion of the component of brachium conjunctivum destined for suprasegmental levels, i.e., of the nucleus dentatus, of the midbrain tegmentum (Carrea and Mettler, 1955), and of the nucleus ventrolateralis of the thalamus (Walker, 1938). Hypotonia was not mentioned in either of these two studies. Botterell and Fulton (1938) described hypotonia after either removal of the lateral hemispheres or injury to the nucleus dentatus. To the hypotonia which appeared after the first lesion, tremor and ataxia were added after the second. The latter finding agrees with the work of Carrea and Mettler.

The fastigial nucleus is credited with ability to inhibit extensor tone, because Sherrington found that electrical stimulation of the anterior lobe decreased extensor tone of decerebration (cat) and that removal increased decerebrate tone without change in its distribution. Repetition of these experiments has given similar results. The decerebrate state can be produced by removal of the inhibitory activity of the reticular formation by lesions in the midline of the medulla and allowing the facilitatory mechanism to be stimulated by afferents from sensory systems of the spinal cord and medulla oblongata (macaque; Ward, 1947).

Facilitatory activity, on the other hand, can be destroyed by lesions in the middle third of the reticular system in the midbrain or by lateral lesions in more caudal levels. Apparently, in the primate the vestibulospinal tract cannot without the reticulospinal fibers in the ventral division of the spinal cord maintain the hypertonicity of the decerebrate state (Bach and Magoun, 1947).

In the analysis of this state in cats, Sprague and Chambers (1953) describe collapse of decerebrate rigidity as following destruction of the nucleus fastigii and of Deiters' nucleus. The pontobulbar reticular formation remained intact. Apparently, loss of Deiters' nucleus and consequently the majority of the facilitatory activity of the vestibulospinal tract left the reticular formation without adequate afferent stimulation. The results given suggest that the impulses discharged by the fastigial nucleus into the reticular formation are dominantly facilitatory, not inhibitory—an apparent contradiction to the results of Sherrington's original experiments. Or does destruction of these two nuclei release the crossed inhibitors of this hypertonic state?

The activity of the crossed inhibitors of extensor rigidity can be seen in a decerebrate cat in which the right foreleg has become atonic after removal of the caudal pole of the opposite nucleus fastigii. This crossed inhibition of entensor rigidity when removed (Moruzzi and Pompeiano, 1957) by deafferentation of the spastic left leg by cutting the left VIII nerve or the left vestibular nerve, or by hemisection of the spinal cord on the left side between T_{12} and L_1, the atonic right foreleg becomes hypertonic and the left hypertonic foreleg becomes atonic. Apparently the crossed afferents are inhibitory, the ipsilateral afferents facilitatory. Myotactic influences can be removed by cutting the dorsal roots to both forelegs. Both forelegs become atonic, the hind legs tonic. This local modification of the decerebrate state by local removal of the activity of several varieties of afferent inhibitors or facilitators suggests that local inhibitors are able to take precedence over general arousal of the more cephalic facilitators, at least in the cat.

The brain stem, which contains the nervous mechanisms which assure smooth movement and facilitate and inhibit tone or movement, also contains complex nerve mechanisms which when stimulated directly produce complex movements, such as flexion of one extremity, extension of the opposite one, with either extension or flexion of a third. Eyes, ears, and facial muscles were sometimes included. Crosby (1956) explored the tegmentum of the macaque under anesthesia. The movements elicited resembled those which she and her associates had evoked by stimulation of the cortices of the parietal and temporal lobes. She thought that the impulses so aroused traveled to motor centers of the brain stem and spinal cord via multineuron arcs—identified as corti-costriatal (Dusser de Barrene *et al.*, 1940), striatotegmental, or striatoru-bral—thence to the spinal motor cells via the rubrospinal or tegmento-spinal systems.

To Sprague and Chambers (1954), no general facilitators or inhibitors of tone exist, because they have been unable to duplicate these painstak-ing experiments of Magoun and his co-workers, by stimulating the reticu-lar formation in unanesthetized cats. Direct stimulation with implanted electrodes placed medially elicited flexion of the ipsilateral legs and extension of the contralateral legs plus the inhibition appropriate to reciprocal innervation. With the electrodes placed laterally, the pattern of tonic flexion and extension was completely reversed. When the inten-sity of the stimulating current was reduced, the cat circled and came to rest in the normal curled position of sleep.

These interrelated neurons form the substrata in the brain stem and spinal cord which the cortex cerebri utilizes in its control of skeletal muscle. These brain mechanisms adjust the tone in skeletal muscle to fit the task and lend to all movement the smoothness and sequence of contraction that is termed normal. Here also is found the basis for organized movements, such as quadrupedal progression and the mass organization of synergistic activities.

V. The Motor Cortices

The majority of investigators have emphasized the lack of correspon-dence between contraction of skeletal muscle elicited by stimulation of the precentral gyrus and the use of that musculature made by the animal or by man. This discrepancy has been noted particularly by neural surgeons who have stimulated the cerebral cortex of conscious man prior to removal of diseased tissue. Topical localization of contrac-tion of single muscles or parts of muscles and low intensity of the stimu-lating current distinguish the precentral gyrus of all primates.

On the other hand, many movements evoked by the sine wave current used on the precentral gyrus of fetal and infant macaques present a striking similarity to those in use by the young monkey at the time of stimulation. Because Hines and Boynton (1940) could not assign these movements to the extrapyramidal or to the corticospinal systems, they were classified respectively as holokinetic and idiokinetic. The holo-kinetic movements were those of infantile behavior patterns, patterns of progression, movements of girdle and neck musculature, and rhythmical movements of lip, tongue, and upper extremity. In early postnatal development, the idiokinetic movements lacked discreteness. Rather, they were coinnervations resembling those used by the conscious infant at the age they were elicited. Strangely enough, during the last month of gestation and the first two weeks of postnatal life, the cortical electrode was able to elicit such discrete movements as extension of the thumb or flexion of the hallux, which the monkey did not make at the age of stimulation; but, thereafter, the evoked contraction corresponded extraordinarily well with those used by the infant itself.

The recent explorations of the cortex cerebri of monkeys and of men are concerned with elicitation of contraction of skeletal muscle rather than with inhibition or facilitation of its tone. The threshold for excitability of the cortical points in these new motor fields is higher than that of the "motor" cortex in the precentral gyrus of primates and the movements obtained are far more complex. Each of these fields is independent of the "motor" cortex and its corticifugal systems.

A. Motor Area II

The oldest of these recently discovered motor fields (Sugar *et al.*, 1948) joins the most ventral part of the precentral gyrus in the monkey's cortex, follows the posterior frontal operculum, and spreads out over the insula and under the anterior parietal operculum (Fig. 1). The stimulating voltage was extremely high (10–20 V), twice as much as

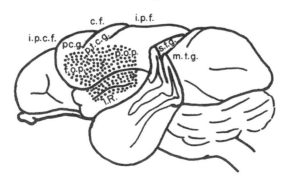

Fig. 1. Drawing of the brain of a macaque. The anterior division of the superior temporal gyrus (s.t.g.) and of the middle temporal gyrus (m.t.g.) have been removed exposing the parietal (pt.op.) and frontal opercula (f.op.) together with the medial surface of the Island of Reil (I.R.). Other abbreviations are: pc.g., precentral gyrus; pt.c.g., postcentral gyrus; c.f., central fissure; and i.p.f., intraparietal fissure. The stippled area outlines the limit of motor cortex II as described by Sugar *et al.* (1948).

that necessary in these monkeys to evoke movement of the thumb by stimulating the precentral motor cortex. Either the type of stimulating current was not adapted for excitation of these cortical areas (including the precentral gyrus), or the resistance of the cortex was very high. Somatotopical localization was ragged. The arm area overlapped that of the face dorsally and that of the leg ventrally. Movements of the thumb and arrest of respiration were generously and without apparent organization peppered over the insula. Posteriorly, this motor cortex overlapped the second somatic area (Woolsey and Fairman, 1946).

B. *The Supplementary Motor Cortex*

The second of these motor fields was first discovered in conscious man by Penfield and Welch (1951) on the medial surface of the superior frontal gyrus and was named the supplementary motor cortex. The region outlined by these authors is extensive, stretching from the leg area in the paracentral lobule to the anterior border of the medial subdivision of Brodmann's area 6. Penfield and Welch are particularly interested in discovering new sites which give inhibition of speech. They uncovered no somatotopical localization on the medial surface for leg, arm, or face. The movements elicited were complex, frequently located ipsilaterally, and the intensity of the stimulating current was greater than the threshold values of excitable loci on the precentral gyrus. Erickson and Woolsey (1951) substantiated the loci but found evidence of some somatotopical localization.

Before this study was published, Bates (1953) had seized the opportunity of stimulating the medial surface of a sound hemisphere, exposed by removal of the opposite cortex cerebri because of infantile hemiplegia. The exploration, done under a general anesthesia after the ablation of the injured hemisphere, was confined to a small region about the medial tip of the central fissure. The resulting movements were complex and the threshold was high. Plotting the primary movements after the manner of Leyton and Sherrington (1917) showed some topical localization. Area 6 was not stimulated. In spite of a general anesthesia, Bates obtained contraction of ipsilateral muscles. One of these sequential complex movements stopped with ipsilateral extension of the ipsilateral hallux—a positive Babinski.

On the medial surface of the macaque's frontal lobe (Erickson and Woolsey, 1951; Woolsey *et al.*, 1952) confluent posteriorly with the precentral motor cortex and ending in the vicinity of the anterior border of area 6, lies the supplementary motor cortex. The two leg areas meet on the dorsal crest of the sulcus cinguli. The anesthesia was deep, the

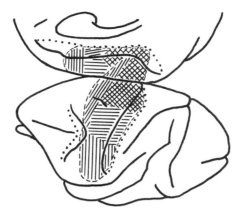

Fig. 2. A diagram of the lateral surface and adjoining medial surface of the precentral motor cortex of the macaque's brain. The fissures are drawn as if they were open in order to reveal the extent of the motor areas therein. The precentral motor cortex occupies the precentral gyrus, the anterior bank of the central fissure, a small area in the posterior bank of the central fissure, a small area in the posterior bank of the inferior precentral fissure, as well as the posterior part of the medial surface of the superior frontal gyrus. The supplementary motor cortex occupies the medial surface on the superior frontal gyrus anterior to the leg area of Brodmann's 4, as well as the dorsal bank of the sulcus cinguli. The anterior limit on the medial surface stops short of the anterior boundary of the homologous area discovered in Cercopithecus by Horsley and Schäfer (1888). Representation of the musculature: Epoxial (diagonals); leg (cross-hatched); arm (horizontals); face (verticals). (Redrawn from a figure made by Woolsey and used by Erickson and Woolsey in a paper presented before the American Neurological Association in 1951.)

threshold was high, and the contractions of skeletal muscles, sequential and complex, were located without exception contralaterally (Fig. 2; cf. Fig. 3).

The more rostrally placed areas on the precentral gyrus required greater strength of current to produce contraction of skeletal muscle than do those areas situated caudally near to or bordering on the central fissure (Woolsey *et al.*, 1952). Kuypers (1960) correlated this finding with his, namely that the more rostral areas synapse with interneurons located in the intermediate grey of the spinal cord, while the more caudal areas synapse directly with the motor cells of the ventral horn in the spinal cord (Kuypers, 1960, 1964).

C. The Postcentral Motor Cortex

Again, stimulation of the postcentral gyrus with 60 cps sine wave current after both of these motor cortices were removed for weeks or

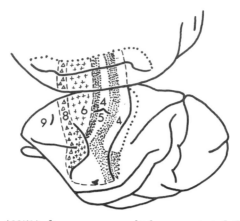

Fig. 3. Woolsey's (1951) diagram, upon which are projected the extent of Brodmann's areas (1903, 1906) as found by the author (Hines, unpublished). Comparisons of the extent of area 4 with that of the precentral motor cortex (Fig. 2) will reveal that the latter extends more anteriorly than does area 4 on the lateral surface of the cortex cerebri; while on the medial surface, the precentral motor cortex does not include the small region turned inward as the superior bank of the sulcul cinguli (cytoarchitectonically area 4). The supplementary motor cortex includes the dorsal bank of the gyrus cinguli in the "leg" area, the anterior border of area 4 on the medial surface and all of the medial division of area 6, together with the dorsal bank of the gyrus cinguli. The extent of these differentiated cytoarchitectonic areas resemble those described for the cortical surface of the frontal lobe by Mettler (1948a). Area 4 has been divided into four strips. The one which passes through the superior precentral fissure (4s) outlines the extent of the cortex, the removal of which is followed by some of the facets of the spastic state—clonus, irradiating brisk tendon reflexes, and the differential distribution of resistance to passive movement—in the contralateral extremities. The paralysis which persisted was confined to the abductors of the toes (Hines, 1936). The posterior clear area is the site of the lesion which gives no permanent paralysis and little or no increase in resistance to passive movement, no clonus, and no brisk tendon reflexes. Initially, there is a real paralysis in the limbs opposite the lesion and muscle tone is greatly decreased. This decrement in tone lasts for years. The paralysis is transient. The cortex, represented by the stippled area, was left intact, so that in the former operation, area 6 would not be injured, and in the latter, the more anterior part of area 4 would be free from the trauma of the operation.

months elicited a detailed representation of contralateral musculature which part for part followed the outline of tactile representation of the surface of the body (Woolsey *et al.*, 1942). The intensity of the stimulating current required to produce these movements was two to three times higher than that necessary to evoke contraction of muscle by stimulation of the precentral gyrus (Woolsey *et al.*, 1953). In man, however, Foerster (1936a) did not elicit movement by stimulation of the postcentral gyrus after the precentral gyrus had been removed.

D. *Additional Motor Areas*

Elizabeth Crosby and her associates called the electrically excitable areas found on the crest of the intraparietal fissure (Fleming and Crosby, 1955) and on the superior temporal gyrus (Schneider and Crosby, 1954) additional motor areas. The results of stimulation of the parietal lobe were an extension of Peele's report (1944). The majority of movements obtained on the parietal lobe was contralateral, while the majority of those evoked from the superior temporal gyrus was ipsilateral. The thresholds were high and varied between 2 and 14 V. Some somatotropic localization occurred. The movements obtained from these areas in the intact cortex were similar to those elicited subsequent to removal of area 4 or of area 4 and the postcentral gyrus. Nonetheless, the presence of these electrically active motor areas did not prevent the development of typical area 4 paralysis (Fig. 4).

Besides these four accessory motor fields, there are others first discovered by Ferrier (1876) and located by him within each lobe of the monkey's cerebral cortex. Ferrier's fields were subsequently lost because of Sherrington's pronouncement that complete flaccidity of skeletal muscle was the prerequisite for electrical stimulation of the motor cortex. Later, they were rediscovered by Vogt and Vogt (1926) for monkeys and by Foerster (1936a,b) for man.

Each lobe of the monkey's cerebral cortex presents its own motor field, far from the origin of the frontospinal tract and electrically active even after the pyramids are surgically divided (Hines, 1943, 1947). Each of these motor fields is located within the cortical area to which thalamic nuclei of association project. Removal of any one of these fields except the one in the parietal lobe (part of area 7) does not visibly change the use of skeletal muscle by the macaque. After removal of area 7, the deficiency in motor performance is related to loss of recognition of objects by somaesthetic sensibility and to hypotonia of skeletal muscles of the upper extremity (Peele, 1944; Kennard and Kessler, 1940). Each of these fields responds to the stimulating current (60 cps sine wave) with adversive movements and quieting (Hines, 1947). The quieting effect causing cessation of spontaneous movement (the ether had been previously lightened) confers upon the macaque an appearance to attentive repose. This effect is more easily obtained from the anterior field on the lateral surface of Brodmann's area 9 than from any one of the three posterior fields. The adversive movements common to these fields were conjugate deviation of the eyes plus, at times, that of the head as well. In the frontal field, these adversive movements

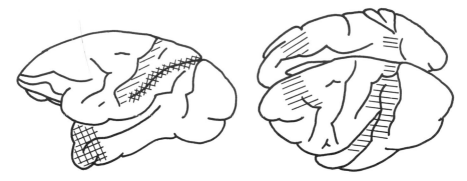

Fig. 4 (left). Drawing of a hemisphere of a macaque's brain upon which is projected the postcentral motor cortex of Woolsey *et al.* (1954) and the two additional motor cortices of Crosby and her associates. The "motor cortex" of the temporal lobe was discovered by Schneider and Crosby (1954), that of the parietal lobe, by Fleming and Crosby (1955). Peele's results (1944) of electrical stimulation of the parietal lobe are similar to those reported by the latter authors.

Fig. 5 (right). Drawing of the extrapyramidal motor areas of the four lobes of the cerebral cortex, as found in the macaque, illustrating the motor areas. These areas resemble those located by Ferrier (1876) outside of the precentral gyrus (Hines, 1947).

sometimes included the axis, the contralateral extremities, and the tail. Orientation of the ears was obtained from three of the four fields (not from the parietal lobe) (Fig. 5).

Three of these fields yielded other movements of the eyes, such as opening and dilation of the pupil from the frontal field (Smith, 1949), convergence from the parietal field (Tower and Hines, 1947; cf. Hines, 1947), closure of the eyes, and constriction of the pupil from the occipital area (Walker and Weaver, 1940). Besides these movements of the eyes, the temporal and occipital fields responded with a complex movement called by Tower and Hines (Hines, 1947) a reaching and grasping act.

In the macaque, but not in man (Foerster, 1931), stimulation of area 17 oriented the eyes to the visual fields represented in the region stimulated (Walker and Weaver, 1940). Area 17 was at one time the only primary sensory projection area of the monkey's cortex which responded with contraction of skeletal muscle.

Outside of these regions, an electrode on the cortex around the caudal end of the fissura principalis (area 8) evoked nystagmus and conjungate deviation of the eyes, whereas one on the cortex dorsal to that fissure elicited deivation of the eyes, either upward or downward.

These are the areas of cortical surface which have been excited by the parameters of the electric current now in use. The results of the early stimulations of the motor cortex have produced many of the interpretations of the fundamental relationships between nervous tissue and skeletal muscle. This multiplication of motor cortices challenges new interpretations; for with the exception of the supplementary motor cortex in the macaque, none of these motor areas is able to substitute for loss of the precentral motor cortex. This search for additional motor cortices is motivated by more than curiosity, for that search hopes to uncover some way of substituting the activity of these new motor areas for that of the precentral gyrus, which is sometimes congenitally and regrettably absent in man.

Electrical recording of recent explorations of the precentral gyrus and its descending systems in primates suggests that certain varieties of excitatory changes characterize the pyramidal system. A temporal triad of facilitation, inhibition, and delayed facilitation evoked by single stimuli to the precentral gyrus of primates is completely abolished by section of a unilateral pyramid (Stewart *et al.*, 1968a). Although the first two of this triad cannot be separated at the cortical level (Preston and Whitlock, 1960), only the third survives removal of the precentral cortex itself. Could late facilitation originate within the cortex of the postcentral gyrus? Some motor neurons in the spinal cord receive excitatory and inhibitory inputs from pathways mediating precentral gyrus effects (Preston and Whitlock, 1961).

The initial facilitation and inhibition induced by cortical stimulation were abolished by large lesions in the lateral funiculus at C_4. The late delayed facilitation remained, to disappear subsequent to hemisection. Only the descending pathways in the dorsolateral quadrant are required for preservation of inhibition, whereas the ventrolateral quadrant may preserve initial facilitation. The final facilitation requires repetitive volleys for excitation and the presence either of an intact dorsolateral or of a ventrolateral quadrant in the spinal cord. Collaterals from the lateral corticospinal tract are capable of exciting pathways in the ventrolateral quadrant (Stewart *et al.*, 1968b). Such collaterals are unknown by any other technique. The fact that the presence of the dorsolateral quadrant is necessary for preservation of inhibition suggests that the corticospinal tract which it contains may possess inhibitory activity. Cortically evoked initial facilitation remains to be abolished by lateral hemisection of the spinal cord.

E. *Participation of the Small Nerve Fiber in Control of Tone and Movement*

Since decerebrate rigidity of the Sherringtonian type is ushered in by uniform rates of firing of small nerve fibers which innervate the intrafusal muscle fibers of the spindles, it was logical for Granit and his co-workers (Granit, 1956) to ask whether the activity of the ventral horn cells fired by the activity of the equatorial ending was organized for cooperation with direct excitation of the extrafusal fibers.

These workers found that the small nerve fiber of the spindle was facilitated or inhibited by the same cephalad areas of brain which had been found previously to be facilitatory or inhibitory of tone and of movement. Stimulation of the motor cortex of the cat (3 msec shock) with a low strength of current elicited small nerve activity only; an increase in current strength added activity of the large nerve fibers to that of the small ones. Stimulation with needle electrodes in the lateral midbrain tegmentum found the large nerve fibers silent, while the rate of discharge in the small nerve fibers gradually rose from 8 to 40 per second and fired at maximum frequency for 20 sec after the current had been removed. A similar stimulation of the pyramids selectively activated the spindle without either the slow recruitment or aftereffect characteristic of facilitation of small nerve fibers which resulted from stimulation of the reticular formation. In some instances, spindle activity was excited by stimulation of the cerebellum.

Complete inhibition of both small and large nerve fiber discharge was obtained by stimulating regions from which inhibition of tone and/or of movement have been obtained by other methods. These regions were the medial reticular formation dorsal to the pyramids, the anterior part of the cingulate gyrus, and the anterior lobe of the cerebellum.

The "double reciprocal innervation" as a principle of reflex organization has not been found by Granit (1956) to be symmetrical in flexors and extensors. In the double sensory innervation of skeletal muscle, the afferents on the spindles facilitate the muscle which contains them and inhibits the antagonist, whereas the Golgi tendon organs inhibit their muscle and facilitate their antagonist. There may sometimes be simultaneous contraction or simultaneous relaxation of both flexors and extensors.

Certain supraspinal lesions appear to shift this reflex organization. Reciprocal innervation, which Hering and Sherrington (1897) obtained by stimulation of the precentral gyrus in monkeys, disappears after the pyramids are severed (Tower and Hines, unpublished; cf. Hines, 1947). Bosma and Gellhorn (1946) showed that extensor and flexor muscles can be caused to contract simultaneously without an initial phase of inhibitory activity of the antagonistic muscle, and Chang et al. (1947) showed that contraction of single muscle can be elicited without concomitant inhibition of the opposing muscle or groups of muscles.

Cortical lesions upset the normal loop control of tone and movement in skeletal muscle so that reciprocal innervation disappears in states of tonic innervation of flexors contralateral to removal of area 6 in the monkey and in tonic innervation of flexors and of extensors created by bilateral removal of area 6 and of the anterior border of area 4 (Hines, 1937, 1943).

Muscular tension stimulates the Golgi endings and they in turn inhibit the synergists. Is it the marked increase in tonic activity of the flexors of the fingers subsequent to bilateral removal of Brodmann's area 4 which causes the loss of dorsoflexion of the wrist when the fingers flex upon an object? Certainly in the stage of development of the infant macaque known as tonic innervation, the flexors of the fingers are accompanied by flexion of the wrist (Hines, 1942). When the synergistic action of the dorsoflexors appears, tonic innervation had disappeared.

VI. Posture

The ability to maintain normal posture is the result of the sum of the activity of many parts of the nervous system. For example, the posture assumed by the infant macaque is related to the distribution of tone in his muscles and the mode of use of musculature which characterizes each stage in his postnatal development (Hines, 1942). Discrete organization of skeletal muscle demands fixation of the muscles of the girdles and trunk and of the more proximal ones of the extremity in order to free the distal ones for the initiation of movement.

A loss of fixation and a decrease in tone of the muscles follow removal of the cerebellar hemispheres in the macaque—never, however, as severe as in the baboon or in the chimpanzee (Botterell and Fulton, 1938). When the nucleus dentatus is included, tremor and ataxia are added (Carrea and Mettler, 1955).

Fixation is lost after bilateral surgical division of the pyramids (Tower, 1940). In standing, the legs are more adducted, the arms less so than in the intact animal. In sitting, the head falls forward (the extensors of the head are hypotonic), the back is extended, the arms hang loosely in extension with a real droop of the shoulders, similar to the condition of the macaque in which the pes pedunculi was partially severed (Cannon *et al.*, 1944).

The hypotonia of total parietal lobe removal presents a posture similar to that assumed by the unilateral pyramidal animal (Peele, 1944). In the writer's experience, the degree of hypotonia seen in the macaque

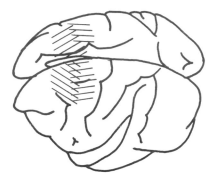

Fig. 6. Drawing of the extrapyramidal motor inhibitory and facilitatory areas in the area frontalis agranularis of the frontal lobe of the macaque. The more anterior field (area 6 plus the anterior boundary of anterior 4) yields flexor synergies, reaching and grasping, and inhibition of flexor tone and of the grasp reflex, when stimulated with the sine wave current (60 Hz). With similar electrical stimulation, the posterior field (anterior 4 plus posterior boundary of 6) yields quadrupedal progression, extensor synergies, standing tone, and inhibition of extensor tone. These phenomena were elicited under light ether anesthesia, not only before, but also after surgical division of the pyramid ipsilateral to the cortex stimulated.

reciprocals of each other, for the removal of the area from which release of the grasp is obtained by electrical stimulation is followed by appearance of the grasp reflex stronger and more enduring in the contralateral hand than in the foot (Richter and Hines, 1934). Removal of the anterior border of area 4, stimulation of which decreased standing tone, is followed by exaggerated standing tone, brisk and irradiating tendon reflexes, hypertonus of the clasp-knife type (distributed in the typical differential manner), clonus, and a minimal residual paralysis (Hines, 1936, 1937, 1943).

Standing tone is decreased after removal of the posterior part of area 4, of a part of or the whole of the parietal lobe, or of the hemisphere of the cerebellum. Surgical division of one pyramid (Tower, 1940) gives a generalized decrease in tone contralateral to the lesion, as does severance of the lateral part of the basis pedunculus—the extensors of the digits (Cannon *et al.*, 1944) are not affected by the latter operation. Cutting bilaterally the facilitators of tone in the lateral part of reticular formation causes a great decrease in tone of skeletal muscle (Ward, 1947). Hemisection of the medulla caudal to the pons (Niemer and Magoun, 1947) or division of the dorsal half of the lateral funiculus which includes the corticospinal ract was followed by decrease in tone of skeletal muscle opposite the lesion (Cannon *et al.*, 1943).

Surgical interventions which produce hypotonia show little differential distribution. The hypotonic states which follow pyramidal lesions, ablations of the cerebellar hemispheres (Botterell and Fulton, 1938), or removal of the parietal lobes effect muscles of the girdles and the proximal muscles of the extremities a little more than they do the distal muscles, except perhaps after the removal of the parietal lobe, when the flexors of the wrists and fingers, and the extensors and the adductors of the fingers are especially hypotonic (Peele, 1948).

The differential distribution of increased tone so characteristic of the spastic state of spastic hemiplegia in man is produced in the monkey by removal not only of the anterior division of Brodmann's area 4 (Hines, 1936), but also of the whole of area 4 (Hines, 1937). The hypertonia after the first lesion is maximal in the flexors of the elbow, the extensors of the knee, and the adductors of the thigh. The removal of total area 4 adds an appreciable hypertonus to the retractors and adductors of the upper arm, to the ventral flexors and the ulnar flexors of the wrist, and to the flexors of the fingers. In the leg, the increase in tone is found in the protractors and adductors of the thigh, in the ventroflexors, and in the evertors of the ankle (Hines, 1937). The ablation of the whole of the supplementary and precentral motor cortices presents a comparable differential distribution of tone contralateral to the lesion, except that resistance to protraction of the leg rather than to retraction was found, and the toes resisted flexion (Travis and Woolsey, 1956). Bilateral lesions increased the degree but did not change the distribution of tone in these muscles on the side of the last ablation.

Lesions involving one supplementary motor cortex only produced increased tone in the above muscles of the arm and sometimes in three groups of muscles of the leg, that is, the protractors and the adductors of the thigh and the ventral flexors of the foot (Travis, 1955b). Again bilateral lesions give a greater increase in tone, bilaterally, than does a unilateral ablation upon contralateral musculature, but excess tone characterizes the flexors, not the extensors of the knee. Indeed, the clasp-knife type of resistance was described only once, and yet it is characteristic of the spastic state in man and in the monkey when the anterior part of or the whole of Brodmann's area 4 is removed.

Clonus did not succeed removal of the precentral motor cortex, of the supplementary motor cortex (Travis and Woolsey, 1956), of the parietal lobe (Peele, 1944), or of the cerebellar hemisphere. None is found after any type of lesion which produces hypotonia, although tremor does occur after removal of the cerebellar hemisphere when the nucleus dentatus is included (Carrea and Mettler, 1955). Clonus is found subsequent to ablation of the anterior half of area 4, of the whole of area 4, or to that of the combined removal of the supplementary motor cortex and the precentral motor cortex. After the last combined operation, clonus appeared in the flexors of the fingers and in the ventroflexors of the ankle, whereas, after a unilateral removal of either one of the former two, clonus could be elicited by sudden and maintained stretch contralateral to the lesion in the quadriceps femoris, the gastrocnemius-soleus group, the adductors of the thigh, the tibialis anterior, the peroneus longus, and the flexors of the toes. In the arm, clonus was more difficult to demonstrate; it was often found in biceps brachii, in the long flexors of the wrist and fingers, and rarely in the triceps brachii.

Tendon reflexes were brisk opposite the removal of the anterior part of area 4, of the total area 4, and of areas 4 and 6. In the former two preparations, the greater the relaxation, the brisker were these reflexes. In the last, the greater the relaxation in a quiet environment, the less brisk these reflexes became. After the former two types of removals of area 4, the tendon reflexes were found to be brisk in the flexors of the fingers, the M. flexor carpi ulnaris, the M. brachioradialis, the Mm. biceps and triceps brachii, the extensors of the knee, the flexors of the knee (i.e., the hamstrings), the ventroflexors of the ankle, and the M. tibialis anterior; and after the removal of anterior area 4 only, briskness in the flexors of the toes was added.

These brisk and hyperactive tendon reflexes frequently recruited others, which after birth involved more proximal muscles. Contraction of M. quadriceps femoris was accompanied by a strong contraction of the adductors of the thigh on the same or opposite side. The flexors of the ankle irradiated to the flexors of the knee, while the tendon reflexes of the flexors of the digits might spread to some or all of the flexors of the extremity concerned. The M. biceps brachii has recruited the adductors of the upper arm.

In general, removals of the precentral arm area and of the precentral leg area plus slight injury to the supplementary leg area (on the lower bank of the sulcus cinguli) produced hyperactive tendon reflexes (triceps and flexors of the fingers) without any sign of hypertonia in the leg. Unilateral removal of the precentral arm area alone was followed by hyperactive tendon reflexes in the triceps and in the flexors of the fingers without palpable resistance to passive movement (Travis, 1955a). On the other hand, increased resistance to retraction of the leg and to extension of the knee was reported present subsequent to lateral excision of both supplementary cortices in one monkey. Only one tendon reflex in the leg was listed as hyperactive, that of the flexors of the toes. Greater spread of resistance to passive movement characterized the upper extremity in this animal. Nevertheless, only the triceps and finger jerks were brisk (Travis, 1955b).

A different partition of the facets of spasticity marked the results of unilateral partial surgical division of the peduncle (Cannon *et al.*, 1943) and of two minute nicks in the white matter of the monkey's spinal cord (Wagley, 1945). Opposite the first lesion, tone was decreased and tendon reflexes were hyperactive without clonus. A cut into the ventral funiculus at T₇ secured on the side of the lesion a minimal increment in tone, clonus or tremor, and brisk irradiating reflexes; one in the ventral part of the lateral funiculus at the same level produced a differential distribution of resistance to passive movement, brisk and irradiating tendon reflexes on the ipsilateral side, and no clonus. A similar partition of the elements of spasticity occurred during the postnatal development of the infant macaque. The differential resistance to passive movement disappeared in time before the tendon reflexes had ceased their irradiation, and recruitment disappeared before all the tendon reflexes had assumed a normal status.

The tendon reflexes are pendular, exaggerated only in the arc of their excursion, after pyramidal section (Tower, 1940), after removal of the parietal lobe (Peele, 1944), and in the hypotonic state subsequent to ablation of the cerebellar hemispheres. In the pyramidal macaque, the tendon reflexes were slow, full, and unchecked. The ankle recruited the flexors of the knee. After removal of the postcentral gyrus and the posterior half of the precentral gyrus (Brodmann's area 4), the tendon reflexes were full, free, pendular, and recruited proximal muscles (Hines, unpublished).

In conclusion, the fact that stimulation of the lateral surface of area 6 or of the anterior border of area 4 evoked, respectively, inhibitory action against tone in the flexors or in the extensors, strongly suggests that localization of inhibitors of tone is not confined to the medial surface of areas 4 and 6 (Travis, 1955b). The addition of ablation of the precentral motor cortex to that of the supplementary motor cortex shifts the distribution of tone in the leg from the flexors to the extensors. Certainly, this finding insinuates that the former cortical area contains inhibitors of tone.

The evidence presented should persuade the most recalcitrant (but it will not) that facilitatory and inhibitory systems, separated from the corticospinal systems, can be activated in the reticular formation. Besides these fiber tracts clearly separated anatomically from the corticospinal tracts, there are others which are not. When the corticospinal systems are severed in the pyramids or spinal cord, either facilitatory systems are divided or facilitation is a part of the motor activity of these descending systems. Although the inhibitors of tone and movement are located in the midbrain tegmentum, a few inhibitory fibers lie dorsal to the pes pedunculi (Cannon *et al.,* 1944), continue caudalward into the pons with the corticospinal tracts (Tower, 1942), and become located in the reticular formation dorsal to the level of the pyramids.

The inhibitors of tone are a system apart from the corticospinal system, for the state of hypotonia caused by surgical division of the pyramids can be transformed by hypertonia by two types of additional operations, one upon the cerebral cortex, removing the anterior half of area 4 or the whole of area 4 (Hines and Tower, see Hines, 1943) and the other upon the spinal cord, cutting preferably the ventral half of the lateral funiculus (Wagley, 1945). For contrary results, see Clark and Ward, 1948, 1949.

IX. Phasic Movements

The phasic activity of experimental animals can be assessed (1) by coordinated responses of the extremities to external stimuli—the placing, stepping, and hopping reactions of Rademaker, and (2) by observation of spontaneous or self initiated movements. The coordinated movements of placing, stepping, and hopping are responses to tactile, proprioceptive, and visual stimuli.

X. Coordinated Movements as Response to Sensory Stimuli

Placing is a quickly executed slight retraction followed by protraction which lifts the leg over the edge of the table. Slight extension at the knee is followed by slight retraction of the leg and flexion at the knee, with almost simultaneous ventral flexion at the ankle. Contact is made on the digital and interdigital pads. The toes are loosely extended and slightly abducted. Placing with the arm requires similar adjustments at the girdle, i.e., retraction and protraction of the upper arm. With the retraction, the elbow is flexed; with the protraction, it becomes extended. The wrist is dorsoflexed and contact is made by the digital and interdigital pads. The thenar and hypothenar eminences do not touch the surface unless the monkey pauses to rest.

The position of the extremities when stepping or hopping has much in common with the final position assumed by the extremity when contact with the surface has been achieved. The normal adult macaque keeps up with the examiner's pace of transporting him through space, maintaining contact with the digital or interdigital pads of hand or foot. In the forward direction, the extremity alternately protracts and retracts in adduction. At maximum protraction or retraction, the second joint is about 160° extended. In between the maxima, the elbow or the knee is flexed and the ankle moves from a 90° to a 40° or less dorsoflexion. The retractors initiate stepping backward. The steps are shorter than in forward stepping and the extremity remains more flexed than extended. The contact of the foot or hand is made when dorsoflexion of these members is great. When the leg is in the protracted position, these distal joints are less dorsoflexed. Stepping to the right or left presents reciprocal images of each other. In stepping to the right, for example, the movement is initiated by the abductors of the leg, without changing the flexion at the second joint or at the distal one. The left leg follows quickly by protracting and adducting the thigh in one movement, so that as the advancing extremity crosses the right, it makes contact lateral and anterior to the locus momentarily occupied by the right foot. The arms follow a similar routine, except that the adducting extremity does not invariably cross over the abducting one.

The sequence of contraction of skeletal muscle which characterizes hopping is similar to that described for stepping, except that the degree of adduction is rarely great enough to meet the midplane of the body.

These reactions must meet three criteria to be judged normal. The movements must (1) be quickly, easily, and smoothly performed, (2) be in correct sequence without disintegration, and (3) make contact with the surface lightly and without hesitation. The number of steps or hops per meter varies with the individual animal. He establishes his own norm for postoperative comparison.

No deviation from the normal mode of placing was reported after lesion of the precentral gyrus or of the supplementary motor cortex (Travis, 1955a,b) nor after a small injury in the ventral funiculus of the spinal cord at T_7 (Wagley, 1945), nor after a complete ventral hemisection at that level (Hines, unpublished). Placing disappeared from the extremities contralateral to removal of Brodmann's area 4 (Hines, 1937), to removal of the frontal lobe (Travis and Woolsey, 1956), and to a cut into the dorsal half of the lateral funiculus (Cannon *et al.*, 1944). After unilateral pyramidal section, the whole movement is never executed by the contralateral extremities, and only fragmentary parts of the movement occur. Subsequent to a unilateral combined lesion of the precentral and supplementary motor cortices, placing was absent from the contralateral extremities until after comparable operations were performed on the other side. After 2 months, placing was "grossly defective" in all four extremities, but greatly facilitated by excitement. After removal of the parietal lobe, tactile placing disappears, proprioceptive placing returns, and visual placing remains. The motor reponse was "slower than normal in initiation and execution and frequently awkward" (Peele, 1944).

Given time for recovery, unilateral lesions of the precentral motor cortex or of the supplementary motor cortex, hopping was reported to be normal. Stepping was not tested (Travis, 1955a,b). Small lesions in the ventral funiculus or in the ventral part of the lateral (Wagley, 1945) did not change the mode of the stepping or of the hopping reactions. But when a lesion in the lateral funiculus transgresses upon the site of the lateral corticospinal tracts, the act of stepping or of hopping is modified by inability of the retractors and adductors of the leg to contract.

Conjoint removals of the precentral and supplementary motor cortices is initially followed by the inability of the contralateral extremities to hop at all; later, however, hopping forward and laterally is accomplished in a crude fashion (Travis and Woolsey, 1956). Unilateral ablation of Brodmann's area 4 gives a similar result, for both stepping and hopping, medially and backward, by the extremities opposite the lesion has vanished. If these removals become bilateral, the arms refuse to step or hop (Hines, 1943).

Unilateral frontal lobectomy caused disapperance of contralateral hopping; in bilateral lobectomy, the hopping forward and laterally reappeared and became eventually similar for each leg, and infrequently for each arm. The manner of use was not described (Travis and Woolsey, 1956).

The effort necessary to innervate contralateral mulculature after surgical division of one pyramid made the initiation of hopping difficult. Like placing, the hopping was fragmentary and ceased after a few efforts because of fatigue. Speeding the translation of the animal over the table improves the sequence of the movements of the legs forward and lateralward. The arms do not attempt to move.

The most severe loss of these coordinated adjustments occurred after cutting bilaterally the nerve fibers in the site of the corticospinal tracts (T_7). All hopping disappeared; only forward stepping remained. The toes turned under and the ankle was incompletely dorsoflexed.

These coordinated adjustments to somaesthetic and visual stimuli are not greatly modified when the extrapyramidal systems in the spinal cord are interrupted. Rather, they become markedly changed whenever the corticospinal systems are severed in the pyramids or in the spinal cord. The loss of large cortical areas does not obliterate either stepping or hopping completely in the lower extremity, although these losses may do just that for the upper extremity. The lower extremity does lose its ability to step or hop backward or toward the medial plane of the body. The performance of the leg in stepping or hopping forward or laterally can be said to be present, but the finer adjustments of dorso-flexion at the ankle, the loose easy extension, and slight adduction of the toes are gone. Contact with the surface is no longer made by the digital and interdigital pads. The heel lands heavily and the toes no longer orient to the surface.

XI. Spontaneous or Self-Initiated Movements

The mass organization of muscle (Tower, 1940, 1949) which survives bilateral surgical division of the pyramids in the monkey depends upon the intactness of extrapyramidal systems. These extrapyramidal motor systems have not been specifically identified as yet. That the use of musculature made by the bilateral 4 and 6 (Brodmann's areas) preparation is similar to that made by the bilateral pyramidal monkey suggests that, aside from the difference in muscle tone in the animals, cutting the parietospinal fibers has not added any factor to the loss of discrete use of skeletal muscle. The temporal lobe and the occipital lobe in the cat have been assigned an appreciable contribution of very finely medullated fibers to the pyramid. Even if these subdivisions of the corticospinal system were present in the monkey (Marchi preparations fail to show them), their contribution cannot be assayed by cutting them in conjunction with those from the frontal and parietal lobes.

Surgical division of the pyramids reduces the adult monkey to an infantile age of 7 days, a time at which the movement which enabled the infant to get hold of an object was a reaching and grasping act. This act is prepared for by retraction of the upper arm and flexion of the elbow, followed by a limited protraction and extension of the elbow, with the forearm in incomplete pronation. The hand is brought down upon the object with fingers partially flexed. Should the position of the desired object be shifted before this act is complete, the movement continues to the end, for a new orientation has to be made and another similar sequence of contraction of skeletal muscles has to be originated. The threshold for initiation of these movements is high, their activity easily fatigued (Tower, 1940, 1949).

In a similar manner, the 4 and 6 monkey (Hines, unpublished), in preparation for taking an object, retracts the upper arm, flexes the elbow, then protracts the upper arm, extends the elbow with the forearm in almost complete pronation with the fingers equally extended, and reaches the object (Fig. 9). The fingers flex about it and the arm returns to the original position. During the progress of these mass movements, the opposite hand grasps the cage bars or any other object for support. If the monkey discovers that readjustment is necessary to reach the desired object, she returns to the original position. She makes her feet secure by grasping the seat, leans slowly forward, changing her grasp to the side of the cage as she progresses, holds the mesh, and takes the object from the observer's hand with her mouth. Slowly she retraces her grasp upon the cage mesh, sits up, back curved, leaning against the cage or steadying herself by grasp upon the cage bars. She takes the object out of her mouth, and holding it between equally flexed fingers and thumb, meets by flexion of the head the limited protraction of the arm. (See also Fig. 10 for difficulty in turning.)

The bilateral area 4 preparation (Brodmann) does not move into position to initiate this act. Rather, that animal is able to begin the movement by slight protraction of the upper arm, followed by extension at the elbow, incomplete pronation of the forearm, equal flexion of the fingers, and flexion of the thumb. After grasping the object, the upper arm is retracted, the elbow flexed, the forearm in the 90° pronation/supination position. If the object is small it will be held between a flexed index finger and an adducted thumb (Fig. 7). The flexion of the elbow will meet a head, flexed upon a flexed trunk. (See Fig. 8 also for Magnus (1924) head and body reflexes.)

Each of these three preparations starts the rhythmic movements of scratching at the girdles. No readjustment is made at the knee, the aim is poor. If the right spot is not met, readjustment must be started again at the girdle. Initiation of all movements is slow, and their execution fatiguing. Beginning in the proximal muscles, the sequence of contraction flows down the arm similar to "the march of movement" described by Hughlings Jackson for man after severe injury to "the motor aspect of the mind" in the frontal lobe.

In these animals, proximal initiation of movement is a condition dependent upon intact extrapyramidal systems (Goldberger, 1969). The muscles no longer play the part of fixers, for they contract visibly and excessively, initiating a synergy which flows down the extremity. This exaggerated synergistic use of muscle by the pyramidal animal has been interpreted as a facet of release (Goldberger, 1969). Moreover, in the normal monkey, this investigator considers that the pyramidal system exerts an inhibitory action that is able to split the synergy of extrapyramidal origin.

In the pyramidal preparations presented by Lawrence and Kuypers (1965, 1968), both pyramids were surgically severed in the medulla oblongata. Frequently the surgical knife cut into that part of the reticular formation found dorsal to the pyramids, thus adding to the former lesion a·second in the extrapyramidal systems.

After the localization which characterizes the pyramidal system from motor cortex to motor neurone has been removed by bilateral division of the pyramids, it is surprising to discover as did Lawrence and Kuypers that the extrapyramidal systems present a comparable relation to the motor neurons of the spinal cord via synapses with tract terminals and interneurons. Cutting into the ventrolateral quadrant of white matter of the spinal cord resulted in impairment of independent movements of the distal extremity and inability to flex the extended limb. On the other hand, the interruption of the ventromedial pathways resulted in severe impairment of axial and proximal muscles.

This localization depends upon the synapses of nerve fibers in the ventrolateral funiculus with interneurons and motor neurons of the lateral division of the ventral horn and upon comparable synapses of the endings of the ventromedial funiculus with a comparable arrangement of nerve cells in the medial division of the ventral horn.

Wagley's (1945) analysis of the results of similarly placed lesions in the spinal cord and of those superimposed contralaterally to a severed pyramid did not resemble the above findings. Indeed, in lesions confined to the spinal cord, the degree of paralysis varied directly with the degree of damage to the corticospinal tract in the dorsal part of the lateral funiculus. When this lesion included extrapyramidal systems ventral to the corticospinal tract, hypertonus appeared. Such hypertonus included all, or only part of these facets of hypertonus, namely, brisk irradiating tendon reflexes, lengthening and shortening reactions, and clonus. Whenever interruption of the corticospinal tract in the spinal cord or at the pyramids was succeeded by that of the extrapyramidal systems in the spinal cord, the hypotonia which followed the first operation became hypertonus. Paralysis remained.

In the developmental maturation of the discrete use of skeletal muscle, the infant macaque moves the proximal muscle into position and then innervates the more distally placed muscles (Hines, 1942). During the time span when this was a common occurrence, it was possible to obtain fixation as a separate event by stimulating the lateral surface of area 6 with the 60 cps sine wave current. Similarly, during this same time in postpartum development, the electric current is able to elicit contraction of a muscle plus fixation of more proximally lying muscles, preceding or accompanying the contraction (Hines and Boynton, 1940). The locus

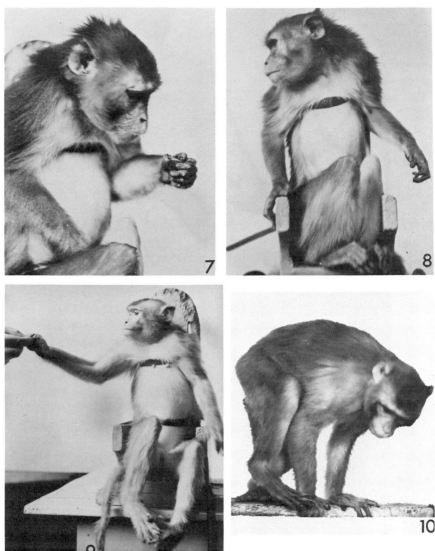

Figs. 7–10. (*See facing page for individual legends.*) Photographs of two macaques (*Macaca mulatta*), H 26 and T 55. H 26 was 12 years of age when the left Brodmann's area 4 (sparing the face area) was removed on December 16, 1936. The right Brodmann's area 4 (sparing the face area) was removed May 11, 1938. T 55 ultimately became a bilateral 4 and 6 preparation (sparing the face areas). The operations were performed as follows: left area 6 removed, June 16, 1939; left area 4, October 19, 1939; and right areas 4 and 6, October 24, 1940. She was able to oppose the extended right thumb to an extended index finger, in spite of the removal of all cortical tissue even to the depth of the central fissure. On the left, she could with a flexed thumb touch a flexed index finger. The remaining

of this stimulus was topically typical of the muscles which contracted actively, not of those which were fixed.

In summary, after each of the above operations that destroy the corticospinal tracts, fixation of proximal musculature disappears and with it goes the discrete use of certain muscles. For example, subsequent to bilateral ablation of area 4 some of these muscles, which were used to initiate movement can be used only in a sequence of movement, others cannot. Still others, which are able to initiate movement, present a limited degree of contraction.

This selective and peculiarly baffling paralysis falls naturally into three categories: (1) complete loss of use of muscle, or paralysis, (2) loss of ability of muscles to act as prime movers although able to participate in generalized movement, and (3) retention of initiation of movement, although the degree of contraction is less than normal. Apposition of the thumb; flexion and extension of individual fingers; flexion and extension of toes, opposition of the hallux; dorsoflexion, ulnar flexion, and radial flexion of the wrist; dorsoflexion and eversion of the ankle; and adduction of the thigh and upper arm are gone. The flexion or extension of all fingers, adduction of the thumb, ventroflexion of the wrist, pronation and supination of the forearm, and ventroflexion and inversion of the foot at the ankle are a part of generalized, sequential movements. A few proximally placed muscles retain the ability to initiate movements,

---

three fingers were also flexed. On August 20, 1942, I opened her skull and found that a minute strip of tissue which at that time lay anterior to the central fissure responded to electrical stimulation with extension of the thumb and extension of the first finger. More dorsally, flexion of the elbow and retraction of the arm were elicited. This strip of tissue was removed and with it varnished the ability to oppose the thumb to an extended index finger.

Fig. 7. Photograph of H 26 taken October 23, 1940. She is holding a small piece of fruit between an adducted left thumb and flexed index finger. The shoulders are raised because of the increased tone in the M. trapezius. Although she could flex the elbow, she flexes her head on a flexed trunk to meet a flexing arm, because protraction of the upper arm is limited.

Fig. 8. Photograph of H 26, taken October 23, 1940, illustrating the associated movement (in reality, the head on the body reflexes of Magnus, 1922) which accompanies voluntary turning of the head to the right. The left arm is abducted and extended. Note that she is able to place her feet on the seat of the examining chair. Her toes are more extended than flexed.

Fig. 9. Photograph of T 55 taken on March 14, 1941, showing that she is able to take a raisin from the hand of an examiner by flexing her thumb and meeting a flexed index finger. The other three fingers are flexed. She prepared for this movement by retraction of the upper arm and flexion at the elbow. She was able to elevate the arm to a greater degree than she could protract it.

Fig. 10. Photograph of T 55, taken May 12, 1941, showing the difficulty experienced in turning to the right on a small base. The right foot is lifted, but the left foot did not move until after the photograph was taken.

although the degree of their contraction is less than normal. These muscles are the extensors and flexors of the arm and leg, protractors of the leg, retractors of the arm, and abductors of the arm and of the leg.

Therefore, the selection of the prime movers, the action of the antagonists and synergists, and fixation at the girdle or trunk seem to be under the control of the descending systems which pass through the pyramids, because these activities disappear whenever the corticospinal systems are bilaterally interrupted. Further evidence for this generalization is given by electrical stimulation of one cortical surface after the ipsilateral pyramid has been cut. Under this condition, the electrode on the motor cortex does not evoke (1) contraction of single muscles or parts of muscles, (2) contractions of flexor or of extensor sheets of muscles, nor (3) sequential contractions which resemble the use patterns characteristic of the animal (Hines, 1940). Although fixation has been elicited by stimulating area 6 as a separate event and by stimulating area 4 as preceding or accompanying a movement, neither Dr. Boynton nor myself was able to obtain any type of fixation subsequent to surgical division of a pyramid. The phasic qualities of the discrete use of skeletal muscle are dependent upon the intactness of the corticospinal systems. Furthermore, without these important descending systems, initiation of movement is difficult and slow, aim is poorly realized, and the monkey is unable to change his mind after a movement has been initiated. Indeed, the extrapyramidal systems which remain are not able to compensate for this loss, even when given years (Fig. 10) to do so.

XII. Conclusions

Skeletal muscle is controlled at the spinal cord level by the activity of two types of endings, the Golgi tendon terminal and the equatorial endings on the intrafusal fibers of the muscle spindle. The Golgi endings have a high threshold to stretch and are sensitive to the tension of contraction. The equatorial ending of the spindle is very sensitive to stretch. This ending can be activated directly by stretch of the muscle which contains it or indirectly by the activity of small nerve fibers which innervate the intrafusal fibers of the spindle. This arrangement is known as the gamma (γ) loop.

The cephalad centers of the central nervous system control skeletal muscle by direct innervation of ventral horn cells or by discharges to the intrafusal fibers which fire the equatorial sensory endings. The gamma loop of the spindle is accessible to electrical stimulation not only of facilitators and inhibitors of tone found in the reticular formation, but also of the motor cortex, of the cerebellum, and of the nerve fibers within the pyramids (cat; Granit, 1956).

The normal regulation of small nerve activity to the spindle of skeletal muscle is the result of a combination of different discharges from the skin, muscle, and possibly the deep structures. This combination of exteroceptive and proprioceptive discharge may be a part of a peculiarly important pattern in the regulation of activity of skeletal muscle (Hunt, 1951).

The results of this analysis of the use of skeletal muscle by the macaque cannot be transferred directly to man. Rather, they are only suggestive. Although, in general, comparable results follow comparable lesions, there are several instances in which supposedly similar lesions do not apparently give similar results. Cutting the lateral two-thirds or four-fifths of the cerebral peduncle produced in the monkey a hypotonic paralysis with brisk tendon reflexes (Cannon *et al.*, 1944). Not so in man, for surgical division of the lateral two-thirds to four-fifths of the cerebral peduncle for hemiballismus or Parkinsonian tremor stopped the pathological movement and left a few of the patients with "little paralysis" (how little, or what was paralyzed, was not mentioned) and with no increase in their previous spastic state (Bucy, 1957; Bucy *et al.*, 1964; Walker, 1952, 1955). At present there is no explanation of the variation of the degree of paralysis produced by apparently similar lesions. This inconsistency could be due to dissimilar preoperative brain damage, to difference in the lesions which resulted from what seemed to be similar lesions, or to variations in the disposition of the corticospinal systems in the peduncle.

In man, as in the monkey (Travis, 1955b), no permanent paralysis follows removal of tissue in the medial surface (area 6) of the frontal lobe (Erickson and Woolsey, 1951); but, unlike the monkey, the change in tone is confined to the tonic innervation of the grasp, and again unlike the monkey, the intact supplementary motor cortex in man is apparently unable to prevent paralysis from following small ablations of the precentral gyrus. "If the hand area is completely removed, the hand becomes paralyzed for any skilled movement whatever" and "if the removal is small, the delicate movements of the fingers and thumb disappeared, although the movements of the digits altogether in flexion or extension and movements of the wrist, elbow and shoulder may be produced" (Penfield and Erickson, 1941). Foerster (1936a) also found that the muscles of the hand or fingers were not implicated where a paralysis of the shoulder, upper arm, and forearm existed. Conversely, a circumscribed loss of use of the muscles of the hand and fingers followed removal of the hand–finger area alone. After years of systematic training, the patient learned to hold a pen and write. Each movement of the right hand, however, was accompanied by a similar one of the left (normal). Foerster considered this relearning to be dependent upon the ipsilateral corticospinal tract. It is possible that the ability to supinate (45°) the forearm remained and if so it could be utilized in the retraining process.

Even a year after the ablation of the whole precentral gyrus, when Marinesco's (1903) patient picked up an object, the arm moved into retraction and flexed as preparation for protraction and extension. In this movement, the wrist was in complete pronation and the fingers were extended. After touching the object, the fingers flexed tightly upon the object without the synergistic action of the dorsoflexors of the wrist.

Since isolation of the precentral gyrus—made by removal of all surrounding cortical area (Penfield, 1954)—is not followed by any deviation from the normal use of skeletal muscle, the "motor" cortex in man seems to be independent of interregional cortical connection. That the cerebellothalamocortical system can give the information about the body necessary to "run" the precentral gyrus is doubtful. The new spinocortical system carrying proprioceptive sensibility to the motor cortex may have to be evoked—if, of course, it is eventually discovered as an ascending system in the pyramids of man.

Strangely enough, large removals of cortical tissue are not counterindicated, for Welch and Penfield (1950) report that ablations of diseased tissue surrounding and even including the precentral gyrus (man) decreased the preoperative spasticity without increasing the paralysis. Similarly, subsequent to removal of a whole hemisphere for infantile hemiplegia, the resulting decrease in spasticity seems to free the extremities for greater use in some patients. (Krynauw, 1950; Bates and McKissoch, 1951). This greater use rarely included a real improvement in that of the muscles of the hand, although many of these patients managed to walk with greater ease after the operation. Similarly, large ablations of normal cortical tissue (frontal lobe plus parietal lobe) from the macaque caused the upper extremity opposite the lesion to be demoted from an organ of exploration of space to that of a moving support incapable of initiation of progression (Travis and Woolsey, 1956).

The final and minute analysis of the ability of the macaque to direct his skeletal muscle toward the achievement of a desired end (object visible or otherwise) awaits completion. We do know, however, that increased or decreased tone and facilitation of movement are contributions of descending systems not located in the medullary pyramids. Selection of prime movers, the activity of antagonists and of synergists are impossible in the absence of the corticospinal systems. Fixation proximal to contracting muscles, that prerequisite of all discrete movement, is dependent upon the existence of the corticospinal systems which stem from the frontal lobe; fixation at the girdles, however, is dependent upon this corticospinal system and another from the parietal lobe, as well as upon the lateral hemispheres of the cerebellum. The mass organization of skeletal muscle innervated proximally survives injury to the corticospinal systems, whereas the discrete organization of that tissue innervation distally does not. Ease of initiation of all movements, the ability to reach for and take the desired object (without movement into position), the ability to stop sequential movements and to start that sequence again toward the same or another end, and the ability to split and utilize the synergies of mass movements are characteristic of the corticospinal systems' contribution to muscle use.

The facilitation of movement so necessary for ease of initiation and the shifts in degrees of tone from one muscle group to another as the intended sequence of movements develops are dependent upon the extrapyramidal systems. These systems control the proximal initiation of contraction of muscles of the whole extremity as flexor or extensor synergies. This use of muscle contributes force and smoothness to the resultant movement (Fig. 10).

In the absence of the extrapyramidal systems, the remaining corticospinal systems are not able to raise the central excitatory state to the level where movement is easy to initiate. After loss of the corticospinal systems, the initiation of all movement is delayed, its progress slowly executed, and its aim rarely realized. Fatigue is indeed great. Consequently, an accompanying emotional state transforms this picture of results of loss of the corticospinal tracts by raising the central excitatory state so that the latent period for beginning of the movement is shortened, the total sequence of the movement more quickly executed, and the aim frequently realized. Fatigue, although present, is less. The hypokinesis of extrapyramidal lesion seems to be restricted only to the reluctance to start a movement. Once begun, the movement appears to be normal in ease of execution, in sequence of muscle contraction, and in manner of use of distal musculature. Fatigue, if present, is not apparent. Certainly, the corticospinal systems are able to utilize the lower centers of the brain stem and spinal cord to produce with the aid of the extrapyramidal systems those movements which distinguish man from other primates and other primates from other mammals.

I have attempted to analyze the relationship of the central nervous system in the primate—particularly of the cortex cerebri via the corticifugal systems—to the control of the use of skeletal muscle. Muscle is that tissue which expresses our slowly and sometimes painfully achieved education and confers upon us, as men, the gift of communication.

REFERENCES

Allen, W. F. (1923). *J. Comp. Neurol.* 35, 275.
Bach, L. M. N., and Magoun, H. W. (1947). *Fed. Proc., Fed. Amer. Soc. Exp. Biol.* 6, 70.
Bates, J. A. V. (1953). *Brain* 76, 405.
Bates, J. A. V., and McKissoch, W. (1951). *J. Physiol. (London)* 115, 51.
Beevor, C. E. (1903). *Lancet* 1, 1715 and 1783.
Bieber, I., and Fulton, J. F. (1938). *Arch. Neurol. Psychiat.* 39, 435.
Bodian, D. (1946). *Proc. Soc. Exp. Biol. Med.* 61, 170.
Bosma, J. F., and Gellhorn, E. (1946). *J. Neurophysiol.* 9, 263.
Botterell, E. H., and Fulton, J. F. (1938). *J. Comp. Neurol.* 69, 63.
Bremer, G. (1935). *C. R. Soc. Biol.* 118, 1241.
Brodal, A. (1956). *Progr. Neurobiol.* 111, 210.
Brodal, A., and Wahlberg, F. (1952). *AMA Arch. Neurol. Psychiat.* 68, 755.
Brodmann, K. (1903). *J. Psychol. Neurol.* 2, 79.
Brodmann, K. (1906). *J. Psychol. Neurol.* 6, 275.
Brongeest, P. Q. (1860). "De Tono Musculorum Voluntati Subditorum," x, p. 91. Utrecht.
Bucy, P. C. (1957). *Brain* 80, 376.
Bucy, P. C., and Klüver, H. (1940). *AMA Arch. Neurol. Psychiat.* 44, 1142.
Bucy, P. C., Keplinger, J. E., and Siqueira, E. B. (1964). *J. Neurosurg.* 21, 385.
Bucy, P. C., Ladpte, R., and Ehrlich, A. (1966). *J. Neurosurg.* 25, 1.
Cannon, B. W., Beaton, L. E., and Ranson, S. W., Jr. (1943). *J. Neurophysiol* 6, 425.
Cannon, B. W., Magoun, H. W., and Windle, W. F. (1944). *J. Neurophysiol.* 7, 425.
Carpenter, M. B. (1956). *J. Comp. Neurol.* 105, 195.
Carpenter, M. B., Whittier, J. R., and Mettler, F. A. (1950). *J. Comp. Neurol.* 92, 293.
Carrea, R. M. E., and Mettler, F. A. (1954). *J. Comp. Neurol.* 101, 565.
Carrea, R. M. E., and Mettler, F. A. (1955). *J. Comp. Neurol.* 102, 151.
Chang, H. T., Ruch, T. C., and Ward, A. A. (1947). *J. Neurophysiol.* 10, 39.
Charcot, J. M., and Pitres, A. (1895). "Les centres moteurs corticaux chez l'homme," Ruess et cie. Paris.
Clark, G., and Ward, J. W. (1948). *Brain* 71, 332.
Clark, G., and Ward, J. W. (1949). *Amer. J. Physiol.* 158, 474.
Crosby, E. C. (1956). *Progr. Neurobiol.* 3, 217.
Dejerine, J. (1901). "Anatomie des centres nerveux," Vol. 2, No. 1, p. 60.
Denny-Brown, D., and Botterell, E. H. (1948). *Res. Publ., Ass. Res. Nerv. Ment. Dis.* 27, 235.
Dusser de Barrene, J. G., Garol, H. W., and McCulloch, W. S. (1940). *Res. Publ., Ass. Res. Nerv. Ment. Dis.* 21, 246.
Eccles, J. C. (1957). "The Physiology of Nerve Cells," p. 270. Johns Hopkins Univ. Press, Baltimore, Maryland.

Erickson, T. C., and Woolsey, C. N. (1951). *Trans. Amer. Neurol. Ass.* **76**, 50.

Ferraro, A., and Barrara, S. E. (1935). *Brain* **58**, 174.

Ferrier, D. (1876). "Functions of the Brain," p. 323. London.

Fleming, J. F. R., and Crosby, E. C. (1955). *J. Comp. Neurol.* **103**, 485.

Foerster, O. (1923). *Deut. Z. Nervenheilk.* **77**, 124.

Foerster, O. (1931). *Lancet* **221**, 309.

Foerster, O. (1936a). "Handbuch der Neurologie," Vol. 6, p. 48. Bumke & Foerster, J. Springer, Berlin.

Foerster, O. (1936b). *Brain* **59**, 135.

Foerster, O., and Gagel, O. (1932). *Z. Gesamte Neurol. Psychiat.* **138**, 1.

Fulton, J. F., and Kennard, M. A. (1932). *Res. Publ., Ass. Res. Nerv. Ment. Dis.* **13**, 158.

Carol, H. W., and Bucy, P. C. (1944). *Arch. Neurol. Psychiat.* **51**, 528.

Gerebtzoff, M. A. (1940). *AMA Arch. Intern. Physiol.* **50**, 59.

Gillilan, L. A. (1941). *J. Comp. Neurol.* **74**, 367.

Goldberger, M. E. (1969). *J. Comp. Neurol.* **135**, 1.

Granit, R. (1956). "Receptors and Sensory Perception," p. 368. Yale Univ. Press, New Haven, Connecticut.

Hering, H. E., and Sherrington, C. S. (1897). *Arch. Gesamte Physiol. Menschen Tiere* **68**, 222.

Hines, M. (1936). *Amer. J. Physiol.* **116**, 76.

Hines, M. (1937). *Bull. Johns Hopkins Hosp.* **60**, 313.

Hines, M. (1940). *J. Neurophysiol.* **3**, 442.

Hines, M. (1942). *Carnegie Inst. Wash. Publ.* **541**, 153.

Hines, M. (1943). *Biol. Rev. Cambridge Phil. Soc.* **18**, 1.

Hines, M. (1947). *Fed. Proc., Fed. Amer. Soc. Exp. Biol.* **6**, 441.

Hines, M. (1949). *In* "Precentral Motor Cortex" (P. C. Bucy, ed.), Chapter 18, p. 359. Univ. of Illinois Press, Urbana.

Hines, M., and Boynton, E. P. (1940). *Contrib. Embryol. Carnegie Inst.* **28**, 309.

Hines, M., and Knowlton, G. C. (1952). *Res. Publ., Ass. Res. Nerv. Ment. Dis.* **30**, 98.

Holmes, G., and May, W. P. (1909). *Brain* **32**, 1.

Horsley, V., and Schäfer, E. A. (1888). *Phil. Trans. Roy. Soc. London* **179**, 1.

Hunt, C. C. (1951). *J. Physiol. (London)* **116**, 456.

Hunt, C. C. (1952). *J. Physiol. (London* **117**, 359.

Hunt, C. C., and Kuffler, S. W. (1951a). *J. Physiol. (London)* **113**, 283.

Hunt, C. C., and Kuffler, S. W. (1951b). *J. Physiol. (London)* **113**, 298.

Jackson, J. S. (1932). *In* "Selected Writings" (J. Taylor, ed.), Vol. 2, pp. 3–39. Hodder, London.

Kennard, M. A., and Kessler, M. M. (1940). *J. Neurophysiol.* **3**, 248.

Krynauw, R. A. (1950). *J. Neurol., Neurosurg. Psychiat.* **13**, 243.

Kuhn, R. A. (1950). *Brain* **73**, 1.

Kuypers, H. G. J. M. (1960). *Brain* **83**, 161.

Kuypers, H. G. J. M. (1964). *Progr. Brain Res.* **11**, 178.

Laporte, Y., Lundberg, A., and Oscarsson, O. (1956). *Acta Physiol. Scand.* **36**, 188.

Lawrence, D. C., and Kuypers, H. G. J. M. (1965). *Science* **148**, 973.

Lawrence, D. C., and Kuypers, H. G. J. M. (1968). *Brain* **91**, 1 and 15.

Lemmen, L. J. (1951). *J. Comp. Neurol.* **95**, 521.

Levin, P. N. (1936). *J. Comp. Neurol.* **63**, 369.

Levin, P. M., and Bradford, F. K. (1937). *J. Comp. Neurol.* **68**, 411.
Leyton, A. S. F., and Sherrington, C. S. (1917). *Quart. J. Exp. Physiol.* **11**, 135.
Liu, E. N., and Chambers, W. W. (1964). *J. Comp. Neurol.* **123**, 252.
Lloyd, D. P. C. (1941). *J. Neurophysiol.* **4**, 115.
Magnus, R. (1922). *Pflueger's Arch. Gesamte Physiol. Menschen Tiere* **193**, 396.
Magnus, R. (1924). Körperstellung, J. Springer Berlin.
Magoun, H. W., and Rhines, R. (1946). *J. Neurophysiol.* **9**, 165.
Magoun, H. W., and Rhines, R. (1947). "Spasticity: The Stretch Reflex and Extra Pyramidal Systems," p. 59. Thomas, Springfield, Illinois.
Marburg, O. (1903). *Arb. Neurol. Inst. Univ. Wien.* **10**, 66.
Marinesco, M. G. (1903). *Sem. Med.* **23**, 325.
Mettler, F. A. (1935). *J. Comp. Neurol.* **61**, 221 and 509.
Mettler, F. A. (1936). *Arch. Neurol. Psychiat.* **35**, 1338.
Mettler, F. A. (1941). *Res. Publ., Ass. Res. Nerv. Ment. Dis.* **21**, 150.
Mettler, F. A. (1943). *J. Comp. Neurol.* **79**, 185.
Mettler, F. A., (1944). *J. Comp. Neurol.* **81**, 105.
Mettler, F. A. (1948a). *J. Comp. Neurol.* **86**, 119.
Mettler, F. A. (1948b). *Res. Publ., Ass. Res. Nerv. Ment. Dis.* **27**, 162.
Meyers, R., Knott, J., Skultety, M., and Imler, R. (1953). *Trans. Amer. Neurol. Ass.* **153**, 189.
Minckler, J., Klemme, R. M., and Minckler, D. (1944). *J. Comp. Neurol.* **91**, 259.
Morin, F. (1953). *Amer. J. Physiol.* **172**, 483.
Morin, F., Schwartz, H. G., and O'Leary, J. L. (1951). *Acta Psychiat. Neurol. Scand.* **26**, 371.
Moruzzi, G., and Pompeiano, O. (1957). *J. Comp. Neurol.* **107**, 1.
Niemer, W. T., and Magoun, H. W. (1947). *J. Comp. Neurol.* **87**, 367.
Orioli, F. L., and Mettler, F. A. (1957). *J. Comp. Neurol.* **107**, 305.
Peele, T. L. (1942). *J. Comp. Neurol.* **77**, 693.
Peele, T. L. (1944). *J. Neurophysiol.* **7**, 269.
Peele, T. L. (1948). Personal communication.
Penfield, W. (1954). *Brain* **77**, 1.
Penfield, W., and Erickson, T. C. (1941). "Epilepsy and Cerebral Localization," x, p. 623. Thomas, Springfield, Illinois.
Penfield, W., and Rasmussen, T. (1952). "The Cerebral Cortex of Man," p. 248. Macmillan, New York.
Penfield, W., and Welch, K. (1951). *AMA Arch. Neurol. Psychiat.* **66**, 289.
Peterson, E. W., Magoun, H. W., McCulloch, W. S., and Lindsley, D. B. (1949). *J. Neurophysiol.* **12**, 371.
Preston, J. B., and Whitlock, D. G. (1960). *J. Neurophysiol.* **23**, 154.
Preston, J. B., and Whitlock, D. G. (1961). *J. Neurophysiol.* **24**, 91.
Ramóny Cajal, S. (1909). "Histologie du système nerveux de l'homme et des vertébrés," Vol. 1. Maloine, Paris.
Ranson, S. W. (1939). *Arch. Neurol. Psychiat.* **41**, 1.
Richter, C. P., and Hines, M. (1934). *Res. Publ., Ass. Res. Nerv. Ment. Dis.* **12**, 211.
Rossi, J. F., and Brodal, A. (1955). *J. Anat.* **90**, 42.
Scheibel, M., Scheibel, A., Mollica, A., and Moruzzi, G. (1955). *J. Neurophysiol.* **18**, 309.
Schneider, R. C., and Crosby, E. C. (1954). *Neurology* **4**, 612.

Smith, W. K. (1949). *In* "Precentral Motor Cortex" (P. C. Bucy, ed.), Chapter XII, p. 301. Univ. of Illinois Press, Urbana.

Spielmeyer, W. (1906). *Muenchen. Med. Wochenschr.* **53**, 1404.

Sprague, J. M., and Chambers, W. W. (1953). *J. Neurophysiol.* **16**, 451.

Sprague, J. M., and Chambers, W. W. (1954). *Amer. J. Physiol.* **176**, 52.

Starzl, T. E., and Whitlock, D. G. (1952). *J. Neurophysiol.* **15**, 449.

Stewart, D. H., and Preston, J. B. (1968). *J. Neurophysiol.* **31**, 938.

Stewart, D. H., Preston, J. B., and Whitlock, D. G. (1968). *J. Neurophysiol.* **31**, 928.

Sugar, O., Chusid, J. G., and French, J. D. (1948). *J. Neuropathol. Exp. Neurol.* **7**, 182.

Szentágothai, J., and Albert, A. (1955). *Acta Morphol. Acad. Sci. Hung.* **5**, 43.

Therman, P. O. (1941). *J. Neurophysiol.* **4**, 153.

Tower, S. S. (1940). *Brain* **63**, 36.

Tower, S. S. (1942). *Anat. Rec.* **82**, 450.

Tower, S. S. (1949). *In* "Precentral Motor Cortex" (P. C. Bucy, ed.), Chapter VI, p. 150. Univ. of Illinois Press, Urbana.

Travis, A. M. (1955a). *Brain* **78**, 155.

Travis, A. M. (1955b). *Brain* **78**, 174.

Travis, A. M., and Woolsey, C. N. (1956). *Amer. J. Phys. Med.* **35**, 273.

van Harreveld, A., and Marmost, I. (1939). *J. Neurophysiol.* **2**, 101.

Verhaart, W. J. C., and Kennard, M. A. (1940). *J. Anat.* **74**, 239.

Vogt, C., and Vogt, O. (1926). *Naturwissenschaften* **14**, 1191.

von Monakow, C. (1909). *Hirn. Anat. Inst. Univ. Zurich* **3**, 49.

von Monakow, C. (1910). *Hirn. Anat. Inst. Univ. Zurich* **5**, 103.

Wagley, P. F. (1945). *Bull. Johns Hopkins Hosp.* **77**, 218.

Wahlberg, F., and Brodal, A. (1953). *Brain* **76**, 491.

Walker, A. E. (1938). "The Primate Thalamus," p. 321. Chicago, Illinois.

Walker, A. E. (1952). *J. Nerv. Ment. Dis.* **116**, 766.

Walker, A. E. (1955). *Surg., Gynecol. Obstet.* **100**, 716.

Walker, A. E., and Weaver, T. A. (1940). *J. Neurophysiol.* **3**, 353.

Walshe, F. M. R. (1929). *Lancet* **1**, 963.

Ward, A. A. (1947). *J. Neurophysiol.* **10**, 89.

Welch, K., and Penfield, W. (1950). *J. Neurosurg.* **7**, 414.

Winkler, C. (1929). "Manuel de Neurologie," Part 1. Haarlem.

Wohlfahrt, S. (1932). *Acta Med. Scand., Suppl.* **46**, 1.

Woolsey, C. N., and Chang, H. T. (1948). *Res. Publ., Ass. Res. Nerv. Ment. Dis.* **27**, 146.

Woolsey, C. N., and Fairman, D. (1946). *Surgery* **19**, 684.

Woolsey, C. N., Marshall, W. H., and Bard, P. (1942). *Bull. Johns Hopkins Hosp.* **70**, 399.

Woolsey, C. N., Settlage, P. H., Sencer, W., Hamuy, T. P., and Travis, A. M. (1952). *Res. Publ., Ass. Res. Nerv. Ment. Dis.* **30**, 238.

Woolsey, C. N., Travis, A. M., Barnard, J. W., and Ostenso, R. S. (1953). *Fed. Proc., Fed. Amer. Soc. Exp. Biol.* **12**, 160.

ELECTROMYOGRAPHY

JOHN V. BASMAJIAN

I. Introduction

Although electromyography (EMG) is almost as old as electroencephalography (EEG) and electrocardiography (ECG), its development in both research and diagnosis was delayed to World War II. The most significant early contribution was that of Weddell *et al.* (1944), who were charged with the task of developing an improved technique for diagnosis and prognosis in traumatic nerve lesions. Out of their work and the simultaneous work of others (most notably Jasper and Forde, 1947; Bauwens, 1948) grew the extensive use of clinical EMG not only for nerve injuries, for but also the many neurological and myogenic diseases. Indeed, the improvement in the medical understanding, classification, and treatment of the myopathies was closely related to the growth of clinical EMG.

Clinical EMG has its own detailed literature and is not the subject of this chapter. Neither will we consider the extensive literature of EMG which directly underpins diagnostic work. Our concern here is with a basic understanding of the subject as well as a broad view of the general subject of muscle function revealed by the technique.

II. Motor Units and Motor Unit Potentials

A. *Structure of Motor Units*

The number of striated muscle fibers that are served by one axon has a wide range. Generally, muscles controlling fine movements (e.g., those of the middle ear, the eyeball, and the larynx) have the smallest number of muscle fibers per motor unit—about ten or fewer—large coarse-acting muscles (e.g., those in the limbs) have large units—as high as 2000 muscle fibers.

The motor units of the sheep extraocular muscles have 3–10 muscle fibers (Tergast, 1873) and those of man have 5–6 (Bors, 1926). More recently, Feinstein *et al.* (1955) reported nine muscle fibers per motor unit in the human lateral rectus and 25 in platysma. The rat diaphragm has 7–17 fibers per motor unit (Krnjević and Miledi, 1958). The size of motor units in the rabbit pharyngeal muscles is also quite small, ranging from as few as 2 to a maximum of only 6 (Dutta and Basmajian, 1960). In contrast, there are 108 in the first lumbrical of the hand, and 2000 fibers per motor unit in the medial head of gastrocnemius (Feinstein *et al.*, 1955).

Van Harreveld (1946) concluded that the fibers in a motor unit of the rabbit sartorius may be scattered and intermingled with fibers of other units. Thus, the individual muscle bundles one sees in cross section in routine histological preparations of normal striated muscles do not correspond to individual motor units as such. Our own studies and those of Buchthal *et al.* (1957) indicate this is true in man as well, and Norris and Irwin (1961) state that in rat muscle the fibers of a motor unit are widely scattered.

B. Motor Unit Potentials

When an impulse reaches the myoneural junction or motor end plate where the axonal branch terminates on the muscle fiber, a wave of contraction accompanying a wave of depolarization spreads over the fiber, resulting in a brief twitch followed by rapid and complete relaxation. During the twitch a minute electrical potential is generated and is dissipated into the surrounding tissues. Since the muscle fibers of a motor unit do not all contract at exactly the same time (some being delayed for several milliseconds), the composite electrical potential developed by the single twitch of all the fibers in the motor unit is prolonged to about 9 msec, with a range of 5–12 msec. The majority of motor unit potentials have a total amplitude of about 0.5 mV. The result is a sharp spike that is often diphasic or triphasic, though it may also have a more complex form partly dependent on the electrode type. Generally, the larger the motor unit potential being registered, the larger is the motor unit that produces it. However, complicating factors, such as the distance of the unit from the electrodes and the types of electrodes and equipment used, determine the final amplitude of individual motor unit potentials (Håkansson, 1956, 1957a,b; Buchthal, 1959).

In the human biceps brachii, the spike potentials of each motor unit are localized to an approximately circular region, with an average diameter of 5 mm to which the fibers of the unit are confined (Buchthal *et al.*, 1957). However, the potentials can be traced in their spread to over 20 mm distance, and the units overlap considerably.

By isolating and stimulating single alpha motor neuron fibers in the nerves to tibialis anterior in rabbits, Close *et al.* (1960) obtained pure recordings of single motor unit action potentials. These remained remarkably consistent over periods of more than an hour. Amplitudes varied from 4 to 8 mV and durations were much briefer than those found with standard techniques in a whole animal. The primary main spike is "almost constant" in duration, lasts only 2 msec, and is followed by a shorter reverse spike.

Under normal conditions, during a voluntary or reflex contraction, the smaller potentials appear first with a slight contraction. As the force is increased, larger and larger potentials are recruited (Henneman *et al.*, 1965). This is called the normal pattern of recruitment. It is absent in cases of partial lower motor neuron paralysis, i.e., the small potentials never appear, apparently because only the larger motor units have survived. Recently, we have shown that man can be trained to suppress the action of small units and deliberately fire the much larger units individually (Basmajian *et al.*, 1965). This is discussed below in Section V,A.

C. Single Muscle Fiber Potentials

A motor unit potential represents the fusion of all accessible individual fiber potentials within a set limit of time (Fleck, 1962). Buchthal and Engbaek (1963) reported that the transmembrane potentials in single frog muscle fibers varies with temperature; at 25°C, the absolute refractory period is 2 msec, and the conduction velocity is 2.8 m/sec. Ekstedt (1964) found the single muscle fiber action potentials in normal man to have median values of 5.6 mV and 70 msec duration.

III. Technology

A. Background

A great deal of controversy has centered around the techniques used in EMG. Many of the earlier studies were rendered useless by inadequate techniques. Because of unavoidable circumstances, many investigators used discarded ECG and EEG equipment. Unfortunately there was too strong a sense of caution in the approach to electrodes and many inappropriate techniques of electrode placement and insertion were employed. We now generally agree that there is a wide selection of different types of electrodes that have their special application; the good electromyographer employs the whole range.

B. Electrodes

The most useful recent methodology employs the fine-wire electrodes. Such inserted electrodes are being widely used in many centers because of their simplicity and lack of trouble for both investigators and human or animal subjects. For kinesiological studies, they are much to be preferred over other types of inserted electrodes. They may be inserted in pairs and in large numbers of pairs. For example, in one of our studies we have worked with fourteen pairs inserted in different areas around the human hip joint for the study of hip kinesiology. In other studies, we have placed a great many in the confined regions of the hand, of the tongue, and of the foot. Concentric needle electrodes (Adrian and Bronk, 1929) and unipolar Teflon-coated needles are used widely in clinical examinations.

Bipolar fine-wire electrodes isolate their pick-up either to the whole muscle being studied or, if it has a multipenate structure, to the confines of the compartment within a muscle. Barriers of connective tissue within a muscle or around it act as insulation, and so one records all the activity as far as such a barrier without interfering with pick-up from beyond the barrier (such as there always is with surface electrodes).

Surface electrodes are usually used where a broad or global pick-up of a number of muscles or a large area of muscle is desired. However, in the case of muscles without internal partitions, the bipolar fine-wire electrodes sample the activity of the whole muscle as broadly as the best surface electrodes. Indeed, investigation reveals that the surface electrodes will miss deeper potentials which the wire electrodes pick up very well; on the other hand, the wire electrodes do not miss any of the potentials that the surface electrodes immediately overlying them do miss.

Any type of insulated wire can be used. Our bipolar fine-wire electrodes (Basmajian and Stecko, 1962) are made from a nylon-insulated or polyurethane-insulated Karma alloy wire only 25 μ in diameter. Jonsson (1970) has shown that this small diameter of wire is not suitable for extremely energetic movements because the wire will occasionally break. Even coarser wire up to 75 μ in diameter is extremely fine and impalpable to human subjects.

C. Apparatus

Electromyographs are simply amplifiers with a frequency range from about 10 to several thousand Herz. An upper limit of 1000 Hz is satisfactory. For kinesiological studies, the best instruments are multichannel. An obvious deficiency of ink-writing equipment is that the pens are too slow to record faster frequencies. This is dodged in some laboratories by the integration of potentials; but the concurrent monitoring of raw EMG potentials is essential to avoid integrating artifacts.

The recording device should either be photographic or employ electromagnetic frequency modulation (FM) tape recording. With multiple channels, one may photograph a row of cathode ray traces on photographic film in a variety of ways. Most convenient is the recording of multiple traces from miniature ultraviolet galvanometers on bromide recording papers. The paper requires no developing giving an immediate display. In recent years, multitrack FM tape recorders have provided a relatively cheap method of storing EMG signals (Basmajian, 1967). The signals can be converted and manipulated by analog-to-digital equipment and appropriate computing devices.

IV. EMG Kinesiology

No attempt can be made to touch upon the thousands of studies reported in the literature, especially on the individual actions of specific muscles in specific movements. These have been dealt with in detail elsewhere (Basmajian, 1967). Instead, several special areas will be discussed.

A. *Human Locomotion and Posture*

Although until recently its contribution has not been as great as it might have been, electromyography has added a new dimension in the latest studies on locomotion. The main reason for this slow start seems to have been that multifactorial studies are difficult and time consuming. Only recently has equipment improved to the point where electromyography gives especially useful results. Excellent multifactorial studies by Radcliffe (1962), Sutherland (1966), Murray *et al.* (1964), Liberson (1965), Battye and Joseph (1966), and many others have been reviewed elsewhere in detail (Basmajian, 1967). Some general findings will be given here.

During walking, the soleus begins to contract before it lifts the heel from the ground; it stops before the great toe leaves the ground. Apparently these are supportive rather than propulsive functions.

Quadriceps femoris contracts as extension of the knee is being completed, not during the earlier part of extension when the action is probably a passive swing. Quadriceps continues to act during the early part of the supporting phase (when the knee is flexed and the center of gravity falls behind it). Quadriceps activity occurs at the end of the supporting phase to fix the knee in extension, probably counteracting the tendency toward flexion imparted by gastrocnemius.

The hamstrings contract at the end of flexion and during the early extension of the thigh, apparently to prevent flexion of the thigh before the heel is on the ground and to assist the movement of the body over the supporting limb. In some persons, the hamstrings also contract a second time in the cycle during the end of the supporting phase; this may prevent hip flexion.

Gluteus maximus shows activity at the end of the swing and at the beginning of the supporting phase. This is contrary to the general belief that its activity is not needed for ordinary walking. Perhaps gluteus maximus contracts to prevent or to control flexion at the hip joint.

Many other studies in the lower limb have been done in the last few years. For example, studies by Basmajian and Bentzon (1954), Joseph and Nightingale (1956), and Gray and Basmajian (1968) have emphasized the factors which might influence the arch support in the human foot. These studies indicate that muscles are not important in the primary maintenance of the arches of the foot in the plantagrade static foot. However, they are very important during locomotion when the extremes of force are applied to the foot. Apparently the first line of defence against flat feet is a ligamentous one, but the added stresses of walking require special mechanisms.

Tibialis anterior has two peaks of activity at heel-strike and toe-off of the stance phase (Battye and Joseph, 1966; Gray and Basmajian, 1968), is inactive during mid-swing and middle of the stance phase, is active at full-foot in flat-footed subjects, and is generally more active during toe-out and toe-in walking.

Tibialis posterior is inactive throughout the swing phase. In flat-footed persons it becomes activated at heel-strike and more active at full-foot during level walking. The toe-out position reduces its activity.

Flexor hallucis longus is most active in mid-stance; during toe-out walking, activity increases in both phases, generally being more active in normal persons than in flatfooted persons.

Peroneus longus is most active at mid-stance and heel-off and generally more active in flat-footed persons.

Abductor hallucis and flexor digitorium brevis are generally more active in flatfooted persons.

An important regular pattern of inversion and eversion during the walking cycle occurs. Contingent arch support by muscles rather than continuous support is the rule, muscles being recruited to compensate for lax ligaments and special stresses during the walking cycle (Gray and Basmajian, 1968).

B. Trunk Muscles

Erector spinae shows two periods of activity during gait (Battye and Joseph, 1966). They occur "at intervals of half a stride when the hip is fully flexed and fully extended at the beginning and end of the supporting phase." Apparently the bilateral activity of the erectores spinae prevents falling forward of the body and also rotation and lateral flexion of the trunk. Sheffield (1962) found the abdominal muscles inactive during walking on a horizontal level.

Trunk musculature received most of the attention of electromyographers in the 1950s. Perhaps the most significant and largely ignored finding was that of Jones *et al.* (1953), who suggested that the intercostal muscles play a part in posture which is more important than their role in respiration. Their role in respiration seems to be the maintenance of a proper distance between the ribs while the rib cage is actively elevated by the neck muscles (scalenes) during inspiration. In quiet breathing, the diaphragm is the chief muscle of respiration in man.

C. Upper Limb

In the upper limb, the classic work of Inman *et al.* (1944) on the shoulder girdle has been followed by a long series of scattered EMG studies. Bearn's (1961) finding, that the activity in the upper fibers of trapezius falls off after a minute or two to disappear completely while the person is upright, is especially significant. It is also rather surprising to find that serratus anterior has only slight activity in an upright posture.

At the shoulder joint, the downward dislocation of the shoulder is resisted by the superior capsule of the shoulder joint and supraspinatus (Basmajian and Bazant, 1959). This finding has also emphasized that muscles which cross a joint longitudinally are not necessarily active when there is distraction on the joint. Other work on the elbow joint has confirmed this finding. The general principle seems to be that capsules and capsular ligaments are sufficient to prevent distraction except where excessive forces are applied.

When muscle is a contributing factor, it often is part of a locking mechanism rather than a source of transarticular forces. On the other hand, during movement such as flexion and extension of a joint, certain muscles are extremely important as a transarticular component to prevent distraction of a joint. Thus, brachioradialis shows little if any activity in maintaining flexed postures even against added loads, but it is very active in either flexion or extension of the elbow. This is its shunt muscle function, i.e., it acts chiefly during rapid movement along the long axis of the moving bone to provide centripetal force. The whole question of spurt and shunt muscles has been thoroughly discussed elsewhere. Of course such muscles are not confined to the upper limb and have widespread significance in the economy of the body. (See MacConaill and Basmajian, 1969, for a fuller discussion.)

V. Neuromuscular Physiology

A. Motor Unit Training

Given a clear response of his motor unit activity on a cathode ray oscilloscope and loudspeaker, and though completely unaware of any movement in the muscle, everyone can achieve notably wilful control over isolated motor-unit contractions in a muscle. We have known for a long time that almost anyone with EMG feedback can learn to relax a whole muscle instantly, on command. More striking, human subjects can recruit the activity of a single motor unit instantaneously and keep it active for a considerable period of time; also, they can deliberately change the frequency of firing of motor units (Basmajian, 1963).

Most persons can be trained to produce specific rhythms on motor units. It is easy for human subjects to gain control over a number of motor units and consciously switch the activity from one to the other. These and other findings on motor unit controls have deep significance in kinesiology. Indeed, they underlie the normal control of movement and muscles. This then would indicate that motor unit controls underlie the very topic of kinesiology itself.

1. LOCAL FACTORS

Moving a neighboring joint while a motor unit is firing is a distracting influence, but most subjects can keep right on doing it in spite of the distraction (Simard and Basmajian, 1967; Basmajian and Simard, 1967). Wagman *et al.* (1965) believe that subjects require our form of motor unit training before they can fire isolated specific motor units with the limb or joints in varying positions. Their subjects reported that "activation depended on recall of the original position and contraction effort necessary for activation." This apparently is a form of proprioceptive memory. Carlsöö and Edfeldt (1963) also concluded that, "Proprioception can be assisted greatly by exeroceptive auxiliary stimuli in achieving motor precision." However, conscious control can be easily maintained, despite the distraction produced by voluntary movements elsewhere in the body (head and neck, upper limbs, and contralateral limb) (Basmajian and Simard, 1967).

2. Effect of Competitive Nerve Stimulation

Contrary to expectation, the superimposition of a massive contraction in a muscle by electrical stimulation of its motor nerve does not significantly alter the regular conscious firing of a motor unit in that muscle. Even the coincidence of the motor unit potential with elements of the electrically induced massive contractions does not abolish the motor unit potential (Scully and Basmajian, 1969a).

3. Effects of Previous Training and Skills

The earliest studies, even though they included hundreds of subjects, failed to reveal any correlation between the ability of subjects to isolate and train individual motor units and such variables as sex, age, academic record, athletic ability, handedness, and general personality traits (Basmajian *et al.*, 1965). However, Scully and Basmajian (1969b), showed that the training time of most of the manually skilled subjects was above the median, although one might expect the opposite. If anticipatory tensions and/or position memory are learned, whether they are integrated at the cerebral level, at the spinal level, or both, these or some other cerebral or spinal mechanisms may be acting temporarily to block the initial learning of new skills. In a sense, perhaps some neuromuscular pathways acquire a habit of responding in certain ways, and it is not until that habit is broken that a new skill can be learned.

4. Effects of Handedness

When a long series of subjects was studied on two occasions using a different hand each time, Powers (1969) found that they always isolated a unit more quickly in the second hand.

B. Relationship of EMG to Force

Isometrically contracting muscle shows a direct relationship between the mechanical tension developed and the integrated EMG (Lippold,

1952; Bigland and Lippold, 1954; Lippold *et al.*, 1959; Bergström, 1959; Close *et al.*, 1960). However, in the muscles of amputees, there is no direct quantitative relationship between inherent power and the EMG (Inman *et al.*, 1944). With studies involving rapid movement, the mechanical tension lags (less than 0.1 sec) behind the main burst of potentials.

C. Tone and Relaxation

EMG has shown conclusively that normal mammalian striated muscle at rest is completely relaxed (Lindsley, 1935; Clemmesen, 1951; Basmajian, 1957). While tone is not determined by continuous neuromuscular activity at rest, there is no denying its existence. Tone is determined to some extent by passive elasticity and tissue turgor, but its chief component is the reflex reaction of the nervous system to stimuli. Even spastic subjects (both men and rabbits) can be made to relax completely (Basmajian and Szatmari, 1955). Clinical examination is deceptive because the very examination itself is a stimulus for reactive contraction. Muscles are rapidly relaxed and remain relaxed until a reflex or volitional demand is made on them. Ironically, the EMG reveals considerable muscular activity in completely atonic denervated muscle (spontaneous fibrillations), but this cannot be detected by ordinary examinations. (A discussion of fibrillation is beyond the scope of this chapter.)

REFERENCES

Adrian, E. D., and Bronk, D. W. (1929). *J. Physiol. (London)* **67**, 119.
Basmajian, J. V. (1957). *Can. Med. Ass. J.* **77**, 203.
Basmajian, J. V. (1963). *Science* **141**, 440.
Basmajian, J. V. (1967). "Muscles Alive: Their Functions Revealed by Electromyography," 2nd ed. Williams & Wilkins, Baltimore, Maryland.
Basmajian, J. V., and Bazant, F. J. (1959). *J. Bone Joint Surg., Amer. Vol.* **41**, 1182.
Basmajian, J. V., and Bentzon, J. W. (1954). *Surg., Gynecol. Obstet.* **98**, 662.
Basmajian, J. V., and Simard, T. G. (1967). *Amer. J. Phys. Med.* **46**, 1427.
Basmajian, J. V., and Stecko, G. (1962). *J. Appl. Physiol.* **17**, 849.
Basmajian, J. V., and Szatmari, A. (1955). *Neurology* **5**, 856.
Basmajian, J. V., Baeza, M., and Fabrigar, G. (1965). *J. New Drugs* **5**, 78.
Battye, C.K., and Joseph, J. (1966). *Med. Biol. Eng.* **4**, 125.
Bauwens, P. (1948). *Brit. J. Phys. Med.* **11**, 130.
Bearn, J. G. (1961). *Anat. Rec.* **140**, 103.
Bergström, R. M. (1959). *Acta Physiol. Scand.* **45**, 97.
Bigland, B., and Lippold, O. C. J. (1954). *J. Physiol. (London)* **125**, 322.

Bors, E. (1926). *Anat. Anz.* **60**, 444.

Buchthal, F. (1959). *Amer. J. Phys. Med.* **38**, 125.

Buchthal, F., and Engbaek, I. (1963). *Acta Physiol. Scand.* **59**, 199.

Buchthal, F., Guld, C., and Rosenfalck, P. (1957). *Acta Physiol. Scand.* **39**, 86.

Carlsöö, S., and Edfeldt, A. W. (1963). *Scand. J. Psychol.* **4**, 231.

Clemmesen, S. (1951). *Proc. Roy. Soc. Med.* **44**, 637.

Close, J. R., Nickle, E. D., and Todd, F. N. (1960). *J. Bone Joint Surg., Amer. Vol.* **42**, 1207.

Dutta, C. R., and Basmajian, J. V. (1960). *Anat. Rec.* **137**, 127.

Ekstedt, J. (1964). *Acta Physiol. Scand.* **61**, Suppl. 226, 96 pp.

Feinstein, B., Lindegård, B., Nyman, E., and Wohlfart, G. (1955). *Acta Anat.* **23**, 127.

Fleck, H. (1962). *Arch. Phys. Med. Rehabil.* **43**, 99.

Gray, E. G., and Basmajian, J. V. (1968). *Anat. Rec.* **161**, 1.

Håkansson, C. H. (1956). *Acta Physiol. Scand.* **37**, 14.

Håkansson, C. H. (1957a). *Acta Physiol. Scand.* **39**, 291.

Håkansson, C. H. (1957b). *Acta Physiol. Scand.* **41**, 199.

Henneman, E., Somjen, G., and Carpenter, D. O. (1965). *J. Neurophysiol.* **28**, 599.

Inman, V. T., Saunders, J. B. deC. M., and Abbott, L. C. (1944). *J. Bone Joint Surg.* **26**, 1.

Jasper, H. H., and Forde, W. O. (1947). *Can. J. Res.* **25**, 100.

Jones, D. S., Beargie, R. J., and Pauly, J. E. (1953). *Anat. Rec.* **117**, 17.

Jonsson, B. (1970). Personal communication.

Joseph, J., and Nightingale, A. (1956). *J. Physiol. (London)* **132**, 465.

Krnjevic, K., and Miledi, R. (1958). *J. Physiol. (London)* **140**, 427.

Liberson, W. T. (1965). *Arch. Phys. Med. Rehabil.* **46**, 37.

Lindsley, D. B. (1935). *Amer. J. Physiol.* **114**, 90.

Lippold, O. C. J. (1952). *J. Physiol. (London)* **117**, 492.

Lippold, O. C. J., Redfearn, J. W. T., and Vučo, J. (1959). *J. Physiol. (London)* **137**, 473.

MacConaill, M. A., and Basmajian, J. V. (1969). "Muscles and Movements: A Basis for Human Kinesiology." Williams & Wilkins, Baltimore, Maryland.

Murray, P. M., Drought, A. B., and Kory, R. C. (1964). *J. Bone Joint Surg.* **46**, 335.

Norris, F. H., Jr., and Irwin, R. L. (1961). *Amer. J. Physiol.* **200**, 944.

Powers, W. R. (1969). Ph.D. Thesis, Queen's University, Kingston, Canada.

Radcliffe, C. W. (1962). *Artif. Limbs* **6**, 16.

Scully, H. E., and Basmajian, J. V. (1969a). *Arch. Phys. Med. Rehabil.* **50**, 32.

Scully, H. E., and Basmajian, J. V. (1969b). *Psychophysiol.* **5**, 625.

Sheffield, F. J. (1962). *Amer. J. Phys. Med.* **41**, 142.

Simard, T. G., and Basmajian, J. V. (1967). *Arch. Phys. Med. Rehabil.* **48**, 12.

Sutherland, D. H. (1966). *J. Bone Joint Surg.* **48**, 66.

Tergast, P. (1873). *Arch. Mikrosk. Anat. Entwicklungsmech.* **9**, 36.

van Harreveld, A. (1946). *Arch. Neer. Physiol.* **28**, 408.

Wagman, I. H., Pierce, D. S., and Burger, R. E. (1965). *Nature (London)* **207** 957.

Weddell, G., Feinstein, B., and Pattle, R. E. (1944). *Brain* **67**, 178.

7

PROTEINS OF THE MYOFIBRIL

SETSURO EBASHI and YOSHIAKI NONOMURA

I. Introduction

A. General Aspects

The myofibril is the morphological unit of the contractile mechanism of skeletal muscle. The muscle fiber, which ranges in diameter from 30 to 150 μ, consists of a number of myofibrils of a diameter of about 1 μ surrounded by the network of the sarcoplasmic reticulum. The myofibril is composed almost exclusively of those proteins concerned with contractile processes.

Discovery of the presence of such contractile proteins was first made by Kühne as early as 1864. He found that if frozen frog muscle were pulverized, thawed, and then filtered through linen, a viscous and fairly transparent fluid was extracted, which, after a while, coagulated into an opaque mass which he named "myosin." This material could be dissolved in a strong salt solution and precipitated again upon dilution. This pioneering work was followed by the studies of Danilewski (1881), Halliburton (1887), von Fürth (1895), and others. The observation of Halliburton was so penetrating, that if his work were further persistently extended by him or his successors, the history of muscle protein research would have been changed (see Finck, 1968).

The line of investigation directly connected with present research of muscle proteins was reopened much later in 1920s and pursued by a number of workers, particularly by Weber and Edsall as represented by the familiar term "Weber–Edsall's solution." Von Muralt and Edsall (1930a,b.) demonstrated the flow birefringence of myosin solution; H. H. Weber found a way of making myosin thread (1934) and observed its birefringence (Noll and Weber, 1934). Thus, the nature of living muscle appeared to be reconstituted in *in vitro*.

Search for the energy source of contraction by Meyerhof and his associates led to the discovery of ATP. The connection between ATP and muscle proteins was first made by Engelhardt and Ljubimova and then by Szent-Györgyi. This brilliant work, done around 1940, may be summarized as follows: (1) ATPase activity of myosin (Engelhardt and Ljubimova, 1939); (2) shortening of myosin thread and the superprecipitation of myosin suspension (A. Szent-Györgyi, 1941a, 1943); (3) separation of classic myosin, myosin B, into two components—myosin A and a new protein, actin (Banga and Szent-Györgyi, 1941; A. Szent-Györgyi, 1941b; Straub, 1942). These three great discoveries represent those different types of approaches to the study of the mechanism of muscle contraction still essential for muscle research. Another important approach, developed after these discoveries, is the investigation into the fine structural basis of muscle contraction using electron microscopic and diffraction techniques.

This chapter describes the properties of muscle proteins from the standpoint of contractile mechanism. Although all myofibrillar proteins are related to the contractile mechanism, their functions are distinctly specialized (Tables I and II). The fundamental process of contraction is certainly carried out by myosin and actin, but these two proteins alone cannot bring about the contractile process of living muscle. The recognition of the physiological role of tropomyosin and the discovery of troponin have further stimulated the discoveries of new myofibrillar proteins. Altogether, the idea of regulatory proteins (proteins that enable myosin and actin to perform the contraction-relaxation cycle under physiological conditions) has been established.

Therefore, we will emphasize the clarification of the roles of individual proteins in the contractile processes, referring to physicochemical aspects only if they are deeply involved in the contractile mechanism. The description is rather limited to the properties of skeletal muscle proteins, since skeletal muscle has attained the highest degree of differentiation as a contractile tissue, and therefore, we can see there the essentials of muscle proteins in a very simplified form.

Since, for lack of space, this chapter cannot be as comprehensive

TABLE I

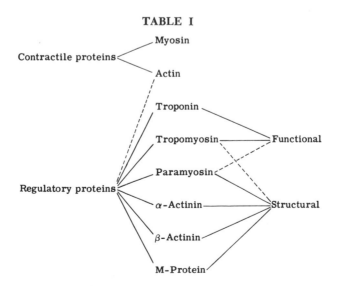

TABLE II

CONTENT OF CONTRACTILE AND
REGULATORY PROTEINS IN
RABBIT MYOFIBRIL[a]

Proteins	Percent myofibril by weight
Myosin	55–60
Actin	20
Tropomyosin	4.5
Troponin	3–5
α-Actinin	2
β-Actinin	~0.5
M-Protein	~0.5

[a] About 60% of proteins of muscle cells are concentrated in myofibrils. These figures indicate an approximate estimate based on theoretical considerations as well as experimental results carried out in many laboratories.

as we would like, we refer the reader to the following pertinent review articles. A comprehensive article by A. G. Szent-Györgyi (1960), which appears in the previous edition of this treatise, covers all important articles published before 1960. Review articles on muscle proteins that appeared in authoritative review books of journals are: Perry (1961), Kielley (1964), Gergely (1966), Stracher and Dreizen (1966), Perry (1967), Young (1969), and Katz (1970). Articles dealing with the regulating mechanism and energetics of muscle contraction though not directly concerned with muscle protein, are: H. E. Huxley (1960), Hasselbach (1964), Davies (1965), A. Weber (1966), Hanson (1968a), Ebashi and Endo (1968), Peachey (1968), H. E. Huxley (1969), Mommaerts (1969), Ebashi *et al.* (1969), Sandow (1970), and Briskey *et al.* (1970). In addition to these, there are a number of books and special issues of general journals that contributed to the research on muscle proteins: Gergely (1964), Ebashi *et al.* (1965), Paul *et al.* (1965), Ernst and Straub (1968), Miller (1968), Bendall (1969), A. F. Huxley and Huxley (1964), Hill *et al.* (1966), Stracher (1967) Laki (1971), and Tonomura (1972). Among them, the book edited by Gergely (1964) should be noted as a thorough collection of work published before 1962; Katz's review (1970) is an extensive survey of recent literature.

B. Terminology and Abbreviations

"Myosin" denotes myosin A or ʟ-myosin. "Myosin B" denotes natural actomyosin, which is directly extracted from muscle mince and contains many proteins. The term "actomyosin" is used when referring only to actin and myosin, irrespective of the presence of other proteins. The complex composed of separately prepared myosin and actin, often called synthetic actomyosin, is here referred to as "reconstituted actomyosin." "Tropomyosin" and "paramyosin" denote tropomyosin B and tropomyosin A, respectively.

Abbreviations used: HMM, H-meromyosin; LMM, L-meromyosin; S-1, subfragment 1 of HMM; S-2, subfragment 2 of HMM; LMP, low molecular weight protein of myosin; PCMB, p-chloromercuribenzoate; NEM, N-ethylmaleimide; EDTA, ethylenediaminetetraacetic acid; SDS, sodium dodecyl sulfate; P_i, inorganic orthophosphate; PP, inorganic pyrophosphate.

II. Myosin

A. General Aspects

Myosin is the major myofibrillar protein, and its interaction with actin in the presence of ATP represents the fundamental molecular basis of muscle contraction. In the sequence of chemical reactions occurring during this interaction, the breakdown of ATP, which is coupled with the conversion of chemical energy of ATP into mechanical energy, is carried out at a specified site on the myosin molecule. It is therefore quite reasonable that a considerable effort of muscle biochemists has been concentrated on the studies on the properties of myosin ATPase under various conditions.

The proposal of the sliding mechanism (A. F. Huxley and Niedergerke, 1954; H. E. Huxley and Hanson, 1954) has stimulated the effort to visualize the fine structure of the myosin molecule by electron microscope or other procedures. This approach has been greatly facilitated by the studies on myosin subfragments produced by digestion by proteolytic enzymes originally used with the intention of isolating the active sites of myosin ATPase. As a result, two types of investigations, one concerned with the enzymic activity of myosin and the other with the structural aspect of the myosin molecule, have been united, and we have now fairly precise knowledge of the fine structure of the myosin molecule.

B. Enzymatic Properties

1. GENERAL PROPERTIES

In spite of elaborate efforts of muscle biochemists, the essential mechanism for converting the chemical energy of ATP to mechanical energy through myosin is not yet understood. It is, however, still the common believe of many muscle scientists that the further investigation into the detailed mechanism of the ATP breakdown would reveal the secret of muscle contraction. If this is true, myosin ATPase must possess unique properties distinct from those of other enzymes. Below are the characteristics of myosin ATPase, which may or may not be related to its physiological functions.

A. WITHOUT EXCHANGE REACTIONS. Myosin ATPase does not catalyze ATP–ADP exchange reaction under usual conditions, in sharp contrast with sodium–potassium- (Skou, 1960) or calcium-transport ATPase (Ebashi and Lipmann, 1962; Hasselbach and Makinose, 1961), which is also involved in the conversion of chemical energy to physical energy like myosin ATPase. This may be related to the fact that the phosphorylated intermediate of myosin in the sense of that of transport ATPases—$E \sim P$ form (Post and Rosenthal, 1962; Yamamoto and Tonomura, 1967; Makinose, 1969)—has not been isolated. Most workers have also failed to demonstrate the presence of ATP–P_i exchange reaction.

B. INITIAL P_i BURST. A. Weber and Hasselbach (1954) first showed that P_i liberation in the presence of magnesium ion in the initial stage is higher than that in the steady state. Tonomura and his colleagues have revealed (Kanazawa and Tonomura, 1965; Imamura *et al.*, 1965; cf. Tonomura, 1972) that the amount of excess P_i liberation has a stoichiometric ratio to the amount of the myosin molecule (i.e., 1 mole P_i per mole myosin); they have concluded that the excess P_i is not due to a higher rate of hydrolytic reaction at the initial stage, but represents the intermediate state of P_i at the site of myosin ATPase. They have proposed that the intermediate state represented by the initial P_i burst reflects the most essential enzymic facet of myosin as the principal contractile protein (for further details, see Tonomura, 1972).

Recent extensive kinetic studies of Taylor and his colleagues on the transient P_i production with reference to the state of enzyme–product complex (Lymn and Taylor, 1970; Taylor *et al.*, 1970) have confirmed the intermediate nature of the excess P_i; i.e., the occurrence of such burst is due to the slow liberation of the product from the enzyme [$k_3 \ll k_2$ in Eq. (1)], and have shown further that the step of ADP liberation, not the P_i liberation, is rate-limiting in this reaction.

$$E + ATP \xrightarrow[k_{-1}]{k_1} E \cdot ATP \xrightarrow{k_2} E \begin{smallmatrix} \nearrow ADP \\ \searrow P_i \end{smallmatrix} \xrightarrow{k_3} E + ADP + P_i \qquad (1)$$

They have given values for k_1 and k_2 at 20°C, 5×10^5 M^{-1} sec^{-1} and 50–75 sec^{-1}, respectively; k_3 is of course the steady state rate constant, about 0.02 sec^{-1}.

The k_1 value is in good agreement with the rate constant for the complex formation of ATP and HMM obtained by Morita (1967, 1969) using a method based on the spectral shift of myosin at 290 mμ induced by ATP or ADP; it should be mentioned that the decay of such induced shift coincides with the rate of steady state. It appears thus to be certain that the finding of Morita is the first clear presentation of the conformational change of myosin induced by its substrates. The direct evidence for the presence of such an intermediate has also been presented by Schliselfeld *et al.* (1970).

According to Lymn and Taylor (1970), excess P_i is 1.8 moles per 5×10^5 gm myosin, i.e., nearly 2 moles per mole. Not only MgATP, but also CaATP and MgITP exhibit a similar amount of initial P_i burst, in agreement with the results based on the ATP-induced spectral shift (Morita, 1967). Such a burst cannot be seen, but there is a definite lag in P_i liberation when potassium ion in the presence of EDTA is used as the activator (see p. 294), probably because of a much higher rate of product liberation than the rate of substrate decomposition as is usually the case with enzymes (Lymn and Taylor, 1970).

C. Initial Proton Liberation. Very rapid proton liberation occurs at the early stage of ATP-myosin interaction (Tokiwa and Tonomura, 1965). This takes place in the presence of magnesium or calcium (0.6–0.7 mole per mole myosin in the case of magnesium ion); MgITP also exhibits the same phenomenon, though the amount of such liberation is much less (Finlayson and Taylor, 1969). The rate of this reaction is of the same order as that of the initial P_i burst, so that both phenomena must arise from the same steps of ATP decomposition.

D. ^{18}O EXCHANGE REACTION. It was found that myosin in the presence of magnesium catalyzes the incorporation, maximally, of two excess oxygen atoms into the phosphate derived from ATP (Levy and Koshland, 1959). Although it was later demonstrated that a considerable part of such exchange is due to the exchange of ^{18}O between $H_2^{18}O$ and P_i in the medium catalyzed by myosin (Dempsey and Boyer, 1961), there still seemed to exist some genuine exchange reaction (Yount and Koshland, 1963). The significance of this phenomenon is not yet understood.

2. EFFECTS OF IONS

A. DIVALENT CATIONS. Divalent cations (those divalent cations that may affect sulfhydryl residues, such as Cu^{2+} or Hg^{2+}, are not included in this discussion) may be classified into three categories: (1) magnesium and manganese ions, which can induce contractile responses, i.e., can give rise to the actin-activated ATPase, superprecipitation, and clearing response; (2) calcium and strontium ions, which activate myosin ATPase but cannot induce contractile responses; and (3) other divalent cations (e.g., nickel, cobalt, zinc, and cadmium ions), which cannot activate myosin ATPase but depress potassium-activated ATPase.

Although conclusive evidence has not yet been presented, it is the tendency to suppose that these ions act on myosin ATPase as the complex with ATP. The following descriptions are principally based on this assumption.

i. Magnesium and manganese ions. Myosin is a very weak enzyme under physiological ionic conditions, i.e., roughly speaking, 10^{-1} M K^+, 10^{-3} M Mg^{2+}, 10^{-4} M (exciting) or 10^{-7} M (resting) Ca^{2+}, and 10^{-7} M H^+. This low ATPase activity is entirely due to the presence of magnesium ion in the myoplasm. Although potassium and calcium ions are activators of myosin ATPase *in vitro*, their concentrations are not high enough to reverse the depressing effect of magnesium ion (A. Szent-Györgyi, 1951; Mommearts and Green, 1954).

It is certain that the ATPase activity of myosin is very low in the presence of magnesium, but it is inaccurate to say that magnesium ion is an inhibitor of myosin ATPase. The low ATPase activity seems due to the stable nature of the intermediate of ATP–myosin interaction (see p. 290) and that this effect may be the basis of the essential role of magnesium ion in the contractile processes (we cannot deny the possibility that, in addition to the role as the complexing agent with ATP, magnesium ion could play another role in the contractile mechanism, probably through its direct effect on myosin). It may be necessary to reconsider all the results concerning the effects of ions and modifiers on myosin ATP-ase in light of the role of magnesium ion in the myosin function.

Manganese ion is the only one which can replace the role of magnesium ion in inducing the contractile responses of actomyosin systems (Watanabe and Sleator, 1957; Fujii and Maruyama, 1970; Ozawa *et al.*, 1970). It should be mentioned that the ion often fails in inducing contractile responses of reconstituted actomyosin system, though it can always exert the dissociating action (see below).

ii. Calcium and strontium ions. Calcium ion has been widely used as an activator of myosin ATPase (A. Szent-Györgyi, 1951; Mommaerts and Green, 1954). In spite of this, its nature as the activator is not yet well understood. The degree of activation of myosin ATPase by calcium ion is very low at physiological concentrations (10^{-7}–10^{-4} M); furthermore, calcium-activated ATPase is very strongly depressed by magnesium ion. Therefore, calcium-activated ATPase has practically no significance in physiological processes.

Calcium and strontium ions cannot replace the role of magnesium ion in the interaction of myosin and actin, unlike manganese ion. As shown in Table III, calcium and strontium ions have much lower depressing actions on potassium-activated myosin ATPase than magnesium and manganese ions. This indicates that the complexes of both ions with ATP or ADP have weaker affinities for the site of myosin ATPase. Since the step of enzyme substrate formation is essentially the same for both calcium and magnesium ions (Lymn and Taylor, 1970), it is very likely that in contrast with the cases of magnesium and manganese ions, cal-

TABLE III
SOME PROPERTIES OF DIVALENT CATIONS CONCERNING MYOSIN ATPASE[a]

Cation	Maximum ATPase activity (relative)	K_i (M^{-1})	Ionic radii
Mg^{2+}	0.02	4.6×10^6	0.65
Mn^{2+}	0.34	6.5×10^5	0.80
Ca^{2+}	(1.0)	1.3×10^4	0.99
Sr^{2+}	>0.2	5.5×10	1.13
Ni^{2+}	~0	1.9×10^6	0.69
Co^{2+}	~0	1.0×10^6	0.72
Zn^{2+}	~0	2.1×10^6	0.74

[a] Values for ATPase activities and K_i were calculated from Figs. 6 and 8 in the article of Seidel (1969c). K_i = the reciprocal of the concentration to inhibit potassium-activated ATPase by 50%.

cium and strontium ions cannot form a stable enzyme–product complex, which would be essential for the interaction of myosin with actin. The observation that MgADP has about twenty times higher affinity for myosin than CaADP (Lowey and Luck, 1969) may be interpreted along a similar line.

Manganese ion, though higher than calcium and strontium ions, exhibits a much weaker inhibiting action than magnesium ion (Table III); in other words, the enzyme–product complex formed with manganese ion is more labile than that with magnesium ion. This may be one of the reasons for the requirement of a high concentration of manganese ion for the activation of the myosin–actin interaction and the frequent failure in inducing the activation of reconstituted actomyosin.

iii. Other divalent cations. Although they cannot activate myosin ATPase, nickel, cobalt, zinc, and cadmium ions are shown to depress potassium-activated ATPase (Seidel, 1969c). The fact that there is a fairly linear relationship between the radii of divalent cations and their concentrations to induce 50% inhibition, irrespective of whether they are activators of myosin ATPase or not (Seidel, 1969c; Table III), suggests that there is a common mechanism among the inhibitory actions of divalent cations.

Zinc and cadmium ions can slightly activate potassium-ATPase (Blum, 1960), but their mode of action may be related to that of modifiers, something like sulfhydryl blocking agents.

B. MONOVALENT CATIONS AND EDTA. While potassium ion activates myosin ATPase, sodium ion cannot, but it depresses the potassium-activated ATPase (see Mommaerts and Green, 1954). Ammonium and rubidium ions behave in a similar way to potassium ion (strictly speaking, effects of ammonium ion are sometimes different from those of potassium and rubidium ions). Lithium ion resembles sodium ion, but it shows a slight but significant activation (Kielley et al., 1956).

It should be mentioned that in addition to myosin, there are a number of enzymes, at least more than ten, that are activated by potassium, rubidium, and ammonium ions but not by sodium ion, e.g., pyruvate kinase (Boyer et al., 1943), surface membrane-bound p-nitrophenylphosphatase (Judah et al., 1962) (probably the partial reaction of Na–K-ATPase), and so on; in other words, it is a fairly common property of enzymes that they can discriminate between sodium and potassium ions.

Friess (1954) and Bowen and Kerwin (1954) have shown that myosin

ATPase is markedly activated by EDTA in the presence of potassium ion at a slightly alkaline pH. The general tendency of the activation by other monovalent cations in the presence of EDTA is essentially the same as that in the absence of chelating agents (Kielley *et al.*, 1956), so that EDTA-activated ATPase is considered a monovalent cation-activated ATPase.

The activation by EDTA seems to be due to the deprivation of magnesium ion from myosin or the depression of the formation of MgATP complex, since the degree of activation by various kinds of chelating agents parallels their magnesium-binding capacities (Fig. 1) (Ebashi *et al.*, 1960; Martonosi and Meyer, 1964).

The effects of monovalent ions may have two aspects; one is related to the ionic strength produced by that ion, and the other its specific action. Whether the latter is based on the complex formation with ATP or direct action on myosin is not yet clear. Even potassium ion, which has least complex-forming property, shows some affinity for ATP (Botts *et al.*, 1965); we cannot deny the possibility that monovalent cations also complex with ATP. However, it should be noted that hydrolytic

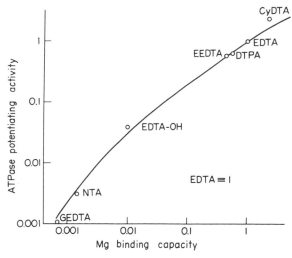

Fig. 1. The relationship between magnesium-binding capacities of various chelating agents and the degrees of activation of myosin ATPase by them. Apparent binding constants of chelating agents under the experimental conditions—0.6 *M* potassium chloride at pH 8.5—and the reciprocals of their concentrations that produce half maximum activations of myosin ATPase [derived from Fig. 2 of the article of Ebashi *et al.* (1960)] are listed in values relative to that of EDTA. For the abbreviations of other chelating agents and other experimental conditions, refer to the original article.

process induced by potassium ion is qualitatively different from that by divalent cations (see p. 292).

C. pH. Using calcium or potassium ion as an activator, myosin ATPase shows two peaks, at a pH around 6 and one higher than 9.0 (see Mommaerts and Green, 1954). If the pH is raised above the latter peak, the activity sharply falls to zero with the concomitant deterioration of the protein. Thus, the latter peak is somewhat different from the ordinary optimum; the subsequent fall seems to be correlated with the loss of light components from main myosin chains (see p. 303).

The depressed state of ATPase activity at pH 7–7.5 appears to be somehow related to the stabilizing effect of magnesium. Indeed, this characteristic profile of pH activity relationship is not well presented by some modified myosin such as sulfhydryl-blocked myosin (Morales and Hotta, 1960), which is barely responsive to the ATPase-inhibiting action of magnesium ion and the dissociating action of MgATP (see p. 311); furthermore, potassium ATPase in the presence of EDTA (K-EDTA-ATPase) shows a typical sigmoid curve increasing with increase in pH (Bowen and Kerwin, 1954).

D. ADDITIONAL REMARKS. Usual anions have no particular influences on myosin ATPase and myosin-related functions (Seidel, 1969a). The effects of ions other than magnesium ion on myosin ATPase mentioned above have no direct physiological significance. In this respect, these ions are similar to chemical modifiers such as sulfhydryl-blocking agents. However, we can discriminate most divalent cations from modifiers, since they seem to act as a part of a substrate for myosin ATPase by forming a complex with ATP.

3. MODIFIERS

The interesting but complicated effects of PCMB on myosin ATPase found by Kielley and Bradley (1956) have introduced a new approach to the studies on myosin ATPase. Since then, a number of agents have been reported to modify the myosin ATPase in various ways. However, the results obtained with sulfhydryl-blocking agents are still most informative and interesting, so their effects will be first described and then other various modifiers will be reviewed briefly.

A. SULFHYDRYL-BLOCKING AGENTS. i. Effects of PCMB. The effects of PCMB on myosin ATPase are illustrated in Fig. 2. The most remarkable point is that the ATPase in the presence of calcium is maximally acti-

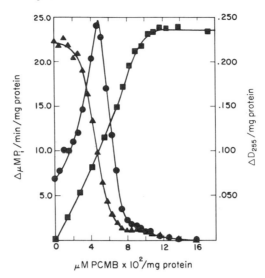

Fig. 2. Titration of myosin with PCMB. ATPase was determined in a system containing 1 mM ATP and either 5 mM calcium chloride and 0.05 M potassium chloride or 1 mM EDTA and 0.4 M potassium chloride at pH 7.6. For the correction for the protein concentrations, refer to the Fig. 1 in the article of Kielley, (1961). ▲ = ATPase activity with EDTA, ● = ATPase activity with calcium, ■ = $\triangle D_{255}$ versus PCMB. From Kielley and Bradley (1956).

vated when nearly half the sulfhydryl residue, about 4 moles per 10^5 gm myosin, are titrated by PCMB. This activation is quite remarkable at 0.6 M potassium chloride, but not so much at 0.05–0.1 M; this is in sharp contrast with the case of ordinary myosin, which shows the highest CaATPase activity at a potassium chloride concentration around 0.05 M (Mommaerts and Green, 1954). The ATPase activity in the presence of magnesium ion behaves in a similar manner, although the maximal activity is still lower than the corresponding calcium-activated ATPase activity by about thirty times (Blum, 1960). In contrast to CaATPase, MgATPase hardly appears to be influenced by the ionic strength (Perry and Cotterill, 1965). It is interesting that K-EDTA-ATPase is almost inversely decreased with increase in CaATPase. The pH–activity relationship is also altered (see p. 296). Actomyosin made of such myosin is not sensitive to the dissociating action of MgATP; it hardly shows the clearing response, only superprecipitation (see p. 312) under the condition where ordinary actomyosin responds to MgATP with clearing.

With further increase in the amount of PCMB, both Ca- and MgATPase activities decrease rather abruptly and reach almost zero when all sulfhydryl groups are blocked. It should also be mentioned that the inhibition by magnesium ion of calcium-activated ATPase becomes weaker with increase in the blocking of sulfhydryl residues (Morales and Hotta, 1960).

If a sulfhydryl-protecting agent is added in excess to remove PCMB at the point before the maximal activation is attained, the properties of myosin regarding the ATPase activity are restored nearly to the original state. However, if all sulfhydryl residues are titrated, although sulfhydryl-protecting agents can completely remove PCMB and restore a part of ATPase activity, the quality of myosin has already irreversibly changed. Magnesium ion does not strongly inhibit the ATPase of such myosin. Although it can react with actin in the presence and absence of ATP, the actomyosin thus formed becomes much less sensitive to the dissociating action of MgATP (Tonomura *et al.*, 1961).

ii. Sulfhydryl residues involved in modification. Kielley and Bradley (1956) have shown that NEM affects myosin similarly to PCMB with respect to the activating effect; complete inhibition cannot be obtained by NEM. Utilizing the slow reacting and covalent bond-forming properties of NEM, Sekine *et al.* (1962) have shown that only one specified sulfhydryl residue per 2×10^5 gm, named S_1, is responsible for the activation. Using ^{14}C-labeled NEM, the amino acid sequences around this sulfhydryl residue has also been determined (Yamashita *et al.*, 1964). The dependence of the CaATPase activity of S_1-blocked myosin by NEM upon potassium chloride concentration is more marked than that blocked by PCMB or phenylmercuric acid; it is activated at 0.6 M but depressed at ionic strength lower than 0.3 (Perry and Cotterill, 1965).

Sekine and Yamaguchi (1963) have further shown that NEM can induce a marked depression of once-activated myosin ATPase (S_1-blocked myosin) in the presence of ATP or ADP. Again, it has been shown that only one sulfhydryl residue per 2×10^5 gm is involved in this depression; they named it S_2. It should be mentioned that this inhibition is qualitatively different from the completely inhibited state of myosin ATPase with PCMB. It exerts definitely higher Ca- and MgATPase than the intact myosin (Sekine and Yamaguchi, 1966); it can exhibit superprecipitation, though not marked, if combined with actin. It should be noted that like S_1, S_2 is a kind of regulatory site, of which the blocking does not lead to the abolition of the ATPase activity. If S_2 alone is blocked, the ATPase is not inhibited but rather accelerated (Seidel, 1969a).

Stracher (1964) has introduced the disulfide–sulfhydryl interchange reaction with bis-β-carboxyethyl disulfide, $(-S-CH_2COOH)_2$, to the modification of myosin molecules, and shown that only one sulfhydryl residue per subunit of myosin is responsible for the complete inhibition of myosin ATPase by iodoacetamide. The amino acid composition around this sulfhydryl residue (Stracher and Dreizen, 1966) is certainly different from that of S_2 (Yamashita *et al.*, 1965).

In this connection, it should be mentioned that the amino acid sequence around sulfhydryl residues including those mentioned above have been extensively studied (Kimura and Kielley, 1966; Weeds and Hartley, 1968). These results may act as useful markers in the attempt to determine the complete amino acid sequences of myosin.

iii. Other sulfhydryl-blocking agents. In view of the results obtained with PCMB and NEM, the modifications of myosin ATPase may fall in the following categories: (1) activation of Ca- and MgATPase (S_1-blocked); (2) depression of once activated Ca- and MgATPase (S_1, S_2-blocked); (3) depression of K-EDTA–ATPase (S_1-blocked); (4) Inactivation of all types of ATPases (blocked by iodacetamide). PCMB may exert all of these actions; NEM usually shows the actions of (1) and a part of (2), but typical (2) under a particular condition.

Other sulfhydryl-blocking agents also show a part of the above actions. The effects of phenylmercuric acetate (Greville and Needham, 1955; Perry and Cotterill, 1965) are not indistinguishable from those of PCMB; Cu^{2+} also behaves like PCMB (Blum, 1962). S-Aminoethylisothiuronium and cysteine ethyl ester exhibit mainly acceleration phase (1), but not the depression (2) (Morales *et al.*, 1957; Morales and Hotta, 1960). Iodoacetamide does not show any accelerating phase but solely depression (4) (Bárány and Bárány, 1959).

In addition to these sulfhydryl-blocking agents, there are several agents that exert similar influences on myosin ATPase to those of PCMB or NEM—some organic solvents, especially *n*-butanol (Ebashi and Ebashi, 1959) and DNP (Greville and Needham, 1955; Chappell and Perry, 1955). It is rather difficult to consider that these agents, particularly the former, would combine with the sulfhydryl residue itself; perhaps, they would induce a similar conformational change of the myosin molecule induced by sulfhydryl blocking.

B. Modifiers Affecting Amino Acid Residues Other than Sulfhydryl Residues. *i. Lysyl-blocking agents.* Since the report that benzaldehyde might affect lysyl residues (Kuschinsky and Turba, 1950), several agents have been reported to modify the ATPase by combining with crucial lysyl residues, e.g., trinitrobenzene sulfonate (Kubo *et al.*, 1960; Kitagawa *et al.*, 1961; Fabian and Mühlrad, 1968), dimethylamino-naphthalene-5-sulfonyl chloride (Kasuya and Takashina, 1965), succinyl anhydride (Oppenheimer *et al.*, 1966), and salicylaldehyde (Mühlrad *et al.*, 1970). Kubo *et al.* (1960) first presented the evidence that lysyl residue is involved in this chemical modification; the amino acid sequence around the lysyl residue responsible for the trinitrophenylation has been determined (Tokuyama *et al.*, 1966).

Although the effect of lysyl blocking on myosin ATPase is somewhat different from agent to agent, the common features of the modified ATPase due to this blocking are: (1) activation of MgATPase; (2) depression of K-EDTA–ATPase; (3) depression of CaATPase with or without a preceeding slight activation of CaATPase. Thus, MgATPase, which behaves in a similar way to CaATPase in the case of sulfhydryl blocking, is now shown to be dissociated from the latter.

ii. Histidyl-blocking agent. Involvement of histidyl residue in ATPase activity is first suggested by Morales and Hotta (1960). Photooxidation in the presence of methylene blue abolishes the ATPase activity of sulfhydryl-protected myosin, probably by affecting histidyl residue (Stracher and Chan, 1964; Sekiya *et al.*, 1965).

iii. Agents blocking other amino acid residues. Provided sulfhydryl residues are protected, diazonium-1*H*-tetrazol has been shown to abolish ATPase by affecting tyrosine residue (Shimada, 1970). *p*-Nitrothiophenol appears to affect a glutamyl residue (Imamura *et al.*, 1965).

C. Other Modifiers. *i. Diazonium compounds.* The interesting point of the modification by these agents is that either CaATPase or K-EDTA–ATPase can be suppressed without significant change in the other ATPase (Yamashita *et al.*, 1967, 1969). It is certain that diazonium compounds affect particular sulfhydryl residues, which may be different from those mentioned above (Yamashita *et al.*, 1967), but further evidence seems necessary for the conclusion that the modification is certainly due to the blocking of such sulfhydryl residues.

ii. Other agents. The effects of 1-fluoro-2,4-dinitrobenzene have been extensively studied, but the target amino acid residue has not yet been specified (Bailin and Bárány, 1968).

D. General Considerations. The chemical modifications of myosin have revealed the following features of myosin ATPase: (1) Mg-, Ca-, and K-EDTA–ATPase can behave fairly independently of the other ATPases; (2) K-EDTA–ATPase is never activated, but is depressed by any modifications; (3) CaATPase is liable to be influenced by the change in ionic strength, but MgATPase is not.

Generally speaking, MgATPase and K-EDTA–ATPase seem more suited for representing the state of myosin than CaATPase (see p. 319).

It may not be so bold to assume that the decrease of K-EDTA–ATPase and the increase of MgATPase are mainly related to the depression of enzyme substrate formation and the acceleration of enzyme product decomposition, respectively, and that CaATPase is dependent on the balance of these two features of myosin ATPase.

One purpose of chemical modifications of myosin is to provide information as to what reaction or what aspect of the hydrolytic processes is crucial to the contractile mechanism. From this standpoint, it is important to examine the interaction of such modified myosins with actin in detail. An important feature of modified ATPase will be also discussed in a later section (p. 316).

C. Subunits and Subfragments

There have been two kinds of approaches to the clarification by chemical procedures of the fine structure of the myosin molecule. One is to separate myosin into subunits by using agents or procedures that can break noncovalent bonds. This is a common approach in protein chemistry, but the procedures used in this approach are largely drastic, so that the fine characteristics of the original protein are lost in the course of separation and the recombination of separated components does not restore the original activity. In this respect, myosin was not exceptional. Its ATPase activity had been found to be irreversibly inactivated by such procedures, so that the studies along this line had been largely confined to physicochemical aspects. Recent success in restoring the ATPase activity by recombination of once separated components, however, has changed the situation, and this approach has become one of the key subjects in the myosin research.

The other approach is to split the myosin molecule into several fragments by proteolytic enzymes. In usual proteins, this approach has a disadvantage of localizing the separated fragments in the entire chain of the molecule. In the case of myosin, however, it has brought about enormous success because of the unique structure of the myosin molecule. The localization of split products in the whole chain of the myosin molecule is easily estimated and has now been correlated with the electron microscopic profiles.

1. COMPONENTS OBTAINED BY BREAKAGE OF NONCOVALENT BONDS

Tsao (1953) was the first to suggest that the myosin molecule was not a single chain but was composed of smaller subunits. He determined the molecular weight of myosin in 5.7 M guanidine hydrochloride and found it to be about 1.65×10^5 (in addition to this, he pointed out the presence of smaller subunits as described below). This was followed by similar reports which showed the molecular weight of the myosin subunit to be 1.8–2.06×10^5 (Kielley and Harrington, 1960; Small et al., 1961). At that time, there were rather controversial discussions as to whether the molecular weight of whole myosin molecule would be around 5×10^5 or 6×10^5 (see Dreizen et al., 1967). This difference had an important meaning in answering to the question as to whether myosin is composed of two strands or three. There is now general agreement that the molecular weight of myosin is less than 5×10^5 (Table IV), and therefore, myosin is composed of two identical chains. Conclusive evidence for this has been supplied by recent electron microscopic observations (see p. 307).

Another important aspect concerning the subunits of the myosin molecule is the presence of small subunits. Tsao (1953) was again the pioneer in pointing out in the ultracentrifugation diagram the presence of a light component of molecular weight 1.6×10^4 in 5.7 M guanidine hydrochloride. The presence of such a light component was confirmed by a few workers (Kominz et al., 1959; Wetlaufer and Edsall, 1960), but not much attention was paid to these works. About 1966, a renewed interest was raised in this problem. Several research groups have confirmed previous results and succeeded in isolating such light components using high pH, 4 M lithium chloride, succinylation, or acetylation (Dreizen et al., 1966; Gershman et al., 1966; Oppenheimer et al., 1966; Locker and Hagyard, 1967; Gaetjens et al., 1968). According to the results of these workers, 1 mole myosin contains one to three small

TABLE IV
MOLECULAR WEIGHT OF MYOSIN

Molecular weight	Method	Author
470,000	Osmotic pressure	Tonomura et al. (1966a)
510,000	Sedimentation equilibrium	
510,000	High speed equilibrium method (Yphantis, 1964)	Chung et al. (1967)
430,000 ± 30,000	Molecular weight of heavy chain and LMP (high speed equilibrium method)	Gazith et al. (1970)
458,000	High speed equilibrium method	Godfrey and Harrington (1970)
465,000	High speed equilibrium method	Rossomando and Piez (1970)

components of molecular weight 2–3×10^4. It has also been shown that HMM and S-1 contain such small subunits (Gaetjens *et al.*, 1968; Trotta *et al.*, 1968). These small components are called by various names such as light component, light chain, low molecular weight component, or G subunit; but since many people now use LMP (low molecular weight protein), we follow this usage hereafter.

It is widely known that myosin loses its ATPase activity at pH above 10. In view of the separation of LMP under this condition, a question was raised as to whether the irreversible disappearance of the ATPase activity of myosin at high pH might be due not to mere deterioration of myosin, but due to the separation of such LMP from myosin. Indeed, Frederiksen and Holtzer (1968) have shown the restoration of ATPase activity of myosin in a solution of pH 11 after confirmation of the separation of LMP in the ultracentrifugation diagram.

Furthermore, Dreizen and Gershman (1970) have succeeded in recombination experiments after complete separation of LMP from heavy chains using a 4 M lithium chloride–potassium citrate system; Stracher (1969) has obtained the same success with S-1 by recombining components once separated completely by Sephadex G-200 chromatography. It should be noted here that in any case the restoration of ATPase activity requires immediate recombination after separation.

According to SDS–polyacrylamide gel electrophoresis, rabbit skeletal myosin contains at least two LMP, of which the molecular weights are 17,000 and 25,000 (Weeds, 1969; Lowey, 1970; Paterson and Strohman, 1970). The 17,000 component is removed from myosin without affecting ATPase activity by 5,5′-dithio-bis(2-nitrobenzoic acid) treatment (Weeds, 1969; Gazith *et al.*, 1970), but the elimination of the heavier LMP is always accompanied by the loss of ATPase.

In addition to this, the presence of another LMP has been noted (Weeds, 1970; Perrie and Perry, 1970). Analysis of amino acid sequence has shown that this lightest LMP has a close relationship to the 25,000 LMP (Weeds, 1970).

One mole of myosin seems to contain 2 moles of the 17,000 component and 2 moles of the sum of the other two LMP (Weeds and Lowey, 1971). Physicochemical properties of heavy chains of myosin without LMP have been studied by Gazith *et al.* (1970); their molecular weight is estimated to be 194,000.

It seems thus almost established that the LMP is indispensable for the ATPase activity of the myosin molecule. It is one of the most fascinating problems in muscle research of nowadays to clarify the interaction of LMP with heavy chains.

2. Subfragments Obtained by Proteolytic Enzymes

Pioneering work in this direction was independently made by Gergely (1950) and Perry (1951) under different intentions. They both showed that ATPase activity became solubilized by trypsin treatment. Further elaborate work of A. G. Szent-Györgyi and Mihályi (Mihályi and Szent-Györgyi, 1953; A. G. Szent-Györgyi, 1953) have defined the conditions for trypsin treatment and shown that the treatment separates the myosin molecule into two distinct portions, H-meromyosin (HMM) and L-meromyosin (LMM).

A. H- and L-Meromyosin. These subfragments are obtained by brief trypsin treatment of the myosin molecule. One fragment has a larger molecular weight of about 3.4×10^5 and retains almost every property of the original myosin concerning ATPase and actin-binding activities. The other is of a smaller molecular weight about 1.4×10^5 and possesses the fibrous and globulin-like nature of myosin, i.e., insoluble at low ionic strengths but solubilized at higher ionic strengths.

It was found that essentially the same components were obtained by other proteolytic enzymes such as chymotrypsin (Gergely, 1953) or subtilisin (Middlebrook, 1958). These facts may indicate that there is a loose structure between the portion containing the ATPase site (i.e., HMM) and the fibrous part (i.e., LMM) and that this part is easily attacked by any proteolytic enzyme (Mihályi and Harrington, 1959). There is no doubt that proteolytic enzymes would also attack other parts of myosin molecule in addition to the specified loose part between HMM and LMM. Indeed, a considerable number of small peptides are separated during the process of digestion, indicating that myosin is split at various points of myosin chains (Segal *et al.*, 1967).

B. S-1. Mueller and Perry (1961a,b) have shown that HMM prepared by the method of A. G. Szent-Györgyi (1953) consists of two major fractions according to DEAE-cellulose chromatography. One is HMM, but the other is of a smaller molecular weight, retaining full ATPase and actin-binding activities. They named it S-1 (subfragment 1 of HMM).

As indicated by the intrinsic viscosity shown in Table V, HMM is a fibrous protein, but S-1 is almost an ideal globular protein. S-1 is obtained also by other proteolytic enzymes, such as papain (Kominz *et al.*, 1965; Slayter and Lowey, 1967; Nihei and Kay, 1968), chymotrypsin (Jones and Perry, 1966; Hotta and Usami, 1967), subtilisin (Jones and Perry, 1966), and Nagarse (Yazawa and Yagi, 1971). The properties

TABLE V

SOME PHYSICOCHEMICAL PROPERTIES OF MYOSIN AND ITS SUBFRAGMENTS[a]

Fragment	Molecular weight ($\times 10^4$)	Intrinsic viscosity (dl/gm)	$s^0_{20,w}$ (S)	α-Helix (%)	Remarks
Myosin	51	2.1	6.4	57	—
HMM	34	0.49	7.2	46	see Fig. 4
LMM	14	1.2	2.9	90	see Fig. 4
S-1	11.5	0.064	5.8	33	see Fig. 4
S-2	6.2	0.4	2.7	87	see Fig. 4
Rod	22	2.4	3.4	94	S-2 + LMM

[a] These values were derived from the table in the article of Lowey *et al.* (1969). The original table includes the standard error of the mean of each value, but in this table only the mean values are listed for simplicity. With regard to the source of each value, refer to the legend of the original table.

of various S-1 preparations are somewhat different from preparation to preparation.

Further treatment of S-1 by Nagarse results in a much smaller subfragments, named S-n, which retain ATPase activity but lose the actin-binding capacity (Yagi *et al.*, 1967).

C. S-2. The remaining fibrous part of HMM, from which S-1 had been removed, was isolated as a homogeneous protein and named HMM subfragment 2 (Lowey *et al.*, 1967), or S-2, and it has a considerable asymmetry (Table V). Unlike LMM, S-2 is soluble and does not form aggregates at low ionic strengths at neutral pH (see p. 304).

3. ENZYMATIC PROPERTIES OF MYOSIN SUBFRAGMENTS

The characteristics of the ATPase activities of HMM and S-1 are qualitatively the same as those of the original myosin, e.g., they are activated by Ca- or K-EDTA or sulfhydryl-blocking agents and inhibited by magnesium (Perry and Cotterill, 1965; Jones and Perry, 1966; Yagi and Yazawa, 1966).

The quantitative comparison of the ATPase activity of HMM with that of S-1 may suggest (Jones and Perry, 1966; Yagi and Yazawa, 1966; Lowey *et al.*, 1969; Nauss *et al.*, 1969) an answer to the question whether or not the two chains of HMM are functionally identical. So far, the ratio of the ATPase activity of HMM to that of S-1 on a molar basis, assuming their molecular weights to be 3.4×10^5 and 1.15×10^5, respectively, has been shown to range from 1.7 to 2.4, favoring an affirmative answer to this question. However, it is quite possible that nature of myosin ATPase would be altered by such a treatment as proteolytic digestion, so that a mere comparison of the ATPase activities under a specified condition cannot bring about the final conclusion. Furthermore, if we take the actin-activated ATPase (Eisenberg and Moos, 1968; Eisenberg *et al.*, 1968; Lowey *et al.*, 1969), there is a considerable divergence among the results of different authors, ranging from 1.0 to 4.3.

In this connection, there is another important problem along a similar line, i.e., how much ATP can bind to the myosin molecule and its subfragments. Early reports regarding the number of ATP, ADP, or PP bound to myosin or HMM varied. Recent observations in several laboratories, however, agree that myosin and HMM must have two nucleotide binding sites and S-1 only one site, as shown in Table VI.

As a whole, all the evidence is rather consistent with the idea that both chains of HMM are identical, even from enzymic standpoints. How-

TABLE VI

NUMBER OF NUCLEOTIDE-BINDING SITES OF MYOSIN AND ITS SUBFRAGMENTS[a]

Fragment	Myosin (moles per 5×10^5 gm)	HMM (moles per 3.5×10^5 gm)	S_1 (moles per 1.1×10^5 gm)	References[b]
PP	1.8	1.7	0.9	Nauss *et al.* (1969) 1
	—	1.8–2.3	—	Morita (1969) 2
	1.5	—	—	Kiely and Martonosi (1968) 1
ADP	1.9	1.5	—	Lowey and Luck (1969) 1
	1.2–1.6	—	—	Kiely and Martonosi (1969) 3
	2.3	2.6	1.0	Young (1967) 4
ATP	1.8–1.9	1.9–2.1	0.7–0.8	Eisenberg and Moos (1970) 5
	1.4–1.8	1.3–2.0	—	Schliselfeld and Bárány (1968) 6

[a] Figures in parentheses under the names of proteins are the molecular weights assumed for these calculations.

[b] Data determined by: 1. equilibrium dialysis, 2. ultraviolet difference spectrum, 3. equilibrium dialysis and ultracentrifugation with labeled-ADP, 4. ultracentrifugal transport and equilibrium dialysis, 5. ATPase rate, 6. gel filtration chromatography with labeled-ATP.

ever, this does not necessarily mean that both chains act independently of each other in the contractile processes. Perhaps the investigation of the interaction of both chains would be one of the key subjects in the muscle research at this stage.

4. Fine Structure of the Myosin Molecule with Particular Reference to Its Subfragments

Direct visualization of the myosin molecules was made possible by several electron microscopists about 1961–1963 (Rice, 1961; Zobel and Carlson, 1963; H. E. Huxley, 1963). According to the electron micrograph taken by Huxley using shadowing method, the profile of the molecule coincided well with the model one had expected from physicochemical properties of trypsin-treated subfragments and of subunits formed in strong urea or guanidine hydrochloride solution. It had a globular head, which undoubtedly corresponded to HMM, and a long tail from which LMM might have been derived. The length of the molecule was about 1520 Å, which was also in agreement with the data obtained by physicochemical determination.

Using rotatory shadowing technique, Lowey and her colleagues (Slayter and Lowey, 1967; Lowey *et al.*, 1969) succeeded in visualizing the presence of two heads in myosin molecules (Fig. 3). The most remarkable point is that both heads are completely separated from each other and are connected to the main chain with a neck of an appreciable length; this is a fact that cannot be expected from physicochemical studies so far carried out. This finding has stimulated further studies on the subfragments or subunits of the myosin molecule as described above. Altogether, a fairly concrete model of the myosin molecule has been presented as illustrated in Fig. 4.

Another crucial point of this model is that two chains of the myosin molecule are apparently identical. A question is then raised as to whether both chains actually have the same function. Although general agreement has not yet been reached, recent experimental results favor an affirmative answer (see p. 306). Thus, the myosin molecule seems to provide another example to the concept that fibrous proteins of muscle are composed of two almost identical chains.

A more exact value for the length of the molecule has been given by the success in making the segments of the rod part of the myosin molecule (C. Cohen *et al.*, 1970), i.e., the part of myosin from which the globular part is removed (Table V). The length of the rod part is 1450 Å, so that the total length of myosin should be about 1550 Å.

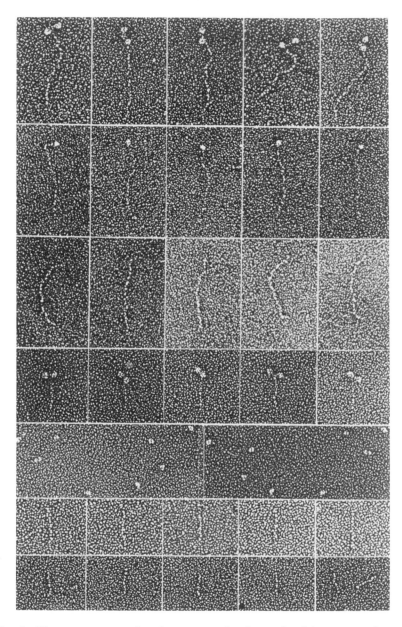

Fig. 3. Electron micrographs of myosin molecules and subfragments of myosin by rotating shadow cast method. *From the top to the bottom:* myosin, single headed myosins, rods, HMM, S-1, LMM, and S-2. From Lowey *et al.* (1969) (an original copy was kindly supplied by Dr. Lowey).

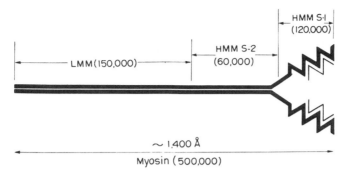

Fig. 4. Model of the myosin molecule. The thin strand connected to HMM indicates LMP. From Lowey *et al.* (1969).

D. Assembly of Myosin and Its Subfragments

1. Filament Formation of Myosin

The intrinsic nature of myosin molecules to form fibrous aggregates of a length of about 1–2 μ, which is similar to that of the thick filament in the sarcomere, was first noticed by using flow birefringent technique (Noda and Ebashi, 1960). Electron microscopic observations have revealed that even the profile of the filament resembles the thick filament separated from fresh muscle (H. E. Huxley, 1963). This strongly suggests that not only the thin filament but also the thick filament of the sarcomere are formed as a result of self assembly.

The formation of such filaments is essentially dependent on the ionic strength, not so much on other factors (Noda and Ebashi, 1960; Kaminer and Bell, 1966). In a solution of a high ionic strength, say 0.6 M, myosin exists as dispersed, almost as a single molecule. If the ionic strength is gradually reduced, the filament formation of myosin starts at an ionic strength about 0.3 M and is completed at about 0.2 M. To the naked eye, myosin molecules thus aggregated appear still solubilized. ATP has a dispersing effect on the filamentous aggregates, but its influence is rather limited. Magnesium ion of a physiological concentration has no particular effect. A more influential factor is the rate of dilution. Gradual decrease in ionic strength tends to form longer aggregates. If the ionic strength is further lowered to about 0.07, the random aggregation of such filaments takes place; the solution becomes suddenly turbid, and soon silky precipitates are gradually sedimented.

On the basis of immunoelectron microscopic studies of the myofibril, Pepe (1966, 1967) has proposed a model for the fine structure of the thick filament. H. E. Huxley and Brown (1967), using a low-angle X-ray diffraction technique, have presented a precise numerical basis for the architecture of the thick filament and also referred to the state in contracting muscle. Since this problem may be discussed in other parts of this treatise in detail, no further reference is made here. It will be interesting to see whether the periodicities 143 Å and 429 Å revealed by these investigations can be shown with reconstituted myosin filaments. Since such periodicities can be observed with paracrystals of myosin subfragments (see below), the reconstituted filaments must have the same periodicities. However, no investigation comparable to above studies has not yet been made.

2. Ordered Aggregates of Myosin and Its Subfragment

Myosin molecules form ordered aggregates, "segment aggregates" (Harrison *et al.*, 1971), but cannot form paracrystals or crystals in the strict sense, probably because of steric hindrance due to the globular shape of HMM part. On the other hand, myosin without HMM, namely LMM, does form crystalline structures (H. E. Huxley, 1963; Lowey *et al.*, 1967; Podlubnaya *et al.*, 1969; King and Young, 1970). Surprisingly, the profiles of paracrystals and crystals of LMM (like the amino acid composition) resemble those of tropomysin (cf. Fig. 9) to a great extent (Table XIII). They show three kinds of periodicities, 400, 430, and 145 Å, the latter two probably corresponding to the periodicities of the same lengths observed with the thick filament *in vivo*.

It is worthy of note that S-2 forms a paracrystal that has a periodicity of 145 Å (Lowey *et al.*, 1967). This may suggest that the S-2 part is also involved in the filament formation of myosin molecules. The segments of the rod part have already been discussed in the previous section.

E. Interaction with Actin in the Presence of MgATP

1. General Aspects

The interaction of myosin with F-actin in the presence of MgATP is nothing but the fundamental process of contraction itself. As is first pointed out by A. Szent-Györgyi, MgATP has two kinds of actions which are apparently opposite to each other. One is to activate the interaction of myosin and actin; this certainly corresponds to the contractile process. The other is to depress the interaction or to keep both proteins in a dissociated state; this is undoubtedly related to the relaxed state.

The activating action of ATP can be replaced only by certain nucleoside triphosphates (see p. 316). No nucleoside diphosphate or inorganic polyphosphate can substitute for ATP. The interaction is always accompanied by the breakdown of the terminal phosphate of that nucleoside triphosphate.

Almost all compounds that have pyrophosphate residues at their terminals show more or less the dissociating action on the actin–myosin system as ATP. No marked splitting of terminal phosphate is necessary for this action.

Roughly speaking, the contraction is the state in which the activating effect of MgATP is predominating over its dissociating effect, and the relaxation is *vice versa*. There are many factors that influence the balance between the activating and depressing effects of ATP. In the case of *in vitro* experiments, the ionic strength and MgATP concentration are the most influential factors, of which the mode of action is illustrated in Fig. 5 (pure actomyosin is most sensitive to these factors, whereas organized contractile systems such as myofibrils or glycerinated fibers are relatively insensitive; the reason for this is not yet clear). The lower the temperature, the more favorable it is for relaxation (Varga, 1950). The influence of pH is complicated and cannot be simply illustrated.

It should be emphasized that these physical or physicochemical factors are constant in living muscle, particularly in skeletal muscle, and, therefore, could not be the physiological regulators.

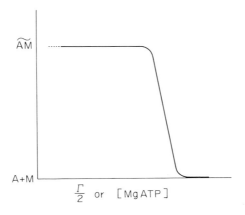

Fig. 5. Schematic representation of the interaction of myosin and actin in the presence of MgATP. $\widetilde{AM}$ denotes the state in which myosin and actin filaments are interacting with the simultaneous breakdown of ATP; in the case of suspension of actomyosin, this state is represented by the superprecipitation. A + M denotes the state in which myosin and actin filaments are dissociated from each other; this state corresponds to the cleared state of actomyosin (see also Figs. 6 and 7).

2. Hydrolytic and Dissociating Sites of Myosin

As described above, the interaction of myosin and actin is based on the activating and dissociating effects of MgATP. It is certain that the activating effect is exerted through the hydrolytic site of the myosin molecule. In this connection, there has been an interesting but controversial problem as to whether or not the hydrolytic site of myosin is identical with the site through which ATP exerts its dissociating effect. Based on the fact that Salyrgan, an organic mercurial, can induce relaxation of glycerinated fibers when it inhibited their ATPase activity, H. H. Weber and Portzehl (1954) suggested the presence of a site for the dissociating action different from the hydrolytic site. The view was confirmed later by kinetic analysis (A. Weber *et al.*, 1964; Eisenberg and Moos, 1965; Levy and Ryan, 1967).

On the other hand, Bárány and Bárány (1959) have shown that if free myosin is treated by iodoacetamide, both hydrolytic and actin-binding activities are lost, but that if actomyosin is subjected to such treatment, only the hydrolytic activity is abolished; this observation, which proves the separate presence of the actin-binding and hydrolytic sites, has later been confirmed in various ways (Perry and Cotterill, 1964; Stracher, 1964). These results might have appeared to indicate that the actin-binding site would be closely related to the dissociating site.

On the basis of kinetic studies, however, it has recently been concluded that there is no ATP-binding site in HMM other than the hydrolytic site (Eisenberg and Moos, 1968), and that the dissociating site of myosin may not be identical with the actin-binding site (Finlayson *et al.*, 1969).

This complicated situation reminds us of a finding in the paper of Bárány and Bárány (1959) mentioned above. The sensitivity of iodoacetamide-treated actomyosin (the actomyosin which has lost its hydrolytic activity but retains actin-binding activity) to the dissociating effect of ATP at high ionic strengths is intensely lowered to the level of PP; the latter exerts a weak but definite dissociating action on actomyosin, irrespective of iodoacetamide treatment. Thus, the mechanism of dissociation seems not so simple as could be explained by assuming a simple "dissociation" site.

3. Superprecipitation and Actomyosin ATPase

As stated above, both myosin and actin form filamentous structures under conditions where the interaction of both proteins takes place;

i.e., the interaction actually occurs between both kinds of filaments. In this respect, the interaction of myosin and actin *in vitro* is not far different from that *in vivo*.

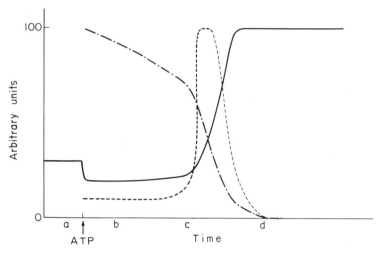

Fig. 6. Schematic representation of the processes of superprecipitation. The scheme represents the case where the response is initiated by clearing. The symbols a, b, c, and d under the abscissa indicate the time where the electron micrographs of Fig. 7 were taken. (——) Optical density; (---) ATPase activity; (-·-) ATP concentration.

The usual *in vitro* experiment is started by adding ATP to the suspension of the mixture of myosin and actin (Fig. 6). Since myosin and actin filaments cannot be kept apart without MgATP, both proteins bind together and form an unphysiological state of aggregates, most of them forming the so-called arrowhead structure[1] (Figs. 6 and 7a; see p. 328); this is the structural basis of actin–myosin complex which has been called actomyosin. At a low ionic strength, say 0.1, the aggregates form a very hydrous gel, of which the suspension is considerably transparent. On addition of ATP, if the conditions favor the dissociating action of MgATP, myosin molecules are detached from the actin filaments and aggregate with one another to form myosin filaments; both filaments now exist separated from each other as in living muscle (Figs. 6 and 7b), so that the suspension becomes quite transparent, almost

[1] The arrowhead structure is obvious at high ionic strength of about 0.6, where myosin molecules are dispersed. The gel formation of actomyosin at low ionic strengths is partly due to the association of LMM part of the myosin molecules. However, most of their heads are still attached to actin, retaining arrowhead structure. Even when actomyosin is formed by adding F-actin to myosin filaments at low ionic strength, careful examination reveals that the arrowhead structure is gradually formed (see also p. 328).

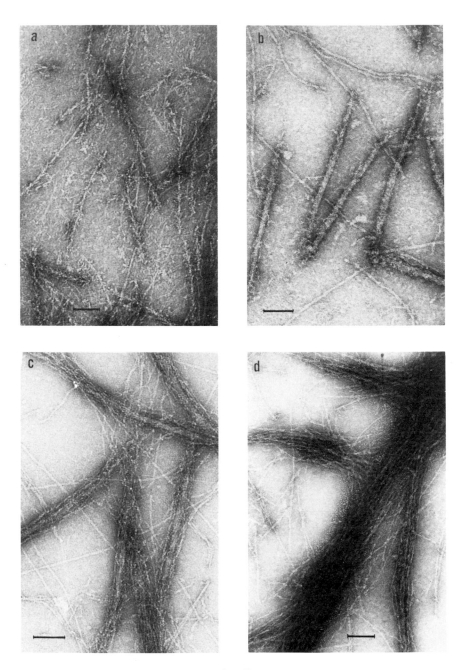

Fig. 7.

a true solution. This state is called the "clear phase" or "cleared state." Although very low, the ATPase activity of the myosin filaments causes ATP to split slowly but steadily (Ebashi, 1961; Eisenberg and Moos, 1967) and finally to reach a critical concentration whereupon actin filaments start to align parallel with myosin filaments (Figs. 6 and 7c) (Ikemoto *et al.*, 1966) and ATP breakdown becomes quite vigorous. This state certainly corresponds to the initiation of actual contraction. At this stage, although the medium still looks clear, the viscosity of the solution is already elevated (Maruyama and Gergely, 1962; Eisenberg and Moos, 1967), corresponding to the change in the alignment of both filaments. No one knows whether the sliding of both filaments would occur in this stage. ATP concentration now decreases very rapidly (Fig. 6). Concurrently, the alignment of actin filaments with myosin filaments becomes more and more extensive (Figs. 6 and 7d) (Ikemoto *et al.*, 1966); finally, the aligned myosin–actin filaments aggregate with one another to form larger aggregates. This process is a kind of nucleation phenomenon and can be promoted by adding nuclear substances (Fig. 6) (Tada and Tonomura, 1967). At this stage, the suspension becomes quite turbid and loses transparency completely; this state is called superprecipitation. ATP is already almost split out, but if ATP is added, it is split rapidly. The aggregates becomes larger and larger and finally precipitate. If the actomyosin suspension is well dispersed before or at the time of the addition of ATP, the aggregate formation stops at a certain stage, and a constant turbid state is retained for a fairly long time. If the conditions do not favor the dissociating action of MgATP, no delay in the turbidity change from the ATPase activity is observed (Watanabe, 1970).

Thus, the contraction *in vitro* may be characterized in two ways. One is the physical change, i.e., the ordered arrangement of both filaments, and the other the chemical change, i.e., the breakdown of ATP. It is not impossible to measure the primary physical change that corresponds to the contractile process, but it is quite impractical to use it as a routine procedure. As stated above, the superprecipitation does not represent the primary physical change itself, but an inevitable conse-

---

Fig. 7. Electron micrographs of superprecipitation of myosin B. Specimens were picked up at points marked in Fig. 6 and stained negatively with uranyl acetate. Bars indicate 0.1 μ. (a) Before the addition of ATP; arrowhead complex filaments are visible. (b) Clear phase; myosin molecules are instantaneously dissociated from the complex filament and form thick (myosin) filaments separately from the thin filaments ($\times$80,000). (c) Starting phase of actin–myosin interaction; thick filaments are surrounded by thin filaments aligning parallel ($\times$80,000). (d) Superprecipitation phase; aligned myosin–actin filaments form larger aggregates ($\times$70,000).

quence of such physical change. It can be measured very conveniently by the increase in optical absorption or turbidity (Ebashi, 1961; T. Yasui and Watanabe, 1965). The increase in absorbancy is due to the increased light scattering caused by the increase in the refractive index of actomyosin gel as a result of its strong shrinking (Rice et al., 1963). Since such scattering is directed to near 0°, the precise alignment of the optical axis is necessary for the determination of absorbancy.

If HMM or S-1 is used in place of myosin, the ATPase activity is the only practical measure of interaction with actin (Perry et al., 1966; Eisenberg and Moos, 1968; Eisenberg et al., 1968). In the cases of certain kinds of modified myosin (see p. 298) or smooth muscle myosin (see D. M. Needham and Schoenberg, 1967), where the activation of ATPase by actin is not pronounced, ATPase activity is not a reliable index; superprecipitation should be observed simultaneously. This is also the case in examination of ATP analogs to determine whether they behave like ATP. However, it should be mentioned that *in vitro* phenomena do not represent all the features of contraction. If it is a question of whether there is actually contraction, the real contractile material (e.g., glycerinated fibers) should be used as the testing system (see p. 317).

3. SUBSTITUTES FOR ATP

A. GENERAL CONSIDERATION

ATP has a dual action on the interaction of myosin and actin—activating and depressing (see p. 310). In addition to this, ATP has different features as a substrate of myosin ATPase from those in the presence of actin (see p. 290). Thus, ATP has at least three facets: (1) as the substrate of myosin ATPase; (2) as the activator of the interaction of myosin and actin, i.e., as the inducer of contractile response and as the substrate of actin-activated myosin ATPase; and (3) as the depressor of the interaction of myosin and actin or as the dissociating agent of actomyosin. Corresponding to these different facets, the substitutes for ATP fall into different groups, although there is considerable overlap. To examine the degree to which an agent resembles ATP, the criteria for each facet of ATP must be established.

i. MYOSIN ATPASE. Myosin ATPase—usually calcium (1–10 mM)-activated ATPase in a solution of high ionic strength, e.g., 0.6 M potassium chloride, at neutral or slightly alkaline pH—is adopted as the index of myosin ATPase activity. Even if we use myosin B or myofibrils, the ATPase measured under these conditions should be considered to reflect the myosin ATPase. As will be described later, there is some doubt whether CaATPase is a pertinent index of ATPase activity.

ii. CONTRACTION OR ACTIN-ACTIVATED MYOSIN ATPASE. As far as ATP is concerned, actomyosin-type P_i liberation, i.e., P_i liberation in the presence of magnesium at a relatively low ionic strength, is considered as representing the contractile response. However, this is not always a reliable measure (see p. 316). Actomyosin-type hydrolytic reaction can be used only when P_i liberation by myosin in the presence of magnesium ion is very low. Even if this condition is satisfied, we should be cautious in correlating the actin-activated P_i liberation with contraction when we are dealing with a new substitute of which the chemical structure is definitely different from ATP. In this case, the superprecipitation of myofibrils or actomyosin may provide more information. In any case, it is a safer way to examine the real contractile response of glycerinated fibers.

iii. DISSOCIATING ACTION. The viscosity drop of actomyosin at a high ionic strength (higher than 0.4)—the phenomenon found by J. Needham *et al.* (1941) was the first evidence of the physical effect of ATP on actomyosin system prior to the discovery of actomyosin thread contraction—is usually adopted as the criterion. This method is simple so that it is suited for large scale experiments, but there the question might be raised whether the effects at a high ionic strength represent those at a relatively low ionic strength. This is a reasonable doubt. On the other hand, the relaxation of glycerinated fibers or the dissociation of actomyosin at low ionic strengths may strongly be interfered with by the activating effect of the substrate. Thus, no satisfactory method of examining the dissociating action has so far been established. This may correspond to the fact that the mechanism of dissociation seems more complicated than had been thought (see p. 312).

B. SPECIAL ASPECTS

As for the natural analogs of ATP, Hasselbach (1956) made fundamental studies and Blum (1955) made a kinetic analysis of some analogs. Tonomura and his associates, (Ikehara *et al.*, 1961; Tonomura *et al.*, 1965, 1967) extensively investigated the activities of various synthetized analogs. The following outline of the structure–activity relationship of ATP analogs is drawn mainly from the work of the above authors, especially Tonomura and his colleagues.

i. Myosin ATPase. So far, almost all the compounds mentioned, except a few cases which have terminal triphosphoric acid, are more or less substrates of myosin ATPase. It is interesting that even such an inorganic compound as tripolyphosphate can be split by myosin (Dainty *et al.*, 1944). Generally speaking, however, the compounds that are rather poor substrates cannot be the agent of the next category.

ii. Contraction. It is an absolute prerequisite that the agent of contraction should contain both triphosphoric acid and base. The structural requirements for the base are (see Tonomura *et al.*, 1965): (1) a nitrogen or oxygen atom should be attached to the 6′-position of the base; whether the nitrogen is methylated or not or whether the oxygen is hydrogenated or not is not essential for the activity; (2) the triphosphoric acid and the base must be connected by carbohydrate, not necessarily by ribose or deoxyribose, but the length of the chain should be the same as that of ribose and the chain must contain an oxygen atom at 3′-position.

iii. Dissociating Action. So far, almost all compounds that have a terminal pyrophosphate residue, including inorganic pyrophosphate itself (Mommaerts, 1948), have been shown to be dissociating. However, the degree of such activity is very much influenced by the structure of carbohydrate chain and base. Whether it is diphosphate- or triphosphate-type is sometimes not so important, although triphosphate-types always have a stronger activity than the corresponding diphosphate.

TABLE VII

RATES OF HYDROLYSIS OF ATP AND ITS ANALOGS BY MYOSIN

Analog	Calcium ion[a]			EDTA[b]	
	Intact myosin (a)	PCMB-titrated[c] myosin (b)	(b)/(a)	Potassium ion	Ammonium ion
ATP	1.3	3.7	2.8 (6.7)	3.0	17.9
ITP	7.2	2.7	0.37 (0.3)	0.0	0.0
GTP	5.5	1.9	0.34 (0.3)	0.0	1.0
CTP	0.52	0.82	1.6 (3.7)	1.1	2.1
UTP	2.0	2.8	1.4 (1.1)	0.4	2.7

[a] Calculated from Fig. 2 in the article of Kielley (1961); figures in parentheses were derived from the data in Fig. 1 in the article of Blum (1960) with myosin B.

[b] Derived from Table II in the article of Kielley *et al.* (1956).

[c] The activities of myosin, of which the half of sulfhydryl residues are titrated by PCMB, where nearly maximum activation is observed with ATP.

The above is a superficial survey. If we inquire into quantitative differences, the situation is more complicated. In Table VII, the mode of actions of ATP and its four analogs on Ca- and K-EDTA-nucleotidase are described. It might give an impression that nucleotidase activities, particularly K-EDTA-nucleotidase, would not indicate anything of biological significance but a mere *in vitro* artifact. However, if these enzymatic activities are compared with the activating and depressing actions on myosin–actin interaction (see p. 310), we can find a rather good coincidence among the activating effect on the myosin–actin interaction, K-EDTA-nucleotidase and NEM-activated calcium nucleotidase; this is not the case with PCMB-activated nucleotidase, but if the ratios of the degrees of activation by PCMB are taken into consideration, the order becomes the same as that of NEM-activated nucleotidase (Table VIII). It seems reasonable that the order of dissociating actions of ATP analogs has no relationship with any types of ATPases, but it is rather unexpected that ordinary CaATPase has no correlation with any other category, suggesting its complicated nature. The question as to whether or not CaATPase is suited to represent myosin ATPase should be reconsidered in the light of facts mentioned above as well as the discussions made in the previous section (see p. 300).

TABLE VIII

COMPARISON OF EFFECTS OF ATP AND ITS ANALOGS ON VARIOUS TYPES OF MYOSIN ATPASE AND MYOSIN–ACTIN INTERACTIONS[a]

Analog	Activation of myosin–actin interaction (contraction)[b]	Dissociation of actomyosin (relaxation)[b]	CaATPase[c]	K-EDTA-[c] ATPase	NEM-activated CaATPase[d]	Degree of activation of CaATPase by PCMB[c]
ATP	1	1	4	1	1	1
CTP	2	4	5	2	2	2
UTP	3	4	3	3	3	3
ITP	4	3	1	4	4	4
GTP	5	2	2	4	5	5
ADP	—	4	—	—	—	—
PP	—	4	—	—	—	—

[a] Figures in the columns indicate the order of analogs in respective actions.

[b] Derived from Tables I and II in the paper of Hasselbach (1956); glycerinated muscle fibers were used as the testing material.

[c] See Table VII.

[d] Derived from Fig. 4 in the paper of Sekine and Kielley (1964).

III. Actin

A. General Properties and Preparative Aspects

Actin is the major constituent of the thin filaments. Its elementary unit is a globular protein of a molecular weight of less than 50,000 (see below) called G-actin. G-actin has a tendency in itself to form a long chain, i.e., actin filament or F-actin.

Actin has two important functions. One is to activate myosin and to cooperate with it in carrying out the reaction to convert the chemical energy contained in ATP into mechanical energy. The other is to provide a place for regulatory proteins (i.e., troponin–tropomyosin system) through which the physiological triggering factor, calcium ion, can control the interaction of actin and myosin.

Recent progress in column chromatography has contributed to the purification of actin to a great deal. Even with recent gel electrophoretic technique using various denaturing agents, G-actin seems to be a single protein, having no subunit (see p. 324).

Isolation of actin can be made by two methods different in principle. One is the classic method, i.e., to decompose the intermolecular linkage in F-actin and then to extract it as G-actin. Myosin must be removed in advance. As represented by the brilliant success of Straub (1942; cf. A. Szent-Györgyi, 1951), acetone is so far the most favorable agent for breaking the link without denaturing actin itself. Another agent that induces such action is potassium iodide (A. G. Szent-Györgyi, 1951).

The other method is to isolate F-actin directly from myofibrils. The idea itself is simple, but it is not so easy a task as might be supposed. The presence of myosin is a more serious hindrance to the extraction than the above case, so that it must be removed completely in advance, e.g., by using Hasselbach–Schneider solution (1951). A further obstacle to this method is the attachment of F-actin to the Z band, which makes the separation of F actin very difficult. Taking advantage of trypsin-resistant nature of actin (see p. 324) and trypsin–sensitive nature of the Z band (see p. 348), Hama *et al.* (1965) have succeeded in isolating F-actin in a large scale and named it natural F-actin. Further problems of preparation methods, will be discussed in the next section.

Every kind of muscle, of course, contains actin as a main contractile protein. Generally speaking, actin has a very low species and organ specificity compared with myosin. The amino acid compositions of actin from various kinds of muscles of vertebrates are almost identical with each other, irrespective of organs or species (Carsten and Katz, 1964).

This may be reflected in the fact that it is extremely difficult to prepare the antibody against actin (the antibody reported as that against actin is now considered to be the antibody against native tropomyosin or troponin).

It is of great interest that an actin-like preparation that can interact with rabbit myosin has been isolated from slime mold by Hatano and his colleagues (Hatano and Oosawa, 1966; Hatano et al., 1967). In this connection, attention should be placed on the report of Ishikawa et al. (1969) that even noncontractile cells contain HMM-binding filament, although it is not yet clear whether it is actually actin-related protein or not.

B. G–F Transformation

1. General Aspects

As stated above, the unit of actin (G-actin) forms a fibrous arrangement, F-actin (Straub, 1943). This phenomenon is generally called the G–F transformation. Interesting enough, G-actin contains one exchangeable ATP per molecule, and this ATP is split into ADP and P_i during the filament formation (Straub and Feuer, 1950). P_i is liberated, but ADP is retained in the chain of actin and becomes nonexchangeable under usual conditions (Martonosi et al., 1960). In view of the important role of ATP and actin in the contractile mechanism, it is quite reasonable that many people have tried to correlate this phenomenon directly with the contractile process. However, so far, every attempt has failed. It is now the common understanding of muscle scientists that G–F transformation itself is not involved in the contractile process at any stage (this does not necessarily mean that the actin filament always remains a rigid rod without substantial conformational change; see p. 340).

The transformation is a typical condensation phenomenon, being promoted markedly by a minute amount of F-actin fragments as a nuclear substance (Asakura et al., 1958; Oosawa et al., 1965). It requires the presence of cations (Straub, 1943). Although there are differences in their activities, almost all kinds of cations, unless they are denaturing, have such actions (Kasai and Oosawa, 1968; Strzelecka-Golaszewska and Drabikowski, 1968). Potassium or magnesium ion is most commonly used as the agent to bring about this transformation.

G-actin usually contains one exchangeable calcium per molecule, in addition to ATP (Maruyama and Gergely, 1961; Bárány et al., 1962). This calcium is retained during the course of G–F transformation induced by potassium or other monovalent cations, and once transformed, it becomes nonexchangeable like ADP. If magnesium ion is used as the polymerizing agent, the situation is complicated. Under the conditions where free calcium ion is extremely low, magnesium ion is involved in the actin filament instead of calcium ion.

There is some difference, though not essential, in physicochemical properties between magnesium and calcium-containing F-actin. Generally speaking, the former is more stable than the latter. F-actin in muscle is now considered to be of the magnesium type, not the calcium type (A. Weber, 1966; Kasai, 1969).

F-actin is converted back to G-actin at a low ionic strength (than 0.01) in the presence of ATP; ADP in F-actin is replaced by ATP from the surrounding medium, giving rise to G-actin–ATP. Thus, the cycle of G–F transformation is completed at the sacrifice of 1 mole ATP per mole G-actin.

Since the depolymerization of F-actin to G-actin requires the reduction of ionic strength, the G–F transformation does not take place at a fixed ionic strength. However, there is a remarkable fact that F-actin behaves like ATPase when it is subjected to sonication (Asakura, 1961) or the temperature of its solution is elevated (Asai and Tawada, 1966). The detailed mechanism of this phenomenon is not yet fully clarified, but it is certain that the breakdown of ATP occurs as a result of the interaction between one end of F-actin filament and one ATP-containing actin molecule, whether the latter be a monomer or a terminal member of a filament; in other words, a similar process to that which takes place in G–F transformation is involved in this phenomenon.

ATP is thus deeply involved in the conversion of the morphological state of actin. Furthermore, if ATP is removed, G-actin rapidly and irreversibly denatures with concomitant loss of bound calcium (Straub and Feuer, 1950); removal of calcium from G-actin also induces the loss of ATP together with the loss of almost every physiological property of actin (Martonosi *et al.*, 1960). The specificity of ATP for this phenomenon is very high compared with that as the substrate of myosin ATPase (see p. 316). For example, deoxyATP cannot be the substitute (Tonomura *et al.*, 1966b; Bárány *et al.*, 1966), only ITP and GTP, in that order, can replace ATP to some extent (Bárány *et al.*, 1966).

Thus, G-actin without ATP no longer seemed to be actin. However, Hayashi and Rosenbluth (1960) have shown the G–F transformation using G-actin in which ATP has been replaced by ADP. Furthermore, Kasai *et al.* (1965) have succeeded in demonstrating the polymerization of G-actin from which the nucleotide is completely removed in a sucrose solution. F-actin thus formed can interact with myosin in the presence of ATP in a usual manner. Since ATP (or ADP) rapidly incorporates into such F-actin, this might not be evidence that F-actin without ADP would interact with myosin. However, the fact that such F-actin interact with myosin like usual F-actin in the presence of deoxyATP (Tonomura *et al.,* 1966b; Bárány *et al.,* 1966), which cannot be incorporated into F-actin, has conclusively indicated that ADP in F-actin does not have primary importance from functional standpoints.

However, it is worthy of note that the rate of polymerization of G-actin without ATP is considerably lower than that of ATP-containing G-actin (Oosawa *et al.,* 1965). ATP may act as an indispensable regulating factor in the process of filament formation of actin in living muscle.

The effect of myosin on the G–F transformation will be referred to in a later section (see p. 327).

2. G–F Transformation from Preparative Standpoints

G–F transformation of actin has given rise to various problems to the preparation of actin. The fact that actin can easily convert between a small molecule, about 3 S, and a large sedimentable aggregate, about 40 S, is a very favorable condition for the purification, so that this phenomenon has widely been used as the routine purification procedure (Mommaerts, 1952).

Apart from such a practical standpoint, the facts related to G–F transformation have raised important questions as to the nature of the actin filament in myofibrils: (1) If the residue from which myosin and other structural proteins have been removed is immersed in a low ionic strength solution with ATP, G-actin should have been extracted. However, this is not true. Appreciable amounts of G-actin can be obtained only after acetone or potassium iodide treatment. It is highly probable that acetone or potassium iodide causes some change in the properties of actin, and as a result, actin become convertible to G-actin. (2) If "natural" F-actin is dialyzed against ATP at a low ionic strength, i.e., the conditions where F-actin is converted to G-actin, it is also depolymerized, but the resulting monomer has lost the polymerizatility. After acetone treatment, ordinary G-actin can be obtained from such F-actin (Hama *et al.,* 1967). (3) There is good evidence that the nucleotide in the acetone-treated muscle residue still exists as ADP. When extracted, it becomes ATP (Tsuboi, 1963). It has been suggested that ADP is converted to ATP by adenylate kinase during extraction (Tsuboi and Hayashi, 1963). If so, the addition of ADP or ATP (together with adenylate kinase) should have increased the yield of actin. This is not the case.

So far, no plausible explanations have been given to these puzzles. They may indicate that there is a considerable difference between actin filament *in vivo* and F-actin transformed from G-actin and that some important points are still missing with regard to the nature of actin.

C. Physicochemical Properties

G-actin is one of the most typical globular proteins. Its intrinsic viscosity is 0.037 dl/gm (L. B. Cohen, 1966), which is very close to the ideal value of a sphere of that molecular weight. The molecular weight of actin had been considered to be about 60,000 for a fairly long time, but recent determinations based on different methods using chromatographically purified preparations all agree with the value less than 50,000 listed in Table IX.

G-actin does not have any subunit nor any disulfide linkage. It shows a single band according to gel electrophoretic pattern in the presence of a denaturing agent such as urea (Rees and Young, 1967). Taking this advantage, the attempt to reveal the amino acid sequence of the entire molecule is now being made in two laboratories (Adelstein and Kuehl, 1970; Elzinga, 1970). Although there is no space available to introduce these works, it is expected that the whole sequence will be clarified in a few years.

Actin, either in G-form or F-form, is very resistant to trypsin digestion (Mihályi, 1953). A significant amount of amino acid is released from G-actin on trypsin treatment but the remaining main part still retains the ability to polymerize and to react with myosin (Laki, 1964).

TABLE IX
MOLECULAR WEIGHT OF G ACTIN

Molecular weight	Method	Author
47,400	Sedimentation equilibrium	Adelstein *et al.* (1963)
46,000	High speed equilibrium method	Rees and Young (1967)
47,300	Amount of bound ATP (Yphantis, 1964)	
47,600	Content of 3-methylhistidine	Johnson *et al.* (1967)
45,000	Amount of bound ATP	Tsuboi (1969)
42,000	Amount of bound calcium	
44,000	Content of cysteine	Lusty and Fasold (1969)
43,000	Light scattering	Sakakibara and Yagi (1970)
43,000	Gel electrophoresis	
43,000	Amount of bound ADP	

The sulfhydryl residues of actin have less important meaning than those of myosin. Titration of rapidly reacting sulfhydryl groups with PCMB does not exert any substantial effect on the properties of actin, including those related to myosin. If all sulfhydryl residues are titrated by PCMB, however, actin loses its polymerizability (Tonomura and Yoshimura, 1962; Katz and Mommaerts, 1962). The disappearance of the polymerizability is not due to the loss of bound ATP, since ATP is released at much slower rate on PCMB treatment (Drabikowski and Gergely, 1963).

The length of F-actin polymerized under ordinary conditions ranges from 0.1 to 20 μ, showing a Poisson type of distribution with a peak at about 1–2 μ (Kawamura and Maruyama, 1970a). F-actin exhibits every property expected as a filamentous protein, e.g., high viscosity, strong flow-birefringence, thixotropy, and so on. The thixotropic properties of F-actin so far reported are partly due to the contamination of α-actinin in classic actin preparations; in spite of complete elimination of α-actinin, however, F-actin shows a definite thixotropic nature (see p. 330). This is in sharp contrast with the cases of tropomyosin or polyelectrolytes.

D. Fine Structure of F-Actin and Its Paracrystals

Prior to the direct visualization of the fine structure of the actin filament, Selby and Bear (1956) and Oosawa and Kasai (1962) had suggested the helical nature of molecular arrangement of F-actin. The former group gave two alternative interpretations, helical or planar, to the X-ray diffraction pattern obtained with dried intact muscle, whereas the latter group has proposed an idea on the basis of theoretical consideration of the G–F transformation that F-actin should not be a simple linear polymer but a helical one.

In 1963, Hanson and Lowy showed the double helical structure of the actin filament by examining the negatively stained specimens in the electron microscope. The model presented by them, consisting of a helical arrangement of globular subunits of about 55 Å in diameter (i.e., G-actin), has widely been accepted as representing the essentials of the actin filament. The pitch of the helix has been investigated by X-ray diffraction analysis of several living muscles and was shown to range from 2×360 Å to 2×370 Å (H. E. Huxley and Brown, 1967; Lowy and Vibert, 1967). The sense of the double helix is right-handed according to shadowed profile of F-actin (Depue and Rice, 1965).

F-Actin has structural polarity. This is clearly visualized when myosin,

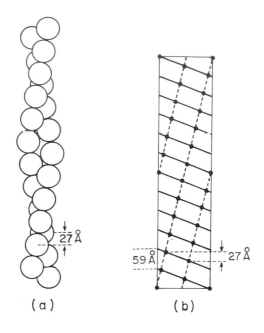

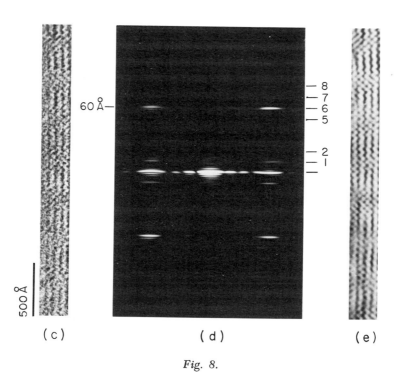

Fig. 8.

HMM, or S-1 binds to F-actin and forms a characteristic structure, called arrowheads, as first shown by H. E. Huxley (1963). Recently, Moore *et al.* (1970) applied the technique of three-dimensional reconstruction developed by De Rosier and Klug (1968) to the analysis of arrowhead structure with S-1. According to their results, to each actin molecule is attached one S-1 molecule with a constant angle to the axis of the actin filament. This finding has finally confirmed the view that the two strands of F-actin are parallel, not antiparallel.

Actin can form a paracrystal in the presence of divalent cations (Hanson, 1968b). It consists of closely packed actin filaments running parallel and in identical phase. The discovery of this paracrystal structure has made it possible to analyze the detailed molecular arrangement of actin (Fig. 8) and its relation to the troponin–tropomyosin system (O'Brien *et al.*, 1971; Moore *et al.*, 1970; Ohtsuki and Wakabayashi, 1972;. Recently, a different type of paracrystal, which shows rhombic networks, has been reported (Kawamura and Maruyama, 1970b).

Most of structural studies using electron microscopic and X-ray diffraction techniques have so far been based on the tacit assumption that the thin and actin filaments are almost identical with each other. In view of the introduction of the troponin–tropomyosin system, however, the detection of a subtle difference between the thin and actin filaments and also between activated and inactivated thin filaments will become the crucial problem of the future; this trend has already been indicated in some of the papers mentioned above, especially that of Moore *et al.* (1970).

E. Interaction with Myosin

The most fascinating interaction of actin with myosin is that in the presence of MgATP, to which reference was made in a previous section. In this section, the discussion will be confined to the phenomena in the absence of ATP. It is a well known fact that if myosin is added

---

Fig. 8. Structure of F-actin filament. (a) Model of F-actin filament according to Hanson and Lowy (1963). The filament consists of helically arranged globular subunits of about 55 Å in diameter. (b) Radial projection of the centers of subunits of F-actin in (a). (c) Electron micrograph of Hanson-type F-actin paracrystal ($\times 280,000$). (d) Optical diffraction pattern of (c). (e) Optically filtered image of (c). Removal of noise from the image reveals the helical structure of each F-actin filament, running parallel and in register with each other. (a) and (b) from Hanson and Lowy (1963); (c), (d), and (e) from Ohtsuki and Wakabayashi (1972).

to F-actin in a solution of a high ionic strength, a striking viscosity increase is produced, indicating the formation of a complex known as actomyosin. On electron microscopic observation, it is found that actin and myosin together forms a unique arrowhead structure (H. E. Huxley, 1963), to which references have already been made (see p. 313 and Fig. 7a). Since this structure is easily decomposed by ATP at any ionic strength, it is impossible to see this structure under physiological conditions.

However, the steric relationship of the myosin head to the actin filament in the rigor state is somewhat similar to that in the arrowheads. On the other hand, at the end of superprecipitation, i.e., after the decomposition of ATP, an arrowheadlike structure can be seen in many parts of the actin–myosin aggregates (Takahashi and Yasui, 1967). Furthermore, in insect flight muscles, a state similar to the rigor state (and therefore to the arrowhead structure) appears to be utilized for their physiological function (Reedy, 1968). These facts may afford some grounds for the idea that a near rigor state is involved in the cyclic interaction of myosin and actin molecules in the contractile mechanism.

There is another type of the myosin–actin interaction. Myosin has a marked accelerating effect on the rate of G–F transformation (A. Szent-Györgyi, 1944; Laki and Clark, 1951). HMM shows essentially the same effect (Martonosi and Gouvea, 1961). This effect is not due to the enzymic properties of myosin or HMM, but due to its complex formation with G-actin (Yagi *et al.*, 1965). In other words, the affinity between myosin and actin is the basis of this interesting phenomenon, and therefore, at the final stage, myosin exists as a form of actomyosin. S-1 prepared under specified conditions can also exhibit this accelerating effect (Onodera and Yagi, 1971).

Thus, all kinds of interactions of myosin and actin, including those started by the addition of ATP, finally reach the state of actomyosin characterized by the arrowhead structure.

IV. Regulatory Proteins

A. *Tropomyosin*

1. GENERAL ASPECTS

Tropomyosin, found by Bailey in 1946, is a very stable protein that has unique properties as a fibrous protein. Therefore, it has been a suitable material for physicochemical studies not directly related to muscle research. As a result, tropomyosin is the most thoroughly investigated protein among muscle proteins from the standpoint of protein chemistry. However, its physiological function had been left unrevealed until 1963, when a protein factor resembling tropomyosin, tentatively called "native tropomyosin," was found to be responsible for the calcium sensitivity of the contractile system (Ebashi, 1963). Regarding the physiological role of tropomyosin, the details are described in the section on troponin.

From the standpoint of preparation, tropomyosin has following characteristics: (1) Tropomyosin is resistant to denaturing effect of organic solvents such as alcohol, acetone, or ether (Bailey, 1948). (2) Tropomyosin can be separated from muscle residue at a very low ionic strength (Perry and Corsi, 1958), but with difficulty at higher ionic strengths, unless actin is denatured. (3) Crude tropomyosin preparation is contaminated by troponin.

Bailey's method (1948), which is based on properties (1) and (2), is still widely used with slight modifications. The main point of the modification is to repeat precipitation at pH 4.6 in 1 M potassium chloride instead of a low ionic strength solution. This is suited for the elimination not only of troponin but also of 260 mμ-absorbing material.

2. Size and Shape of the Tropomyosin Molecule

Bailey and his colleagues have made it clear that the tropomyosin molecule is an asymmetrical protein of a high axial ratio. Although they presented the value 53,000–54,000 (Tsao *et al.*, 1951; Kay and Bailey, 1960) for its molecular weight, a slightly larger value, e.g., 68,000 (Woods, 1967), is now widely accepted. The length of the molecule was estimated to be 385 Å by Tsao *et al.* (1951). This is in good agreement with the values determined on electron micrographs of the ordered structure of tropomyosin molecules, i.e., 402 Å (Caspar *et al.*, 1969) or 400 Å (Higashi-Fujime and Ooi, 1969). This is now accepted as the minimum length of tropomyosin, which may be close to the real length.

Tropomyosin has been known to have nearly 100% α-helix (Urnes and Doty, 1962). Using this property, the molecular weight can be estimated from the length of the molecule; according to Caspar *et al.* (1969) it is 63,000.

3. Subunits

Woods (1967) has proposed that tropomyosin is dissociated into two identical subunits in 8 M urea in the presence of β-mercaptoethanol and that a disulfide linkage is involved in the association of two subunits. The molecular weight of the subunit is 33,500 according to sedimentation equilibrium, which is just half of the molecular weight of the whole molecule, 68,000. This result is essentially confirmed with SDS–poly-acrylamide gel electrophoresis (K. Weber and Osborn, 1969).

4. Aggregates

The most remarkable property of tropomyosin is the formation of linear aggregates under a low ionic strength (Tsao *et al.*, 1951). The characteristic properties of the aggregates are: (1) The degree of aggregation sharply decreases with increase in ionic strength (Bailey, 1948). (2) It decreases with decrease in protein concentration (Tsao *et al.*, 1951). (3) The sedimentation constants of the aggregates are not related to the degree of aggregation, i.e., it is always not far different from that of monomer, i.e., about 2.5 S. (4) The aggregates do not show any thixotropic nature.

These properties are in sharp contrast with those of actin aggregates (i.e., F-actin) (see p. 325). The interaction of each tropomyosin mole-cule seems essentially dependent on the electrical charge localized in each molecule. Tropomyosin aggregates are composed mainly of linear head-to-tail association principally based on such charge (Kay and Bailey, 1960), though the presence of side-by-side association cannot be denied (Ooi *et al.*, 1962). The aggregates thus formed may not be a monodisperse system of a certain molecular weight, but a polydisperse system in which each unit of aggregates is in rapid equilibrium with one another (Ooi *et al.*, 1962).

5. Crystals and Paracrystals

Tropomyosin is an easily crystallized protein, even though it is not so easily purified. It was already crystallized by Bailey (1948) when he discovered this protein. Since then, several pieces of electron micro-graphs of the crystal have been reported (Hodge, 1959; H. E. Huxley, 1963; Caspar *et al.*, 1969; Higashi-Fujime and Ooi, 1969) (see Fig. 10). Caspar *et al.* have also carried out X-ray diffraction analysis and provided precise informations of the parameters of the crystal. Higashi-Fujime and Ooi studied the peripheral structure of the crystal and sug-gested the location of the link between two adjacent tropomyosin mole-cules. On the basis of such observations, both groups have independently presented very similar models for the crystal. The essential point is that every strand of the unit structure of the crystal consists of two tropomyosin filaments that run in antiparallel.

Tropomyosin also shows different types of ordered aggregates, i.e., the paracrystal (Tsao *et al.*, 1965; C. Cohen and Longley, 1966; Nonomura *et al.*, 1968; Caspar *et al.*, 1969; Higashi-Fujime and Ooi, 1969; Millward and Woods, 1970) or square and hexagonal nets (Caspar *et al.*, 1969; Higashi-Fujime and Ooi, 1969).

There is one feature common to the crystal and various types of paracrystals or ordered aggregates, i.e., there is a periodicity of about 400 Å that corresponds to the length of tropomyosin molecule.

6. INFLUENCES OF TROPOMYOSIN ON THE PROPERTIES OF F ACTIN

A. PHYSICOCHEMICAL ASPECTS. Tropomyosin undergoes a parallel association with F-actin (Maruyama, 1964), but this association is not as firm (Martonosi, 1962; Drabikowski *et al.*, 1968b) as that between actin and myosin in the absence of ATP or troponin and tropomyosin. It also should be noted that tropomyosin does not provide substantial change in the property of F-actin from physicochemical standpoints (Laki *et al.*, 1962).

The following facts depict the essential features of the binding: (1) The flow birefringence of F-actin increases on addition of such an amount of tropomyosin as by itself does not show any birefringence (Maruyama, 1964). The weight ratio of maximally bound tropomyosin to F-actin estimated from the increase in birefringence is 1:4 (molar ratio, 1:6.5) (Laki *et al.*, 1962; Maruyama, 1964). (2) In spite of the increase in flow birefringence, there is essentially no change in the extinction angle, indicating together with the above facts that tropomyosin makes parallel arrangement with F-actin (Maruyama, 1964). (3) The viscosity of tropomyosin–F-actin complex is fairly equal to the sum of that of each component (Ebashi and Kodama, 1966). (4) The association of tropomyosin and F-actin requires the presence of specified concentrations of certain cations (Martonosi, 1962), e.g., 0.8 mM magnesium ion or 20 mM potassium ion. This property can be utilized for the removal of tropomyosin from F-actin.

B. INFLUENCES OF TROPOMYOSIN ON THE INTERACTION OF F-ACTIN WITH MYOSIN. Tropomyosin exerts a definite influence, though not dramatic, on the capability of F-actin to interact with myosin. Katz (1964) has worked out the effects of tropomyosin on superprecipitation as well as on actomyosin type ATPase. Generally speaking, the interaction of myosin and actin is promoted by tropomyosin if the conditions are favorable for the interaction, whereas it is depressed by tropomyosin under the conditions where actomyosin is apt to dissociate. Thus, tropomyosin enhances both contractile and dissociating responses of the myosin–actin interaction; as a result, tropomyosin would endow the myosin–actin interaction with a property to behave in an all-or-none manner. This may be related to the physiological role of tropomyosin in the control mechanism through the calcium–troponin–tropomyosin system. The accelerating action of tropomyosin is a function of the temperature; if the temperature becomes lower than 15°C, the accelerating action is converted to a depressing one (Fujii and Maruyama, 1971).

Schaub *et al.* (1967) have reported a strange phenomenon. At a low ionic strength, calcium ion of a fairly high concentration, i.e., 1–10 mM, strongly stimulates the ATPase of troponin–tropomyosin-free actomyosin, and this calcium-activated ATPase is depressed by tropomyosin. This may be somehow related to the unique effect of tropomyosin mentioned above.

B. Troponin

1. General Aspects

The establishment of the role of calcium ion in regulating the physiological contraction–relaxation cycle renewed interest in the fact that pure actomyosin does not respond to calcium ion. This has lead to the discovery of the protein system responsible for the regulation by calcium ion (see A. Weber, 1966; Ebashi and Endo, 1968; Ebashi *et al.*, 1969).

First, the protein system was found as a protein complex that resembled tropomyosin from physicochemical standpoints but possessed a unique property providing actomyosin with the responsiveness to calcium ion, the property not shared by tropomyosin (Ebashi, 1963). The protein, called native tropomyosin (Ebashi and Ebashi, 1964), was then shown to consist of tropomyosin and a new globular protein named troponin (Ebashi and Kodama, 1965, 1966).

Before inquiring into the properties of troponin, it may be pertinent to emphasize that calcium–troponin–tropomyosin is concerned with the regulation of the fundamental process of contraction carried out by myosin, actin, ATP, and magnesium ion, but that the physiological contraction–relaxation cycle can never be performed without such regulatory protein system.

2. PREPARATIVE ASPECTS

There have been reported several methods for the preparation of troponin (Ebashi and Ebashi, 1964; B. Yasui *et al.*, 1968; Ebashi *et al.*, 1968; Arai and Watanabe, 1968; Hartshorne and Mueller, 1969; Schaub and Perry, 1969). All of them isolate native tropomyosin first and then separate troponin from it. A method of isolating troponin directly from muscle residue is described in Table X (Ebashi *et al.*, 1971).

Generally speaking, troponin is a fairly stable protein, particularly to usual denaturing agents. However, it is much more sensitive to proteolytic enzymes than myosin or tropomyosin. Therefore, caution should be exercised against putrefaction during preparation; the whole preparation should be finished in a short period. Following the method described in Table X, it does not require more than 10 hr until the step before the elimination of ammonium sulfate.

The protein fraction thus obtained is sometimes already fairly pure, but usually contains small amounts of other proteins, which may be identified by SDS–polyacrylamide gel electrophoresis with tropomyosin, β-actinin, γ-component, or α-actinin. The total amount of these contaminating proteins, however, does not exceed 5%.

Although native tropomyosin can be prepared from cardiac and smooth muscles (Ebashi *et al.*, 1966), routine procedure for preparing tropomyosin-free troponin has not yet been established.

3. PHYSICOCHEMICAL PROPERTIES

In usual solution, troponin behaves as a single molecule according to various kinds of chromatography. Sometimes it shows extra bands, but they are mainly composed of aggregates of troponin (Wakabayashi and Ebashi, 1968). Troponin is a globular protein, of which the molecular weight has been reported to be from 44,000 to 86,000 (Ebashi *et al.*, 1968; Arai and Watanabe, 1968). Troponin has two distinct characteristics—strong affinities for calcium and tropomyosin. Since its relation with tropomyosin will be described later, properties concerning calcium binding will be discussed here.

As shown in Table XI troponin binds about 4–5 moles calcium per 10^5 grams of protein (Ebashi *et al.*, 1967; B. Yasui *et al.*, 1968; Fuchs and Briggs, 1968; Arai and Watanabe, 1968). This binding property is fairly specific for calcium and is not influenced by the presence of excess magnesium ion at all. As expected, troponin also has some affinity for strontium. Interestingly, the affinity of troponin for strontium relative to that for calcium varies from tissue to tissue, e.g., cardiac troponin

TABLE X
PREPARATION OF TROPONIN

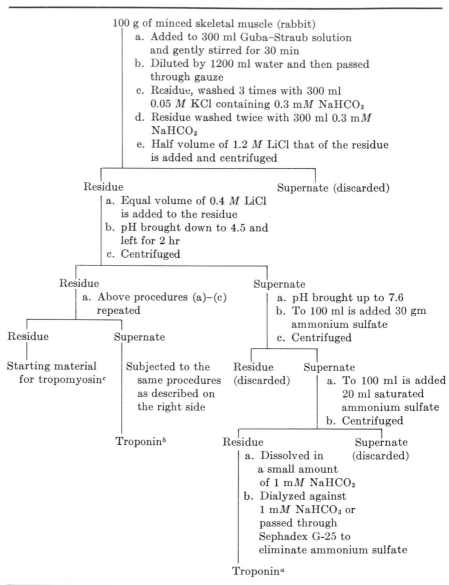

100 g of minced skeletal muscle (rabbit)
  a. Added to 300 ml Guba–Straub solution and gently stirred for 30 min
  b. Diluted by 1200 ml water and then passed through gauze
  c. Residue, washed 3 times with 300 ml 0.05 M KCl containing 0.3 mM NaHCO$_3$
  d. Residue washed twice with 300 ml 0.3 mM NaHCO$_3$
  e. Half volume of 1.2 M LiCl that of the residue is added and centrifuged

Residue
  a. Equal volume of 0.4 M LiCl is added to the residue
  b. pH brought down to 4.5 and left for 2 hr
  c. Centrifuged

Supernate (discarded)

Residue
  a. Above procedures (a)–(c) repeated

Supernate
  a. pH brought up to 7.6
  b. To 100 ml is added 30 gm ammonium sulfate
  c. Centrifuged

Residue
Starting material for tropomyosin[c]

Supernate
Subjected to the same procedures as described on the right side

Residue (discarded)

Supernate
  a. To 100 ml is added 20 ml saturated ammonium sulfate
  b. Centrifuged

Troponin[b]

Residue
  a. Dissolved in a small amount of 1 mM NaHCO$_3$
  b. Dialyzed against 1 mM NaHCO$_3$ or passed through Sephadex G-25 to eliminate ammonium sulfate

Supernate (discarded)

Troponin[a]

[a] The yield is about 120–200 mg per 100 gm at this stage. Contaminating proteins can be removed to some extent by certain procedures (Ebashi *et al.*, 1971).

[b] The yield is about one-third that of the above preparation.

[c] After washing once with 0.6 M LiCl, the residue is neutralized and then subjected to the preparation of tropomyosin.

TABLE XI
SOME PROPERTIES OF TROPONIN FROM RABBIT SKELETAL MUSCLE

Properties	Values	Authors
Calcium-binding capacity (moles/10^5 gm) and binding constant (M^{-1})	3.8; 1.3×10^6 and 0.5×10^5 (pH 6.8)	Ebashi *et al.* (1968)
	5.2; 2.6×10^6 (pH 7.0)	B. Yasui *et al.* (1968)
	2.1; 2.2×10^6 (pH 7.0)	Fuchs and Briggs (1968)
	3.0; 3.7×10^5 (pH 7.0)	Arai and Watanabe (1968)
Molecular weight, ($\times 10^4$) (methods)	5.3 (light scattering)	cf. Ebashi *et al.* (1968)
	8.6 ($s°_{20,w}$ and viscosity)	
	9.1 (gel filtration)	} Arai and Watanabe (1968)
	4.4 (gel filtration; 0.1 M dithiothreitol)	
Intrinsic viscosity (dl/gm)	0.1	Arai and Watanabe (1968)
	0.05	cf. Maruyama and Ebashi (1970)
Helical content (%) (methods)	40 (optical rotatory dispersion)	} Staprans and Watanabe (1970)
	33–41 (circular dichromism)	
ϵ_{278} ($\epsilon280/\epsilon260$)	3.5 (1.5)	Arai and Watanabe (1968)
ϵ_{278} ($\epsilon278/\epsilon255$)	4.0 (2.1)	cf. Maruyama and Ebashi (1970)

(a crude preparation) has much higher affinity for strontium than skeletal troponin. These affinities are reflected in the sensitivities to strontium ion of the contractile systems that contain respective troponins (Ebashi *et al.*, 1968, 1969).

Using the exchangeability of troponin-bound ^{45}Ca during gel filtration as the index, Fuchs *et al.* (1970) have shown that troponin has practically no affinities for manganese, cobalt, nickel, zinc, and lanthanum ions. This extraordinarily high selectivity of troponin for metal ions may provide a unique material for investigation from the standpoint of chelate chemistry.

Troponin has 5–5.5 moles sulfhydryl residues per 10^5 grams (Ebashi *et al.*, 1968; Arai and Watanabe, 1968), but PCMB or NEM treatment does not produce marked effect on its calcium-sensitizing activity or calcium-binding capacity (Staprans *et al.*, 1968; Ebashi *et al.*, 1968; Hartshorne and Daniel, 1970).

Troponin has a tendency to form aggregates upon addition of salt. This aggregating tendency is modified by the change in the concentration of calcium ion, the range of which coincides with that in the regulation of contraction–relaxation of muscle; the presence of calcium ion abolishes the aggregating tendency and the removal of it promotes the aggregation

(Wakabayáshi and Ebashi, 1968). Strontium ion has the same tendency to such an extent as expected from its activity relative to calcium ion.

4. Subunits of Troponin

Separation of troponin into its components was first attempted by Hartshorne and Mueller (1968; Hartshorne *et al.*, 1969). Troponin can be separated at pH 1 in 1.2 *M* potassium chloride (in the presence of dithiothreitol) into two components—troponin A, precipitable under above conditions but very soluble at neutral pH, and troponin B, soluble under above conditions but practically insoluble at neutral pH at low ionic strengths. The addition of troponin A made troponin B soluble. Since troponin A was capable of binding calcium ion and troponin B was inhibitory of actomyosin ATPase, it was considered that the inhibitory action of troponin B would be removed by troponin A. Schaub and Perry (1969) presented a method using SE–Sephadex column chromatography at pH 6 in 6 *M* urea for separating troponin into components—calcium-sensitizing factor and inhibitory factor, which, according to Schaub and Perry, corresponded to troponin A and troponin B, respectively. The introduction of SDS–polyacrylamide gel electrophoresis has revealed, however, that both troponin A and troponin B are also not homogenous.

Troponin A is a relatively homogeneous fraction; its major component has a molecular weight of about 17,000 (hereafter this component will be called troponin A). Troponin A represents most of the calcium-binding capacity of the original troponin; it binds 20 moles calcium or more per 10^5 gm. Some of the function or troponin A can be replaced by certain polyanions (Hartshorne, 1970; Ebashi *et al.*, 1971).

Troponin B consists of two major components, of which the molecular weights are estimated to be around 40,000 and 22,000, respectively (Ebashi *et al.*, 1971) (in addition to these, troponin B also contains a small but a definite amount of troponin A). The larger component, troponin I, represents the tropomyosin-binding property of the original troponin. The smaller component, troponin II, seems to play the main role in the inhibitory action of troponin B on the myosin–actin–ATP interaction in the presence of tropomyosin. Troponin B also depresses the interaction in the absence of tropomyosin; this inhibition is mainly due to troponin I. Combination of troponin II with troponin A provides some calcium-sensitivity to actomyosin (Schaub and Perry, 1971), but troponin I is required for the complete restoration of the calcium-sensitizing activity of the original troponin (Ebashi *et al.*, 1971; Greaser and Gergely,

1971). The precise role of each component in the physiological processes, however, remains to be studied.

Both troponin I and troponin II are very sensitive to trypsin and other proteolytic enzymes. This is the reason trypsin treatment can abolish the calcium sensitivity of myosin B and myofibrils (Ebashi and Ebashi, 1964). In contrast with this, troponin A is fairly stable to tryptic digestion.

5. Interaction with Other Proteins

A. Physicochemical Aspects. Troponin has a strong affinity for tropomyosin (Ebashi and Kodama, 1965). Flow birefringent properties of troponin have revealed (Maruyama and Ebashi, 1970) that troponin not only promotes the polymerization of tropomyosin, but also provides tropomyosin with thixotropic properties. This may suggest that the end-to-end aggregates of tropomyosin are stabilized by troponin so that each unit of tropomyosin aggregates is no more in rapid equilibrium with one another; in other words, the dissociation and reassociation of tropomyosin in aggregates cannot take place easily. The fact that tropomyosin bound to F actin is easily exchangeable with tropomyosin in the outer medium but no more exchangeable in the presence of troponin may be compatible with the above explanation (Drabikowski *et al.*, 1968b).

Troponin or troponin I can form coprecipitates with F actin (Drabikowski and Nonomura, 1968; Fujii *et al.*, 1972), but this phenomenon may not be related to the physiological role of troponin.

As stated before (p. 331), tropomyosin does not exert any marked effect on the physicochemical properties of F-actin. However, if we add troponin to F-actin–tropomyosin complex, a remarkable change takes place in the properties of F-actin (Ebashi and Kodama, 1966; Maruyama and Ebashi, 1970). Most pronounced is the appearance of the rheopexic property, i.e., F-actin–tropomyosin–troponin complex forms large aggregates under a low shear force but disperses at a high shear rate. However, it is not clear whether or not this is related to the physiological action of troponin.

It is worthy to mention that the ordinary filamentous structure of F-actin cannot be observed under the electron microscope with purified F-actin from which troponin and tropomyosin had been completely removed. Addition of tropomyosin and troponin to such actin preparations restores the usual profile (cf. Ebashi and Endo, 1968).

B. Morphological Aspects. In view of strong interaction between troponin and tropomyosin and the inherent property of tropomyosin

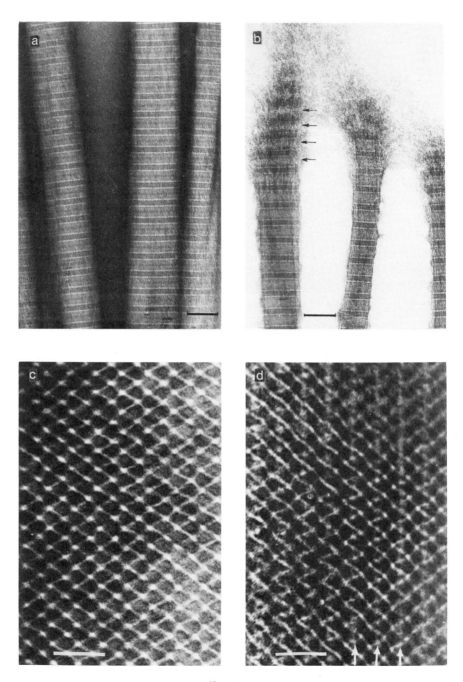

Fig. 9.

molecules to form aggregates of highly ordered arrangement (see p. 330), it is quite natural to suppose that the fine profile of the state of actual binding of troponin to tropomyosin would be visualized. Higashi and Ooi (1968) first showed that troponin binds to the particular site of tropomyosin crystals, i.e., the middle of the longer arm of the unit network as shown in Fig. 10. Essentially the same structure can be seen by crystallizing the native tropomyosin, i.e., troponin–tropomyosin complex (Fukazawa *et al.*, 1970).

A similar observation has been made by Nonomura *et al.* (1968) with tropomyosin paracrystals. Troponin binds to the middle of the slightly dark line in the broad light band (Fig. 9). Since tropomyosin molecules are precisely arranged in these crystalline structures, it is certain that troponin binds to a specified site of the tropomyosin molecules. This is reflected in the finding that antitroponin stains the separated thin filaments as well as those in myofibrils with 400 Å periodicities, which undoubtedly indicates that tropomyosin molecules are distributed along the entire filament with the same periodicity and that the troponin molecule is located at a specific site of the tropomyosin molecule as in the case of paracrystals (Ohtsuki *et al.*, 1967). On the basis of these findings and considerations, a model for the thin filament, the troponin–tropomyosin–actin complex, has been presented (Ebashi *et al.*, 1969; Volume I, p. 355, Fig. 34).

These findings have raised a question as to whether troponin binds to an end of the tropomyosin molecule where the next tropomyosin molecule is linked. The evidence so far obtained indicates that troponin may not be located at one end but some other part in the molecule, probably one third from one end of the molecule (Ohtsuki, 1971). This indicates that the stabilizing effect of troponin on the aggregates of tropomyosin molecules would be exerted in an allosteric way, not through the direct binding to the linking point of two adjacent tropomyosin molecules.

Fig. 9. Paracrystals (a and b) crystals (c and d) of tropomyosin (a and c) and troponin–tropomyosin complex (b and d). Specimens were stained negatively with uranyl acetate. Bars indicate 0.1 μ (a and b) and 500 Å (c and d). (a) Paracrystals were formed in magnesium chloride–tris–hydrochloric acid (pH 7.6) solutions. The periodicity of 395 Å is visible, repeating the broad light band and the narrow dark band. The slightly dark line is noticeable in the middle of the broad light band ($\times 82,000$). (b) Notice the binding sites of troponin in the middle of the broad light band (arrows). (c) Crystals were formed by dialyzing against potassium chloride–sodium acetate (pH 5.6) solutions. The network structure showing kite-like shape is composed of long (220 Å) and short (180 Å) strands ($\times 270,000$). (d) Notice the localization of troponin in the center of the long strand (arrows) ($\times 270,000$). (c) and (d) by the courtesy of Ohtsuki (1968).

6. MECHANISM OF REGULATION OF THE CONTRACTILE PROCESS
   BY THE CALCIUM–TROPONIN–TROPOMYOSIN SYSTEM

Based on the various findings and considerations, it has been suggested (Ebashi *et al.*, 1968, 1969) Fig. 10: (1) In the absence of calcium ion troponin exerts a depressing effect on the interaction of myosin, actin and MgATP.[2] (2) This depressing effect is mediated through tropomyosin to F-actin; as a result, F-actin displays a conformational change that makes its interaction with myosin difficult. (3) Calcium ion affects troponin to remove this depression; thus, calcium ion acts as a kind of derepressor.

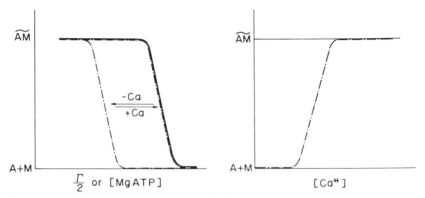

Fig. 10. Schematic representation of the control mechanism of the myosin–actin–MgATP interaction through calcium–troponin–tropomyosin system. Broken lines indicate the response of the complete system consisting of troponin, tropomyosin, actin, and myosin. Solid lines indicate the response of the system consisting of myosin and actin, or myosin, actin, and tropomyosin. As described in the text (p. 331), tropomyosin has a complicated effect, so that the responses of the systems with and without tropomyosin are somewhat different from each other, but their essentials may be represented by the above scheme.

The evidence supporting this mechanism is: (1) Calcium ion can influence the quaternary structure of troponin (see p. 335). (2) Calcium ion exerts an influence on electron spin resonance of NEM-labeled F-actin and tropomyosin only in the presence of troponin (Tonomura *et al.*, 1969). (3) Calcium ion alters the quasi elastic light scattering of F-actin (Fujime, 1970) only when troponin and tropomyosin are present;

[2] If troponin–tropomyosin complex is added to the actomyosin system, an acceleration of the myosin–actin interaction is often observed in the presence of calcium ion. However, this does not indicate the activating nature of the calcium–troponin–tropomyosin system. The accelerating effect is mainly indebted to the stimulating effect of tropomyosin on the myosin–actin interaction (see p. 331).

reduction of calcium ion in the medium removes the flexibility of actin filaments, which is restored by the addition of calcium ion (Ishiwata and Fujime, 1971).

The last finding has shed an illuminating light on the finer mechanism of calcium–troponin–tropomyosin regulation. Since the interaction with actin filaments of HMM or S-1, which cannot form filamentous structure, can be regulated by the calcium–troponin–tropomyosin system (Kominz, 1966; Eisenberg and Kielley, 1970; Parker *et al.*, 1970; Spudich and Watt, 1971), the above fact may not be interpreted that calcium ion would increase the accessibility of actin to myosin heads. Perhaps the increase in the flexibility of each actin molecule in the chain may have a more important meaning.

It should be noted that there is an opinion that troponin is directly involved in the site of interaction of actin with myosin. In view of the contents of troponin and tropomyosin in muscle based on the yield from muscle (Table II) as well as the densities in SDS–polyacrylamide gel electrophoresis, the amounts of associable troponin and tropomyosin to F-actin (Maruyama and Ebashi, 1970), and the ratios of troponin–tropomyosin complex to actin that produce the maximum calcium-sensitizing activity (Ebashi *et al.*, 1968; Spudich and Watt, 1971), this idea seems to have little possibility.

C. Paramyosin

1. General Properties

Paramyosin is a fibrous protein contained in those muscles that display so-called catch mechanism or holding mechanism. It is a very unique protein which exists in living muscle in a kind of crystalline form. It forms a large filament of a diameter from 150 to 600 Å and a length from 10 to 30 μ, to the surface of which myosin molecules are attached. Thus, paramyosin is the major component of the thick filament, and under particular conditions, the thick filament appears almost as a pure paramyosin filament. Therefore, it was reasonable that its presence in muscle was first made by electron microscopic studies and X-ray diffraction analyses (Hall *et al.*, 1945). The extraction of the protein and its identification with that noticed by above techniques were made much later. The extraction of the protein was made by Hodge already in 1952, but the substantial success in preparing the protein of an excellent quality was first made in 1956–1957 as a tropomyosin (tropomyosin B)-like protein in invertebrate muscles (Bailey, 1957; Laki, 1957; Kominz *et al.*, 1958).

In view of this historical background, the protein was first named paramyosin by people working with electron microscope or X-ray diffraction, but termed by biochemists as tropomyosin A. However, tropomyosin and paramyosin are functionally entirely different from each other; their only common feature is that they are both fibrous proteins of nearly 100% α-helix (C. Cohen and Szent-Györgyi, 1957). Even as a fibrous protein, they do not have much similarity from the physicochemical standpoint. Therefore, the term paramyosin is preferable not to give a misleading impression.

Paramyosin is a rod-shaped molecule of about 1350 Å in length and 220,000 in molecular weight (Lowey et al., 1963). The proposal based on the X-ray diffraction analysis that paramyosin is composed of a two-chain α-helical coiled coil (C. Cohen and Holmes, 1963) has been substantiated by the fact that its molecular weight is decreased to half by a denaturing agent in the presence disulfide-reducing agents (Olander et al., 1967). Paramyosin is very soluble at high ionic strengths of neutral to slightly alkaline pH. In such a solution, a considerable part of paramyosin molecules exist as aggregates, but the degree of aggregation is not related to the ionic strength, unlike the aggregation of tropomyosin. If the ionic strength of the solution is lowered to less than 0.3 at the pH slightly acid ($\sim$6.5) precipitates in a form of needle-shaped crystal are formed. They present periodicities characteristic to myofilaments in vivo. In the absence of salt, paramyosin forms a hydrous gel (Hodge, 1952). Thus, paramyosin bears more resemblance to myosin in its physicochemical properties than tropomyosin.

Paramyosin has now been isolated from many invertebrates, not only mollusks and annelids, but also horseshoe crab (Ikemoto and Kawaguti, 1967). There may be no doubt that the protein serves as the so-called holding mechanism of certain invertebrate muscle. However, its precise mechanism and the role of paramyosin in it remain to be elucidated.

2. Fine Structure of Paramyosin Aggregates

There are two kinds of crystalline aggregates of paramyosin molecules. One is those in living muscle, which are the core structure of the thick filament in catch muscles (see above). The filament is characterized by the checker-board type structure with distinct periodicity of 145 Å. Elliott and Lowy (1970) have analyzed this structure in detail using electron microscopic and diffraction techniques. According to them, the filament is a rolled up assembly of a flat sheet. As stated above, paramyosin once solubilized forms needle-like precipitates. These precipitates often show the periodicity of 145 Å (Hodge, 1952).

Another type of ordered arrangement of paramyosin can be formed by precipitating the molecules with divalent cations (Kendrick-Jones *et al.*, 1969) as in the case of tropomyosin. The paracrystals thus formed are more similar in appearance to those of tropomyosin rather than to those of paramyosin filaments mentioned above. Although the periodicities and general profiles of these paracrystals are different from one another depending upon the sources of paramyosin and the conditions in which the aggregates are formed, there can be found among such paracrystals a common periodicity, i.e., 725 Å (Hall *et al.*, 1945: Bear and Selby, 1956), which is tacitly contained in the natural filament. One of the merits of these paracrystals is that they suggest the minimum length of the paramyosin molecule, 1275 Å.

The interesting problem from the physiological standpoint is whether the 145 Å periodicity, which coincides with the periodicity of myosin assembly, has a physiological meaning for the attachment of myosin molecules to the paramyosin filament.

D. Actinins

1. GENERAL CONSIDERATIONS

At the time of the discoveries of α-actinin (original α-actinin) (Ebashi and Ebashi, 1965; Maruyama and Ebashi, 1965) and β-actinin (Maruyama, 1965a,b), it appeared that amino acid compositions of these proteins closely resembled that of actin, so they were considered to be somehow related to actin in its biogenesis. However, functionally defined α-actinin was later shown to be a rather minor constituent of crude α-actinin (Nonomura, 1967; Masaki and Takaiti, 1969) (the major component has tentatively been called as the 10 S component[3]) and to have a different amino acid composition from that of actin. A less pronounced but significant difference in amino acid composition was

---

[3] The 10 S component of original α-actinin was later shown to be a protein of which the amino acid composition, peptide map, and behavior in SDS–polyacrylamide gel electrophoresis are not distinguishable from those of actin except that there are a few different peptides in the peptide map. Therefore, it is quite natural to think it as a kind of denatured actin. However, the 10 S component has a higher rate of amino acid incorporation than actin by about 70% (Table XIV) and shows a high sensitivity to trypsin, in sharp contrast with actin and its denatured forms (see p. 324). Therefore, even if the 10 S component would be nothing but a type of denatured actin, there remains an unsettled problem as to the nature of actin in myofibrils (see p. 323).

also noticed between β-actinin and actin. Therefore, the term actinin seems to have lost its original meaning. However, it is still true that their amino acid compositions are more or less similar to that of actin compared with other structural proteins (Table XIV). Furthermore, their peptide maps (Horiuchi et al., 1970) show a fairly close similarity to one another. Therefore, the idea first proposed by Maruyama may be worthy of further consideration.

2. α-ACTININ (6 S COMPONENT OF ORIGINAL α-ACTININ)

A. PREPARATIVE ASPECTS. α-Actinin can get rid of myofibrils at low ionic strengths (less than 0.003). The removal of myosin prior to extraction facilitates the release of α-actinin. Under these conditions, α-actinin can be extracted from myofibrils together with troponin, tropomyosin, and the 10 S component of crude α-actinin (Ebashi and Ebashi, 1965). Differential centrifugation method can be used to separate the 6 S component of crude α-actinin (Nonomura, 1967; Seraydarian et al., 1967), but the separation can be made more easily by ammonium sulfate fractionation (Masaki and Takaiti, 1969; Arakawa et al., 1970; Robson et al., 1970). However, the range of ammonium sulfate fractionation for α-actinin is not far different from the range for the 10 S component, and once α-actinin is precipitated together with the 10 S component, the two are difficult to separate. Therefore, the step of ammonium sulfate fractionation should be made carefully as indicated.

Another point to be mentioned is the liability of α-actinin preparation to be contaminated by M-protein (Masaki et al., 1967). If once contaminated, it is almost impossible to eliminate M protein from α-actinin. Since M-protein is very antigenic, even a small amount of contamination would spoil the immunochemical investigation. Therefore, M-protein must be removed together with myosin prior to the isolation of α-actinin. For this purpose, Guba–Straub solution (1943) is not suited; only repeated extraction with Hasselbach–Schneider solution (1951) can remove all the M-protein together with myosin (Masaki and Takaiti, 1969).

B. GENERAL PROPERTIES. The molecular weight of α-actinin and some other physicochemical properties are listed in Table XII. One of the interesting properties of α-actinin is its tendency to form aggregates upon addition of salts. Even at such a low ionic strength as 0.01, the ratio of aggregated molecules to the total reaches nearly maximum (Masaki and Takaiti, 1969).

α-Actinin has an affinity for F-actin, not for G-actin, and exerts marked biological effects (Ebashi and Ebashi, 1965; Maruyama and Ebashi,

TABLE XII

SOME PHYSICOCHEMICAL PROPERTIES OF ACTININS AND M-PROTEIN[a]

	Sedimentation coefficients $s_{20,w}$ (S)	Intrinsic viscosity (dl/gm)	Molecular weight ($\times 10^4$)	Helical content (%)	ϵ_{278} ($\epsilon 278/\epsilon 255$)
α-Actinin (6 S)	6.8 ($s^0_{20,w}$)	0.14	16 (light scattering) 9.5 (SDS gel electrophoresis)	41	10.0 (1.9)
β-Actinin	4.5	0.05	6.5 (high speed equilibrium, Yphantis 1964)	13	9.3 (2.2)
M-protein	8	0.06	15.5 (SDS-gel electrophoresis)	—	10.4 (2.1)

[a] These data were derived from: Maruyama (1965b), Nonomura (1967), Masaki and Takaiti (1969, 1972), and Maruyama and Ebashi (1970).

1965; Briskey *et al.*, 1967a,b): (1) It promotes the superprecipitation of actomyosin free of α-actinin. (2) It accelerates the tendency of F-actin to form gel; in excess, it produces coprecipitates. The relation of these two phenomena is not yet clear.

The superprecipitation-promoting action may include two aspects: (1) α-Actinin enlarges the ionic strength of Mg ATP concentration at which the superprecipitation can take place (Ebashi and Ebashi, 1965; Seraydarian *et al.*, 1967). In other words, α-actinin depresses the dissociating tendency of actomyosin due to MgATP (actomyosin ATPase is also enhanced, but this is due to the decrease in the degree of dissociation of actomyosin and not due to the activation of actomyosin ATPase itself). Consequently, α-actinin cannot exert its effect under a condition where the dissociating action of MgATP is hardly effective, e.g., at low concentrations of MgATP (Ebashi, 1968) or when certain sulfhydryl groups of myosin are blocked (Seraydarian *et al.*, 1968) (S_1-blocked; see p. 298). (2) In addition to the effects mentioned above, α-actinin has a unique action that promotes the degree of shrinking of actomyosin in the presence of ATP, as indicated by the increase in the turbidity of the superprecipitated actomyosin. In this connection, it is interesting the α-actinin promotes the tension development of actomyosin thread (Maruyama and Kimura, 1971).

Under the electron microscope, F-actin gel induced by α-actinin forms bundles due to the lateral association of F-actin filaments, but the association is loose and irregular (Kawamura *et al.*, 1970), not so compact as the coprecipitates of F-actin and troponin (see p. 337). This gel formation is inhibited by tropomyosin (Drabikowski and Nowak, 1968; Drabikowski *et al.*, 1968a; Kawamura *et al.*, 1970).

In spite of these remarkable actions, α-actinin is now considered not to be directly involved in the contractile mechanism, but to be a protein localized at the Z band (see p. 348). However, since its superprecipitation-promoting action is very unique, though not physiological, and cannot be shown by any other agents, it can be a useful material for the investigation of the interaction of actin and myosin.

3. β-ACTININ

β-Actinin, discovered by Maruyama (1965a,b), is a protein contained in potassium iodide-extracted solution (see p. 323). Although β-actinin, like α-actinin, is partially extracted by solutions of a low ionic strength, a large part remains in the residue from which α-actinin and other regulatory proteins have almost been extracted (Ebashi and Maruyama, 1965). Preparation method was described by Maruyama in detail (1965a), but the β-actinin preparation thus obtained was contaminated by other proteins. To obtain much purer preparation, he has recently recommended washing muscle mince several times in Hasselbach–Schneider solution (1951) before the potassium iodide extraction step. In the previous method, Perry's myofibrils (1952) were used as the starting material, but muscle mince was found to be more suitable for the purpose.

The actions of β-actinin on F-actin are: (1) It inhibits the intramolecular interaction of F-actin (Maruyama, 1965a); this is most remarkably demonstrated by measuring the dynamic rigidity and viscosity (Maruyama, 1971). (2) It prevents the reassociation of mechanically fragmented F-actin (Maruyama, 1966; Kawamura and Maruyama, 1970a).

β-Actinin was first considered as the factor settling the length of F actin filament (Maruyama, 1966). However, detailed study has revealed (Kawamura and Maruyama, 1970a) that β-actinin does reduce the average length of F-actin, but does not provide the uniformity; as far as *in vitro* system is concerned, myosin is the most efficient agent in settling the length of F-actin near 1 μ.

β-Actinin binds to the top of actin paracrystals (Maruyama, 1971). According to fluorescent antibody technique, β-actinin is located around the A-I junction region of the myofibril (Masaki and Maruyama, 1971). β-Actinin is a globular protein of a molecular weight around 60,000

(Table XII). As stated above, its amino acid composition and its peptide map are similar to that of actin (Table XIV).

E. *Other Myofibrillar Proteins*

1. M PROTEIN

The protein of which the antibody specifically stains M line has been isolated (Masaki *et al.*, 1968) and purified (Masaki and Takaiti, 1972). The protein, M-substance or M - protein, has been shown to be homogeneous with SDS–polyacrylamide gel electrophoresis; its molecular weight is about 155,000 (Table XII).

Crude M protein has an aggregation-promoting action on myosin filaments, but purified preparation does not possess this action.

2. γ-COMPONENT

Lee and Watanabe (1970) have isolated a protein component of a molecular weight about 80,000 from crude troponin preparation. Its amino acid composition is different from any known structural protein of muscle. The antibody against this protein stains the entire A band. Its physiological function is not yet known.

3. MISCELLANEOUS

A. TROPOMYOSIN EXTRACTED AT $\mu = 1$ (Hamoir, 1955). This protein may be identical with native tropomyosin.

B. METIN (A. Szent-Györgyi and Kaminer, 1963). This was later shown to be the mixture of tropomyosin and native tropomyosin (Azuma and Watanabe, 1965).

C. Δ PROTEIN (Amberson *et al.*, 1957). The nature of this protein is not yet well understood, but it perhaps contains native tropomyosin as a component.

D. FIBRILLIN (Guba *et al.*, 1968). After the removal of all structural proteins, a protein fraction can be isolated. Its amino acid composition is similar to that of actin. The gel electrophoretic pattern of the protein is the same as that of actin.

V. Some Aspects of Myofibrillar Proteins

A. *Protein Compositions of Particular Structures in the Myofibril*

1. THICK FILAMENT

There is no doubt that myosin is the main constituent of the thick filament, and it is rather a common belief that the thick filament is composed exclusively of myosin (M line is considered the extra-thick filament structure; see p. 347). However, we must take the report into consideration that γ-component seems to be located at the whole A band region (see p. 347).

In the case of so-called catch muscle, paramyosin forms the core of the thick filaments, which is surrounded by myosin molecules (see p. 341). In addition to this, there is evidence that the calcium-receptive site, which is associated with the thin filament in vertebrate muscle, is located at the thick filament in all kinds of molluscan muscles, either smooth or striated (Kendrick-Jones *et al.*, 1970).

2. THIN FILAMENT

The main constituent of the thin filament is, of course, actin, but tropomyosin and troponin are also important members. As described above, it is reported that calcium receptive site is not associated with the thin filament in the case of molluskan muscles. Tropomyosin seems not to be firmly associated with F-actin unlike the case of skeletal muscle (Kendrick-Jones *et al.*, 1970).

3. Z BAND

The presence of the Z band is the characteristic feature of striated muscles; many muscle scientists had tried to ascribe some important roles to this structure, although most of them are no more considered to be the functions of the Z band. It was once assumed that tropomyosin might be an important constituent of the Z band. However, the observations using immunofluorescent microscopic technique do not favor this idea, although it is very probable that the end of the tropomyosin molecule is located very close to the Z band (Endo *et al.*, 1966; Pepe, 1966).

It is now widely accepted that α-actinin (6 S component) is one of the Z band materials (Fig. 11): (1) The antibody against α-actinin stains only the Z band (Masaki *et al.*, 1967). (2) The structure of the Z band is easily destroyed by trypsin, and at the same time, α-actinin-like protein is released from myofibrils (Goll *et al.*, 1969). (3) The Z band

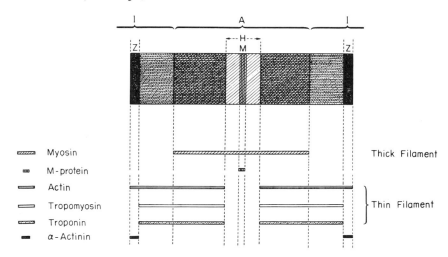

Fig. 11. Scheme illustrating the localization of myofibrillar proteins in the sarcomere.

structure is removed by extracting myofibrils by a solution of a low ionic strength (Samosudova, 1966); the same procedure eliminates α-actinin-like substance from myofibrils (Stromer *et al.*, 1969).

However, reconstitution experiments using α-actinin and those myofibrils of which the Z band have been removed is not yet successful [some success has been reported by using α-actinin-containing crude fraction (Stromer *et al.*, 1967), but the degree of reconstitution is not yet satisfactory]. Since the content of α-actinin does not afford the estimated amount of proteins in the Z band, even if we assume that the concentration of actin in the Z band is twice that of the I band, it is conceivable that there is another protein(s) in this region. The fact that the architecture of the Z band is very sensitive to trypsin (Ashley *et al.*, 1951), in spite of the trypsin-resistant nature of α-actinin and actin, is another evidence for this possibility.

In this connection, there is a muscle disease, called nemaline myosis, in which anomalous dense structures develop from the Z band (Shy *et al.*, 1963). These structures have a periodicity of about 200 Å. The protein responsible for this anomalous structure is not yet identified; it is neither tropomyosin nor α-actinin. With regard to the fine structure of the Z band, several models have been presented (Knappeis and Carlsen, 1962; Franzini-Armstrong and Porter, 1964; Reedy, 1964).

4. M Line

M-protein (p. 347) is a constituent of the M line (Fig. 11). As for the reconstitution experiment, some success has been obtained using a crude fraction which may contain M-protein (Stromer *et al.*, 1969), but the attempt with pure M-protein is seldom successful. It is possible that another substance(s) is also necessary to form the architecture of the M line. The electron microscopic structure of the M line has been studied in detail (Knappeis and Carlsen, 1968; Kundrat and Pepe, 1971).

B. Amino Acid Composition and Related Problems

1. AMINO ACID COMPOSITION

The amino acid composition of myofibrillar proteins are listed in Tables XIII and XIV. As represented by the case of actin (see p. 324), general

TABLE XIII

AMINO ACID COMPOSITION OF MYOSIN GROUP AND FIBROUS
PROTEINS FROM RABBIT SKELETAL MUSCLE

Amino acid	Composition (moles/10^5 gm)[a]								
	Myosin		HMM		S_1		LMM	Tropo-myosin	Para-myosin
	1	2	3	2	4	2	2	5	6
Lys	94	92	64	86	71	83	94	116	59
His	17	16	13	14	15	18	21	6	4
NH₃	—	—	—	90	93	—	—	—	(110)
Arg	46	43	25	34	32	34	60	43	81
Asp	82	85	88	82	91	85	83	91	114
Thr	36	44	47	44	50	49	33	23	36
Ser	45	39	41	39	43	41	34	39	39
Glu	166	157	130	137	126	117	210	220	169
Pro	20	22	26	32	30	37	0	0–4	15
Gly	41	40	53	50	55	61	18	13	15
Ala	75	78	72	73	67	70	81	113	108
Cys½	8	9	10	7	10	11	4	3–5	—
Val	42	43	38	48	51	55	38	29	28
Met	24	23	27	26	29	28	19	20	11
Ile	42	42	37	44	52	53	39	33	22
Leu	86	81	71	73	73	73	96	99	106
Tyr	17	20	21	21	31	21	9	16	18
Phe	30	29	43	36	46	36	4	4	6
Try	—	—	—	—	—	—	—	—	—

[a] References: 1. Oppenheimer *et al.* (1967); 2. Lowey and Cohen (1962); 3. Mueller (1965); 4. Jones and Perry (1966); 5. Woods (1967); 6. Kominz *et al.* (1958) (from the venus smooth muscle).

interest today is directed toward the entire sequence of amino acids of myofibrillar proteins. The determination of the whole sequence of myosin might be a short cut to the clarification of the secret of muscle contraction. Although very troublesome because of the large molecular weight, it may not be impossible in view of the unique properties of the myosin molecule (see pp. 299 and 304).

2. METHYLATED AMINO ACIDS

Perry and his colleagues first noticed the presence of a methylated amino acid in muscle proteins and are energetically pursuing this investi-

TABLE XIV

AMINO ACID COMPOSITION OF ACTIN AND OTHER GLOBULAR
PROTEINS FROM RABBIT SKELETAL MUSCLE

Amino acid	Composition (moles/10^5 gm)[a]								
	Actin		α-Actinin	β-Actinin	Troponin		Tropo-nin I	Tropo-nin II	Tropo-nin A
	1	2	3	4	5	6	5		5
Lys	47	47	50	52	115	99	138	115	57
His	18	18	22	19	17	13	19	20	5
3-Methyl-histidine	2	2	—	—	—	—	—	—	—
NH₃	—	63	89	75	—	—	—	—	—
Arg	38	43	48	36	72	63	79	74	39
Asp	86	81	85	80	77	92	66	76	129
Thr	62	62	43	54	22	25	19	22	23
Ser	53	53	41	46	33	37	26	42	35
Glu	99	100	116	89	154	173	178	150	146
Pro	46	44	40	46	22	30	27	37	7
Gly	70	67	46	61	39	45	25	40	69
Ala	72	71	67	67	72	81	81	68	68
Cys½	12	12	—	—	5	—	—	—	—
Val	48	48	41	42	35	35	32	32	37
Met	38	37	22	33	30	30	15	39	43
Ile	64	68	50	63	31	31	26	26	45
Leu	65	63	71	65	67	69	62	82	55
Tyr	38	36	22	34	9	13	10	10	8
Phe	30	28	25	27	21	27	14	16	51
Try	10	11	—	—	—	—	—	—	—

[a] References: 1. Adelstein and Kuehl (1970); 2. Johnson and Perry (1968) [corrected for hydrolytic losses according to Adelstein and Kuehl (1970)]; 3. Masaki and Takaiti (1969) (from the chicken skeletal muscle); 4. Maruyama (1965b); 5. Ebashi *et al.* (1971); 6. Hartshorne and Mueller (1968).

TABLE XV

RATIOS OF INCORPORATED AMINO ACIDS
INTO MUSCLE PROTEINS[a]

Protein	Rate ± S.E.
Myosin	73 ± 4
Actin	30 ± 2
Tropomyosin	(100)
Troponin	202 ± 14
α-Actinin	99 ± 7
10 S component of crude α-actinin	53 ± 2
Soluble proteins	124 ± 8

[a] The figures indicate the means of relative radioactivities of each protein of rabbit skeletal muscle 24 hr after injection of 150 μC per kilogram of glycine-^{14}C, leucine-^{3}H, valine-^{3}H, or lysine-^{3}H.

gation. The fact that the actin molecule contains only one 3-methylhistidine has contributed to the determination of the molecular weight of actin to be definitely less than 50,000 (Johnson *et al.,* 1967).

The amount of 3-methylhistidine in myosin is varied. It is interesting that fetal myosin contains no methylhistidine, whereas adult white muscle myosin contains a definite amount, 1.6 moles per mole; red and cardiac muscle myosins have an intermediate amount (Trayer *et al.,* 1968; Johnson and Perry, 1970). In addition to this, myosin contains 1 mole mono-N-methyllysine and 3.3 moles tri-N-methyllysine per 5×10^5 gm (Hardy *et al.,* 1970). They are not contained in actin. It is interesting that no methylated amino acids are present in regulatory proteins (Hardy *et al.,* 1970).

3. RATE OF AMINO ACID INCORPORATION INTO MUSCLE PROTEINS

The ratios shown in Table XV indicate something about the turnover rate of muscle proteins (Koizumi, 1971). It is interesting that a great variance can be observed among myofibrillar proteins; some of them show even faster rates than those of soluble proteins.

REFERENCES

Adelstein, R. S., and Kuehl, W. M. (1970). *Biochemistry* **9,** 1355.
Adelstein, R. S., Godfrey, J. E., and Kielley, W. W. (1963). *Biochem. Biophys. Res. Commun.* **12,** 34.

Amberson, W. R., White, J. I., Bensusan, H. B., Himmelfarb, S., and Blankenhorn, B. E. (1957). *Amer. J. Physiol.* **188**, 205.

Arai, K., and Watanabe, S. (1968). *J. Biol. Chem.* **243**, 5670.

Arakawa, N., Robson, R. M., and Goll, D. E. (1970). *Biochim. Biophys. Acta* **200**, 284.

Asai, H., and Tawada, K. (1966). *J. Mol. Biol.* **20**, 403.

Asakura, S. (1961). *Biochim. Biophys. Acta* **52**, 65.

Asakura, S., Hotta, K., Imai, N., Ooi, T., and Oosawa, F. (1958). *In* "Conference on the Chemistry of Muscular Contraction" (H. Kumagai, ed.), p. 57. Igaku Shoin, Tokyo.

Ashley, C. A., Porter, K. R., Philpott, D. E., and Hass, G. M. (1951). *J. Exp. Med.* **94**, 9.

Azuma, N., and Watanabe, S. (1965). *J. Biol. Chem.* **240**, 3852.

Bailey, K. (1946). *Nature (London)* **157**, 368.

Bailey, K. (1948). *Biochem. J.* **43**, 271.

Bailey, K. (1957). *Biochim. Biophys. Acta* **24**, 612.

Bailin, G., and Bárány, M. (1968). *Biochim. Biophys. Acta* **168**, 282.

Banga, I., and Szent-Györgyi, A. (1941). *Stud. Inst. Med. Chem. Univ. Szeged* **1**, 108.

Bárány, M., and Bárány, K. (1959). *Biochim. Biophys. Acta* **35**, 293.

Bárány, M., Finkelman, F., and Therattil-Antony, T. (1962). *Arch. Biochem. Biophys.* **98**, 28.

Bárány, M., Tucci, A. F., and Conover, T. E. (1966). *J. Mol. Biol.* **19**, 483.

Bear, R. S., and Selby, C. C. (1956). *J. Biophys. Biochem. Cytol.* **2**, 55.

Bendall, J. R. (1969). "Muscles, Molecules and Movement." Heinemann, London.

Blum, J. J. (1955). *Arch. Biochem. Biophys.* **55**, 486.

Blum, J. J. (1960). *Arch. Biochem. Biophys.* **87**, 104.

Blum, J. J. (1962). *Arch. Biochem. Biophys.* **97**, 309.

Botts, J., Chashin, A., and Young, H. L. (1965). *Biochemistry* **4**, 1788.

Bowen, W. J., and Kerwin, T. D. (1954). *J. Biol. Chem.* **211**, 237.

Boyer, P. D., Lardy, H. A., and Phillips, P. H. (1943). *J. Biol. Chem.* **149**, 529.

Briskey, E. J., Seraydarian, K., and Mommaerts, W. F. H. M. (1967a). *Biochim. Biophys. Acta* **133**, 412.

Briskey, E. J., Seraydarian, K., and Mommaerts, W. F. H. M. (1967b). *Biochim. Biophys. Acta* **133**, 424.

Briskey, E. J., Cassens, R. G., and Marsh, B. B., eds. (1970). "The Physiology and Biochemistry of Muscle as a Food," Vol. 2. Univ. of Wisconsin Press, Madison.

Carsten, M. E., and Katz, A. M. (1964). *Arch. Biochem. Biophys.* **90**, 534.

Caspar, D. L. D., Cohen, C., and Longley, W. (1969). *J. Mol. Biol.* **41**, 87.

Chappell, B. J., and Perry, S. V. (1955). *Biochim. Biophys. Acta* **16**, 285.

Chung, C. S., Richards, E. G., and Olcott, H. S. (1967). *Biochemistry* **6**, 3154.

Cohen, C., and Holmes, K. (1963). *J. Mol. Biol.* **6**, 423.

Cohen, C., and Longley, W. (1966). *Science* **152**, 794.

Cohen, C., and Szent-Györgyi, A. G. (1957). *J. Amer. Chem. Soc.* **79**, 248.

Cohen, C., Lowey, S., Harrison, R. G., Kendrick-Jones, J., and Szent-Györgyi, A. G. (1970). *J. Mol. Biol.* **47**, 605.

Cohen, L. B. (1966). *Arch. Biochem. Biophys.* **117**, 289.

Dainty, M., Kleinzeller, A., Lawrence, A. S. C., Miall, M., Needham, J., Needham, D. M., and Shen, S. C. (1944). *J. Gen. Physiol.* **27**, 355.

Danilewski, A. (1881). *Z. Physiol. Chem.* **5**, 158.
Davies, R. E. (1965). *Essays Biochem.* **1**, 29.
Dempsey, M. E., and Boyer, P. D. (1961). *J. Biol. Chem.* **236**, PC6.
Depue, R. H., Jr., and Rice, R. V. (1965). *J. Mol. Biol.* **12**, 302.
De Rosier, D. J., and Klug, A. (1968). *Nature (London)* **217**, 130.
Drabikowski, W., and Gergely, J. (1963). *J. Biol. Chem.* **238**, 640.
Drabikowski, W., and Nonomura, Y. (1968). *Biochim. Biophys. Acta* **160**, 129.
Drabikowski, W., and Nowak, F. (1968). *Eur. J. Biochem.* **5**, 209
Drabikowski, W., Nonomura, Y., and Maruyama, K. (1968a). *J. Biochem. (Tokyo)*
    63, 761.
Drabikowski, W., Kominz, D. R., and Maruyama, K. (1968b). *J. Biochem. (Tokyo)*
    63, 802.
Dreizen, P., and Gershman, L. C. (1970). *Biochemistry* **9**, 1688.
Dreizen, P., Hartshorne, D. J., and Stracher, A. (1966). *J. Biol. Chem.* **241**, 443.
Dreizen, P., Gershman, L. C., Trotta, P. P., and Stracher, A. (1967). *J. Gen.
    Physiol.* **50**, 85.
Ebashi, S. (1961). *J. Biochem. (Tokyo)* **50**, 236.
Ebashi, S. (1963). *Nature (London)* **200**, 1010.
Ebashi, S. (1968). *In* "Symposium on Muscle" (E. Ernst and F. B. Straub eds.),
    p. 77. Akadémiai Kiadó, Budapest.
Ebashi, S., and Ebashi, F. (1959). *J. Biochem. (Tokyo)* **46**, 1255.
Ebashi, S., and Ebashi, F. (1964). *J. Biochem. (Tokyo)* **55**, 604.
Ebashi, S., and Ebashi, F. (1965). *J. Biochem. (Tokyo)* **58**, 7.
Ebashi, S., and Endo, M. (1968). *Progr. Biophys. Mol. Biol.* **18**, 123.
Ebashi, S., and Kodama, A. (1965). *J. Biochem. (Tokyo)* **58**, 107.
Ebashi, S., and Kodama, A. (1966). *J. Biochem. (Tokyo)* **60**, 733.
Ebashi, S., and Lipmann, F. (1962). *J. Cell Biol.* **14**, 389.
Ebashi, S., and Maruyama, K. (1965). *J. Biochem.* **58**, 20.
Ebashi, S., Ebashi, F., and Fujie, Y. (1960). *J. Biochem. (Tokyo)* **47**, 54.
Ebashi, S., Oosawa, F., Sekine, T., and Tonomura, Y., eds. (1965). "Molecular
    Biology of Muscular Contraction." Igaku Shoin, Tokyo.
Ebashi, S., Iwakura, H., Nakajima, H., Nakamura, R., and Ooi, Y. (1966). *Biochem.
    Z.* **345**, 201.
Ebashi, S., Ebashi, F., and Kodama, A. (1967). *J. Biochem. (Tokyo)* **62**, 137.
Ebashi, S., Kodama, A., and Ebashi, F. (1968). *J. Biochem. (Tokyo)* **64**, 465.
Ebashi, S., Endo, M., and Ohtsuki, I. (1969). *Quart. Rev. Biophys.* **2**, 4.
Ebashi, S., Wakabayashi, T., and Ebashi, F. (1971). *J. Biochem. (Tokyo)* **69**,
    441.
Eisenberg, E., and Kielley, W. W. (1970). *Biochem. Biophys. Res. Commun.* **40**,
    50.
Eisenberg, E., and Moos, C. (1965). *Arch. Biochem. Biophys.* **110**, 568.
Eisenberg, E., and Moos, C. (1967). *J. Biol. Chem.* **242**, 2945.
Eisenberg, E., and Moos, C. (1968). *Biochemistry* **7**, 1486.
Eisenberg, E., and Moos, C. (1970). *Biochemistry* **9**, 4106.
Eisenberg, E., Zobel, C. R., and Moos, C. (1968). *Biochemistry* **7**, 3186.
Elliott, A., and Lowy, J. (1970). *J. Mol. Biol.* **53**, 181.
Elzinga, M. (1970). *Biochemistry* **9**, 1365.
Endo, M., Nonomura, Y., Masaki, T., Ohtsuki, I., and Ebashi, S. (1966). *J. Biochem.
    (Tokyo)* **60**, 605.
Engelhardt, W. A., and Ljubimova, M. N. (1939). *Nature (London)* **144**, 668.

Ernst, E., and Straub, F. B., eds. (1968). "Symposium on Muscle." Akadémiai Kiadó, Budapest.

Fabian, F., and Mühlrad, A. (1968). *Biochim. Biophys. Acta* 162, 596.

Finck, H. (1968). *Science* 160, 332.

Finlayson, B., and Taylor, E. W. (1969). *Biochemistry* 8, 802.

Finlayson, B., Lymn, R. W., and Taylor, W. (1969). *Biochemistry* 8, 811.

Franzini-Armstrong, C., and Porter, K. R. (1964). *Z. Zellforsch. Mikrosk. Anat.* 61, 661.

Frederiksen, D. W., and Holtzer, A. (1968). *Biochemistry* 7, 3935.

Friess, E. T. (1954). *Arch. Biochem. Biophys.* 51, 17.

Fuchs, F., and Briggs, F. N. (1968). *J. Gen. Physiol.* 51, 655.

Fuchs, F., Reddy, Y., and Briggs, F. N. (1970). *Biochim. Biophys. Acta* 221, 407.

Fujii, T., and Maruyama, K. (1970). *Sci. Pap. Coll. Gen. Educ., Univ. Tokyo* 20, 163.

Fujii, T., and Maruyama, K. (1971). *Sci. Pap. Coll. Gen. Educ., Univ. Tokyo.* 21, 45.

Fujii, T., Kawamura, K. and Maruyama, K. (1972). Personal communication.

Fujime, S. (1970). *J. Phys. Soc. Jap.* 29, 751.

Fukazawa, T., Briskey, E., and Mommaerts, W. F. H. M. (1970). *J. Biochem.* (*Tokyo*) 67, 147.

Gaetjens, E., Bárány, K., Bailin, G., Oppenheimer, H., and Bárány, M. (1968). *Arch. Biochem. Biophys.* 123, 82.

Gazith, J., Himmelfarb, S., and Harrington, W. F. (1970). *J. Biol. Chem.* 245, 15.

Gergely, J. (1950). *Fed. Proc., Fed. Amer. Soc. Exp. Biol.* 9, 176.

Gergely, J. (1953). *J. Biol. Chem.* 200, 543.

Gergely, J., ed. (1964). "Biochemistry of Muscle Contraction." Little, Brown, Boston, Massachusetts.

Gergely, J. (1966). *Annu. Rev. Biochem.* 35, 628.

Gershman, L. C., Dreizen, P., and Stracher, A. (1966). *Proc. Nat. Acad. Sci. U.S.* 56, 966.

Godfrey, J. E., and Harrington, W. F. (1970). *Biochemistry* 9, 894.

Goll, D. E., Mommaerts, W. F. H. M., Reedy, M. K., and Seraydarian, K. (1969). *Biochim. Biophys. Acta* 175, 174.

Greaser, M. L., and Gergely, J. (1971). *J. Biol. Chem.* 246, 4226.

Greville, G. D., and Needham, D. M. (1955). *Biochim. Biophys. Acta* 16, 284.

Guba, F., Straub, F. B. (1943). *Stud. Inst. Med. Chem. Univ. Szeged.* 3, 40.

Guba, F., Harsányi, V., and Vajda, E. (1968). *Acta Biochim. Biophys.* 3, 353.

Hall, C. E., Jakus, M. A., and Schmitt, F. O. (1945). *J. Appl. Phys.* 16, 459.

Halliburton, W. D. (1887). *J. Physiol.* (*London*) 8, 133.

Hama, H., Maruyama, K., and Noda, H. (1965). *Biochim. Biophys. Acta* 102, 249.

Hama, H., Maruyama, K., and Noda, H. (1967). *Biochim. Biophys. Acta* 133, 251.

Hamoir, G. (1955). *Arch. Int. Physiol. Biochim.* 63, Suppl.

Hanson, J. (1968a). *Quart. Rev. Biophys.* 1, 177.

Hanson, J. (1968b). *In* "Symposium on Muscle" (E. Ernst and F. B. Straub eds.), p. 99. Akadémiai Kiadó, Budapest.

Hanson, J., and Lowy, J. (1963). *J. Mol. Biol.* 6, 46.

Hardy, M., Harris, I., Perry, S. V., and Stone, D. (1970). *Biochem. J.* 117, 44.

Harrison, R. G., Lowey, S. and Cohen, C. (1971). *J. Mol. Biol.* **59**, 531.

Hartshorne, D. J. (1970). *J. Gen. Physiol.* **55**, 585.

Hartshorne, D. J., and Daniel, J. L. (1970). *Biochim. Biophys. Acta* **223**, 214.

Hartshorne, D. J., and Mueller, H. (1968). *Biochem. Biophys. Res. Commun.* **31**, 647.

Hartshorne, D. J., and Mueller, H. (1969). *Biochim. Biophys. Acta* **175**, 301.

Hartshorne, D. J., Theiner, M., and Mueller, H. (1969). *Biochim. Biophys. Acta* **175**, 320.

Hasselbach, W. (1956). *Biochim. Biophys. Acta* **20**, 355.

Hasselbach, W. (1964). *Progr. Biophys. Mol. Biol.* **14**, 167.

Hasselbach, W., and Makinose, M. (1961). *Biochem. Z.* **333**, 518.

Hasselbach, W., and Schneider, G. (1951). *Biochem. Z.* **321**, 462.

Hatano, S., and Oosawa, F. (1966). *Biochim. Biophys. Acta* **127**, 488.

Hatano, S., Totsuka, T., and Oosawa, F. (1967). *Biochim. Biophys. Acta* **140**, 109.

Hayashi, T., and Rosenbluth, R. (1960). *Biol. Bull.* **119**, 294.

Higashi, S., and Ooi, T. (1968). *J. Mol. Biol.* **34**, 699.

Higashi-Fujime, S., and Ooi, T. (1969). *J. Microsc. (Paris)* **8**, 535.

Hill, A. V., and 66 authors. (1966). *Biochem. Z.* **345**, 1–426.

Hodge, A. J. (1952). *Proc. Nat. Acad. Sci. U.S.* **38**, 850.

Hodge, A. J. (1959). *Rev. Mod. Phys.* **31**, 409.

Horiuchi, S., Hama, H., Yamasaki, M., and Maruyama, K. (1970). *Sci. Pap. Coll. Gen. Educ., Univ. Tokyo* **20**, 69.

Hotta, K., and Usami, Y. (1967). *J. Biochem. (Tokyo)* **61**, 415.

Huxley, A. F., and Huxley, H. E. ed. (1964). *Proc. Roy. Soc. Ser. B* **160**, 433–542.

Huxley, A. F., and Niedergerke, R. (1954). *Nature (London)* **173**, 971.

Huxley, H. E. (1960). *In* "The Cell" (J. Brachet and A. E. Mirsky eds.), IV, p. 365. Academic Press, N.Y.

Huxley, H. E. (1963). *J. Mol. Biol.* **7**, 281.

Huxley, H. E. (1969). *Science* **164**, 1356.

Huxley, H. E., and Brown, W. (1967). *J. Mol. Biol.* **30**, 383.

Huxley, H. E., and Hanson, J. (1954). *Nature (London)* **173**, 973.

Ikehara, M., Ohtsuka, E., Kitagawa, S., Yagi, K., and Tonomura, Y. (1961). *J. Amer. Chem. Soc.* **83**, 2679.

Ikemoto, N., and Kawaguchi, S. (1967). *Proc. Jap. Acad.* **43**, 974.

Ikemoto, N., Kitagawa, S., and Gergely, J. (1966). *Biochem. Z.* **345**, 410.

Imamura, K., Kanazawa, T., Tada, M., and Tonomura, Y. (1965). *J. Biochem. (Tokyo)* **57**, 627.

Ishikawa, H., Bischoff, R., and Holtzer, H. (1969). *J. Cell Biol.* **43**, 312.

Ishiwata, S., and Fujime, S. (1971). *J. Phys. Soc. Jap.* **30**, 303.

Johnson, P., and Perry, S. V. (1968). *Biochem. J.* **110**, 1967.

Johnson, P., and Perry, S. V. (1970). *Biochem. J.* **119**, 293.

Johnson, P., Harris, C. I., and Perry, S. V. (1967). *Biochem. J.* **105**, 361.

Jones, J. M., and Perry, S. V. (1966). *Biochem. J.* **100**, 120.

Judah, J. D., Ahmed, K., and McLean, A. E. M. (1962). *Biochim. Biophys. Acta* **65**, 742.

Kaminer, B., and Bell, A. L. (1966). *J. Mol. Biol.* **20**, 391.

Kanazawa, T., and Tonomura, Y. (1965). *J. Biochem. (Tokyo)* **57**, 604.

Kasai, M. (1969). *Biochim. Biophys. Acta* **172**, 171.

Kasai, M., and Oosawa, F. (1968). *Biochim. Biophys. Acta* **154**, 520.

Kasai, M., Nakano, E., and Oosawa, F. (1965). *Biochim. Biophys. Acta* **94**, 494.
Kasuya, M., and Takashina, H. (1965). *Biochim. Biophys. Acta* **99**, 452.
Katz, A. M. (1964). *J. Biol. Chem.* **239**, 3304.
Katz, A. M. (1970). *Physiol. Rev.* **50**, 63.
Katz, A. M., and Mommaerts, W. F. H. M. (1962). *Biochim. Biophys. Acta* **65**, 82.
Kawamura, K., and Maruyama, K. (1970a). *J. Biochem.* (*Tokyo*) **67**, 437.
Kawamura, M., and Maruyama, K. (1970b). *J. Biochem.* (*Tokyo*) **68**, 885.
Kawamura, M., Masaki, J., Nonomura, Y., and Maruyama, K. (1970). *J. Biochem.* (*Tokyo*) **68**, 577.
Kay, C. M., and Bailey, K. (1960). *Biochim. Biophys. Acta* **40**, 149.
Kendrick-Jones, J., Cohen, C., Szent-Györgyi, A. G., and Longley, W. (1969). *Science* **163**, 1196.
Kendrick-Jones, J., Lehman, W., and Szent-Györgyi, A. G. (1970). *J. Mol. Biol.* **54**, 313.
Kielley, W. W. (1961). *In* "The Enzymes" (P. D. Boyer, H. Lardy, and K. Myrbäck, eds.), 2nd ed., vol. 5, p. 159. Academic Press, New York.
Kielley, W. W. (1964). *Annu. Rev. Biochem.* **33**, 403.
Kielley, W. W., and Bradley, L. B. (1956). *J. Biol. Chem.* **218**, 653.
Kielley, W. W., and Harrington, W. F. (1960). *Biochim. Biophys. Acta* **41**, 401.
Kielley, W. W., Kalckar, H. M., and Bradley, L. B. (1956). *J. Biol. Chem.* **219**, 95.
Kiely, B., and Martonosi, A. (1968). *J. Biol. Chem.* **243**, 2273.
Kiely, B., and Martonosi, A. (1969). *Biochim. Biophys. Acta* **172**, 158.
Kimura, M., and Kielley, W. W. (1966). *Biochem. Z.* **345**, 188.
King, M. V., and Young, M. (1970). *J. Mol. Biol.* **50**, 491.
Kitagawa, S., Yoshimura, J., and Tonomura, Y. (1961). *J. Biol. Chem.* **236**, 902.
Knappeis, G. G., and Carlsen, F. (1962). *J. Cell Biol.* **13**, 323.
Knappeis, G. G., and Carlsen, F. (1968). *J. Cell Biol.* **38**, 202.
Koizumi, T. (1971). Personal communication.
Kominz, D. R. (1966). *Arch. Biochem. Biophys.* **115**, 583.
Kominz, D. R., Saad, F., and Laki, K. (1958). *In* "Conference on the Chemistry of Muscular Contraction" (H. Kumagai, eds.), p. 66. Igaku Shoin, Tokyo.
Kominz, D. R., Carroll, W. R., Smith, E. N., and Mitchell, E. R. (1959). *Arch. Biochem. Biophys.* **79**, 191.
Kominz, D. R., Mitchell, E. R., Nihei, T., and Kay, C. M. (1965). *Biochemistry* **4**, 2373.
Kubo, S., Tokura, S., and Tonomura, Y. (1960). *J. Biol. Chem.* **235**, 2835.
Kühne, W. (1864). "Untersuchungen über das Protoplasma und die Kontraktilität." Engelmann, Leipzig.
Kundrat, E., and Pepe, F. A. (1971). *J. Cell Biol.* **48**, 340.
Kuschinsky, G., and Turba, F. (1950). *Experientia* **6**, 103.
Laki, K. (1957). *Arch. Biochem. Biophys.* **67**, 240.
Laki, K. (1964). *In* "Biochemistry of Muscular Contraction" (J. Gergely ed.), p. 135. Little, Brown, Boston, Massachusetts.
Laki, K., ed. (1971). "Contractile Proteins and Muscle." Dekker, N.Y.
Laki, K., and Clark, A. (1951). *Arch. Biochem. Biophys.* **30**, 187.
Laki, K., Maruyama, K., and Kominz, D. R. (1962). *Arch. Biochem. Biophys.* **86**, 16.
Lee, L., and Watanabe, S. (1970). *J. Biol. Chem.* **245**, 3004.
Levy, H. M., and Koshland, D. E., Jr. (1959). *J. Biol. Chem.* **234**, 1102.
Levy, H. M., and Ryan, E. M. (1967). *J. Gen. Physiol.* **50**, 2421.

Locker, R. H., and Hagyard, C. J. (1967). *Arch. Biochem. Biophys.* **120**, 454.

Lowey, S. (1970). *Proc. Int. Congr. Biochem., 8th, 1969* p. 28.

Lowey, S., and Cohen, C. (1962). *J. Mol. Biol.* **4**, 293.

Lowey, S., and Luck, S. M. (1969). *Biochemistry* **8**, 3195.

Lowey, S., Kucera, J., and Holtzer, A. (1963). *J. Mol. Biol.* **7**, 234.

Lowey, S., Goldstein, L., Cohen, C., and Luck, S. M. (1967). *J. Mol. Biol.* **23**, 287.

Lowey, S., Slayter, H. S., Weeds, A. G., and Baker, H. (1969). *J. Mol. Biol.* **42**, 1.

Lowy, J., and Vibert, P. J. (1967). *Nature (London)* **215**, 1254.

Lusty, C. J., and Fasold, H. (1969). *Biochemistry* **8**, 2933.

Lymn, R. W., and Taylor, E. W. (1970). *Biochemistry* **9**, 2975.

Makinose, M. (1969). *Eur. J. Biochem.* **10**, 74.

Martonosi, A. (1962). *J. Biol. Chem.* **237**, 2795.

Martonosi, A., and Gouvea, M. A. (1961). *J. Biol. Chem.* **236**, 1345.

Martonosi, A., and Meyer, H. (1964). *J. Biol. Chem.* **239**, 640.

Martonosi, A., Gouvea, M. A., and Gergely, J. (1960). *J. Biol. Chem.* **235**, 1707.

Maruyama, K. (1964). *Arch. Biochem. Biophys.* **105**, 142.

Maruyama, K. (1965a). *Biochim. Biophys. Acta* **94**, 208.

Maruyama, K. (1965b). *Biochim. Biophys. Acta* **102**, 542.

Maruyama, K. (1966). *Biochim. Biophys. Acta* **126**, 389.

Maruyama, K. (1971). *J. Biochem. (Tokyo)* **69**, 369.

Maruyama, K., and Ebashi, S. (1965). *J. Biochem. (Tokyo)* **58**, 13.

Maruyama, K., and Ebashi, S. (1970). *Sci. Pap. Coll. Gen. Educ., Univ. Tokyo* **20**, 171.

Maruyama, K., and Gergely, J. (1961). *Biochem. Biophys. Res. Commun.* **6**, 245.

Maruyama, K., and Gergely, J. (1962). *J. Biol. Chem.* **237**, 1095.

Maruyama, K., and Kimura, S. (1971). *J. Biochem. (Tokyo)* **69**, 983.

Masaki, T., and Takaiti, O. (1969). *J. Biochem. (Tokyo)* **66**, 637.

Masaki, T., and Takaiti, O. (1971). Personal communication.

Masaki, T., and Takaiti, O. (1972). *J. Biochem. (Tokyo)* **71**, 355.

Masaki, T., Endo, M., and Ebashi, S. (1967). *J. Biochem. (Tokyo)* **62**, 630.

Masaki, T., Takaiti, O., and Ebashi, S. (1968). *J .Biochem. (Tokyo)* **64**, 909.

Middlebrook, W. R. (1958). *Proc. Int. Congr. Biochem., 4th, 1958* Vol. 17, p. **84**.

Mihályi, E. (1953). *J. Biol. Chem.* **201**, 197.

Mihályi, E., and Harrington, W. F. (1959). *Biochim. Biophys. Acta* **36**, 447.

Mihályi, E., and Szent-Györgyi, A. G. (1953). *J. Biol. Chem.* **201**, 211.

Miller, P. L. ed. (1968). "Aspects of Cell Motility," *Sympos. Soc. Exp. Biol. XXII,* Cambridge Univ. Press, Cambridge.

Millward, G. R., and Woods, E. F. (1970). *J. Mol. Biol.* **52**, 585.

Mommaerts, W. F. H. M. (1948). *J. Gen. Physiol.* **31**, 361.

Mommaerts, W. F. H. M. (1952). *J. Biol. Chem.* **198**, 445.

Mommaerts, W. F. H. M. (1969). *Physiol. Rev.* **49**, 427.

Mommaerts, W. F. H. M., and Green, I. (1954). *J. Biol. Chem.* **208**, 833.

Moore, P. B., Huxley, H. E., and De Rosier, D. J. (1970). *J. Mol. Biol.* **50**, 279.

Morales, M. F., and Hotta, K. (1960). *J. Biol. Chem.* **235**, 1979.

Morales, M. F., Osbahr, A. J., Martin, H. L., and Chambers, R. W. (1957). *Arch. Biochem. Biophys.* **72**, 54.

Morita, F. (1967). *J. Biol. Chem.* **242**, 4501.

Morita, F. (1969). *Biochim. Biophys. Acta* **172**, 319.
Mueller, H. (1965). *J. Biol. Chem.* **240**, 3816.
Mueller, H., and Perry, S. V. (1961a). *Biochim. Biophys. Acta* **50**, 599.
Mueller, H., and Perry, S. V. (1961b). *Biochem. J.* **80**, 217.
Mühlrad, A., Ajta, K., and Fabian, F. (1970). *Biochim. Biophys. Acta* **205**, 342.
Nauss, K. M., Kitagawa, S., and Gergely, J. (1969). *J. Biol. Chem.* **244**, 755.
Needham, D. M., and Shoenberg, C. F. (1967). In "Cellular Biology of Uterus" (R. M. Wynn, ed.), p. 291. North-Holland Publ., Amsterdam.
Needham, J., Shen, S. C., Needham, D. M., and Lawrence, A. S. C. (1941). *Nature (London)* **147**, 466.
Nihei, T., and Kay, C. M. (1968). *Biochim. Biophys. Acta* **160**, 46.
Noda, H., and Ebashi, S. (1960). *Biochim. Biophys. Acta* **41**, 386.
Noll, D., and Weber, H. H. (1934). *Pfluegers Arch. Gesamte Physiol. Menschen Tiere* **235**, 234.
Nonomura, Y. (1967). *J. Biochem. (Tokyo)* **61**, 796.
Nonomura, Y., Drabikowski, W., and Ebashi, S. (1968). *J. Biochem. (Tokyo)* **64**, 419.
O'Brien, E. J., Bennett, P. M., and Hanson, J. (1971). *Phil. Trans. Roy. Soc. London, Ser. B* **261**, 201.
Ohtsuki, I. (1968). Personal communication.
Ohtsuki, I. (1971). Personal communication.
Ohtsuki, I., and Wakabayashi, T. (1972). *J. Biochem. (Tokyo)* **72**, 369.
Ohtsuki, I., Masaki, T., Nonomura, Y., and Ebashi, S. (1967). *J. Biochem. (Tokyo)* **61**, 817.
Olander, J., Emerson, M. F., and Holtzer, A. (1967). *J. Amer. Chem. Soc.* **89**, 3058.
Onodera, M., and Yagi, K. (1971). *J. Biochem. (Tokyo)* **69**, 145.
Ooi, T., Mihashi, K., and Kobayashi, H. (1962). *Arch. Biochem. Biophys.* **98**, 1.
Oosawa, F., and Kasai, M. (1962). *J. Mol. Biol.* **4**, 10.
Oosawa, F., Asakura, S., Higashi, S., Kasai, M., Kobayashi, S., Nakano, E., Ohnishi, T., and Taniguchi, M. (1965). In "Molecular Biology of Muscular Contraction" (S. Ebashi *et al.*, eds.), p. 56. Igaku Shoin, Tokyo.
Oppenheimer, H., Bárány, K., Hamoir, G., and Fenton, J. (1966). *Arch. Biochem. Biophys.* **115**, 233.
Oppenheimer, H., Bárány, K., Hamoir, G., and Fenton, J. (1967). *Arch. Biochem. Biophys.* **120**, 108.
Ozawa, I., Fujii, T., Kaneko, N., and Maruyama, K. (1970). *Sci. Pap. Coll. Gen. Educ., Univ. Tokyo* **19**, 91.
Parker, L., Pyun, H. Y., and Hartshorne, D. J. (1970). *Biochim. Biophys. Acta* **223**, 453.
Paterson, B., and Strohman, R. C. (1970). *Biochemistry* **9**, 4094.
Paul, W. M., Daniel, E. E., Kay, C. M., and Monckton, G., eds. (1965). "Muscle." Pergamon, Oxford.
Peachey, L. D. (1968). *Annu. Rev. Physiol.* **30**, 201.
Pepe, F. A. (1966). *J. Cell Biol.* **28**, 505.
Pepe, F. A. (1967). *J. Mol. Biol.* **27** ,203.
Perrie, W. J., and Perry, S. V. (1970). *Biochem. J.* **119**, 31.
Perry, S. V. (1951). *Biochem. J.* **48**, 257.
Perry, S. V. (1952). *Biochem. J.* **51**, 495.
Perry, S. V. (1961). *Ann. Rev. Biochem.* **30**, 473.

Perry, S. V. (1967). *Progr. Biophys. Mol. Biol.* **17**, 325.
Perry, S. V., and Corsi, A. (1958). *Biochem. J.* **68**, 5.
Perry, S. V., and Cotterill, J. (1964). *Biochem. J.* **92**, 603.
Perry, S. V., and Cotterill, J. (1965). *Biochem. J.* **96**, 224.
Perry, S. V., Cotterill, J., and Hayter, D. (1966). *Biochem. J.* **100**, 289.
Podlubnaya, Z. A., Kalamkarova, M. B., and Nankina, V. P. (1969). *J. Mol. Biol.* **46**, 591.
Post, R. L., and Rosenthal, A. S. (1962). *J. Gen. Physiol.* **45**, 614A.
Reedy, M. K. (1964). *Proc. Roy. Soc., Ser. B* **160**, 458.
Reedy, M. K. (1968). *J. Mol. Biol.* **31**, 155.
Rees, M. K., and Young, M. (1967). *J. Biol. Chem.* **242**, 4449.
Rice, R. V. (1961). *Biochim. Biophys. Acta* **53**, 29.
Rice, R. V., Asai, H., and Morales, M. F. (1963). *Proc. Natl. Acad. Sci. U.S.* **50**, 549.
Robson, R. M., Goll, D. E., Arakawa, N., and Stromer, M. H. (1970). *Biochim. Biophys. Acta* **200**, 296.
Rossomando, E. F., and Piez, K. A. (1970). *Biochem. Biophys. Res. Commun.* **40**, 800.
Sakakibara, I., and Yagi, K. (1970). *Biochim. Biophys. Acta* **207**, 178.
Sandow, A. (1970). *Annu. Rev. Physiol.* **32**, 87.
Samosudova, N. V. (1966). *Electron. Microsc., Proc. Int. Congr., 6th, 1966* Vol. 2, p. 691.
Schaub, M. C., and Perry, S. V. (1969). *Biochem. J.* **115**, 993.
Schaub, M. C., and Perry, S. V. (1971). *Biochem. J.* **123**, 367.
Schaub, M. C., Perry, S. V., and Hartshorne, D. J. (1967). *Biochem. J.* **105**, 1235.
Schliselfeld, L. H., and Bárány, M. (1968). *Biochemistry* **7**, 3206.
Schliselfeld, L. H., Conover, T. E., and Bárány, M. (1970). *Biochemistry* **9**, 1133.
Segal, D. M., Himmelfarb, S., and Harrington, W. F. (1967). *J. Biol. Chem.* **242**, 1241.
Seidel, J. C. (1969a). *J. Biol. Chem.* **244**, 1142.
Seidel, J. C. (1969b). *Biochim. Biophys. Acta* **180**, 216.
Seidel, J. C. (1969c). *Biochim. Biophys. Acta* **189**, 162.
Sekine, T., and Kielley, W. W. (1964). *Biochim. Biophys. Acta* **81**, 336.
Sekine, T., and Yamaguchi, M. (1963). *J. Biochem.* (*Tokyo*) **54**, 196.
Sekine, T., and Yamaguchi, M. (1966). *J. Biochem.* (*Tokyo*) **59**, 195.
Sekine, T., Barnett, L. M., and Kielley, W. W. (1962). *J. Biol. Chem.* **237**, 2769.
Sekiya, K., Mii, S., and Tonomura, Y. (1965). *J. Biochem.* (*Tokyo*) **57**, 192.
Selby, C. C., and Bear, R. S. (1956). *J. Biophys. Biochem. Cytol.* **2**, 71.
Seraydarian, K., Briskey, E. J., and Mommaerts, W. F. H. M. (1967). *Biochim. Biophys. Acta* **133**, 399.
Seraydarian, K., Briskey, E. J., and Mommaerts, W. F. H. M. (1968). *Biochim. Biophys. Acta* **162**, 424.
Shimada, T. (1970). *J. Biochem.* (*Tokyo*) **67**, 185.
Shy, G. M., Engel, W. K., Somers, J. E., and Wanko, T. (1963). *Brain* **86**, 793.
Skou, J. C. (1960). *Biochim. Biophys. Acta* **42**, 6.
Slayter, H. S., and Lowey, S. (1967). *Proc. Nat. Acad. Sci. U.S.* **58**, 1611.
Small, P. A., Harrington, W. F., and Kielley, W. W. (1961). *Biochim. Biophys. Acta* **49**, 462.
Spudich, J. A., and Watt, S. (1971). *J. Biol. Chem.* **246**, 4866.

Staprans, I., and Watanabe, S. (1970). *J. Biol. Chem.* **245**, 5962.

Staprans, I., Arai, K., and Watanabe, S. (1968). *J. Biochem. (Tokyo)* **64**, 65.

Stracher, A. (1964). *J. Biol. Chem.* **239**, 1118.

Stracher, A. (1967). *J. Gen. Physiol.* **50**, No. 6, Part 2.

Stracher, A. (1969). *Biochem. Biophys. Res. Commun.* **35**, 519.

Stracher, A., and Chan, P. C. (1964). In "Biochemistry of Muscle Contraction" (J. Gergely, ed.), p. 106. Little, Brown, Boston, Massachusetts.

Stracher, A., and Dreizen, P. (1966). *Curr. Top. Bioenerg.* **1**, 154.

Straub, F. B. (1942). *Stud. Inst. Med. Chem. Univ. Szeged* **2**, 3.

Straub, F. B. (1943). *Stud. Inst. Med. Chem. Univ. Szeged* **3**, 23.

Straub, F. B., and Feuer, G. (1950). *Biochim. Biophys. Acta* **4**, 455.

Stromer, M. H., Goll, D. E., and Roth, L. E. (1967). *J. Cell. Biol.* **34**, 431.

Stromer, M. H., Hartshorne, D. J., Mueller, H., and Rice, R. V. (1969). *J. Cell Biol.* **40**, 167.

Strzelecka-Golaszewska, H., and Drabikowski, W. (1968). *Biochim. Biophys. Acta* **162**, 581.

Szent-Györgyi, A. (1941a). *Stud. Inst. Med. Chem. Univ. Szeged* **1**, 17.

Szent-Györgyi, A. (1941b). *Stud. Inst. Med. Chem. Univ. Szeged* **1**, 67.

Szent-Györgyi, A. (1943). *Stud. Inst. Med. Chem. Univ. Szeged* **3**, 86.

Szent-Györgyi, A. (1944). "Studies on Muscle." Inst. Med. Chem., Univ. Szeged.

Szent-Györgyi, A. (1951). "Chemistry of Muscular Contraction," 2nd rev. ed. Academic Press, New York.

Szent-Györgyi, A., and Kaminer, B. (1963). *Proc. Nat. Acad. Sci. U.S.* **50**, 1033.

Szent-Györgyi, A. G. (1951). *J. Biol. Chem.* **192**, 361.

Szent-Györgyi, A. G. (1953). *Arch. Biochem. Biophys.* **42**, 305.

Szent-Györgyi, A. G. (1960). In "The Structure and Function of Muscle" (G. H. Bourne, ed.), Vol. 2, p. 1. Academic Press, New York.

Tada, M., and Tonomura, Y. (1967). *J. Biochem. (Tokyo)* **61**, 123.

Takahashi, K., and Yasui, T. (1967). *Proc. Int. Congr. Biochem., 7th, 1967* Vol. 1, p. 32.

Taylor, E. W., Lymn, R. W., and Moll, G. (1970). *Biochemistry* **9**, 2984.

Tokiwa, T., and Tonomura, Y. (1965). *J. Biochem. (Tokyo)* **57**, 616.

Tokuyama, H., Kubo, S., and Tonomura, Y. (1966). *J. Biochem. (Tokyo)* **60**, 701.

Tonomura, Y. (1972). "Muscle Proteins, Muscle Contraction and Cation Transport." Univ. of Tokyo Press, Tokyo.

Tonomura, Y., and Yoshimura, J. (1962). *J. Biochem. (Tokyo)* **51**, 259.

Tonomura, Y., Yoshimura, J., and Kitagawa, S. (1961). *J. Biol. Chem.* **236**, 1968.

Tonomura, Y., Kubo, S., and Imamura, K. (1965). In "Molecular Biology of Muscular Contraction" (S. Ebashi *et al.*, eds.), p. 11. Igaku Shoin, Tokyo.

Tonomura, Y., Appel, P., and Morales, M. (1966a). *Biochemistry* **5**, 515.

Tonomura, Y., Tokiwa, T., and Shimada, T. (1966b). *J. Biochem. (Tokyo)* **59**, 322.

Tonomura, Y., Imamura, K., Ikehara, M., Uno, H., and Harada, F. (1967). *J. Biochem. (Tokyo)* **61**, 460.

Tonomura, Y., Watanabe, S., and Morales, M. (1969). *Biochemistry* **8**, 2171.

Trayer, I. P., Harris, C. I., and Perry, S. V. (1968). *Nature (London)* **217**, 452.

Trotta, P. P., Dreizen, P., and Stracher, A. (1968). *Proc. Nat. Acad. Sci. U.S.* **61**, 659.

Tsao, T. C. (1953). *Biochim. Biophys. Acta* **11**, 368.

Tsao, T. C., Bailey, K., and Adair, G. S. (1951). *Biochem. J.* **49**, 27.
Tsao, T. C., Kung, T.-H., Peng, C.-M., Chang, Y.-S., and Tsou, Y.-S. (1965). *Sci. Sinica* **14**, 91.
Tsuboi, K. K. (1963). *Biochim. Biophys. Acta* **74**, 359.
Tsuboi, K. K. (1969). *Biochim. Biophys. Acta* **160**, 420.
Tsuboi, K. K., and Hayashi, T. (1963). *Arch. Biochem. Biophys.* **100**, 313.
Urnes, P., and Doty, P. (1962). *Advan. Protein Chem.* **16**, 401.
Varga, L. (1950). Cf. A. Szent-Györgyi (1951, p. 98).
von Fürth, O. (1895). *Arch. Exp. Pathol. Pharmakol.* **36**, 231.
von Muralt, A., and Edsall, J. T. (1930a). *J. Biol. Chem.* **89**, 315.
von Muralt, A., and Edsall, J. T. (1930b). *J. Biol. Chem.* **89**, 351.
Wakabayashi, T., and Ebashi, S. (1968). *J. Biochem. (Tokyo)* **64**, 731.
Watanabe, S. (1970). *J. Biochem. (Tokyo)* **68**, 913.
Watanabe, S., and Sleator, W., Jr. (1957). *Arch. Biochem. Biophys.* **68**, 81.
Weber, A. (1966). *Curr. Top. Bioenerg.* **1**, 203.
Weber, A., and Hasselbach, W. (1954). *Biochim. Biophys. Acta* **15**, 237.
Weber, A., and Osborn, M. (1969). *J. Biol. Chem.* **244**, 4406.
Weber, A., Herz, R., and Reiss, I. (1964). *Proc. Roy., Soc., Ser. B* **160**, 489.
Weber, H. H. (1934). *Pfluegers Arch. Gesamte Physiol. Menschen Tiere* **235**, 206.
Weber, H. H., and Portzehl, H. (1954). *Progr. Biophys. Biophys. Chem.* **4**, 60.
Weeds, A. G. (1969). *Nature (London)* **223**, 1362.
Weeds, A. G. (1970). *Proc. Int. Congr. Biochem. 8th, 1969* p. 29.
Weeds, A. G., and Hartley, B. S. (1968). *Biochem. J.* **107**, 531.
Weeds, A. G., and Lowey, S. (1971). *J. Mol. Biol.* **61**, 701.
Wetlaufer, D. B., and Edsall, J. T. (1960). *Biochim. Biophys. Acta* **43**, 132.
Woods, E. F. (1967). *J. Biol. Chem.* **242**, 2859.
Yagi, K., and Yazawa, Y. (1966). *J. Biochem. (Tokyo)* **60**, 450.
Yagi, K., Mase, R., Sakakibara, I., and Asai, H. (1965). *J. Biol. Chem.* **240**, 2448.
Yagi, K., Yazawa, Y., and Yasui, T. (1967). *Biochem. Biophys. Res. Commun.* **29**, 331.
Yamamoto, T., and Tonomura, Y. (1967). *J. Biochem. (Tokyo)* **62**, 558.
Yamashita, T., Soma, Y., Kobayashi, S., Sekine, T., Titani, K., and Narita, K. (1964). *J. Biochem. (Tokyo)* **55**, 576.
Yamashita, T., Soma, Y., Kobayashi, S., and Sekine, T. (1965). *J. Biochem. (Tokyo)* **57**, 460.
Yamashita, T., Kabusawa, I., and Sekine, T. (1967). *J. Biochem. (Tokyo)* **63**, 608.
Yamashita, T., Kobayashi, S., and Sekine, T. (1969). *J. Biochem. (Tokyo)* **65**, 869.
Yasui, B., Fuchs, F., and Briggs, F. N. (1968). *J. Biol. Chem.* **243**, 735.
Yasui, T., and Watanabe, S. (1965). *J. Biol. Chem.* **240**, 98.
Yazawa, Y., and Yagi, K. (1971). Personal communication.
Young, M. (1967). *J. Biol. Chem.* **242**, 2790.
Young, M. (1969). *Annu. Rev. Biochem.* **38**, 711.
Yount, R. G., and Koshland, D. E., Jr. (1963). *J. Biol. Chem.* **238**, 1708.
Yphantis, D. A. (1964). *Biochemistry* **3**, 297.
Zobel, C. R., and Carlson, F. (1963). *J. Mol. Biol.* **7**, 78.

Addendum in Proof

For progress made after completion of this manuscript, we refer the reader to *Cold Spring Harbor Symposia on Quantitative Biology*, Volume XXXVII, 1973

8

BIOCHEMISTRY OF MUSCLE

D. M. NEEDHAM

I. Energy Provision in Muscle

A. The Lactic Acid Period

The energy used by muscle in performance of work and maintenance of tension is ultimately derived from chemical reactions going on within it. The nature of these reactions was the subject of much experimentation and controversy during the latter half of the nineteenth century, but quantitative and reproducible results were not obtained until the classic work of Fletcher and Hopkins (1907). They showed clearly for the first time that fatigue and death rigor are accompanied by lactic acid production. This great step forward was due to their recognition of the need to reduce to a minimum any stimulation of the muscle during fixation and extraction. Later, Parnas and Wagner (1914) showed that the lactic acid is derived from the muscle glycogen. Now it was known that the formation of lactic acid from glycogen is a reaction going on with output of heat—the difference in the heats of combustion of the two substances was 16,300 cal per gram molecule lactic acid formed, according to Meyerhof and Meier (1924). It seemed reasonable to suppose—and the assumption has proved correct—that here was a reaction providing energy necessary for contraction. The long series of studies in Meyerhof's laboratory afforded support for this view, showing as they did the proportionality during anaerobic contraction between lactic acid formation and tension production or work done (Meyerhof, 1920, 1921).

The experiments on heat production in frog muscle during anaerobic contraction and relaxation (the initial heat) showed that the heat production was much greater than the expected amount—about 35,100 cal per gram molecule of lactic acid formed. A part of this excess could be explained by neutralization in the tissue, but a discrepancy of some 45% of the initial heat remained until it was explained by the metabolism of phosphagen. The significant fact was discovered that the amount of the initial heat (Weizäcker, 1914), its distribution in time (A. V. Hill and Hartree, 1920), and its relation to tension production (A. V. Hill, 1928b) are the same when the muscle contracts in oxygen as when it contracts in nitrogen. Thus, the chemical reactions underlying contraction must be nonoxidative; later, D. K. Hill (1940a) showed for frog muscle at 0°C with a short tetanus that even when conditions are from the beginning aerobic, increased oxygen consumption does indeed only start after activity (contraction and relaxation) is over. The aerobic recovery–heat production is large, almost equal to the initial heat (A. V.

Hill, 1928b); but if the conditions are maintained anaerobic, the period after activity shows only small and variable heat production, averaging about 20% of the initial heat spread over about 30 min (see D. K. Hill, 1940b).

B. Contraction and Phosphocreatine

For twenty years, the idea of lactic acid formation from carbohydrate as the only energy-providing reaction remained unchallenged. Then Lundsgaard (1930a) found that lactic acid formation in muscle could be stopped by poisoning with iodoacetate, but that nevertheless contraction could go on. He showed that in these poisoned muscles, tension production was proportional to breakdown of phosphocreatine. This substance (phosphagen) had been discovered in muscle by P. Eggleton and Eggleton (1927a,b) and independently by Fiske and Subbarow (1927), who first elucidated its structure (Fiske and Subbarow, 1929a). Eggleton and Eggleton connected phosphocreatine metabolism with contraction, for they found it to decrease in amount during contraction, while resynthesis occurred on recovery in oxygen. Nachmansohn (1928, 1929) showed that even in nitrogen there was rapid resynthesis of about 30% of the phosphocreatine in the 30 sec immediately after relaxation. At the same time, Meyerhof and Lohmann (1928a,b) found phosphocreatine hydrolysis to be an exothermic reaction, about 12,000 cal per gram molecule of inorganic phosphate being liberated in the conditions of their experiments. Nachmansohn (1928, 1929), comparing tension development with phosphocreatine disappearance, found that during a succession of contractions, this disappearance was much greater during the early contractions than during the later ones.

These facts concerning phosphocreatine necessitated a review of the lactic acid theory of contraction. In the first place, all experiments up to that time had seemed to show proportionality between tension production, heat liberation, and lactic acid formation (see A. V. Hill, 1928a); but if phosphocreatine hydrolysis, an exothermic reaction, was going on to a greater degree at the beginning of a contraction series, a greater heat production at this time was to be expected. Again, some explanation was needed for the rapid anaerobic resynthesis of phosphocreatine immediately after contraction. Here was an endothermic reaction going on during a short space of time when heat exchange was negligible and certainly no equivalent heat absorption could be detected (Hartree and Hill, 1928). The suggestion was made at the time that phosphocreatine was merely "unstabilized" *in vivo*, some change rendering the compound unusually liable to be broken down by the chemical treatment used in estimation (see A. V. Hill, 1928a; Nachmansohn, 1928; Meyerhof et al., 1930).

Then Lipmann and Meyerhof (1930) made the significant observation that, during the first few of a series of short tetani, there is a change in the muscle pH not to the acid but toward the alkaline side. These pH changes were studied in intact uninjured muscle (thin frog sartorii) lying in a bath of bicarbonate Ringer solution with a nitrogen–carbon dioxide atmosphere above. They were related by Lipmann and Meyerhof to an earlier study (Meyerhof and Lohmann, 1928b) of the titration curves of phosphocreatine and of an equimolecular mixture of free creatine and inorganic phosphate. Comparison of these curves showed that between pH 3 and 7.5 hydrolysis of phosphocreatine is accompanied by liberation of base; it was striking that the degree of pH change in the muscle on contraction varied according to the pH of the bathing medium and closely paralleled the amount of base liberation to be expected at different pH values from the titration curves. Thus it was shown that phosphocreatine breakdown in the muscle is a reality, and the way was prepared for the results of Lundsgaard.

Lundsgaard (1930a,b) took the point of view that, in muscle poisoned with iodoacetate (which prevents lactic acid formation by inhibiting glyceraldehydephosphate dehydrogenase), phosphocreatine breakdown supplies the energy for contraction; he went on to suggest that this might be the normal role of phosphocreatine hydrolysis in unpoisoned muscle also, the role of carbohydrate breakdown being to supply energy for resynthesis of phosphocreatine. In further experiments, he confirmed the results of Nachmansohn in finding in normal muscle a rise in the ratio of tension production to inorganic phosphate formation as the number of contractions increased, but he also found that in these circumstances there was greater lactic acid formation per unit of tension. Thus although the heat–tension ratio remains constant, the chemical reactions responsible for the heat production vary with time in a contraction series (Lundsgaard, 1931).

With regard to the anaerobic recovery, Lehnartz (1931) brought convincing evidence that much of the lactic acid production takes place after the contraction is over. Such claims made earlier by the Embden school had been disregarded, since with the strength of direct stimuli used, a pathological condition of some fibers supervened, with prolonged relaxation time and incomplete recovery. Lehnartz now used stimulation through the nerve and established that 20–30% more lactic acid was formed in the 5 min after relaxation. Indeed, at low temperatures, more than half the lactic acid formation may take place after relaxation

(Meyerhof, 1931). In this lactic acid production, we have the energy source for the anaerobic resynthesis of phosphocreatine.

Phosphocreatine is found in all vertebrate muscle examined (G. P. Eggleton and Eggleton, 1929–1930) and also in some invertebrates (Needham *et al.,* 1932; Baldwin and Yudkin, 1950). Phosphoarginine is the most widely distributed phosphagen in the rest, being characteristic of most invertebrate muscle (Meyerhof and Lohmann, 1928a,b). However, the discovery has been made that certain invertebrates contain new phosphagens; thus in the annelids, glycocyamine phosphate has been identified in *Nereis diversicola* and taurocyamine phosphate in *Arenicola* (Thoai *et al.,* 1953; Hobson and Rees, 1955). Hobson and Rees (1957) have shown that in these animals phosphokinases are present that can bring about the phosphorylation of the bases in question by means of ATP. Thoai and Robin (1954) have isolated guanidyl-ethylseryl phosphate from the earthworm; chromatographic examination showed the presence of the corresponding phosphagen in its muscle.

Creatine	$H_2NC(:NH)\cdot N(CH_3)\cdot CH_2\cdot COOH$
Arginine	$H_2N\cdot C(:NH)\cdot NH\cdot (CH_2)_3\cdot CH(NH_2)\cdot COOH$
Glycocyamine	$H_2N\cdot C(:NH)\cdot NH\cdot CH_2\cdot COOH$
Taurocyamine	$H_2N\cdot C(:NH)\cdot NH\cdot CH_2CH_2\cdot SO_3H$
Guanidylethyl- seryl phosphate	$H_2N\cdot C(:NH)\cdot NH\cdot (CH_2)_2\cdot O\cdot (O)P(OH)\cdot O\cdot CH_2\cdot CH(NH_2)\cdot COOH$

Later two more phosphagens were identified—those derived from hypotaurocyamine in the muscles of the gephyreans *Phascolosoma blainvilli* and *elongatum* (Robin and Thoai, 1962; Roche *et al.,* 1962), and those from guanidinoethylmethyl phosphoric acid in the Polychaete *Ophelia neglecta* (Thoai *et al.,* 1963).

Hypotaurocyamine	$H_2N\cdot C(:NH)\cdot NH\cdot CH_2\cdot CH_2\cdot SO_2H$
Guanidoethylmethyl phosphoric acid	$H_2N\cdot C(:NH)\cdot NH\cdot CH_2\cdot CH_2\cdot O(O)P\cdot OH\cdot OCH_3$

In each phosphagen, one hydrogen of the terminal amino group is replaced by the phosphate group, with formation of the energy-rich phosphoamide linkage.

C. Discovery of Adenosine Triphosphate

Soon after the isolation of phosphagen, there was discovered in muscle, independently by Lohmann (1929) and by Fiske and Subbarow (1929b), the substance adenosine triphosphate (ATP). Lohmann's studies (1931, 1932, 1935) indicated the structure shown, which has been confirmed by synthesis (Baddiley *et al.*, 1948). Two observations of importance were made; in the first place, that ATP acted as a coenzyme of glycolysis (Lohmann, 1929), though the details of its participation were not worked out till later; in the second place, that hydrolysis of the two terminal phosphate groups led to liberation of heat—about 12,000 cal per gram molecule of phosphate, according to Meyerhof and Lohmann (1932).

The realization of the importance of ATP hydrolysis for the contraction process itself came only later, and arose out of the work of Lohmann (1934) on hydrolysis of phosphocreatine in dialyzed cell-free muscle extracts. Such extracts cannot cause splitting off of phosphate from phosphocreatine; this only happens if adenylic compounds are present and the reaction occurs in two stages.

$$\text{Phosphocreatine} + \text{ADP} \rightarrow \text{creatine} + \text{ATP} \tag{1}$$

$$\text{ATP} \rightarrow \text{ADP} + \text{H}_3\text{PO}_4 \tag{2}$$

The first is a transfer of phosphate to adenosine diphosphate (ADP), then hydrolysis of the ATP thus formed goes on to give ADP again. It should be mentioned here in parentheses that, in all the earlier work, adenosine monophosphate (AMP) was used in reaction systems which, as we now know, require ADP. The adequacy of the AMP is explained by the presence in such systems of the enzyme myokinase which enabled the AMP to react with traces of ATP (Colowick and Kalckar, 1943).

$$\text{ATP} + \text{AMP} \rightleftharpoons 2\,\text{ADP} \tag{3}$$

This work of Lohmann had consequences of great significance. He deduced that before phosphocreatine breakdown can yield energy, ATP hydrolysis must have occurred; this reaction thus became the energy-yielding reaction closest to contraction. Further, this was the first detailed study of phosphate transfer and involved two compounds each containing what we now call an energy-rich phosphate bond (Lipmann, 1941). Transfer to phosphate between such molecules without formation of inorganic phosphate is a mechanism for conservation of free energy which has turned out to be of enormous physiological importance. At the time, Lohmann pointed out that the phosphate transport in Reaction

(1) (Lohmann's reaction) went on with very little heat exchange; he expressly remained noncommittal about free energy changes.

To provide for resynthesis of ATP, then, is the role of phosphocreatine in muscle metabolism. Parnas (Parnas *et al.*, 1934) next initiated an enquiry into the mechanism whereby carbohydrate breakdown provided energy and phosphate for the rephosphorylation, perhaps of creatine, perhaps of adenosine diphosphate. The stages in glycogen breakdown, already known in the nineteen thirties, are summarized in Table I.

It was known that hydrolysis of phosphopyruvic acid was an exo-thermic reaction (about 9000 cal being liberated per gram molecule of phosphoric acid released) and this seemed a likely stage. It was in fact found (Ostern *et al.*, 1935; Needham and van Heyningen, 1935; Meyerhof and Lehmann, 1935) in muscle extracts that phosphopyruvate (like phosphocreatine) transferred phosphate to adenylic compounds and was not dephosphorylated when these were absent.

$$\text{Phosphopyruvate} + \text{ADP} \rightarrow \text{pyruvate} + \text{ATP} \qquad (4)$$

In the presence of creatine as well as a catalytic amount of ADP, phos-phocreatine synthesis took place, but no direct reaction between phos-phopyruvate and creatine was found. It follows that Reaction (1) must be reversible, and this reversibility has been directly demonstrated (Leh-mann, 1936). The equilibrium point depends on the pH, more alkaline reactions favoring phosphocreatine synthesis. It seems, then, that in the

TABLE I
REACTIONS IN MUSCLE GLYCOLYSIS

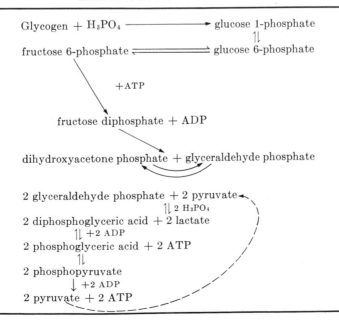

Glycogen + H_3PO_4 ⟶ glucose 1-phosphate

⇅

fructose 6-phosphate ⇌ glucose 6-phosphate

+ATP

fructose diphosphate + ADP

dihydroxyacetone phosphate + glyceraldehyde phosphate

2 glyceraldehyde phosphate + 2 pyruvate

⇅ 2 H_3PO_4

2 diphosphoglyceric acid + 2 lactate

⇅ +2 ADP

2 phosphoglyceric acid + 2 ATP

⇅

2 phosphopyruvate

↓ +2 ADP

2 pyruvate + 2 ATP

(Why Cr pool is important)

recovering muscle, once the stimulation to ATP breakdown has ceased, the phosphate from carbohydrate intermediates is transferred through ATP mediation to free creatine to rebuild the phosphocreatine store.

Now Lundsgaard (1931) had found that for every molecule of lactic acid formed in the anaerobic recovery period, about two molecules of phosphocreatine were resynthesized. The reaction we have discussed could account for only half this resynthesis. But there is another exothermic reaction going on in glycolysis, the oxidoreduction between glyceraldehyde phosphate and pyruvate, giving as end products phosphoglyceric and lactic acids. This was known to be accompanied by esterification of inorganic phosphate (Meyerhof and Kiessling, 1935); it was now found that, if adenylic acid is added to the oxidoreduction system, stoichiometric synthesis of ATP occurs, one molecule of phosphate being esterified for every molecule of lactic acid produced (Needham and Pillai, 1937; Meyerhof et al., 1937). We must further remember that one molecule of ATP is needed in the phosphorylation of fructose mono- to diphosphate. Thus, in vitro, during formation of two molecules of lactic acid, three molecules of phosphocreatine may be resynthesized (see Table I). Whether the difference is real between this figure and Lundsgaard's figure for results in vivo has not been further investigated.

The coupled esterification and oxidoreduction needs a further word of explanation. Although it has been known for some years that 1,3-diphosphoglyceric acid is formed in this coupled esterification and acts as the phosphate donor to ADP (Warburg and Christian, 1939; Negelein and Brömel, 1939), the sequence of reactions was only made clear in the nineteen fifties (Racker and Krimsky, 1952; Segal and Boyer, 1953). It involves reaction of the aldehyde group of the glyceraldehyde phosphate with the sulfhydryl group of the enzyme glyceraldehydephosphate dehydrogenase; the oxidation of this hemimercaptal with formation of an energy-rich bond; and then phosphorolysis of the enzyme–acyl compound by means of inorganic phosphate to form diphosphoglyceric acid, which can transfer its acyl phosphate to ADP.

The studies made by Meyerhof and his colleagues (Meyerhof and Lohmann, 1932; Meyerhof and Schulz, 1935; Meyerhof et al., 1938) on the heats of hydrolysis of compounds involved in muscle metabolism led to the early distinction between compounds containing the guanidino-, pyro-, or enolphosphate linkage on the one hand and the phosphate ester linkage on the other. Hydrolysis of the last was found to liberate only about 3000 cal per gram molecule of phosphate. It should be noted here that the heat of hydrolysis of ATP has been re-evaluated by several workers during recent years and considerably reduced. For instance, the value found by enzymic attack in vitro in buffers of known heats of ionization takes into account the heat of neutralization of H+ ions produced during the reaction and amounts to only 4700 cal per gram molecule of phosphoric acid set free (Podolsky and Morales, 1956). The important aspect of these reactions is, of course, the free energy and not the heat change; this was realized at the time, but since methods were not then available for measuring the free energies, the heats of reactions were taken as a rough guide. Since the treatment of this subject by Lipmann (1941) and Kalckar (1941), much effort by many workers has been put into the important task of finding true values for the free energy of hydrolysis of these compounds. The figures given for $\Delta F°$ of the energy-rich phosphate bond (the change in free energy on hydrolysis under standard conditions) lie between —7000 and —9000 cal (Levintov and Meister, 1954; Burton, 1955; Robbins and Boyer, 1957). The review by Huennekens and Whiteley (1960) gives a comprehensive account of the behavior of phosphoric acid anhydrides and other energy-rich compounds.

We have, then, this picture of the sequence of events after the stimulus reaches the muscle. First is dephosphorylation of ATP, which though masked by resynthesis in moderate contraction, is the essential reaction. As we shall see, there is good reason for postulating a direct reaction between the ATP and the myofibrillar protein actomyosin. This is followed at once by reaction between phosphocreatine and ADP; later, rephosphorylation of ADP by phosphopyruvate and diphosphoglycerate from carbohydrate breakdown becomes quantitatively more important.

Attention has been concentrated here on contraction under anaerobic conditions, depending ultimately on glycolysis; this is because the reactions concerned are easier to study in cell-free extracts than are many of the oxidative reactions concerned, and so gave the first insight into problems of energy provision. But although anaerobic contraction must sometimes happen *in vivo*, conditions are much more usually aerobic, and oxidative rephosphorylation of ADP is far more efficient. This is discussed in another chapter.

II. Interaction of Adenosine Triphosphate and Actomyosin as the Basis of Muscle Contraction

A. *The Early Experiments*

During the years that saw this preoccupation with the energy sources of contraction, a parallel and quite independent line of work had been pursued in the study of the muscle proteins. As early as the middle of the nineteenth century, Kühne (1864) had extracted from muscle a protein capable of gel formation, which he called myosin. This protein was soon recognized as belonging to the class of globulins—soluble in sodium or potassium chloride solutions, but precipitated by dilution with large volumes of water. Edsall (1930) and von Muralt and Edsall (1930) purified the protein and studied its physicochemical properties, particularly the double refraction of flow of its solutions. E. C. Smith (1933–1934) investigated the dependence of the solubility of myosin on salt concentration and on pH. He concluded that under the conditions prevailing in resting muscle (pH about 7 and salt concentration equivalent to about 0.18 M potassium chloride), at least 90% of the myosin must be in the gel form. About the same time, H. H. Weber (1935) was making myosin filaments by squirting the solution in 0.5 M potassium chloride through fine extruders into a large volume of water, so that dilution precipitation took place.

It was not until some years later that Engelhardt and Ljubimova (1939) brought the two lines of work together by their discovery that myosin, prepared and purified by the classic methods, is a specific ATPase; when activated by certain divalent ions, it splits off the terminal phosphate group to give ADP.

This was very quickly followed by the finding that there is a reciprocal action of the substrate on the enzyme protein. Thus, Needham *et al.* (1941) (see also Dainty *et al.*, 1944) observed that addition of ATP to a solution of myosin (made by many hours of extraction of the muscle with salt solution) led to a striking fall in viscosity and double refraction of flow (see Fig. 1)—changes which are to be interpreted as a diminution in the axial ratio of anisometric molecules in the solution (Lawrence *et al.*, 1944). These changes were reversed as the ATP was destroyed by the enzymic activity of the myosin. The tentative suggestion was made that a shortening of the myosin molecule took place and might be the basis of contraction. But in Szent-Györgyi's laboratory in Hungary, where intensive study of these questions was also going on, it was shown that two proteins are concerned: (1) the protein which we now term myosin, extractable from minced muscle by short treatment (20 min) with salt solution, and (2) the protein actin, extractable from the dried residue (Banga and Szent-Györgyi, 1941–1942; Straub, 1942). Each of these proteins alone can give a solution of low viscosity and without double refraction of flow. On mixture of the two, a solution of high viscosity and strong double refraction of flow was obtained. The conclusion was drawn that the two proteins enter into some form of combination, giving actomyosin, and that the changes seen on adding ATP are due to dissociation of the complex, with formation of the two types of protein molecule of smaller axial ratio.

The relevance of the actomyosin–ATP relationship to contraction was

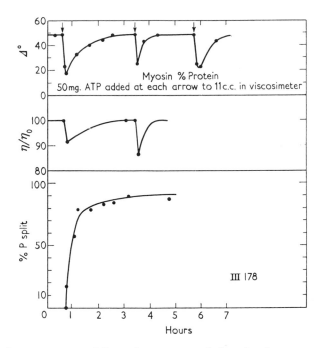

Fig. 1. Three successive falls and recoveries of flow birefringence of a myosin sol treated three times with ATP, with two successive falls and recoveries of relative viscosity (expressed in percentage of initial value) and estimations of inorganic phosphate liberated by ATPase activity of the myosin during the first cycle (Dainty *et al.*, 1944).

more clearly brought out by the experiments of Szent-Györgyi (1941–1942) in which ATP was added to actomyosin at low ionic strengths, in fact to actomyosin gels. Actomyosin threads (prepared by Weber's method) were used, and an isodimensional contraction to about 10% of the original length was obtained. If the micelles in the thread are oriented by partial drying and stretching (Buchthal *et al.*, 1947), the effect of added ATP is to make the thread become shorter and wider, as happens in contraction of a muscle fiber.

B. Interaction of ATP and Actomyosin in Solutions

At the basis of most thinking on the contraction–relaxation mechanism, we find the conception outlined above—that addition of ATP to an actomyosin solution leads to dissociation of the two proteins, while removal of the ATP is followed by recombination. Direct evidence for this dissociation conception was, however, for a long time lacking, but A. Weber (1956) showed that if actomyosin is centrifuged (under conditions in which ATPase activity is inhibited) with ATP for 3 hours at 100,000 g, pure myosin could be recovered from the upper half of the supernate and identified by its ATPase characteristics, its reaction with actin, and its sedimentation constant. The pellet on extraction yielded actin of characteristic behavior. Gergely (1956) has come to the same conclusion from light-scattering experiments. The study of changes in light scattering of a system provides a rapid method of following changes in the size, shape, and interaction of the particles of the light-scattering material (see Oster, 1948). By application of the extrapolation method of Zimm (1948), particle weight can be determined independently of any assumptions as to particle shape. In this way, Gergely obtained evidence of a large fall in molecular weight upon ATP addition to actomyosin solutions, consistent with dissociation.

Mommaerts (1947) has shown that the viscosity drop on ATP addition can be obtained without any accompanying enzymic dephosphorylation; the change is therefore due to combination of enzyme and substrate without breakdown of the substrate. Thus 10^{-3} M Mg^{2+}, which inhibits ATPase activity of the actomyosin solution in 0.5 M potassium chloride, allows the unimpaired viscosity fall, but there is no recovery. More recently, he has studied the same change in the protein particles by the more sensitive light-scattering method and has shown that Ca^{2+}, while accelerating the ATPase activity, greatly decreases the light-scattering fall (Mommaerts, 1956). The high rate of ATP disappearance may have been partly responsible for the diminished fall; but the results of Baranyi et al. (1951), using a method of very rapid measurement of viscosity changes, show that there really is a very marked activating effect of Mg^{2+} ions and a smaller inhibitory effect of Ca^{2+} ions on the dissociability of the actomyosin (see Table II).

TABLE II
INFLUENCE OF MAGNESIUM AND CALCIUM ON RATE OF VISCOSITY FALL
WHEN ATP IS ADDED TO ACTOMYOSIN SOLUTIONS[a]

No activator	Mg^{2+}			Ca^{2+}			0.004 M Mg^{2+} plus 0.003 M Ca^{2+}
	0.001 M	0.01 M	0.004 M	0.001 M	0.002 M	0.004 M	
9.5	0.63	0.47	0.54	—	—	—	—
6.0	—	—	—	9.5	7.9	3.0	—
—	—	—	—	—	—	—	0.66

[a] Results expressed as half-time of the fall in seconds. The relative viscosity of the actomyosin solution was 2.5; the ATP concentration 0.5×10^{-4} M (Representative figures selected from the tables of Baranyi et al., 1951).

Straub (1943) and Mommaerts (1947) further showed that inorganic pyrophosphate, which is not hydrolyzed (Bailey, 1942), and inorganic triphosphate, which is only very slowly hydrolyzed (Dainty *et al.*, 1944), can in certain circumstances show the unreversed viscosity fall. A high concentration of Mg^{2+} (0.01 M) is necessary here, and Ca^{2+} is ineffective.

C. Interaction of ATP and Actomyosin in Gels

1. Types of System Used

We have already mentioned the use of the actomyosin thread after orientation by partial drying in a stretched state as a model of the muscle fiber. An even more useful preparation, introduced by Szent-Györgyi (1949), is the glycerinated fiber bundle. A strip of muscle about 2 mm in diameter is removed, being kept at the resting length; from the psoas of the rabbit, for example, a bundle of parallel fibers some 8 cm long can easily be obtained. The fiber bundles are kept in 50% glycerol at 0°C for some days. In this way, the removal of water from the muscle is very gradual and the water content remains uniform throughout the bundle; about 50% of the soluble proteins and most of the crystalloids are extracted. An important characteristic is that, owing to destruction of membranes, the fibers show no response to electrical stimulation but are permeable to ATP. Such fiber bundles can be kept for many weeks; they can be washed free from glycerol when needed, dissected to smaller dimensions, and then used for the study of tension production upon addition of ATP. The tension production and degree of shortening are quantitatively very similar to the responses of living muscle to electrical stimulation (H. H. Weber and Portzehl, 1952).

Another model used for the study of the contractile actomyosin–ATP system has been the superprecipitating actomyosin gel. At certain concentrations of potassium chloride and ATP, sol to gel transformation takes place; under certain other conditions the gel contracts, squeezing out water and becoming readily sedimentable on centrifugation. The degree of contraction can be expressed in a semiquantitative manner. (See, e.g., Spicer, 1951.)

2. Stages in the Interaction and the Dual Role of ATP

The first investigation of the effect of ATP on such muscle models was made by Engelhardt *et al.* (1941) on actomyosin threads. These contained only 2% of protein but showed a considerable amount of tensile strength. When a load of some milligrams was applied by means of a torsion balance to the thread immersed in a bath, extensibility could be measured. Addition of ATP to the bath led to an increase in extensibility by some 50–100%. At first sight, these results seem to be in contradiction to those of Szent-Györgyi made a little later in which, as we have seen, the effect of ATP was to cause contraction. Some years later, Buchthal *et al.* (1947) showed that the same threads (in this case 20% protein) would give both effects; if loaded (e.g., with 200 mg), they showed extension on ATP application, but unloaded they contracted. By very carefully controlled drying and stretching, it is possible to prepare actomyosin threads that will develop considerable tension (Portzehl, 1951), but the point made here is that the extensibility increase with ATP indicates that the first effect on the gel is a loosening of linkages, as with the solution. There is much evidence, as we have seen, that in the case of the solution, myosin and actin are dissociated from one another; while the same degree of dissociation cannot take place in the gel, it is quite possible that the same bonds are affected. Szent-Györgyi (1949), using glycerol-extracted muscle fibers, emphasized this dual role of ATP. After treatment with ATP, the contracted fiber bundle was hard and opaque, but on renewal of the ATP it became momentarily soft and flexible before hardening again as the fresh ATP was used up. Bozler (1951) (see Fig. 2) was the first to get a full cycle of contraction and relaxation. Here, with the glycerinated fiber bundle, a high ATP concentration was used (0.02 M) and the contraction was brief. A concentration of 0.003 M ATP caused only contraction.

A better understanding of the dual role of ATP in muscle began with the work of Marsh (1952). He used fresh muscle homogenates in which the fragments consisted of fiber bundles, and he followed their changes in size by centrifuging and measuring the volume of the solid layer. When a brei from fresh, actively glycolyzing muscle was used, the fiber volume remained constant for a considerable period, depending upon the glycogen content of the muscle. After 20–50 min, there was a rapid fall in volume, then again a steady state. He considered that the fiber shrinkage corresponded to contraction, the water loss which

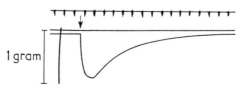

Fig. 2. Brief contraction of glycerinated fiber bundle produced by 1.5% ATP (about 0.03 M). Temperature 21°C; time marks every 5 sec; tension 1 gm. (Bozler, 1951.)

occurred from the fibers being a consequence and not the cause of the diminution in volume. Experiments in which the behavior of the fiber fragments was examined under the microscope bore out this point of view. With fresh homogenates, ATP addition usually led first to an increase in volume and length of the fibers (relaxation); only later, after a time during which diminution of the ATP concentration within the fibers (through their own low ATPase activity and that of soluble ATPase) would have occurred, did the fibers shorten. The rise and fall in volume could be repeated several times.

Marsh was led to emphasize the importance of ATPase activity by the observation that if the fibers were washed twice with salt solution, the only response to ATP added to the suspension was decrease in volume, never increase. At the same time, the ATPase activity of the fibers rose some tenfold as a result of the washing. Marsh deduced the presence in the original brei of a factor responsible for these effects and we shall return to this later. He also deduced that dissociated actin and myosin in the relaxed muscle must change to actomyosin before contraction; for this change and the shortening, both an increased rate of energy liberation from ATP and a fall of ATP concentration at the active sites on the protein are necessary. He did indeed find that the contraction only occurred when the ATP concentration within the fibers had fallen (Marsh, 1951).

In summarizing the effect of ATP on a glycerinated fiber or actomyosin thread, we may picture the series of events thus: first, we have the loosening of linkages; second, ATPase activity reduces the ATP concentration at the active sites. This makes formation of new linkages possible and also supplies energy for the contraction process. From all the evidence of Section I, we must conclude that continuous ATP breakdown is necessary as long as tension production goes on. ATPase activity will continue until all the ATP is used up, and the fiber then remains in the shortened form, but with disappearing tension. Under certain conditions fresh ATP can cause relaxation.

3. PROBLEMS OF DIFFUSION

In experiments such as those of Marsh, the fresh fiber fragments can for a time keep up their own internal ATP concentration by carbohydrate metabolism; with experiments with the artificial thread or the glycerinated fiber, there is dependence from the beginning on the balance between the rate of diffusion inwards of ATP on the one hand, and the rate of hydrolysis by the actomyosin on the other. In the steady state, the concentration of ATP on the outside (C) is related to the concentration at the center (I) according to the Meyerhof–Schulz formula (Meyerhof and Schulz, 1927)

$$C = Ar^2/4D + I$$

where A = rate of splitting, r = radius of the fiber, and D = diffusion constant of ATP within the fiber.

Calculation shows that with a physiological concentration of ATP (about 5×10^{-3} M), the diameter of a fiber must not exceed about 6 μ if the concentration is to be the same in the center (Hasselbach and A. Weber, 1955). With a fiber bundle 500 μ in diameter, the concentration at the center would be zero. In such a case, the fibers of a large central core remain stiff and play no part in the contraction; indeed, they even hinder it (A. Weber, 1951). Tension measurements have been successfully carried out with thin single fibers, for example by Briggs and Portzehl (1957); here the fibers were 50–60 μ in diameter. They showed a much higher tension production per square centimeter, about twice as great as with fiber bundles, presumably because the inactive core is much smaller. Even so, it was calculated that the ATP-free core would be 20–50% of the cross section; but it is much less likely than in a thicker fiber to be in a condition of rigor, since it has available ample supply of ADP, a good plasticizer (see Engelhardt, 1946) in the presence of enough Mg^{2+}. It will be obvious that correlations of tension production and ATPase activity will present difficulties when fiber bundles are used, since the exact conditions of diffusion will vary from one bundle to another.

D. ATPase Activity of Myosin and Actomyosin

1. EFFECTS OF DIVALENT IONS

Purified myosin free from actin has ATPase activity, but actin has none. It is therefore assumed that the ATPase activity of actomyosin, though it shows some characteristic differences, is due to sites on the myosin component. An important difference between actomyosin and myosin ATPase is that the former is activated by Mg^{2+} and Ca^{2+}, the latter by Ca^{2+} but not by Mg^{2+} (Banga, 1941–1942; Kuschinsky and Turba, 1950). Hasselbach (1952) made a detailed study of these Mg^{2+} and Ca^{2+} effects (at pH 7 and ATP concentration about 10^{-3} M). Actomyosin gel at low ionic strength (0.1 M or less) was activated by Mg^{2+} up to a concentration of 3×10^{-4} M; with higher concentrations of Mg^{2+}, activity fell off. Myosin and actomyosin sols were inhibited by Mg^{2+}, and were both more highly activated by Ca^{2+} than was the actomyosin gel. It is interesting that the activation energy of the Mg^{2+}-activated ATPase is greater than that of the Ca^{2+}-activated. Bendall (1961) found the former always greater than 15 kcal/mole, while the latter was only about 11 kcal/mole.

2. INHIBITION BY SULFHYDRYL REAGENTS

The inhibitory effect of sulfhydryl reagents on myosin ATPase activity was early recognized (see, e.g., Singer and Barron, 1944). Then Bailey and Perry (1947) tested myosin after treatment with various of these reagents for ATPase activity and (by means of viscosity measurements) for ability to combine with actin. They found, under their conditions, that the two properties fell off in parallel, and suggested that at certain centers containing sulfhydryl groups, either actin or ATP could combine. The first indication that different sulfhydryl groups are concerned in interaction with ATP and with actin came with the observations of Turba and Kuschinsky (1952) on actomyosin sol treated with oxarsan in absence of ATP; this resulted in failure of superprecipitation when ATP was added and the sol was diluted. Inhibition of the combination between myosin and actin was thus implied, but the ATPase activity was unaffected. Similarly Bárány and Bárány (1959), using dithioglycollate, iodoacetamide, or N-ethylmaleimide, distinguished two types of sulfhydryl group. One, concerned in actin combination, was affected by the reagent when applied to myosin, but not when applied to actomyosin. The sulfhydryl site involved in ATP-ase activity could be protected from the reagent by the presence of ATP. The results of Stracher (1964, 1965) with dithiopropionate also show up clearly this specificity of the different sulfhydryl groups. Out of the fifteen such groups contained in 200,000 gm myosin, two were necessary for ATPase activity; three were essential for actin combination. One of these groups took part in both phenomena.

3. HIGH INITIAL RATE OF ATPASE ACTIVITY

A. Weber and Hasselbach (1954), using glycerinated fibers 30 μ thick showed that the magnesium–ATPase activity was at least twice as great during the first 15 sec of reaction at room temperature as later (after about 100 sec), when a constant rate was reached. Accumulation of end products, decreasing concentration of ATP, impurity in the ATP, and contraction of the originally relaxed fibers were all considered as ruled out as causes of the decrease in rate. The effect could be repeated if the hydrolysis was interrupted by Salyrgan inhibition and then reactivated by cysteine; it was in fact only obtained with enzyme newly beginning to cause splitting. The general features of this curious phenomenon have been confirmed by several workers, but its mechanism has been a matter of discussion. Some explanations involve phosphorylation of the enzyme with ready release of the phosphate on fixation (see the detailed studies by Tonomura and his collaborators to which many references may be found in Tonomura *et al.*, 1969), but this is disputed in others (Bendall, 1961; Bowen *et al.*, 1963; Sartorelli *et· al.*, 1966). Taylor *et al.* (1970) have brought evidence for formation of a myosin–product complex, $MADPP_i$, during the enzymic activity. They explain the phosphate burst as due to formation of this complex at a greater rate than its breakdown, the complex being broken down on fixation.

Other observations taken to indicate phosphorylation of myosin during ATPase activity of myosin or actomyosin in presence of Mg^{2+} are those of Levy and Koshland (1958, 1959). The reaction proceeded in presence of ^{18}O-labeled water, and it was found that the number of atoms of ^{18}O per molecule of inorganic phosphate formed was greater than one; an exchange reaction could thus have taken place between the phosphate (presumably during its attachment to the enzyme) and the water of the medium. However, it seems to remain possible that the exchange concerns the terminal phosphate of ATP in the enzyme–substrate complex.

E. Contraction and ATPase Activity in Vitro

Weber and his collaborators have brought forward much evidence showing the close correlation between tension production in glycerinated fibers and their ability to dephosphorylate ATP. Thus, A. Weber and Weber (1951) (see also H. H. Weber, 1951) and Heinz and Holton (1952) showed the effect of changing ATP concentration upon both tension production and ATPase activity; the effects ran parallel both at room temperature and at 0°C. Ulbrecht and Ulbrecht (1963), with glycerinated fibers from smooth adductor muscles of *Anodonta*, found that the mechanical effects of ATP application and the ATPase activity showed the same temperature dependence. As we have seen, many reagents reacting with sulfhydryl groups are known to inhibit the ATPase activity of myosin. Their effect on contraction of actomyosin threads or fibers has always been found to be similarly or even more strongly inhibitory (Turba and Kuschinsky, 1952; H. H. Weber and Portzehl, 1954). The connection of the ATP utilization with the contraction phase is well brought out by the experiments of Portzehl (1952). In the presence of the mercurial Salyrgan, ATP brought about complete and lasting relaxation of glycerinated fibers, Salyrgan alone having no effect. Only after addition of large excess of cysteine did contraction again become possible.

In all these investigations, Mg^{2+} was supplied as ATPase activator in the medium of the contracting fibers. Hasselbach (1952) mentioned the highly significant finding by H. H. Weber that calcium ions, though they could mediate ATPase activity, could not mediate tension production. Further such evidence was brought forward by Ashley *et al.* (1956); they used glycerinated myofibrils and measured their shortening under the microscope. After dialysis, the fibrils did not contract, though they could split ATP. On addition of ATP and Ca^{2+}, there was increased hydrolysis but no contraction. Addition of ATP and Mg^{2+} instead brought about immediate contraction. Bendall (1961) also saw instant shortening of glycerinated myofibrils on addition of Mg^{2+} to the ATP-containing medium, but none with Ca^{2+} addition. As we shall see (Section IV,C), it later became evident that Ca^{2+} is necessary as well as Mg^{2+} for ATPase activity and contraction, but in such small concentration (about 10^{-6} M) that only with very highly purified material can its effect be identified.

TABLE III

COMPARISON OF NUCLEOTIDE TRIPHOSPHATES IN THEIR EFFECT ON TENSION PRODUCTION, AND IN. BEHAVIOR WITH THE MARSH FACTOR[a]

Type	Tension in presence of 10^{-3} M Mg^{2+}	Tension in presence of 10^{-2} M Mg^{2+}	Minimum concnetration of NTP needed to get relaxatoin with Marsh factor (M)
ATP	100	100	2.5×10^{-3}
CTP	80	95	7.5×10^{-3}
UTP	—	75[b]	No effect up to 10^{-2}
ITP	30	50	No effect up to 10^{-2}
GTP	15	30	No effect up to 10^{-2}

[a] The tension development with ATP under optimal conditions is put equal to 100; temperature 21°C; ionic strength 0.1; nucleotide triphosphate (NTP) concentration in columns 2 and 3, 5×10^{-3} M. Single fibers or fiber bundles 160 μ in diameter were used. (Hasselbach, 1956.)
[b] 5×10^{-3} M Mg^{2+}.

Other nucleotide triphosphates can also supply energy for contraction of glycerinated fibers. Hasselbach (1956) puts them in the order ATP > CTP > UTP > ITP > GTP for rate of hydrolysis, ability to produce tension, and plasticising effect[*] (Table III). The pre-eminence of ATP

[*] CTP = cytidine triphosphate; UTP = uridine triphosphate; ITP = inosine triphosphate; and GTP = guanosine triphosphate.

as plasticiser was specially marked at normal Mg^{2+} concentration (about 10^{-3} M); with higher Mg^{2+} concentration (10^{-2} M), the efficacy of the other nucleotide triphosphates was somewhat increased.

III. The Regulation of Carbohydrate Metabolism for Energy Supply

The presence of ADP may be considered an adequate trigger at pH values around 7 for the breakdown of phosphocreatine; its importance in regulating oxidative activity will be considered in another chapter. But it is not so clear what mechanism starts the rapid breakdown of glycogen which begins on stimulation. As C. F. Cori (1956) has emphasized, the increment in rate here is enormous. In studying the complicated patterns of regulatory mechanism we shall find, besides the well known influence of concentration of reactants and products, such controlling factors as rapid changes in enzyme concentration and allosteric effects. We shall consider the key enzymes concerned in glycogen and glucose utilization and in the synthesis of glycogen from glucose and of glucose from lactic acid.

A. *Activation and Inactivation of Phosphorylase*

The earliest observations on regulation of glycogen breakdown by control of phosphorylase activity were those of C. F. Cori and Cori (1936) on the stimulating effect of adenylic compounds on the formation of hexose monophosphate in frog muscle. A little later, G. T. Cori *et al.* (1938) showed that adenylic acid was specifically concerned here. Then G. T. Cori and Green (1943) observed that the enzyme existed in two forms now known as phosphorylase a and b. The former had about 70% of maximum activity in absence of adenylic acid; the latter was quite inactive unless adenylic acid was present. An enzyme brought about the change from the a to the b form; in this process, halving of the molecular weight of the enzyme from about 495,000 to about 242,000 occurred (Keller and Cori, 1953).

Madsen and Cori (1957) found that four molecules of adenylic acid were bound by phosphorylase a, while phosphorylase b bound only two. About the same time, Baranowski *et al.* (1957) reported that no less than eight atoms of phosphorus were present per molecule of phosphorylase a, and of these, four were identified as contained in pyridoxal 5-phosphate. Phosphorylase b contained only 2 moles of this compound per mole of enzyme (C. F. Cori and Illingworth, 1957). It was necessary for activity of the enzyme. Its removal (by ammonium-sulfate precipitation at pH 3.6) left the enzyme inactive, but activity could be restored by incubation of phosphorylase a with pyridoxal 5-phosphate or of phosphorylase b with this compound plus adenylic acid. Experiments with ^{32}P-labeled inorganic phosphate or glucose 1-phosphate gave no indication of the participation of the phosphate groups of either adenylic acid or of pyridoxal 5-phosphate in the enzymic mechanism (Cohn and Cori, 1948; Illingworth *et al.*, 1958).

The next step was the finding by E. H. Fischer and Krebs (1955; E. G. Krebs and Fischer, 1955) that it was chiefly the b form that was present in resting muscle, and that the change to the a form went on in muscle extracts if ATP and Mg^{2+} as well as a certain protein constituent of the extract were present. With crystalline phosphorylase b and ^{32}P-labeled ATP, E. G. Krebs *et al.* (1958) found that during the conversion of b to a by the enzyme, which they named phosphorylase b kinase, four phosphate groups entered the enzyme molecule and four molecules of ADP appeared. The work of Wosilait (1958) and of E. H. Fischer *et al.* (1959) showed that this newly introduced phosphate was contained in phosphoserine. Sutherland and Wosilait (1955), using liver phosphorylase a, had earlier observed release of inorganic phosphate during the enzymic inactivation and conversion to the b form. Krebs and his collaborators found no exchange of inorganic phosphate with phosphoserine phosphate during the enzymic phosphorylation of glycogen.

We come now to a further regulatory mechanism, the activation of phosphorylase b kinase itself by the hormone adrenaline. In this work of Rall *et al.* (1957) with cell-free liver homogenates, an active, heat stable, and dialyzable factor was formed in the particulate fraction, and this stimulated conversion of phosphorylase b to a·in the supernatant fraction. This factor was identified by Rall and Sutherland (1957, 1962) as 3′,5′-adenosine monophosphate or cyclic adenylic acid. It was formed (also in skeletal muscle and heart particulate fractions) from ATP, with setting free of pyrophosphate; presence of magnesium ions was essential. Rall and Sutherland called the enzyme concerned adenyl cyclase. That still more complicated mechanisms were concerned in this phosphorylase b kinase activation appeared from the work of E. G. Krebs *et al.* (1959). They extracted the kinase from muscle in a form inactive at pH 7 and found that it could be activated in two ways: by short incubation with either ATP plus Mg^{2+}, or with 10^{-3} M Ca^{2+}. Later work showed (a) that the former effect could be enhanced by adding a trace of cyclic AMP (ineffective alone); phosphorylation of the kinase seemed to be involved (E. G. Krebs *et al.*, 1964); (b) that a protein factor (kinase-activating factor) was concerned in the latter effect with calcium ions, though there was no evidence that it was of enzymic nature (Meyer *et al.*, 1964). Then Ozawa *et al.* (1967) reported that calcium ions were also required in the activation by ATP, but in much smaller amount—only 10^{-6} M. Table IV attempts a summary of these complicated relationships.

The relative potencies of adrenaline and other catecholamines with regard to their stimulation of cyclic AMP formation in dog heart (Murad *et al.*, 1962) resemble their relative powers to increase the contractile force of the heart and its phosphorylase content (Mayer and Moran, 1960). But there has been much discussion as to the timing of the succession of events. For example, Drummond *et al.* (1964) (see also Williamson and Jamieson, 1966) found with small adrenaline doses that the rise in contractile force preceded the rise in phosphorylase a concentration by about 10 sec. In more recent work with perfused rat hearts and use of the maximum physiological dose of adrenaline, Drummond

TABLE IV

SOME FACTORS AFFECTING THE ACTIVITY OF PHOSPHORYLASE a AND PHOSPHORYLASE b KINASE[a]

Phosphorylase b	Phosphorylase a
Inactive without AMP	70% active without AMP
Inhibited by ATP	—
Molecular weight 242,000	Molecular weight 495,000
Loosely binds 2 moles AMP	Loosely binds 4 moles AMP
Contains 2 moles pyridoxal phosphate	Contains 4 moles pyridoxal phosphate

Phosphorylase b $\xrightarrow[\text{ATP, Mg}^{2+}]{\text{phosphorylase b kinase}}$ Phosphorylase a + ADP (4 moles serine → phosphorylserine)

Phosphorylase b + $4H_3PO_4$ $\xleftarrow[\text{(PR enzyme)}]{\text{phosphatase}}$ Phosphorylase a

Inactive phosphorylase b kinase $\xrightarrow[\substack{\text{or preincubation with} \\ Ca^{2+}10^{-3} M + \text{protein factor}}]{\substack{\text{Preincubation with ATP,} \\ Mg^{2+}, \text{ and cyclic AMP}}}$ Active phosphorylase b kinase

ATP $\xrightarrow[Mg^{2+}; \text{stimulated by epinephrine}]{\text{adenyl cyclase}}$ Cyclic AMP + pyrophosphate

Cyclic 3,5-AMP $\xrightarrow{\text{Cyclic phosphodiesterase}}$ 5-AMP

[a] Needham, 1971.

et al. (1966) reported activation of phosphorylase b kinase within 1 sec. Simultaneously there was rise in cyclic AMP concentration, but the inotropic peak was reached only at about 10 sec. They suggest that the phosphorylase a and phosphorylase b kinase may be structurally separated. The former might be free in the cytoplasm, while the latter could be situated in the membrane, where it would be readily accessible to the adenyl cyclase. There is evidence that the cyclase is membrane-bound (Davoren and Sutherland, 1963). Cyclic AMP has no direct effect on actomyosin superprecipitation or ATPase activity (Sutherland and Robison, 1966); Williamson (1966) has tentatively suggested that it might (presumably via a change in the permeability of the sarcoplasmic reticulum or cell membrane) increase the ionized calcium content of the heart and thus increase the activity of the contractile protein.

It became clear by 1962 that changes in concentration of adenine nucleotides might exercise control on phosphorylase activity in other ways than through the b → a transformation. Thus, Parmeggiani and Morgan (1962) found that under anaerobic conditions there was more rapid glycolytic breakdown than was to be explained by phosphorylase a formation. Cornblaeth *et al.* (1963) had similar experience, and their results suggested that increased tissue levels of AMP and inorganic phosphate were responsible. Morgan and Parmeggiani (1964a,b) ascertained the levels of ATP, AMP, ADP, and inorganic phosphate in the aerobic and anaerobic conditions with perfused hearts. With tests *in vitro* corresponding to aerobic conditions, ATP strongly inhibited phosphorylase b; AMP stimulated both forms of the enzyme. With tests under anaerobic conditions, both forms of the enzyme were inhibited by ATP and stimulated by AMP. It thus seems that in resting muscle, with the enzyme mainly in the b form, the activity is maintained at a low level by ATP inhibition and also by very low concentration of the substrate inorganic phosphate. In anoxia, the rise in AMP and inorganic phosphate can counteract the ATP inhibition. These workers, as well as Helmreich and Cori (1964) found that the AMP activation of phosphorylase in the b form was due to increased affinity of the enzyme for both its substrates, inorganic phosphate and glycogen. Parmeggiani and Morgan found that ATP inhibited phosphorylase b by lowering its affinity for AMP.

B. Activation and Inhibition of Phosphofructokinase

C. F. Cori (1941) suggested that phosphofructokinase might act as a limiting factor in muscle glycolysis since both glucose and fructose

6-phosphate can accumulate in muscle without any increase in lactic acid formation. Lardy and Parks (1956) found beef liver phosphofructo-kinase to be inhibited by ATP if present in excess of the Mg^{2+} concentration. Then Passonneau and Lowry (1962) showed that this inhibition with muscle phosphofructokinase could be overcome on adding cyclic AMP, AMP, or inorganic phosphate. Inhibition of the enzyme from heart muscle by ATP and activation by cyclic AMP were also noted by Mansour et al. (1962). In further experiments of Passonneau and Lowry, anaerobic conditions were imitated by increasing the inorganic phosphate level and lowering the ATP; AMP and ADP were present also in low concentrations. A thirtyfold increase in phosphofructokinase activity resulted. The reaction product fructose diphosphate was also an activator. They suggested (see also Lowry and Passonneau, 1966) that there may be two binding sites for ATP on the enzyme, one the active center, the other inhibitory. Deinhibitors or activators like AMP and fructose diphosphate might function by displacing the ATP from the inhibitory site. It was shown in this work and also in that of Pogson and Randle (1966) that a factor in the inhibition by increased concentration of ATP was the increase in the K_m of the enzyme for fructose 6-phosphate. Further evidence for phosphofructokinase as a rate-limiting step came from the work of Randle et al. on glucose metabolism in rat heart and diaphragm (Newsholme and Randle, 1961; Garland and Randle, 1964). Glycolysis was accelerated by anaerobiosis, and in such conditions there was decrease in hexose monophosphate content and increase in fructose diphosphate content. Thus, it seemed that on change from aerobic to anaerobic conditions, phosphorylation of fructose monophosphate increased to a much greater degree than formation of the monophosphate or further breakdown of the diphosphate.

Very interesting interrelationships between glycolysis and the citric acid cycle were disclosed by the results of Randle and his collaborators (Newsholme et al., 1962; Garland and Randle, 1963; Garland et al., 1963). They were using hearts from diabetic or starved rats and here they found glycolysis and phosphofructokinase activity much slower than in normal hearts, although there was no consistent fall in AMP or inorganic phosphate. In such hearts, fatty acids are set free for oxidation and perfusion of the hearts with fatty acids, ketone bodies, or pyruvate was found to cause similar inhibition of lactic acid formation and of phosphofructokinase activity. Study of the effects of members of the citric acid cycle revealed a specific inhibitory action of citrate on phosphofructokinase. At the same time, this observation was also made by Passonneau and Lowry (1963) using liver and brain and by Parmeggiani and Bowman (1963) using diabetic hearts. Passonneau and Lowry (1964) described phosphofructokinase as the valve controlling the flow of fructose 6-phosphate to pyruvate. In aerobic conditions, the flow of pyruvate into the citrate cycle is kept low; in anoxia, increase in inorganic phosphate and AMP can open the valve to the full.

Binding of a number of metabolites to phosphofructokinase from skeletal muscle has been studied by Kemp and Krebs (1967). They also observed that increased pH, or increased concentration of AMP, ammonium ions, or phosphate ions can all heighten the affinity of the kinase for fructose 6-phosphate.

That large fall in pH can limit glycolysis is an old observation. Ui (1966) has recently found that remarkably small pH fall still within the alkaline range can bring about phosphofructokinase inhibition in the presence of 3 mM or higher ATP. This may have important consequences; since any glucose 6-phosphate accumulating during phosphofructokinase inhibition would inhibit hexokinase (see below), any slight pH rise relieving this inhibition could have a very marked effect on glucose breakdown. Table V summarizes some of these effects.

C. Control of Glucose Metabolism in Muscle

It has long been known that entry of glucose from the blood stream into the cell is the rate-limiting factor in utilization of glucose by muscle and that administration of insulin hastened this penetration. The mechanism of this facilitation was not understood; a usual explanation was that insulin, by increasing the rate of glucose 6-phosphate formation in the tissue, decreased glucose concentration there and so created a

TABLE V

SOME FACTORS AFFECTING ACTIVITY OF PHOSPHOFRUCTOKINASE[a]

Inhibitors	Activators and deinhibitors
ATP—this inhibition is pH-dependent, being greater at lower pH values	ADP, AMP, cyclic AMP, and inorganic phosphate overcome ATP inhibition
Citrate	Increased pH, ammonium ions, and phosphate ions raise the affinity of the enzyme for fructose 6-phosphate
	Fructose diphosphate
	Sulfate ions

[a] Needham, 1971.

favorable gradient. It was not until some fifteen years after the discovery of insulin that a well grounded explanation was afforded by Lundsgaard (1939). He showed that the intracellular glucose concentration of normal muscle was extremely low and therefore that the effect of insulin in causing a threefold increase in rate of penetration could not be due to increased concentration gradient between blood and tissue. He suggested that insulin had a direct effect on an active process in the cell membrane responsible for the glucose transfer. This work published in an Uppsala medical journal remained unnoticed, and the discovery had to be made again by Levine and her collaborators (see Levine and Goldstein, 1955, for references) nearly ten years later. They were concerned with entry of nonmetabolizable sugars, but Park et al. (1956) could later show a similar action by insulin on membrane transport using glucose itself. The results of Randle and his collaborators (Randle and Smith, 1958; Morgan et al., 1959) must be remembered here. In perfused heart and isolated rat diaphragm, either anoxia or uncouplers of oxidative phosphorylation could activate membrane transport of sugars; there was no damage to the membrane. Thus, availability of high-energy phosphate seemed to impede glucose entry into the cell, and the effect of insulin could be to overcome this restraint. Morgan and his collaborators (Morgan et al., 1961; Post et al., 1961) showed that inward passage of glucose as a function of glucose concentration outside conformed to Michaelis–Menten kinetics. Thus, its transport could be visualized as the first enzymic reaction in glucose metabolism. After addition of insulin, phosphorylation of glucose by hexokinase is the limiting step (see, e.g., Randle and Smith, 1958). Crane and Sols (1953) using skeletal muscle extracts, had shown the inhibitory effect of the product glucose 6-phosphate on hexokinase activity.

D. Regulation of Carbohydrate Breakdown in Anaerobic Contraction

The recent investigations of the changes going on in frog muscle contracting under strictly anaerobic conditions and the relation of these to regulation of glycogen breakdown are of great interest (Danforth et al., 1962). Only some 5% of the phosphorylase is present in resting frog muscle in the a form; this percentage rises rapidly to about 50 on 1 sec stimulation to isometric contraction, to more than 80 on 2.5 sec stimulation. Return to the resting value had taken place within about 1 min after the stimulation ceased. Danforth and Helmreich (1964) used isotonically contracting sartorius; here the phosphorylase a content increased with frequency of stimulation up to about 8 shocks per second. Changes in the amount of work done had no effect on the activity of this enzyme. A striking difference is perceived in these experiments between the effect of adrenaline treatment and of stimulation, the increase in phosphorylase a activity being much greater and more rapid in the latter case. Cyclic AMP could be detected within 60 sec of the adrenaline administration. It is important to remember that Posner *et al.* (1965) could find no increase in cyclic AMP on electrical stimulation of rat or frog muscle. Drummond *et al.* (1969) confirmed this and also showed that the phosphorylase b to a conversion took place without any detectable change in activity of phosphorylase b kinase. Thus, the mechanism of activation of phosphorylase in contraction seems basically different from that in which adrenaline is involved.

In a study of lactic acid formation under anaerobic conditions, Karpatkin *et al.* (1964) found this proportional, during a 30 min period, to the rate of stimulation up to a rate of 48 stimuli per minute. The increase in rate was more than one hundred times. At stimulation frequencies up to about 24 shocks per minute, there was no rise in hexose monophosphate; above this rate, this ester began to rise in concentration, no doubt due to limited action of phosphofructokinase. The changes found in AMP and ATP over this range of stimulation rate were also examined, but bore little relevance to what is known of their effects on glycolysis (Helmreich and Cori, 1965). This may well be explained by their compartmentation in the cell. Helmreich and Cori brought up the possibility that release of calcium on stimulation might, besides activating actomyosin ATPase, also activate the two key enzymes phosphorylase and phosphofructokinase. However, in recent tests Vaughan and Newsholme (1970) could find no effect of calcium ions (10^{-6} to 10^{-9} M) on the activities of phosphofructokinase or hexokinase in a variety of muscles.

E. Synthesis of Glycogen and Glucose in Muscle

In the early work on phosphorylase, it became clear that the equilibrium point with the enzyme was very unfavorable for glycogen synthesis under conditions *in vivo* (see, e.g., Sutherland and Cori, 1951). Then it was discovered by Leloir and Cardini (1957) that a different mechanism was involved in this synthesis. Leloir and his collaborators (Caputto *et al.*, 1949) had earlier isolated from yeast and mammalian tissues the compound uridine diphosphate glucose (UDPG) in which uridine 5-phosphate and glucose 1-phosphate were connected by a pyrophosphate bridge. It was now discovered that UDPG can act, in presence of a liver enzyme and with a trace of glycogen as primer, as a glucose donor in glycogen synthesis; UDP was formed at the same time. Villar-Palasi and Larner (1958, 1960), using rat diaphragm and skeletal muscle, then studied two enzymic reactions:

$$\text{UTP + glucose 1-phosphate} \rightleftharpoons \text{UDPG + pyrophosphate} \qquad (5)$$

$$\text{UDPG + glycogen} \rightleftharpoons \text{[glycogen + one glucose residue] + UDP} \qquad (6)$$

They showed that with this system under physiological conditions glycogen synthesis was greatly favored.

It soon became apparent that control mechanisms were present for this synthesis, as well as for glycogen breakdown. Thus, Villar-Palasi and Larner (1961) found that incubation of rat diaphragm with insulin led to stimulation of the glycogen synthetase in the extracts by about 35%, indeed to a level near that of the maximum rate. If glucose 6-phosphate was present when the enzyme was tested, its activity was the same whether the extracts were made from diaphragms incubated with or without insulin. Friedman and Larner (1962, 1963) went on to find that two types of the synthetase activity could be distinguished—one type, D, is dependent on presence of glucose 6-phosphate; the other, I, is independent. In presence of ATP and magnesium ions, type I decreased, while the total activity remained constant. With the help of ^{32}P-labeled ATP, two reactions were demonstrated:

$$\text{Synthetase I} + n \text{ ATP} \rightarrow \text{synthetase D} + n \text{ ADP} \qquad (7)$$

$$\text{Synthetase D} \rightarrow \text{synthetase I} + n \text{ inorganic phosphate} \qquad (8)$$

Thus, with the synthetase as with the phosphorylase, one form of the enzyme is phosphorylated, but in the case of the synthetase, the phosphorylated form is the one dependent on the effector. The effect of adrenaline is to cause a decrease in synthetase activity (Belocopitow, 1961), and Rosell-Perez and Larner (1964) showed that cyclic AMP enhanced phosphorylation of the synthetase.

Another resemblance between the systems concerned in control of the phosphorylase and the synthetase is seen in the effect of calcium ions. Incubation of the synthetase with Ca^{2+} in presence of a protein factor caused increased conversion of the I to the D form. The same protein factor was concerned (Belocopitow *et al.*, 1965; Appleman *et al.*, 1965, 1966) here as that operative in the activation of phosphorylase b kinase in presence of calcium ions. Table VI shows these relationships and brings out the similarity of conditions for inhibition of the synthetase (I → D conversion) and for activation of the phosphorylase b kinase.

A still further regulation method appears in the results of Danforth (1965) who showed that the I form of the synthetase increased

TABLE VI

PROCESSES CONCERNED IN ACTIVATION OF GLYCOGEN
SYNTHETASE AND PHOSPHORYLASE[a]

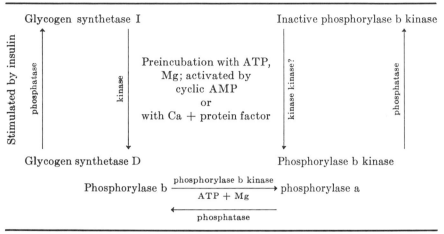

when the glycogen content of rat diaphragms was lowered. Villar-Palasi
and Larner (1966) then showed with muscle extract that the phosphatase converting the D into the I form was inhibited by physiological
concentrations of glycogen. They also found (as had E. G. Krebs *et
al.*, 1964) that glycogen stimulates phosphorylase b kinase. So glycogen
can act in a double feedback mechanism controlling its own breakdown
and synthesis.

Piras *et al.* (1968) showed that the dependent form of the synthetase
can be markedly inhibited by certain metabolites, e.g., ATP, ADP, inorganic phosphate. After a detailed study of glycogen resynthesis *in
vitro* and *in vivo*, Piras and Staneloni (1969) concluded that the regulation of synthetase activity during contraction and recovery can be explained by interconversion of the I and D forms together with the differential action of effectors on the two forms of the enzyme.

Synthesis of carbohydrate from lactic acid does not take place in
mammalian muscle, but this process was well attested for frog muscle
by Meyerhof and his collaborators in the 1920s (see, e.g., Meyerhof
and Lohmann, 1926). The actual pathway of synthesis was at that time
obscure, but H. A. Krebs (1964) recently emphasized that among the
reactions known to occur in glycolysis, three present energy barriers
to reversal. These are the formation of glucose 1-phosphate from glycogen, formation of fructose monophosphate from the diphosphate, and
the formation of phosphopyruvate from pyruvate (see Table I). The

uridine diphosphate pathway can overcome the first difficulty, presence of a specific fructose diphosphatase would be needed to surmount the second, while the third could be dealt with by the following reactions which have been studied by Utter and Keech (1960, 1963; Keech and Utter, 1963):

Pyruvate carboxylase:

$$\text{Pyruvate} + CO_2 + \text{ATP} \rightleftharpoons \text{oxaloacetate} + \text{ADP} + H_3PO_4 \qquad (9)$$

Phosphopyruvate carboxykinase:

$$\text{Oxaloacetate} + \text{GTP} \rightleftharpoons \text{phosphopyruvate} + CO_2 + \text{GDP} \qquad (10)$$

H. A. Krebs and Woodford (1965) found fructosediphosphatase in many kinds of skeletal muscle, but since Keech and Utter had found little or no phosphopyruvate carboxykinase and very little pyruvate carboxylase in skeletal muscle, Krebs and Woodford assumed that carbohydrate synthesis in muscle could only take place from phosphorylated three-carbon intermediates (such as 2-glycerophosphate) and not from lactate or pyruvate.

But the recent work of Bendall and Taylor (1970) has upheld Meyerhof's findings. With much improved methods they could show in frog sartorii oxidation quotients

$$\frac{\text{molecules of lactic acid} \rightarrow \text{glycogen}}{\text{molecules of lactic acid oxidized}}$$

comparable to Meyerhof's—about 6. They point out that although Opie and Newsholme (1967) had found very low pyruvate carboxylase even in frog muscle, this difficulty might be overcome by the presence of the "malic enzyme" (Ochoa *et al.*, 1948), known to be widely distributed. Malate would thus be provided and would then be oxidized to oxalacetate:

$$CH_3COCOOH + H_2CO_3 \underset{\text{NADP}}{\overset{\text{NADPH}}{\rightleftharpoons}} COOHCH_2CHOHCOOH \qquad (11)$$

$$COOHCH_2CHOHCOOH \underset{\text{NADH}}{\overset{\text{NAD}}{\rightleftharpoons}} COOHCH_2COCOOH \qquad (12)$$

Phosphopyruvate carboxykinase was found in all the muscles tested by Opie and Newsholme except the heart. For the synthesis of carbohydrate to occur in the muscle, conditions must be provided permitting the production of pyruvate from lactate by the action of the lactate dehydrogenase. This means that high pH and high oxygen tension are requisite.

IV. Relaxation

A. *Relaxing Factor*

Our consideration so far of the contractile machinery has led to the conclusion that high ATP content and low ATPase activity are characteristic of the resting muscle. If this is accepted, the question arises in acute form, what keeps the unstimulated muscle in the relaxed condition? We have already discussed the results of Marsh, with their demonstration of the presence in muscle extract of a factor, the "relaxing factor," concerned in inhibition of the fiber ATPase and therefore of contraction. This factor was heat labile and nondialyzable; its action was completely prevented by presence of 0.002 M Ca^{2+}.

Further work on this factor quickly followed in several laboratories. Bendall (1952, 1953) and Hasselbach and H. H. Weber (1953) confirmed its inhibitory effects, as well as the antagonistic effect of calcium ions using as test system the ATPase activity and tension production of glycerinated fiber bundles. Hasselbach and Weber emphasized that in experiments on ATPase and superprecipitation with actomyosin preparations, the effect of the factor could be seen as a sensitization to the inhibitory effect of overoptimal ATP concentration. Presence of a low ATP concentration, however, was essential for its effect; Bendall also found both ATP and Mg^{2+} (1–8 mM) to be necessary for the factor action.

At this time, a number of suggestions were made that the role of the relaxing factor was to facilitate rephosphorylation of ADP, and it was variously proposed that the important element in it might be creatine phosphokinase (Goodall and Szent-Györgyi, 1953; Lorand, 1953), or pyruvate phosphokinase (Lorand, 1953), or myokinase (Bendall, 1954). Relaxing effects with these enzyme systems in presence of their substrates were indeed found.

However, it became evident that something more was concerned than providing within the fiber a concentration of ATP supraoptimal for ATPase activity. Kumagai *et al.* (1955) prepared two fractions from muscle extract and showed that both were involved in relaxation. Fraction A was precipitated by 10–20 gm of ammonium sulfate per 100 ml; the other, fraction B, by 30–40 gm per 100 ml. Either alone gave a fairly good relaxing effect in the presence of 5 mM ATP and 10 mM Mg^{2+} on relatively fresh single fibers. But with fibers preserved in 50%

glycerol for weeks and then exhaustively washed, addition of both factors was necessary. Fraction B contained myokinase and creatine phosphokinase, and fraction A plus myokinase gave relaxation with moderately washed fibers. Nevertheless, something more seemed to be contained in fraction B, since A and B together caused relaxation with old well washed fibers for which the combination A plus myokinase was ineffective. They suggested that the fibers used by previous workers had all contained enough of A and B to give relaxation under their conditions.

The situation became clearer with the contributions of Portzehl (1957a) and Briggs and Portzehl (1957). These workers concluded from their observations on the ATPase activity and the degree of shortening of glycerinated myofibrils (not more than 2 μ in diameter and very thoroughly washed) that the ATP-providing enzymic systems described above cannot be equated with the Marsh factor, for under conditions where such fibrils shortened and hydrolyzed ATP unaffected by the presence of creatine phosphokinase plus phosphocreatine, addition of calcium-free dialyzed muscle extract (prepared by homogenizing the muscle in media containing 5 mM potassium oxalate) could abolish both phenomena (see Fig. 3). The relaxing effect of myokinase and of the other enzymic systems can be explained by the use in their testing of fibers still containing traces of Marsh factor, and so thick that in the interior of the bundle the ATP concentration was zero, or at any rate so low, that even in the presence of Marsh factor (which itself required ATP for its effectiveness) it was inadequate for relaxation. In these circumstances, addition of a system that provided for ADP rephosphorylation in the center of the bundle could elicit relaxation.

Portzehl (1957b) found that the whole of the factor activity is contained in the particulate fraction of the muscle extract. This would be in agreement with the findings of Kumagai et al. (1955), since the cell particles are readily precipitated by ammonium sulfate addition. The Japanese workers indeed had pointed out that their fraction A had a high lipid content and ATPase activity and might well contain the Kielley–Meyerhof ATPase believed to be associated with the microsomal fraction of the muscle cell (Kielley and Meyerhof, 1948). Later Ebashi (1958a) showed that granules prepared by centrifuging between 11,000 and 20,000 g had factor activity; so also had an ATPase preparation made by the Kielley–Meyerhof method. However, he showed that the relaxing activity could be eliminated by appropriate treatment. Portzehl's results were confirmed by Bendall (1958), who used washed granules centrifuged down between 18,000 and 80,000 g. With the centrifugal method of Marsh (1952), he compared their effectiveness with the ineffectiveness of myokinase on the synaeresis of washed myofibrils.

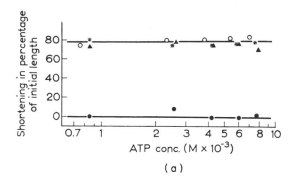

(a)

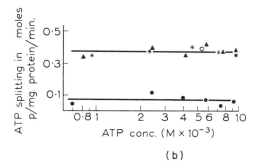

(b)

Fig. 3. (a) Influence of the relaxing factor and of the phosphocreatine–creatine phosphokinase system on the dependence of fibrillar contraction on the ATP concentration. Ionic strength about 0.17; pH about 6.3; oxalate about 4×10^{-3} M; Mg^{2+} 2×10^{-3} M; fibril concentration 4 mg/ml. Symbols: $\blacktriangle$ = with ATP; $\ast$ = with ATP + 0.01 M phosphocreatine + 3–13 mg creatine phosphokinase per milliliter; $\bullet$ = with ATP + relaxing factor; $\bigcirc$ = with ATP + relaxing factor + 4.5 $\times$ 10^{-3} M Ca^{2+}. (b) Influence of the relaxing factor and of the phosphocreatine–creatine phosphokinase system on the dependence of ATP splitting on ATP concentration. Conditions and symbols as in (a) except that pH was 7.0 and the creatine phosphokinase concentration was 0.03–10 mg/ml. (Portzehl, 1957a).

B. The Calcium Pump

The time was now almost ripe for the discovery of the calcium pump. A. Weber (1959), who had evidence for a part played by traces of Ca^{2+} as well as by Mg^{2+} in activating actomyosin ATPase, suggested that the relaxing factor might function by binding calcium. Then Ebashi (1960) arguing from his own results on chelators and from the results of Weber, as well as from his work with Lipmann (not published till two years later), came to the conclusion that the particulate fraction of muscle concentrates calcium in an ATP-coupled reaction and thereby removes it from the medium. Hasselbach and Makinose (1961), using tests on myofibrillar ATPase, showed that the inhibition of the relaxing factor activity by Ca^{2+} was reversible if small amounts of calcium were used and could be repeated several times. The abolition of the Ca^{2+} activation depended on removal of the Ca^{2+} from the medium. In the presence of 5 mM oxalate, its storage within the particles as the crystalline salt could be seen. One liter of granules could store as much as 0.6 moles of calcium. This calcium uptake was accompanied by great increase in ATPase activity of the particles; poisoning by Salyrgan caused both inhibition of the ATPase and suspension of the calcium uptake. The term "calcium pump" could now be used. Hasselbach and Makinose (1962) went on to find that the extra ATP hydrolysis (see Fig. 4) stimulated by calcium (unlike the basic ATP hydrolysis) was

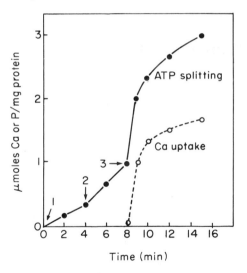

Fig. 4. ATP-splitting and calcium uptake. Preparation: fractionated vesicles from rabbit skeletal muscle 40 hr after isolation. Before the addition of calcium, ATP is split by the basic ATPase (1) (2.10^{-4} M EGTA present). After the addition of calcium (2) (1.2×10^{-4} M), a small amount of calcium is taken up (0.1 μ moles/mg protein) reducing the calcium concentration in the solution by 5%, and the rate of ATP splitting increases by 0.1 μ moles phosphorus per milligram protein per minute. According to the transport ratio, this increase corresponds to a rate of calcium uptake of 0.1 to 0.2 μ moles calcium per milligram protein per minute, which may compensate for a calcium efflux of the same magnitude. Under comparable conditions similar efflux rates have been observed by A. Weber et al., 1964. On the addition of oxalate (3) 5×10^{-3} M, the concentration of calcium inside the vesicles is reduced, calcium is taken up with a high rate, and at the same time the extra ATPase is maximally activated. (Hasselbach, 1966).

accompanied by an ATP-ADP exchange reaction, and they proposed a pumping mechanism based on current theories of active transport. This could involve phosphorylation of an unknown carrier on the outer surface of the vesicular membrane, with the result that the substance acquired greatly increased affinity for calcium. When the calcium complex diffused to the inner surface of the membrane, conditions for splitting off of phosphate had to be postulated, with consequent loss of calcium to the interior, whether or not a high calcium concentration was already present there. Ebashi and Lipmann (1962) also brought evidence for the enormous concentration of calcium in the particulate fraction centrifuged down between 10,000 and 38,000 g; this sequestration was dependent on ATPase activity of the particles and on an ATP–ADP exchange reaction carried out by them. Their electron micrographs showed that the material was made up (to the extent of about 80%) of spherical or flattened vesicles 60–200 mμ in diameter; these appeared to be healed fragments of the sarcoplasmic reticulum.

Strict correlation between the calcium uptake on the one hand and the ATPase activity and exchange reaction on the other was emphasized by Hasselbach and Makinose (1963) (see also Martonosi and Ferretos, 1964a,b). An external concentration of at least 10^{-8} M calcium was necessary. With concentrations greater than 10^{-7} M calcium, 2 moles of calcium were taken up per mole of ATP broken down; the ratio might be rather lower at calcium concentrations below this. The final calcium concentration inside the vesicles might be 500 times as great as the outside concentration. Martonosi and Ferretos also investigated an interesting phenomenon, that of calcium efflux, detected by means of exchange of stored calcium with ^{45}Ca in the medium. This efflux was seen in the absence of oxalate, phosphate, or pyrophosphate, no doubt because the presence of these ions, leading to precipitation of calcium inside the vesicles, kept the concentration there low. Makinose and Hasselbach (1965) also observed efflux of calcium when loaded vesicles were transferred to a medium of lower calcium content; inactivation of the pump, either by lack of ATP or by Salyrgan poisoning led to greatly increased efflux. Thus it seems that a slight activity of the pump is still needed after a steady state of filling is reached to compensate for this efflux by diffusion. A. Weber *et al.* (1966) did indeed find that in absence of any calcium-precipitating agent, the steady state of calcium concentration within depended on the calcium concentration outside; and that during this steady state of filling ATPase activity, about 15% of the initial rate, was still needed.

Hasselbach and Seraydarian (1966) showed that 10^5 gm of vesicular protein contained seven sulfhydryl equivalents of which three reacted readily with N-ethylmaleimide without impairment of calcium transport or of ATPase activity. Loss of the extra ATP splitting associated with calcium intake, as well as loss of calcium transport and storage, followed blockage of the other four sulfhydryl equivalents. The electron microscope observations of Hasselbach and Elfvin (1967) using the sulfhydryl reagent mercury phenyl azoferritin showed that the sulfhydryl groups were located on the outer surface of the transporting membrane.

The interesting observation was made by Balzer *et al.* (1968a,b) that the calcium-dependent ATPase could be inhibited in certain conditions where the calcium-dependent exchange reaction remained unaffected. This happened on treatment of the vesicular fragments with 3×10^{-5} M reserpine, prenylamine, or chlorpromazine. The ATPase was 50% inhibited and the calcium uptake and storage were diminished to the same degree. Thus, these drugs do not affect the membrane phosphorylation, but do interfere with hydrolysis of the phosphorylated membrane constituent.

C. Role of Calcium Ions and Their Removal

We must turn now to discuss in more detail the part played by calcium in MgATPase activity. As we have seen, the enzyme was well known to be inhibited by high ATP concentration. Perry and Grey (1956) had also made the interesting observation that MgATPase was inhibited when the metal chelator EDTA was added in a concentration only a small fraction of the concentration of Mg^{2+} present. They suggested that another metal in traces was necessary. A. Weber (1959) now suggested that free Ca^{2+} in very low concentration might be needed for optimal rate of breakdown of ATP by actomyosin at low ionic strength, and so for contraction. She pointed out that the results just mentioned and many others in the literature could be due to removal of calcium ions by ATP or EDTA. She now showed that, with 8 mM ATP or 0.05 mM EDTA in the medium, actomyosin ATPase was activated by 1–2 μM free magnesium, but at higher concentration, the magnesium inhibited unless free calcium was present. With a calcium ion concentration of 6×10^{-6} M, the rate of hydrolysis depended only on the concentration of $MgATP^{2-}$. She concluded that $MgATP^{2-}$ was the enzyme substrate and that the enzyme must be in equilibrium with a very low calcium ion concentration. In further work (A. Weber and Winicur, 1961), superprecipitation of actomyosin was used as a model of contraction, the actomyosin being made from the two very highly purified proteins from which the last traces of calcium had been removed. These showed low ATPase activity and superprecipitated very slowly unless 0.1 mM Ca^{2+} was added. A. Weber and Herz (1963), using [45]Ca, found that at least 1–2 μmoles of bound calcium per gram of myosin was necessary for superprecipitation. This bound calcium could be partly removed by treatment of the actomyosin with 4 mM MgATP in 0.1 M potassium chloride, and mostly removed by addition of 2 mM EGTA. Then ATPase activity and power to superprecipitate were lost. Ebashi (1960) had found that the relative chelating powers of a number of calcium-chelating agents were in good correlation with their relaxing effect on glycerinated muscle fibers. A. Weber et al. (1962, 1964) reported that the relaxing factor (0.1 mg N/ml) could lower the calcium content of myofibrils as effectively as did 2 mM EGTA, the best calcium-chelating agent known. There was a close parallelism between the lowering of the fibrillar calcium content and the loss of power to superprecipitate and split ATP.

Portzehl et al. (1964) injected calcium EGTA buffers into the large intact muscle fibers of the crab *Maia squinado*. For contraction, a concentration of calcium ions of 0.3–1.5 μM was needed, a result in good agreement with the threshold reported by Weber and her collaborators for ATPase activation and superprecipitation in actomyosin preparations. In the experiments of Podolsky and Costantin (1964), isolated frog fibers from which the sarcolemma had been dissected away were used; for contraction, a calcium ion concentration of 10^{-5} M was needed.

In the early work on the relaxing factor, difficulty was often felt in visualizing close contact between the particles and the fibrils. Evidence was brought forward from time to time that seemed to indicate participation of a soluble calcium-sensitive relaxing factor or of a soluble cofactor (see, e.g., Nagai et al., 1962; Briggs and Fuchs, 1960; Gergely et al., 1959). However, later work, such as that of Seidel and Gergely (1963) and of Seidel (1964) have given other explanations for such observations.

A. Weber et al. (1964) have calculated that the calcium storage capacity of the muscle is adequate, and Hasselbach (1964) calculated that there is a wide safety margin in the rate of sequestration as compared with the rate of relaxation. With regard to the rate of release of the calcium, Jöbsis and O'Connor (1966) have investigated this using toads previously injected intraperitoneally for several days with murexide solution. The course of calcium ion release in stimulated sartorius muscle could then be followed by an optical method depending on calcium murexide formation. The free calcium content of the muscle began to rise within 5 msec of stimulation; the concentration fell again while the tension was still high, and this was explained as probably due to its interaction with the contractile proteins and also its chelation with ATP.

Thus, it seems clear that the contraction is triggered by release of calcium ions from the sarcoplasmic reticulum, while relaxation depends on the active return of this calcium. The mechanism of the excitation–contraction coupling concerned in the calcium release will be considered elsewhere (Chapter 1).

It has become apparent in recent work that the mechanism of the effect of calcium ions in stimulating the magnesium-activated actomyosin ATPase is complicated; three proteins besides actin and myosin are involved, and there is still much to be elucidated (see Chapter 7).

V. Biochemical Effects on the Muscle Equipment of Repeated Stimulation, Denervation, and Cross Innervation

Before discussing these questions it is necessary to say something about the different types of skeletal muscles. Broadly speaking, these muscles can be divided into white and red types, the former contracting rapidly, the latter more slowly but usually capable of maintaining much greater tension. The work of the last twenty years has made clear in detail the great differences in metabolism between the white and red muscles—differences already glimpsed in the early years of this century. (See Needham, 1926, 1971, for reviews.)

Red muscle is characterized by its dependence on oxidative metabolism. R. Hill (1936) and Millikan (1936) described the special oxygen-combining properties of the red pigment, muscle hemoglobin, and made clear its function in providing an oxygen store. Lawrie (1952, 1953a) showed a correlation between muscle redness and activity of the cytochrome system, a correlation carried further by Chappell and Perry (1953) in their finding of the much greater mitochondrion content of red muscle. Lawrie had also pointed out the inverse relationship between myoglobin content and glycolytic activity as indicated by pH fall on contraction.

Histological examinations had early shown that the fibers of white muscles are usually larger and clearer than those of red muscles, the narrow fibers of the latter containing many granules. Very thorough histological studies by George and his collaborators (George and Jyoti, 1955a,b; George and Scaria, 1958; George and Naik, 1958a,b; George and Talesara, 1961) and by Ogata (1958a,b,c), bore out this mitochondrial arrangement. They showed the high succinic dehydrogenase,

NADH and NADPH diaphorase and cytochrome oxidase content of the red fibers, while the white fibers were distinguished by high glycogen content. The presence of fat globules and of lipase in the narrow fibers suggested that here fat oxidation was important in energy provision.

It must be realized that the same muscle may contain both types of fiber, sometimes well segregated, and also that intermediate types occur. Much further biochemical work has been done, often on carefully separated red and white fibers, which bears out the distinction between the two modes of energy provision. These studies are too numerous to give in detail here, but we may mention the finding by Ogata (1960) and by Pette and Bücher (1963) of the much greater activity in white muscle of several enzymes concerned in glycolysis; and the greater activity in red fibers of cytochrome oxidase, malic oxidase, and lactic, glutamic, and isocitric dehydrogenases shown by the work of George and Talesara (1961), Pette and Bücher (1963), and Domonkos and Latzkovits (1961).

A further point must be made here about the evidence for the higher state of development of the large clear fibers. Thus, Denny-Brown (1929) noted that in kittens at birth all the fibers were small and densely packed with granules, but at 2 weeks of age the gastrocnemius, a typical white muscle when adult had 20% of large clear fibers. Dubowitz (1963) later found that increase in size of the fibers in developing muscles of several animals ran parallel with increase in phosphorylase and decrease in certain oxidative enzymes. Close (1964) showed that the speed of shortening of the two kinds of muscle is about the same in early stages of development of the rat. Close and Hoh (1967) later found with the kitten that the intrinsic speed of the red soleus muscle did not change between the time of birth and adulthood. The white flexor digitalis longus had the same intrinsic speed as the soleus in the newborn kittens, but this increased during growth.

A. Effects of Stimulation on Enzyme Content

This question of enzymic adaptation in the muscle to contractile activity was controversial for a number of years, and it is only in fairly recent times that evidence has accumulated for its reality. Gutmann and his collaborators (Bass *et al.*, 1955; Vrbová and Gutmann, 1956) showed activation of glycogen synthesis as the result of some minutes of stimulation of rat muscles. 4 hours after the stimulation, the glycogen content was some 30% greater than in the unstimulated controls. They also found (Žak *et al.*, 1957; Gutmann and Žak, 1961a,b) that after a preliminary fall there was a rise during some hours in the RNA content and in the content of noncollagenous protein nitrogen. The increase in RNA (about 20%) preceded the rise in protein content which amounted to about 5%. Kendrick-Jones and Perry (1965, 1967) found, also with rats, that after 4–12 hours exercise there was an increase of 30–50% in creatine phosphokinase activity and also in aldolase activity. They tried the effect of certain inhibitors upon this increased activity of the kinase. Puromycin, inhibitor of protein synthesis, prevented the increase, as did also the amino acid analogs ethionine and *p*-fluorophenylalanine; but actinomycin D, 5-bromouracil, and 8-azaguanine, inhibitors of RNA synthesis, were ineffective. In all these cases the sartorii contracted with normal heat and tension development in presence of the inhibitors. These results suggest that the increased kinase activity is due to synthesis of new enzyme protein and that the messenger RNA controlling it is stable during the hours of the experiment.

The experiments of Hultman and Bergström (1967; Bergström and Hultman, 1966) on themselves are relevant here. Strenuous exercise was performed by one leg, the other being kept at rest. If a high carbohydrate diet was taken, biopsies showed a rapid glycogen synthesis in the exercised limb, the content rising to twice that in the control. The increased synthesis lasted several days. Then Lamb *et al.* (1969) found the supercompensation effect with the guinea pig in both red and white muscles; the glycogen content was about 75% higher after intensive training than in the resting controls. In the hearts of the trained animals, there was increased glycogen synthetase activity, and in the skeletal muscles, hexokinase activity was about twice the control value. Such changes were not found in the liver.

Holloszy (1967) also found effects of exercise (strenuous running on a treadmill) on several oxidative enzymes in rat gastrocnemius. The total mitochondrial protein was increased by about 60% over the controls; the mitochondria had high respiratory activity and oxidative phosphorylation was tightly coupled. This means, of course, increased capability to synthesize ATP. In connection with oxidative metabolism, we may remember the much earlier experiments of Lawrie (1953c). Long-continued training of rats led to increased myoglobin content of the muscles, and similarly, there was a difference in myoglobin content between the gastrocnemii of free-ranging and confined fowls.

Thus, it seems that conditions can be found in which stimulation to contraction leads to increased activity of certain enzymes; the indications are that in some cases, at any rate, there is synthesis of the enzyme protein. As we shall see, cutting off of the nerve supply may abolish these effects. Further work in this whole field will be of much interest.

B. Biochemical Effects of Denervation

The protein wastage in muscle after denervation is well known; E. Fischer and Ramsey (1945–1946) were among the first workers here. Weinstock *et al.* (1958) had seen increase in cathepsin activity in a number of wasting conditions, and Syrový *et al.* (1966) showed that in denervated muscle (the mixed muscles of the rat leg), this enzyme both in the free state and contained in lysosomes was increased, as was also peptidase. The effects are complicated, however, since certain muscles undergo hypertrophy after denervation (Bajusz, 1964).

The stimulation we have already described of glycogen and protein synthesis by contractile activity was found by Gutmann and Žak (1961a,b; Žak and Gutmann, 1960) not to take place in denervated muscle, although some increase in RNA was still produced. It has also been shown, mainly by histochemical studies, that denervation results in a general diminution in the enzymes concerned in energy supply, both in white and red fibers (see, e.g., Nachmias and Padykula, 1958; B. Smith, 1965).

In most of this work, it appears that in the changes after denervation the red and white muscles become much more alike. The changes are principally in the white muscle, and this loss of differentiation has naturally suggested that a return toward the fetal type is involved. However, Drahota and Gutmann (1963) found that the equalization in potassium content, which occurs after denervation, was brought about entirely by rise in the concentration in the red muscles, while the equalization in myoglobin content concerns fall in the one type and rise in the other type of muscle, as was shown by Maleknia *et al.* (1966).

C. Biochemical Effects of Cross Innervation

Very surprising results were reported by Buller *et al.* (1960); with kittens and cats, they performed operations whereby the nerve going to the fast digitorum longus was made instead to innervate the slow soleus and vice versa. After some weeks, the slow muscle tended in time course of its contraction toward the behavior of the fast muscle,

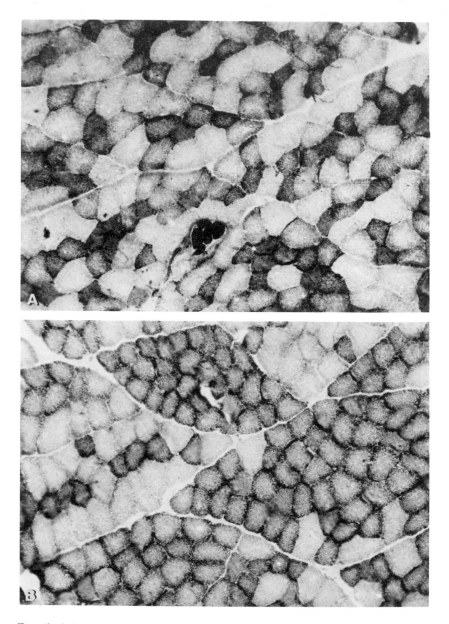

Fig. 5. (a) Normal flexor hallucis longus from adult cat, showing the distribution of fibers giving a strong and a weak reaction for NADH$_2$ diaphorase. (b) Cross-innervated flexor hallucis longus from adult cat, showing large areas with greatly increased proportion of fibers giving a strong reaction for NADH$_2$ diaphorase. (Unpublished photographs by courtesy of Dr. V. Dubowitz).

while changes in the opposite sense took place in the fast muscle. The indications seemed to be that some substance able to affect the muscle fibers passed down the axon and entered the muscle cell.

Biochemical investigations soon followed. Drahota and Gutmann (1963) showed that some months after denervation of the soleus and its reinnervation by the peroneal nerve, its characteristic low glycogen and potassium content had risen to that normal in the fast extensor digitorum. Romanul and van der Meulen (1966) confirmed the physiological results of cross innervation on rats and cats. On histological examination, reversal of the enzyme profiles characteristic of the fast and slow muscles was seen. Dubowitz (1967) had similar results with rats and rabbits, and Dubowitz and Newman (1967) with cats. (See also Fig. 5). Prewitt and Salafsky (1967), using young cats, found increased activity of certain glycolytic enzymes in the cross innervated soleus, while activity of malic and isocitric dehydrogenases decreased. Converse effects were seen in the cross innervated flexor digitorum longus.

The mechanism concerned in the biochemical effects of cross innervation is still a matter of conjecture. Guth and Watson (1967) have suggested that some trophic factor, possibly messenger RNA, is produced or transmitted by the nerve, the rate of this activity depending on the strength and frequency of the nerve impulses. Vrbová's results (1966) are relevant here. In rabbits, after spinalization and transection of the Achilles tendon, electrodes were implanted in the soleus. Stimulation was carried out 8–10 hours a day for 2–3 weeks. Stimulation at 40 impulses per second led to marked increase in the rate of rise to the contraction peak; stimulation at 10 impulses per second had no such effect.

Recent experiments of Korr *et al.* (1967) are concerned with the nature of substances that might be transmitted by the nerve. ^{32}P-labeled inorganic phosphate and ^{14}C-labeled amino acids were applied in rabbits to the tip of the floor of the fourth ventricle. After several days, the substances appeared in the muscle cells of the tongue, apparently after traveling down the hypoglossal nerve.

Sexton and Gersten (1967) have shown with glycerinated fibers from the soleus and gastrocnemius of the rat that there are significant differences in the contractile performance—the red fibers develop higher tension, but much more slowly than the white. *In vivo*, as we have seen, it appears that the preferred energy supply, whether glycolytic or oxidative, depends on the nerve supply. But with the glycerinated fibers, ATP is provided in the medium, so that any difference in its rate of resynthesis in the two muscle types is not relevant. Only the ATP–actomyosin system can be in play. Bárány *et al.* (1965) found similar concentrations of myosin and of actin in red and white muscle of the rabbit, but the ATPase activity of the white muscle actomyosin was two to three times as high as that of the red.

It seems, then, that some difference in the nature of the active centers and in the nature or number of the cross bridges must be involved. Recently, Buller *et al.* (1969), using cat muscles, showed that the myosin and myofibrillar ATPase fell in the cross innervated flexor digitorum longus, becoming nearly the same as that in the normal soleus. There was only slight increase in the ATPase activity of the cross innervated soleus.

VI. Conclusion

Much of this chapter has been concerned with work of the last ten years, in particular with two main themes—the elucidation of the relaxation mechanism and the regulation of energy supply for contraction. With regard to relaxation, a clear story depending on the sequestration of calcium ions has emerged. No doubt there is further knowledge still to be gathered, such as the details of the translocation process in the sarcoplasmic membrane; though here, too, recent work has shed considerable light (see Balzer *et al.*, 1968a,b).

When we consider the interrelated controls of carbohydrate breakdown and synthesis, integrated as these metabolic paths are with the needs of contraction and recovery, we find a truly amazing pattern. There are the extremely rapid changes in make-up of certain enzymes, leading to their altered activity, such changes being themselves the result of enzyme action. Then there are the effectors formed or removed during activity controlling the affinity of the key enzymes for their substrates or producing allosteric effects; slight changes in pH with far-reaching consequences; the action of hormones; and changes in membrane permeability. Progress has been made in the difficult task of following the play of certain of these controlling factors during actual contraction in the living muscle and interpretation of the results in the light of information gained from experiments *in vitro*. Much of this regulation could no doubt go on in a homogeneous medium, but it seems likely that compartmentation (of a more or less subtle kind) must be needed in the tissue.

The effect of prolonged and strenuous exercise in eliciting increased activity of certain enzymes important in energy provision is now well attested. This is a long-term process needing some hours, and there are indications that it depends on new enzyme formation.

Finally, we have touched on the challenging puzzle of the trophic influence of the nerve on muscle. Of the many aspects of this influence (see Gutmann and Hník, 1963), we have mentioned only a few examples concerned with enzymes supplying energy, and in particular, with effects on the fundamental actomyosin ATPase.

Much biochemical work on muscle is now concerned with these problems of metabolic regulation, both in its immediate and in its more prolonged manifestations; the next ten years may well bring a great increase in understanding.

REFERENCES

Appleman, M. M., Birnbaumer, L., Belocopitow, E., and Torres, H. N. (1965). *Fed. Proc., Fed. Amer. Soc. Exp. Biol.* 24, 537.

Appleman, M. M., Birnbaumer, L., and Torres, H. N. (1966). *Arch. Biochem. Biophys.* 116, 39.

Ashley, C. A., Avacimavicius, A., and Hass, G. M. (1956). *Exp. Cell Res.* 10, 1.

Baddiley, J., Michelson, A. M., and Todd, A. R. (1948). *Nature (London)* 161, 761.

Bailey, K. (1942). *Biochem. J.* 36, 121.

Bailey, K., and Perry, S. V. (1947). *Biochim. Biophys. Acta* 1, 506.

Bajusz, E. (1964). *Science* 145, 938.

Baldwin, E., and Yudkin, W. H. (1950). *Proc. Roy. Soc., Ser. B* 136, 614.

Balzer, H., Makinose, M., and Hasselbach, W. (1968a). *Naunyn-Schmiedebergs Arch. Pharmakol. Exp. Pathol.* 260, 444.

Balzer, H., Makinose, M., Fiehn, W., and Hasselbach, W. (1968b). *Naunyn-Schmiedebergs Arch. Pharmakol. Exp. Pathol.* 260, 456.

Banga, I. (1941–1942). *Stud. Inst. Med. Chem. Univ. Szeged* 1, 27.

Banga, I., and Szent-Györgyi, A. (1941–1942). *Stud. Inst. Med. Chem. Univ. Szeged* 1, 5.

Baranowski, T., Illingworth, B., Brown, D. H., and Cori, C. F. (1957). *Biochim. Biophys. Acta,* 25, 16.

Bárány, M., and Bárány, K. (1959). *Biochim. Biophys. Acta* 35, 293.

Bárány, M., Bárány, K., Reckard, T., and Volpe, A. (1965). *Arch. Biochem. Biophys.* 109, 185.

Baranyi, E. H., Edman, K. A. P., and Palis, A. (1951). *Acta Physiol. Scand.* 24, 361.

Bass, A., Gutmann, E., and Vodicka, Z. (1955). *Physiol. Bohemoslov.* 4, 267.

Belocopitow, E. (1961). *Arch. Biochem. Biophys.* **93**, 457.

Belocopitow, E., Appleman, M. M., and Torres, H. N. (1965). *J. Biol. Chem.* **240**, 3473.

Bendall, J. R. (1952). *Nature (London)* **170**, 1058.

Bendall, J. R. (1953). *J. Physiol. (London)* **121**, 232.

Bendall, J. R. (1954). *Proc. Roy. Soc., Ser. B* **142**, 409.

Bendall, J. R. (1958). *Nature (London)* **181**, 1188.

Bendall, J. R. (1961). *Biochem. J.* **81**, 520.

Bendall, J. R., and Taylor, A. A. (1970). *Biochem. J.* **118**, 887.

Bergström, J., and Hultman, E. (1966). *Nature (London)* **210**, 309.

Bowen, W. J., Stewart, L. C., and Martin, H. L. (1963). *J. Biol. Chem.* **238**, 2926.

Bozler, E. (1951). *Amer. J. Physiol.* **167**, 276.

Briggs, F. M., and Fuchs, F. (1960). *Biochim. Biophys. Acta* **42**, 519.

Briggs, F. N., and Portzehl, H. (1957). *Biochim. Biophys. Acta* **24**, 482.

Buchthal, F., Deutsch, A., Knappeis, C. G., and Munch-Petersen, A. (1947). *Acta Physiol. Scand.* **13**, 167.

Buller, A. J., Eccles, J. C., and Eccles, R. M. (1960). *J. Physiol. (London)* **150**, 417.

Buller, A. J., Mommaerts, W. F. H. M., and Seraydarian, K. (1969). *J. Physiol. (London)* **205**, 581.

Burton, K. (1955). *Biochem. J.* **59**, 44.

Caputto, R., Leloir, L. R., Trucco, R. E., Cardini, C. E., and Paladini, A. (1949). *J. Biol. Chem.* **179**, 497.

Chappell, J. B., and Perry, S. V. (1953). *Biochem. J.* **55**, 586.

Close, R. (1964). *J. Physiol. (London)* **173**, 74.

Close, R., and Hoh, F. J. Y. (1967). *J. Physiol. (London)* **192**, 815.

Cohn, M., and Cori, G. T. (1948). *J. Biol. Chem.* **175**, 89.

Colowick, S. P., and Kalckar, H. M. (1943). *J. Biol. Chem.* **148**, 117.

Cori, C. F. (1941). *In* "A Symposium on Respiratory Enzymes," p. 175. Univ. of Wisconsin Press, Madison.

Cori, C. F. (1956). *In* "Enzymes: Units of Biological Structure and Function" (O. H. Gaebler, ed.) Chapt. 27 Academic Press, New York.

Cori, C. F., and Cori, G. T. (1936). *Proc. Soc. Exp. Biol. Med.* **34**, 702.

Cori, C. F., and Illingworth, B. (1957). *Proc. Nat. Acad. Sci. U.S.* **43**, 547.

Cori, G. T., and Green, A. A. (1943). *J. Biol. Chem.* **151**, 31.

Cori, G. T., Colowick, S. P., and Cori, C. F. (1938). *J. Biol. Chem.* **123**, 381.

Cornblaeth, M., Randle, P. J., Parmeggiani, A., and Morgan, H. E. (1963). *J. Biol. Chem.* **238**, 1592.

Crane, R. K., and Sols, A. (1953). *J. Biol. Chem.* **203**, 273.

Dainty, M., Kleinzeller, A., Lawrence, A. S. C., Miall, M., Needham, J., Needham, D. M., and Shen, S-C. (1944). *J. Gen. Physiol.* **27**, 355.

Danforth, W. H. (1965). *J. Biol. Chem.* **240**, 588.

Danforth, W. H., and Helmreich, E. (1964). *J. Biol. Chem.* **239**, 3133.

Danforth, W. H., Helmreich, E., and Cori, C. F. (1962). *Proc. Nat. Acad. Sci. U.S.* **48**, 1191.

Davoren, P. R., and Sutherland, E. W. (1963). *J. Biol. Chem.* **238**, 3016.

Denny-Brown, D. E. (1929). *Proc. Roy. Soc., Ser. B* **104**, 371.

Domonkos, J., and Latzkovits, L. (1961). *Arch. Biochem. Biophys.* **95**, 138.

Drahota, Z., and Gutmann, E. (1963). *In* "The Effect of Use and Disuse on

Neuromuscular Function" (E. Gutmann and P. Hník, eds.), p. 143. Publ. House Czech. Acad. Sci., Prague.

Drummond, G. I., Valadares, J. R. E., and Duncan, L. (1964). *Proc. Soc. Exp. Biol. Med.* **117**, 307.

Drummond, G. I., Duncan, L., and Hertzman, E. (1966). *J. Biol. Chem.* **241**, 5899.

Drummond, G. I., Harwood, J. P., and Powell, C. A. (1969). *J. Biol. Chem.* **244**, 4235.

Dubowitz, V. (1963). *Nature (London)* **197**, 1215.

Dubowitz, V. (1967). *J. Physiol. (London)* **193**, 481.

Dubowitz, V., and Newman, D. L. (1967). *Nature (London)* **214**, 840.

Ebashi, S. (1958a). *Arch. Biochem. Biophys.* **76**, 410.

Ebashi, S. (1960). *J. Biochem. (Tokyo)* **48**, 150.

Ebashi, S., and Lipmann, F. (1962). *J. Cell Biol.* **14**, 389.

Edsall, J. T. (1930). *J. Biol. Chem.* **89**, 289.

Eggleton, G. P., and Eggleton, P. (1929–1930). *J. Physiol. (London)* **68**, 15.

Eggleton, P., and Eggleton, G. P. (1927a). *Biochem. J.* **21**, 190.

Eggleton, P., and Eggleton, G. P. (1927b). *J. Physiol. (London)* **63**, 155.

Engelhardt, W. A. (1946). *Advan. Enzymol.* **6**, 147.

Engelhardt, W. A., and Ljubimova, M. N. (1939). *Nature (London)* **144**, 668.

Engelhardt, W. A., Ljubimova, M. N., and Meitina, R. (1941). *Dokl. Akad. Nauk SSSR* **30**, 644.

Fischer, E., and Ramsey, V. W. (1945–1946). *Amer. J. Physiol.* **145**, 571.

Fischer, E. H., and Krebs, E. G. (1955). *J. Biol. Chem.* **216**, 121.

Fischer, E. H., Graves, D. J., Snyder, E. R., Crittenden, E. R. S., and Krebs, E. G. (1959). *J. Biol. Chem.* **234**, 1698.

Fiske, C. H., and Subbarow, Y. (1927). *Science* **65**, 401.

Fiske, C. H., and Subbarow, Y. (1929a). *J. Biol. Chem.* **81**, 629.

Fiske, C. H., and Subbarow, Y. (1929b). *Science* **70**, 381.

Fletcher, W. M., and Hopkins, F. G. (1907). *J. Physiol. (London)* **35**, 247.

Friedman, D. L., and Larner, J. (1962). *Fed. Proc., Fed. Amer. Soc. Exp. Biol.* **21**, 206.

Friedman, D. L., and Larner, J. (1963). *Biochemistry* **2**, 669.

Garland, P. B., and Randle, P. J. (1963). *Nature (London)* **199**, 381.

Garland, P. B., and Randle, P. J. (1964). *Biochem. J.* **93**, 678.

Garland, P. B., Randle, P. J., and Newsholme, E. A. (1963). *Nature (London)* **200**, 169.

George, J. C., and Jyoti, D. (1955a). *J. Anim. Morphol. Physiol.* **2**, 1.

George, J. C., and Jyoti, D. (1955b). *J. Anim. Morphol. Physiol.* **2**, 10.

George, J. C., and Naik, J. M. (1958a). *Nature (London)* **181**, 709.

George, J. C., and Naik, J. M. (1958b). *Nature (London)* **181**, 782.

George, J. C., and Scaria, K. S. (1958). *Nature (London)* **181**, 783.

George, J. C., and Talesara, C. L. (1961). *J. Cell. Comp. Physiol.* **58**, 253.

Gergely, J. (1956). *J. Biol. Chem.* **220**, 917.

Gergely, J., Kaldor, G., and Briggs, F. N. (1959). *Biochim. Biophys. Acta* **34**, 218.

Goodall, M. C., and Szent-Györgyi, A. G. (1953). *Nature (London)* **172**, 84.

Guth, L., and Watson, P. K. (1967). *Exp. Neurol.* **17**, 107.

Gutmann, E., and Hník, P., eds. (1963). "The Effect of Use and Disuse on Neuromuscular Functions." Publ. House Czech. Acad. Sci., Prague.

Gutmann, E., and Žak, R. (1961a). *Physiol. Bohemoslov.* **10**, 493.
Gutmann, E., and Žak, R. (1961b). *Physiol. Bohemoslov.* **10**, 501.
Hartree, W., and Hill, A. V. (1928). *Proc. Roy. Soc., Ser. B* **103**, 207.
Hasselbach, W. (1952). *Z. Naturforsch.* **7b**, 163.
Hasselbach, W. (1956). *Biochim. Biophys. Acta* **20**, 355.
Hasselbach, W. (1964). *Progr. Biophys. Mol. Biol.* **14**, 169.
Hasselbach, W. (1966). *Ann. N.Y. Acad. Sci.* **137**, 1041.
Hasselbach, W., and Elfvin, L.-G. (1967). *J. Ultrastruct. Res.* **17**, 598.
Hasselbach, W., and Makinose, M. (1961). *Biochem. Z.* **333**, 518.
Hasselbach, W., and Makinose, M. (1962). *Biochem. Biophys. Res. Commun.* **7**, 132.
Hasselbach, W., and Makinose, M. (1963). *Biochem. Z.* **339**, 94.
Hasselbach, W., and Seraydarian, K. (1966). *Biochem. Z* **345**, 159.
Hasselbach, W., and Weber, A. (1955). *Pharmacol. Rev.* **7**, 97.
Hasselbach, W., and Weber, H. H. (1953). *Biochim. Biophys. Acta* **11**, 160.
Heinz, E., and Holton, F. (1952). *Z. Naturforsch.* **7b**, 386.
Helmreich, E., and Cori, C. F. (1964). *Proc. Nat. Acad. Sci. U.S.* **51**, 131.
Helmreich, E., and Cori, C. F. (1965). *Advan. Enzyme Regul.* **3**, 91.
Hill, A. V. (1928a). *Proc. Roy. Soc., Ser. B* **103**, 163.
Hill, A. V. (1928b). *Proc. Roy. Soc., Ser. B* **103**, 183.
Hill, A. V., and Hartree, W. (1920). *J. Physiol. (London)* **54**, 84.
Hill, D. K. (1940a). *J. Physiol. (London)* **98**, 207.
Hill, D. K. (1940b). *J. Physiol. (London)* **98**, 460.
Hill, R. (1936). *Proc. Roy. Soc., Ser. B* **120**, 472.
Hobson, G. E., and Rees, K. R. (1955). *Biochem. J.* **61**, 459.
Hobson, G. E., and Rees, K. R. (1957). *Biochem. J.* **65**, 305.
Holloszy, J. O. (1967). *J. Biol. Chem.* **242**, 2278.
Huennekens, F. M., and Whiteley, H. R. (1960). *In* "Comparative Biochemistry" (M. Florkin and H. S. Mason, eds.) Vol. 1, p. 107. Academic Press, New York.
Hultman, E., and Bergström, J. (1967). *Acta Med. Scand.* **182**, 109.
Illingworth, B., Janszi, H. S., Brown, D. H., and Cori, C. F. (1958). *Proc. Nat. Acad. Sci. U.S.* **44**, 1180.
Jöbsis, F. F., and O'Connor, M. J. (1966). *Biochem. Biophys. Res. Commun.* **25**, 246.
Kalckar, H. M. (1941). *Chem. Rev.* **28**, 71.
Karpatkin, S., Helmreich, E., and Cori, C. F. (1964). *J. Biol. Chem.* **239**, 3139.
Keech, D. B., and Utter, M. F. (1963). *J. Biol. Chem.* **238**, 2609.
Keller, P. J., and Cori, G. T. (1953). *Biochim. Biophys. Acta* **12**, 235.
Kemp, R. G., and Krebs, E. G. (1967). *Biochemistry* **6**, 423.
Kendrick-Jones, J., and Perry, S. V. (1965). *Nature (London)* **208**, 1068.
Kendrick-Jones, J., and Perry, S. V. (1967). *Nature (London)* **213**, 406.
Kielley, W. W., and Meyerhof, O. (1948). *J. Biol. Chem.* **176**, 591.
Korr, I. M., Wilkinson, P. N., and Chornock, F. W. (1967). *Science* **155**, 342.
Krebs, E. G., and Fischer, E. H. (1955). *J. Biol. Chem.* **216**, 113.
Krebs, E. G., Kent, A. B., and Fischer, E. H. (1958). *J. Biol. Chem.* **231**, 73.
Krebs, E. G., Graves, D. J., and Fischer, E. H. (1959). *J. Biol. Chem.* **234**, 2867.
Krebs, E. G., Love, D. S., Bratwold, G. E., Trayser, K. A., Meyer, W. L., and Fischer, E. H. (1964). *Biochemistry* **3**, 1022.
Krebs, H. A. (1964). *Proc. Roy. Soc., Ser. B* **159**, 545.
Krebs, H. A., and Woodford, M. (1965). *Biochem. J.* **94**, 436.

Kühne, W. (1864). "Untersuchungen über das Protoplasma und die Contractilität." Engelmann, Leipzig.

Kumagai, H., Ebashi, S., and Takeda, F. (1955). *Nature (London)* **176**, 166.

Kuschinsky, G., and Turba, F. (1950). *Experientia* **6**, 103.

Lamb, D. R., Peter, J. B., Jeffress, R. N., and Wallace, A. H. (1969). *Amer. J. Physiol.* **217**, 1628.

Lardy, H. A., and Parks, R. E. (1956). *In* "Enzymes: Units of Biological Structure and Function" (O. H. Gaebler, ed.), p. 584. Academic Press, New York.

Lawrence, A. S. C., Needham, J., and Shen, S-C. (1944). *J. Gen. Physiol.* **27**, 201.

Lawrie, R. A. (1952). *Nature (London)* **170**, 122.

Lawrie, R. A. (1953a). *Biochem. J.* **55**, 298.

Lawrie, R. A. (1953b). *Biochem. J.* **55**, 305.

Lawrie, R. A. (1953c). *Nature (London)* **171**, 1069.

Lehmann, H. (1936). *Biochem. Z.* **286**, 336.

Lehnartz, E. (1931). *Klin. Wochenschr.* **10**, 27.

Leloir, L. F., and Cardini, C. E. (1957). *J. Amer. Chem. Soc.* **79**, 6340.

Levine, R., and Goldstein, M. S. (1955). *Progr. Horm. Res.* **11**, 343.

Levintov, L., and Meister, A. (1954). *J. Biol. Chem.* **209**, 265.

Levy, H. M., and Koshland, D. E. (1958). *J. Amer. Chem. Soc.* **80**, 3614.

Levy, H. M., and Koshland, D. E. (1959). *J. Biol. Chem.* **234**, 1102.

Lipmann, F. (1941). *Advan. Enzymol.* **1**, 99.

Lipmann, F., and Meyerhof, O. (1930). *Biochem. Z.* **227**, 84.

Lohmann, K. (1929). *Naturwissenschaften* **17**, 624.

Lohmann, K. (1931). *Biochem. Z.* **233**, 460.

Lohmann, K. (1932). *Biochem. Z.* **254**, 381.

Lohmann, K. (1934). *Biochem. Z.* **271**, 264.

Lohmann, K. (1935). *Biochem. Z.* **282**, 120.

Lorand, L. (1953). *Nature (London)* **172**, 1181.

Lowry, O. H., and Passonneau, J. V. (1966). *J. Biol. Chem.* **241**, 2268.

Lundsgaard, E. (1930a). *Biochem. Z.* **217**, 162.

Lundsgaard, E. (1930b). *Biochem. Z* **227**, 51.

Lundsgaard, E. (1931). *Biochem. Z.* **233**, 322.

Lundsgaard, E. (1939). *Upsala Läekarefoeren. Föerh.* **45**, 143.

Madsen, N. B., and Cori, C. F. (1957). *J. Biol. Chem.* **224**, 899.

Makinose, M., and Hasselbach, W. (1965). *Biochem. Z.* **343**, 360.

Maleknia, N., Ebersolt, E., Schapira, G., and Dreyfus, J. C. (1966). *Bull. Soc. Chim. Biol.* **48**, 905.

Mansour, T. E., Clague, M. E., and Beernink, K. D. (1962). *Fed. Proc., Fed. Amer. Soc. Exp. Biol.* **21**, 238.

Marsh, B. B. (1951). *Nature (London)* **167**, 1065.

Marsh, B. B. (1952). *Biochim. Biophys. Acta* **9**, 247.

Martonosi, A., and Ferretos, R. (1964a). *J. Biol. Chem.* **239**, 648.

Martonosi, A., and Ferretos, R. (1964b). *J. Biol. Chem.* **239**, 659.

Mayer, S. E., and Moran, N. C. (1960). *J. Pharmacol. Exp. Ther.* **129**, 271.

Meyer, W. L., Fischer, E. H., and Krebs, E. G. (1964). *Biochemistry* **3**, 1033.

Meyerhof, O. (1920). *Pfluegers Arch. Gesamte Physiol. Menschen Tiere* **182**, 232.

Meyerhof, O. (1921). *Pfluegers Arch. Gesamte Physiol. Menschen Tiere* **191**, 128.

Meyerhof, O. (1931). *Klin. Wochenschr.* **10**, 214.

Meyerhof, O., and Kiessling, W. (1935). *Biochem. Z.* **281**, 449.

Meyerhof, O., and Lehmann, H. (1935). *Naturwissenschaften* 23, 337.

Meyerhof, O., and Lohmann, K. (1926). *Biochem. Z.* 171, 421.

Meyerhof, O., and Lohmann, K. (1928a). *Biochem. Z.* 196, 22.

Meyerhof, O., and Lohmann, K. (1928b). *Biochem. Z.* 196, 49.

Meyerhof, O., and Lohmann, K. (1932). *Biochem. Z.* 253, 431.

Meyerhof, O., and Meier, R. (1924). *Biochem. Z.* 150, 233.

Meyerhof, O., and Schulz, W. (1927). *Pfluegers Arch. Gesamte Physiol. Menschen Tiere* 217, 547.

Meyerhof, O., and Schulz, W. (1935). *Biochem. Z.* 281, 292.

Meyerhof, O., McCullagh, R. D., and Schulz, W. (1930). *Pfluegers Arch. Gesamte Physiol. Menschen Tiere* 224, 230.

Meyerhof, O., Schulz, W., and Schuster, P. (1937). *Biochem. Z.* 293, 309.

Meyerhof, O., Ohlmeyer, P., and Möhle, W. (1938). *Biochem. Z.* 297, 113.

Millikan, G. A. (1936). *Proc. Roy. Soc., Ser. B* 120, 366.

Mommaerts, W. F. H. M. (1947). *J. Gen. Physiol.* 31, 361.

Mommaerts, W. F. H. M. (1956). *J. Gen. Physiol.* 39, 821.

Morgan, H. E., and Parmeggiani, A. (1964a). *J. Biol. Chem.* 239, 2435.

Morgan, H. E., and Parmeggiani, A. (1964b). *J. Biol. Chem.* 239, 2440.

Morgan, H. E., Randle, P. J., and Regen, D. M. (1959). *Biochem. J.* 73, 573.

Morgan, H. E., Henderson, M. J., Regen, D. M., and Park, C. R. (1961). *J. Biol. Chem.* 236, 253.

Murad, F., Chi, Y.-M., Rall, T. W., and Sutherland, E. W. (1962). *J. Biol. Chem.* 237, 1233.

Nachmansohn, D. (1928). *Biochem. Z.* 196, 73.

Nachmansohn, D. (1929). *Biochem. Z.* 208, 357.

Nachmias, V. T., and Padykula, H. A. (1958). *J. Biophys. Biochem. Cytol.* 4, 77.

Nagai, T., Uchida, K., and Yasuda, M. (1962). *Biochim. Biophys. Acta* 56, 205.

Needham, D. M. (1926). *Physiol. Rev.* 6, 1.

Needham, D. M. (1971). "Machina Carnis; The Biochemistry of Muscle in its Historical Development." Cambridge Univ. Press, London and New York.

Needham, D. M., and Pillai, R. K. (1937). *Biochem. J.* 31, 1837.

Needham, D. M., and van Heyningen, W. E. (1935). *Biochem. J.* 29, 2040.

Needham, D. M., Needham, J., Baldwin, E., and Yudkin, J. (1932). *Proc. Roy. Soc., Ser. B* 110, 260.

Needham, J., Shen, S-C., Needham, D. M., and Lawrence, A. S. C. (1941). *Nature (London)* 147, 766.

Negelein, E., and Brömel, H. (1939). *Biochem. Z.* 303, 132.

Newsholme, E. A., and Randle, P. J. (1961). *Biochem. J.* 80, 655.

Newsholme, E. A., Randle, P. J., and Manchester, K. L. (1962). *Nature (London)* 193, 270.

Ochoa, S., Mehler, A. H., and Kornberg, A. (1948). *J. Biol. Chem.* 174, 979.

Ogata, T. (1958a). *Acta Med. Okayama* 12, 216.

Ogata, T. (1958b). *Acta Med. Okayama* 12, 228.

Ogata, T. (1958c). *Acta Med. Okayama* 12, 233.

Ogata, T. (1960). *J. Biochem. (Tokyo)* 47, 726.

Opie, L. H., and Newsholme, E. A. (1967). *Biochem. J.* 103, 391.

Oster, G. (1948). *Chem. Rev.* 43, 319.

Ostern, P., Baranowski, T., and Reis, J. (1935). *Biochem. Z.* 279, 85.

Ozawa, E., Hosoi, K., and Ebashi, S. (1967). *J. Biochem. (Tokyo)* 61, 531.

Park, C. R., Post, R. L., Kalman, C. F., Wright, J. H., Johnson, L. H., and Morgan, H. E. (1956). *Ciba Found. Colloq. Endocrinol.* 9, 240.

Parmeggiani, A., and Bowman, R. H. (1963). *Biochem. Biophys. Res. Commun.* **12**, 268.
Parmeggiani, A., and Morgan, H. E. (1962). *Biochem. Biophys. Res. Commun.* **9**, 252.
Parnas, J. K., and Wagner, R. (1914). *Biochem. Z.* **61**, 387.
Parnas, J. K., Ostern, P., and Mann, T. (1934). *Biochem. Z.* **272**, 64.
Passonneau, J. V., and Lowry, O. H. (1962). *Biochem. Biophys. Res. Commun.* **7**, 10.
Passonneau, J. V., and Lowry, O. H. (1963). *Biochem. Biophys. Res. Commun.* **13**, 372.
Passonneau, J. V., and Lowry, O. H. (1964). *Advan. Enzyme Regul.* **2**, 265.
Perry, S. V., and Grey, T. C. (1956). *Biochem. J.* **64**, 5P.
Pette, D., and Bücher, T. (1963). *Hoppe-Seyler's Z. Physiol. Chem.* **331**, 180
Piras, R., and Staneloni, R. (1969). *Biochemistry* **8**, 2153.
Piras, R., Rothman, L. B., and Cabib, E. (1968). *Biochemistry* **7**, 56.
Podolsky, R. J., and Costantin, L. L. (1964). *Fed. Proc., Fed. Amer. Soc. Exp. Biol.* **23**, 933.
Podolsky, R. J., and Morales, M. F. (1956). *J. Biol. Chem.* **218**, 945.
Pogson, C. I., and Randle, P. J. (1966). *Biochem. J.* **100**, 683.
Portzehl, H. (1951). *Z. Naturforsch.* **6b**, 355.
Portzehl, H. (1952). *Z. Naturforsch.* **7b**, 1.
Portzehl, H. (1957a). *Biochim. Biophys. Acta* **24**, 474.
Portzehl, H. (1957b). *Biochim. Biophys. Acta* **26**, 373.
Portzehl, H., Caldwell, P. C., and Rüegg, J. C. (1964). *Biochim. Biophys. Acta* **79**, 581.
Posner, J. B., Stern, R., and Krebs, E. G. (1965). *J. Biol. Chem.* **240**, 982.
Post, R. L., Morgan, H. E., and Park, C. R. (1961). *J. Biol. Chem.* **236**, 269.
Prewitt, M. A., and Salafsky, B. (1967). *Amer. J. Physiol.* **213**, 295.
Racker, E., and Krimsky, I. (1952). *J. Biol. Chem.* **198**, 731.
Rall, T. W., and Sutherland, E. W. (1957). *J. Biol. Chem.* **232**, 1065.
Rall, T. W., and Sutherland, E. W. (1962). *J. Biol. Chem.* **237**, 1228.
Rall, T. W., Sutherland, E. W., and Berthet, J. (1957). *J. Biol. Chem.* **224**, 463.
Randle, P. J., and Smith, G. H. (1958). *Biochem. J.* **70**, 501.
Robbins, E. A., and Boyer, P. D. (1957). *J. Biol. Chem.* **224**, 121.
Robin, Y., and Thoai, van N. (1962). *Biochim. Biophys. Acta* **63**, 481.
Roche, J., Robin, Y., di Jeso, F., and Thoai, van N. (1962). *C. R. Soc. Biol.* **156**, 830.
Romanul, C. F. A., and van der Meulen, J. P. (1966). *Nature (London)* **212**, 1369.
Rosell-Perez, M., and Larner, J. (1964). *Biochemistry* **3**, 81.
Sartorelli, L., Fromm, H. J., Benson, R. W., and Boyer, P. D. (1966). *Biochemistry* **5**, 2877.
Segal, H. L., and Boyer, P. D. (1953). *J. Biol. Chem.* **204**, 265.
Seidel, J. C. (1964). *Fed. Proc., Fed. Amer. Soc. Exp. Biol.* **23**, 901.
Seidel, J. C., and Gergely, J. (1963). *J. Biol. Chem.* **238**, 3648.
Sexton, A. W., and Gersten, J. W. (1967). *Science* **157**, 199.
Singer, T. P., and Barron, E. S. (1944). *Proc. Soc. Exp. Biol. Med.* **56**, 120.
Smith, B. (1965). *J. Neurol., Neurosurg. Psychiat.* **28**, 99.
Smith, E. C. (1933–1934). *Proc. Roy. Soc., Ser. B* **114**, 494.

Spicer, S. S. (1951). *J. Biol. Chem.* **190**, 257.

Stracher, A. (1964). *J. Biol. Chem.* **259**, 1118.

Stracher, A. (1965). *In* "Muscle" (W. M. Paul *et al.*, eds.), p. 83. Pergamon, Oxford.

Straub, F. B. (1942). *Stud. Inst. Med. Chem. Univ. Szeged* **2**, 3.

Straub, F. B. (1943). *Stud. Inst. Med. Chem. Univ. Szeged* **3**, 38.

Sutherland, E. W., and Cori, C. F. (1951). *J. Biol. Chem.* **188**, 531.

Sutherland, E. W., and Robison, G. A. (1966). *Pharmacol. Rev.* **18**, 145.

Sutherland, E. W., and Wosilait, W. D. (1955). *Nature (London)* **175**, 169.

Syrový, I., Hájek, I., and Gutmann, E. (1966). *Physiol. Bohemslov.* **15**, 7.

Szent-Györgyi, A. (1941–1942). *Stud. Inst. Med. Chem. Univ. Szeged* **1**, 17.

Szent-Györgyi, A. (1949). *Biol. Bull.* **96**, 140.

Taylor, E. W., Lymn, R. W., and Moll, G. (1970). *Biochemistry* **9**, 2984.

Thoai, van N., and Robin, Y. (1954). *Biochim. Biophys. Acta* **14**, 76.

Thoai, van N., Roche, J., Robin Y., and Thiem, N. V. (1953). *C. R. Soc. Biol.* **147**, 1241.

Thoai, van N., di Jeso, F., and Robin, Y. (1963). *C. R. Acad. Sci.* **256**, 4525.

Tonomura, Y., Nakamura, H., Kinoshita, N., Onishi, H., and Shizakawa, M. (1969). *J. Biochem. (Tokyo)* **66**, 599.

Turba, F., and Kuschinsky, G. (1952). *Biochim. Biophys. Acta* **8**, 76.

Ui, M. (1966). *Biochim. Biophys. Acta* **124**, 310.

Ulbrecht, G., and Ulbrecht, M. (1953). *Biochim. Biophys. Acta* **11**, 138.

Utter, M. F., and Keech, D. B. (1960). *J. Biol. Chem.* **235**, PC17.

Utter, M. F., and Keech, D. B. (1963). *J. Biol. Chem.* **238**, 2603.

Vaughan, H., and Newsholme, E. A. (1970). *Biochem. J.* **114**, 81P.

Villar-Palasi, C., and Larner, J. (1958). *Biochim. Biophys. Acta* **30**, 449.

Villar-Palasi, C., and Larner, J. (1960). *Arch. Biochem. Biophys.* **86**, 270.

Villar-Palasi, C., and Larner, J. (1961). *Arch. Biochem. Biophys.* **94**, 436.

Villar-Palasi, C., and Larner, J. (1966). *Fed. Proc., Fed. Amer. Soc. Exp. Biol.* **25**, 583.

von Muralt, A. L., and Edsall, J. T. (1930). *J. Biol. Chem.* **89**, 315.

Vrbová, G. (1966). *J. Physiol. (London)* **185**, 17.

Vrbová, G., and Gutmann, E. (1956). *J. Physiol. (Paris)* **48**, 751.

Warburg, O., and Christian, W. (1939). *Biochem. Z.* **303**, 40.

Weber, A. (1951). *Biochim. Biophys. Acta* **7**, 214.

Weber, A. (1956). *Biochim. Biophys. Acta* **19**, 345.

Weber, A. (1959). *J. Biol. Chem.* **234**, 2764.

Weber, A., and Hasselbach, W. (1954). *Biochim. Biophys. Acta* **15**, 237.

Weber, A., and Herz, R. (1963). *J. Biol. Chem.* **238**, 599.

Weber, A., and Weber, H. H. (1951). *Biochim. Biophys. Acta* **7**, 339.

Weber, A., and Winicur, S. (1961). *J. Biol. Chem.* **236**, 3198.

Weber, A., Herz, R., and Reiss, I. (1962). *J. Gen. Physiol.* **46**, 679.

Weber, A., Herz, R., and Reiss, I. (1964). *Fed. Proc., Fed. Amer. Soc. Exp. Biol.* **23**, 896.

Weber, A., Herz, R., and Reiss, I. (1966). *Biochem. Z.* **345**, 329.

Weber, H. H. (1935). *Pfluegers Arch. Gesamte Physiol. Menschen Tiere* **235**, 205.

Weber, H. H. (1951). *Z. Elektrochem.* **55**, 511.

Weber, H. H. (1956). *Proc. Int. Congr. Biochem., 3rd, 1955* p. 81.

Weber, H. H., and Portzehl, H. (1952). *Advan. Protein Chem.* **7**, 161.

Weber, H. H., and Portzehl, H. (1954). *Progr. Biophys. Biophys. Chem.* **4**, 60.

Weinstock, I. M. (1966). *Ann. N.Y. Acad. Sci.* **138,** 199.

Weinstock, I. M., Epstein, S., and Milhorat, A. T. (1968). *Proc. Soc. exp. Biol. Med.* **99,** 272.

Weizäcker, V. (1914). *J. Physiol. (London)* **48,** 396.

Williamson, J. R. (1966). *Pharmacol. Rev.* **18,** 205.

Williamson, J. R., and Jamieson, D. (1966). *Mol. Pharmacol.* **2,** 191.

Wosilait, W. D. (1958). *J. Biol. Chem.* **233,** 597.

Žak, R., and Gutmann, E. (1960). *Nature (London)* **185,** 766.

Žak, R., Gutmann, E., and Vrbová, G. (1957). *Experientia* **13,** 80.

Zimm, B. (1948). *J. Chem. Phys.* **16,** 1099.

BIOCHEMISTRY OF MUSCLE MITOCHONDRIA*

E. J. DE HAAN, G. S. P. GROOT, H. R. SCHOLTE,
J. M. TAGER, and E. M. WIT-PEETERS

* *Abbreviations used:* $NAD(P)^+$, $NAD(P)H$ = oxidized and reduced nicotinamide adenine nucleotide (phosphate); LS_2, $L(SH)_2$ = oxidized and reduced α-lipoic acid; Q, QH_2 = oxidized and reduced ubiquinone; fp = flavoprotein; FAD = flavine adenine dinucleotide; DPT = diphosphothiamine; CoASH = coenzyme A; ETF = electron transferring flavoprotein; P_i = inorganic orthophosphate; PP_i = inorganic pyrophosphate; subscript "in" or "out" = within or outside the mitochondrial inner membrane; $\sim$ = energy-rich compound; v = rate of reaction.

I. Introduction

It has been known for more than one hundred years, since the publication of the book "Allgemeine Anatomie" by Henle (1841), that the sarcoplasm of striated muscle contains granules. Henle (1841) observed these granules with the light microscope, and reported that a "tiny skin" envelops the granules.

Kölliker (1850, 1857, 1888) made a thorough study of these granules, which he called "interstitial granules," and showed that they were widely distributed in the animal kingdom. In the third of these papers, he described the separation of the granules from insect muscle and discussed the effect of various treatments, such as suspension in a hypotonic medium, on the appearance of the granules under the microscope. This paper is of considerable historical interest, since it foreshadowed by sixty or seventy years much of the subsequent work on the structure of these and related granules. Kölliker's studies were followed by further detailed investigations by a number of distinguished cytologists. Retzius (1890) extended Kölliker's descriptions and introduced the name "sarcosome" in the following words:

> They exist in many animals in different positions and arrangements—frequently in longitudinal rows, as has been shown by Kölliker—and they constitute, as the latter has already pointed out, *"Körperchen sui generis";* they are not fat granules, as several authors have maintained, even recently. Because of their pronounced and characteristic properties (resistance to reagents, specific staining properties, etc.) and also in order to distinguish them clearly from pathologically formed fat granules . . . I have named the normal interstitial, or more correctly intercolumnar *Körperchen,* as "sarcosomes."

Holmgren (1910) and Bullard (1913, 1916) made detailed studies of the granules present in heart muscle. Holmgren classified striated muscles as falling into two types, depending upon the distribution of the granules, and recognized that these two classes correspond to two physiological types. Muscles with granules at the level of the isotropic bands of the myofibrils are those with intermittent activity, such as the skeletal muscle of many vertebrates and invertebrates. Granules at the level of the anisotropic bands are found in muscles required for a sustained activity, such as the flight muscles of birds and insects, and the heart muscle of vertebrates. These types of muscle contain more and larger granules than the first type.

In more recent years, it has become apparent that Holmgren's classification has a biochemical basis. The muscles with granules at the level of the isotropic bands are rich in glycolytic enzymes, sufficient for muscle activity over short periods. The muscles required for sustained activity are predominantly aerobic in their metabolism. Keilin (1925) showed that among all the large number of tissues examined, the thoracic muscles of insects were the ones with the highest concentrations of cytochrome.

Regaud (1909), Holmgren (1910), and Bullard (1913, 1916) showed that in heart muscle, the granules were very regularly arranged between the myofibrils, with one granule opposite each anisotropic disk. They also described the presence of a few granules at the pole of the nucleus.

Between 1916 and 1950, comparatively little attention was paid to muscle mitochondria. After this date, however, it soon became clear that these organelles are the seat of the respiratory activity in muscle as in other tissues and that they play a fundamental role in the energy metabolism of the cell.

II. Function of Muscle Mitochondria

Particulate preparations from muscle have been used for the study of respiratory enzymes since 1912, when they were first introduced by Batelli and Stern. Similar preparations were used by Warburg (1913) to study cellular respiration. In 1925, Keilin discovered that the cytochrome system of the respiratory chain is located in a particulate fraction from heart muscle. The Keilin and Hartree (1947) heart muscle preparation is still used today for the isolation of many respiratory enzymes. It was only later, however, that such preparations were recognized as consisting of mitochondrial fragments (Cleland and Slater, 1953a,b); see also Weijers (1971).

Bensley and Hoerr (1934) and Claude (1946) introduced the technique of differential centrifugation in order to separate various structural components of the cell. Using this technique, Hogeboom et al. (1946) showed that mitochondria isolated from liver contain the respiratory enzymes. It was afterward found that liver mitochondria contain all the enzymes necessary for the complete oxidation of pyruvate to carbon dioxide and water and for coupling this oxidation to the synthesis of ATP.

The technique of differential centrifugation was subsequently applied to isolate particulate fractions with respiratory activity from various types of muscle, including mammalian muscle (Slater, 1950; Green, 1951) and the thoracic muscle of the blowfly (Watanabe and Williams, 1951). These respiratory granules were afterward identified as mitochondria (Harman, 1950; Harman and Feigelson, 1952; Cleland and Slater, 1953a,b; Chappell and Perry, 1953; Harman and Osborne, 1953).

Methods of isolating mitochondria from various types of muscle are described in detail in "Methods of Enzymology," Volume X (Colowick *et al.*, 1967).

The main function of muscle mitochondria is to bring about the aerobic oxidation of the products of the glycolysis of carbohydrate and the products of the hydrolysis of lipids and proteins, together with the synthesis of ATP from ADP and P_i. The ATP synthesized by this process, known as oxidative phosphorylation, is utilized for muscular contraction and for other ATP-requiring processes.

The respiratory activities of different types of muscles were compared with the mitochondrial concentration by Paul and Sperling (1952). The results with α-ketoglutarate as substrate, assembled in Table I, show that white muscle, which contains very few mitochondria, has a low respiratory activity, while red muscle, which is rich in mitochondria, has a high respiratory activity. The red color is due to myoglobin, which assists the transfer of oxygen from the blood to the respiratory enzymes in the mitochondria. In the rabbit, the highest respiratory activity is found in the heart and the diaphragm, while the skeletal muscle has a very low activity. This is the explanation of the lack of stamina of a rabbit. Short periods of muscular activity are maintained by the high concentration of glycolytic enzymes in rabbit muscle, which is widely used by the biochemist for preparing these enzymes. The low respiratory activity of the breast muscle of the nonflying chicken in comparison with that in the pigeon is also noteworthy.

In insect muscle, the respiratory activity is considerably greater than in vertebrate muscle (see Van den Bergh, 1962).

Some generalizations are possible about the metabolism of the different types of muscle tissues that occur in nature. In most skeletal muscles, where a relatively high oxygen debt can be built up, the lactate dehydrogenase of the tissue is predominantly of the M type, lactate production can be high, and there can be considerable transport of lactate from

TABLE I

RESPIRATORY ACTIVITY OF DIFFERENT TYPES OF MUSCLE CORRELATED WITH
MITOCHONDRIAL CONCENTRATION AND WITH THE COLOR OF THE MUSCLE[a]

Animal	Muscle	Color of muscle	Mitochondrial concentration	Respiratory[b] activity (μl O_2/hr/gm wet weight tissue)
Rabbit	Back	White	(very low)	35
Chicken	Breast	White	(very low)	110
Rabbit	Gastrocnemius	White	(very low)	110
Rabbit	Soleus	Reddish	+	85
Rat	Leg	Reddish	+	85
Bat	Forearm	Red	++	185
Rabbit	Diaphragm	Red	++	545
Mallard	Breast	Red	+++	875
Rabbit	Heart	Red	++++	1110
Pigeon	Breast	Red	++++	1830

[a] Calculated from Paul and Sperling, 1952.
[b] α-Ketoglutarate as substrate.

the muscle to other tissues. In this type of muscle, no active mechanism exists for the transfer of hydrogens from the cytosol to the mitochondrion.

In most flight muscles an exclusively aerobic metabolism takes place. Abundant large mitochondria are present, lactate formation is negligible, and an efficient mechanism (the glycerol 3-phosphate shuttle) exists for the transfer of hydrogens from the cytosol to the mitochondrion, thus ensuring that extramitochondrially generated NADH is rapidly oxidized (see Section VI).

Heart muscle is of an intermediate type. Although the glycerol 3-phosphate shuttle is relatively inactive, transfer of reducing equivalents from the cytosol to the mitochondrion may be brought about by the malate–aspartate shuttle (see Section VI).

III. Detailed Mechanisms of Energy-Yielding Reactions in Muscle Mitochondria

No attempt has been made in this section to document by references to the literature statements that are found in many textbooks of biochemistry.

A. Substrates Oxidized by Muscle Mitochondria

Mitochondria from all types of muscle can oxidize pyruvate, the end product of the glycolysis of carbohydrate. Mitochondria from most muscles oxidize fatty acids and glutamate. The flight muscle of insects of the orders Diptera and Hymenoptera form an exception in this respect; mitochondria from the flight muscles of these insects, in which only carbohydrates are used as a source of fuel, can not oxidize fatty acids (Wigglesworth, 1949; Sacktor, 1955; Van den Bergh, 1962). Acetoacetate can be oxidized by mitochondria from several types of muscle, including heart muscle (Stern *et al.*, 1956). β-Hydroxybutyrate is oxidized more slowly.

Finally, muscle mitochondria can oxidize glycerol 3-phosphate to dihydroxyacetone phosphate (Green, 1936; Tung *et al.*, 1952; Sacktor and Cochran, 1957). The enzyme catalyzing this reaction (L-glycerol-3-phosphate:acceptor oxidoreductase; EC 1.1.99.5) is a flavoprotein distinct from the cytosolic enzyme (L-glycerol 3-phosphate:NAD$^+$ oxidoreductase; EC1.1.1.8).

B. Oxidation of Pyruvate and Acetyl-CoA.

Energy production in muscle takes place predominantly by oxidation of pyruvate and of acetyl-CoA, the end product of the β-oxidation of fatty acids. These compounds are oxidized by the reactions of the citric acid cycle [Reactions (1)–(10)].

$$CH_3COCOOH + CoASH + O \rightarrow CH_3COSCoA + CO_2 + H_2O \qquad (1)$$

$$CH_3COSCoA + HOOCCOCH_2COOH \rightarrow CoASH + HOOCCH_2COH(COOH)CH_2COOH \qquad (2)$$

$$HOOCCH_2COH(COOH)CH_2COOH \rightleftharpoons HOOCCH{=}C(COOH)CH_2COOH + H_2O \qquad (3)$$

$$HOOCCH{=}C(COOH)CH_2COOH + H_2O \rightleftharpoons HOOCCHOHCH(COOH)CH_2COOH \qquad (4)$$

$$HOOCCHOHCH(COOH)CH_2COOH + O \rightarrow HOOCCOCH(COOH)CH_2COOH + H_2O \qquad (5)$$

$$HOOCCOCH(COOH)CH_2COOH \rightarrow HOOCCOCH_2CH_2COOH + CO_2 \qquad (6)$$

$$HOOCCOCH_2CH_2COOH + O \rightarrow HOOCCH_2CH_2COOH + CO_2 \qquad (7)$$

$$HOOCCH_2CH_2COOH + O \rightarrow HOOCCH{=}CHCOOH + H_2O \qquad (8)$$

$$HOOCCH{=}CHCOOH + H_2O \rightleftharpoons HOOCCHOHCH_2COOH \qquad (9)$$

$$HOOCCHOHCH_2COOH + O \rightarrow HOOCCOCH_2COOH + H_2O \qquad (10)$$

Sum: $\qquad CH_3COCOOH + 2\tfrac{1}{2}O_2 \rightarrow 3CO_2 + 2H_2O \qquad (11)$

In recent years, research on pyruvate and acetyl-CoA oxidation in mitochondria has dealt mainly with regulation of the process. We will consider the regulation of pyruvate utilization in muscle in some detail.

1. REGULATION OF THE PYRUVATE DEHYDROGENASE COMPLEX

Oxidation of pyruvate to acetyl-CoA is catalyzed by several enzymes that can be isolated as a single complex from pigeon breast muscle (Jagannathan and Schweet, 1952; Schweet *et al.*, 1952), from *Escherichia coli* (Koike *et al.*, 1960), and from pig heart (Hayakawa *et al.*, 1964, 1966). Koike *et al.* (1963) were able to resolve the pyruvate dehydrogenase complex from *E. coli* into three separate enzyme activities: pyruvate dehydrogenase (EC 1.2.4.1.), lipoate acetyl transferase (EC 2.3.1.12), and lipoamide dehydrogenase (EC 1.6.4.3). Later (Hayakawa and Koike, 1967), these enzyme activities were also separated from the enzyme complex from pig heart. Pyruvate dehydrogenase contains diphosphothiamin as prosthetic group (DPT–protein); lipoate acetyl transferase contains lipoic acid covalently bound to the protein (LS$_2$–protein), whereas lipoamide dehydrogenase is a flavoprotein with FAD as prosthetic group. On recombination of the three separated enzymes, the original CoA-dependent, NAD$^+$-linked oxidation of pyruvate is reconstituted. The mechanism of pyruvate oxidation is given in Reactions (12)–(16):

$$CH_3COCOOH + DPT\text{–protein} \rightarrow CH_3CHOH\text{–DPT–protein} + CO_2 \quad (12)$$

$$CH_3CHOH\text{–DPT–protein} + LS_2\text{–protein} \rightleftharpoons \text{L–protein} \begin{smallmatrix} SH \\ \diagup \\ \diagdown \\ SCOCH_3 \end{smallmatrix} + DPT\text{–protein} \quad (13)$$

$$\text{L–protein} \begin{smallmatrix} SH \\ \diagup \\ \diagdown \\ SCOCH_3 \end{smallmatrix} + CoASH \rightleftharpoons CH_3COSCoA + \text{L–protein} \begin{smallmatrix} SH \\ \diagup \\ \diagdown \\ SH \end{smallmatrix} \quad (14)$$

$$\text{L–protein} \begin{smallmatrix} SH \\ \diagup \\ \diagdown \\ SH \end{smallmatrix} + \text{flavoprotein} \rightleftharpoons LS_2 + \text{reduced flavoprotein} \quad (15)$$

$$\text{Reduced flavoprotein} + NAD^+ \rightleftharpoons \text{flavoprotein} + NADH + H^+ \quad (16)$$

$$\text{Sum: } CH_3COCOOH + CoASH + NAD^+ \rightarrow CH_3COSCoA + CO_2 + NADH + H^+ \quad (17)$$

In 1962 Garland *et al.* showed for the first time that in the perfused rat heart pyruvate oxidation is inhibited by adding fatty acids to the perfusion medium. Since then this inhibitory effect of fatty acids or fatty acylcarnitines has been shown repeatedly, not only with rat heart preparations (Bremer, 1965), but also in homogenates or mitochondria from rat liver (Walter *et al.*, 1966; Bremer, 1966; Nicholls *et al.*, 1967; von Jagow *et al.*, 1968).

Several suggestions have been made as to the mechanism of inhibition of pyruvate oxidation by fatty acids. The inhibition could be due to (a) a high reduction level of mitochondrial nicotinamide nucleotides, (b) competition for CoASH, or (c) inhibition of pyruvate dehydrogenase by high levels of acetyl-CoA (for discussion, see Nicholls *et al.*, 1967; von Jagow *et al.*, 1968; Garland *et al.*, 1969a).

The observation of Haslam and Krebs (1963; Haslam, 1966) that in rat liver mitochondria pyruvate oxidation is inhibited by α-oxo-glutarate was considered to stress the importance of the $NAD^+/NADH$ ratio and the CoASH level for rapid pyruvate oxidation. However, recent experiments of Garland and co-workers (1969a) show that in rat liver mitochondria pyruvate oxidation in the presence of palmitoyl-carnitine is not limited by a lack of NAD^+ or of CoASH. On the other hand, the possibility that pyruvate oxidation is inhibited by acetyl-CoA is supported by the observation (Garland and Randle, 1964a; Garland *et al.*, 1969a; Wieland *et al.*, 1969) that partially purified pyruvate dehydrogenase complexes from pigeon or pig heart are inhibited by acetyl-CoA.

Recently, still another regulatory mechanism for pyruvate oxidation has been discovered. Reed and co-workers observed that the purified pyruvate dehydrogenase complex from beef kidney was 90% inhibited by 1 mM ATP (Linn *et al.*, 1969a; Reed, 1969). By using γ-^{32}P-labeled ATP, they showed that the effect of ATP was due to a phosphorylation of the (active) pyruvate dehydrogenase component of the complex; this phosphorylation was ascribed to a "pyruvate dehydrogenase kinase." Activation could be accomplished by dephosphorylation of the phosphorylated enzyme by means of a phosphatase. Both the kinase and the phosphatase were shown to be intrinsic parts of the pyruvate dehydrogenase complex.

Subsequently it was shown (Linn *et al.*, 1969b; Wieland and von Jagow-Westermann, 1969; Wieland and Siess, 1970; Wieland *et al.*, 1970) that the pyruvate dehydrogenase complexes isolated from pig heart, pig liver, and the beef heart behave similarly. In all cases, both the kinase and the phosphatase require Mg^{2+} for maximal activity, the optimal concentrations being about 1 mM and 10 mM, respectively. It was suggested (Linn *et al.*, 1969a) that the activity of the phosphatase and the kinase, thus of pyruvate dehydrogenase, is regulated by the intracellular Mg^{2+} concentration. At rest, when the ATP concentration in the cell is high, most of the Mg^{2+} is bound to ATP (Burton, 1959). Under these conditions (high ATP, low free Mg^{2+} concentration) pyruvate dehydrogenase kinase will still be active. Thus, pyruvate dehydrogenase will be phosphorylated and simultaneously inactivated. Inversely, lowering of the ATP/ADP ratio will lead to increased pyruvate oxidation.

The effect of fatty acids on pyruvate oxidation mentioned above can be explained by assuming that fatty acid oxidation causes an increase of the ATP/ADP ratio in the mitochondria, thereby inhibiting pyruvate dehydrogenase. Two observations support this hypothesis. First, it was found (Nicholls *et al.*, 1967) that the suppression of pyruvate oxidation by fatty acids is sensitive to uncouplers of oxidative phosphorylation. These agents again tend to lower the ATP/ADP ratio and thus promote pyruvate oxidation. Second, it was found recently by Wieland and co-workers (1971) that *in vivo* under conditions of increased fatty acid utilization caused by starvation or alloxan diabetes, only 15% of the total pyruvate dehydrogenase of the heart or kidney was in the active form. This figure was estimated by comparing the enzyme activities in tissue homogenates in the absence of added Mg^{2+} and after complete activation of pyruvate dehydrogenase by the addition of 10 mM Mg^{2+}. In normal fed rats, about 70% of the pyruvate dehydrogenase was in the active form. These observations certainly point to a role of the activation mechanism *in vivo*.

The physiological role of the regulation of pyruvate dehydrogenase by the ATP/ADP ratio seems obvious. When the cell can use carbohydrate as well as fatty acid (or ketone bodies) to meet its energy demand, the latter will be chosen preferentially. Indeed, it has been shown repeatedly that in the intact heart of normal fed mammals the greater part of the respiration is due to fatty acid oxidation (Olson and Piatnek, 1959; Slater, 1960; Shipp *et al.*, 1961; Williamson, 1962; Randle *et al.*, 1964). In addition, ketone bodies (Williamson and Krebs, 1961; Randle *et al.*, 1964) and acetate (Williamson, 1964) can compete efficiently with glucose as fuel for energy production in perfused hearts. Williamson and Krebs (1961), for example, showed that even in the presence of glucose plus insulin, acetoacetate could account for about 60% of the oxygen uptake of the perfused heart.

In gluconeogenic tissues like liver and kidney, inhibition of pyruvate oxidation by fatty acids is accompanied by a simultaneous stimulation of carboxylation of pyruvate. The latter process is absolutely dependent on the presence of acetyl-CoA (for reviews, see Utter, 1969; Utter and Scrutton, 1969). Indeed, in these tissues, stimulation of gluconeogenesis by fatty acids has been found (Struck *et al.*, 1966; Wieland *et al.*, 1970).

However, certainly in muscle, and even in liver, inhibition of pyruvate oxidation by fatty acids cannot be the complete explanation for the glucose-sparing effect of fat utilization. For, without further regulation of glycolysis, conversion of glucose into pyruvate and lactate would be unimpaired, so that vast amounts of pyruvate would accumulate. This, in turn, would probably lead to a release of inhibition of pyruvate dehydrogenase, since it has been found (Linn *et al.*, 1969b) that the enzyme is not inhibited by ATP in the presence of concentrations of pyruvate of 0.5 mM or higher.

Indeed, a second point of regulation of glucose utilization has been discovered at the level of phosphofructokinase. This enzyme has been shown to be inhibited by a high ATP/AMP ratio (Passonneau and Lowry, 1962; Mansour and Mansour, 1962; Mansour, 1963) as well as by citrate (Parmeggiani and Bowman, 1963). Garland and Randle (1964b) showed that citrate accumulated under conditions when glucose utilization in heart muscle was suppressed by starvation, alloxan diabetes, or by addition of fatty acids or ketone bodies to the perfusion medium. Furthermore, an inhibition of glycolysis at the level of the phosphorylation of fructose phosphate could be established (Randle *et al.*, 1964).

So far, most information has been obtained from studies with liver or heart. Some effects of fatty acids or ketone bodies have also been found in rat diaphragm muscle (Randle *et al.*, 1964). Furthermore, a similar regulation of phosphofructokinase has been demonstrated in rat and rabbit skeletal muscle (Ling *et al.*, 1965; Parmeggiani *et al.*, 1966). Since in this type of muscle both glucose and fatty acids may serve as substrate for energy production, regulation of pyruvate dehydrogenase activity is also to be expected. However, direct experimental evidence for phosphorylation and dephosphorylation of this enzyme is lacking.

Still less is known about the regulation of glucose utilization in the highly active flight muscle cells of the house fly (*Musca domestica*) or the blowfly (*Phormia regia*). It is known that in these cells carbohydrate is practically the sole substrate. Sacktor and Wormser-Shavit (1966) and Sacktor and Hurlbut (1966) measured concentrations of the intermediates of glucose metabolism and of the adenine nucleotides in the flight muscle of the blowfly after various periods of flight. At the onset of flight, they observed a transient increase of the fructose 1,6-diphosphate concentration, without, however, a concomitant decrease in the level of hexose 6-phosphates. Simultaneously the ATP/AMP ratio decreased sharply. Sacktor and co-workers suggest that their results, although not conclusive, may provide evidence for the regulation of glycolysis at the level of phosphofructokinase in blowfly flight muscle. However, no evidence whatsoever has been obtained for regulation of pyruvate dehydrogenase in this tissue. One could, of course, question whether this regulation is to be expected, since in these flight muscles fatty acid oxidation is virtually absent. In this respect it would be interesting to compare the pyruvate dehydrogenases from flight muscle of Diptera with that from locust (*Locusta migratoria*), in which fatty acids can be utilized (Bode and Klingenberg, 1965).

2. Regulation of Acetyl-CoA Oxidation

Acetyl-CoA, formed either from pyruvate or from fatty acids (see Section III,C) is oxidized through the tricarboxylic acid cycle. Control points of this process are (a) control of the respiratory chain through the availability of ADP, (b) availability of oxaloacetate, (c) citrate synthase, and (d) isocitrate dehydrogenase.

A. Respiratory Control. Respiratory control will be discussed in Section III,F of this chapter.

B. Availability of Oxaloacetate. Isolated heart mitochondria oxidize acetate and pyruvate rapidly without further addition of Krebs cycle intermediates, provided the right conditions are chosen (Davis, 1965, 1967; Slater *et al.*, 1965). This means that the mitochondria contain sufficient Krebs cycle intermediates to produce the required oxaloacetate. Also, in house fly flight muscle mitochondria, no additional substrate is required for maximal rates of pyruvate oxidation (Van den Bergh and Slater, 1962; Van den Bergh, 1964). In the latter case it has been suggested by Van den Bergh (1962) that the observed impermeability of flight muscle mitochondria for substrates other than pyruvate or glycerol 3-phosphate has the function of preventing any loss of tricarboxylic acid cycle intermediates from the mitochondria (see also Van den Bergh, 1964; Sacktor and Childress, 1967; Chappell and Haarhoff, 1967). It has been found recently (Tulp and Van Dam, 1969; Tulp *et al.*, 1971) that housefly mitochondria are relatively permeable to dicarboxylate ions when the phosphate concentration of the medium is low, but Tulp and co-workers suggested that no rapid transport of tricarboxylic acid cycle intermediates over the mitochondrial membrane takes place *in vivo*. Since heart mitochondria are permeable for most of the Krebs cycle intermediates (see Section VI), this means that these compounds *in vivo* have to be replenished. Carboxylation of pyruvate is virtually absent in muscle mitochondria (Utter, 1969), so the only way to produce some intermediates is by amino acid oxidation (Van den Bergh, 1964; see Section III,D).

C. CITRATE SYNTHESIS. Regualtion of citrate synthesis by effects on citrate synthase has been discussed in detail in recent reviews (Atkinson, 1969; Garland *et al.*, 1968, 1969a). Although citrate synthase from muscle, like the liver enzyme, is inhibited by ATP (Hathaway and Atkinson, 1965; Kosicki and Lee, 1966), the role of this effect *in vivo* has been questioned (Garland *et al.*, 1969a) because of the extremely high enzyme activity in muscle. Furthermore, citrate synthesis in rat heart mitochondria is not influenced by uncouplers of oxidative phosphorylation, as is the case with citrate formation in rat liver mitochondria.

D. ISOCITRATE DEHYDROGENASE. It has been found that NAD^+-linked isocitrate dehydrogenase obtained from different sources is subject to regulation by the level of AMP or ADP, leading to an increase in substrate affinity (Hathaway and Atkinson, 1963; Chen and Plaut, 1963; Sanwal *et al.*, 1964; Goebell and Klingenberg, 1964). Garland *et al.* (1969a) questioned the importance of the NAD^+-linked isocitrate dehydrogenase in isocitrate oxidation because of the low enzyme activity observed in mitochondria. Recently, however, control of isocitrate dehydrogenase activity by the ATP/ADP level has been shown to play a role in regulating the rate of pyruvate oxidation in insect flight muscle mitochondria (Hansford and Chappell, 1968; Tulp and Van Berkel, 1971).

3. PRODUCTS OF PYRUVATE OXIDATION

In muscle mitochondria under normal conditions, the products of pyruvate oxidation will be carbon dioxide and water (Reaction 11), since no accumulation of tricarboxylic acid cycle intermediates can occur without concomitant inhibition of further oxidation of pyruvate. However, incomplete oxidation of pyruvate in muscle has been reported to lead to formation of acetate or acetoacetate (Davis, 1968; Tulp and Van Dam, 1970). This side path of pyruvate oxidation is suggested to be an escape route to provide energy under conditions where the Krebs cycle is inhibited. Krebs (1970) mentions that ketone body production in extra hepatic tissues does not involve the hydroxymethylglutaryl-CoA pathway, but occurs via acetoacetyl-CoA thiolase. He suggests that the physiological significance of this pathway lies in the utilization of ketone bodies rather than in their production.

C. Oxidation of Fatty Acids

In the absence of carbohydrate, fatty acids are mainly used as respiratory substrate in adult mammalian muscle. The fatty acids can be supplied by lipolysis of neutral fats in the tissue itself and can also be absorbed from the blood. Before they can be oxidized, the fatty acids must be activated to the corresponding acyl-CoA esters.

The inner membrane of the mitochondrion is impermeable to acyl-CoA esters. If the fatty acids are activated on the outside of the outer membrane, as is the case with long-chain fatty acids in heart muscle mitochondria, the acyl-CoA esters are transformed into acylcarnitine esters, which do penetrate the inner mitochondrial membrane. In the matrix the acylcarnitine esters are reconverted to the acyl-CoA derivatives, as will be considered in detail.

The β-oxidation system (Knoop, 1904) or Lynen spiral (Lynen, 1955) leads to the complete degradation of the acyl-CoA esters to acetyl-CoA. The acetyl-CoA, in turn, is oxidized via the tricarboxylic acid cycle, so that the final products of fatty acid oxidation in muscle mitochondria are carbon dioxide and water, this process being coupled with the synthesis of ATP.

1. ACTIVATION OF FATTY ACIDS

The activation of a fatty acid by ATP and CoASH is catalyzed by an acyl-CoA synthetase and leads to the formation of the corresponding acyl-CoA ester. The mechanism of the reaction, as first demonstrated for acetyl-CoA synthetase (acetate:CoA ligase (AMP); EC 6.2.1.1) by Berg (1956), Boyer *et al.* (1959), and Webster (1963) is as follows:

$$RCOOH + ATP + enzyme \rightleftharpoons enzyme\ [RCO—AMP] + PP_i \quad (18)$$

$$Enzyme\ [RCO—AMP] + CoASH \rightleftharpoons RCOSCoA + AMP + enzyme \quad (19)$$

Sum: $\quad RCOOH + CoASH + ATP \rightleftharpoons RCOSCoA + AMP + PP_i \quad (20)$

Both soluble and membrane-bound acyl-CoA synthetases have been described. Furthermore, at least six acyl-CoA synthetases can be distinguished on the basis of the chain length of the fatty acids activated. These enzymes also differ in their physicochemical properties. The enzymes are named according to the fatty acids with which they show the greatest activity: acetyl-CoA synthetase (Webster, 1969), propionyl-CoA synthetase (Scholte *et al.*, 1971), butyryl-CoA synthetase (Webster *et al.*, 1965), heptanoyl-CoA synthetase (Mahler *et al.*, 1953), lauroyl-

CoA synthetase, and palmitoyl-CoA synthetase (Kornberg and Pricer, 1953; Aas, 1971). It should be emphasized that most of these enzymes show a rather broad specificity with regard to the fatty acid activated.

The soluble acyl-CoA synthetases, which are responsible for the activation of C_2, C_3, C_4, and C_7 fatty acids, are localized in the mitochondrial matrix (Aas and Bremer, 1968; Scholte *et al.*, 1971). In liver mitochondria, three acyl-CoA synthetases have been shown to occur in the outer membrane (Aas, 1971); in addition, some (10%) palmitoyl-CoA synthetase activity is associated with the inner membrane–matrix fraction (Van Tol and Hülsmann, 1970). In intact heart mitochondria, palmitoyl-CoA synthetase is completely inactivated by Nagarse; since this proteolytic enzyme, like other macromolecules, does not penetrate the outer membrane, it was concluded that palmitoyl-CoA synthetase is situated solely at the outside of the outer membrane in heart mitochondria (De Jong and Hülsmann, 1970; Pande and Blanchaer, 1970). Palmitoyl-CoA synthetase is also located to a small extent (about 13%) in the microsomal fraction of heart (De Jong and Hülsmann, 1970).

It should be emphasized that the enzymes described in this section are not necessarily present in mitochondria from all types of muscle (Bode and Klingenberg, 1965; Aas, 1971). Aas (1971) has shown that rat skeletal muscle lacks the soluble acyl-CoA synthetases, but that rat heart muscle, in contrast, contains all the fatty acid activating enzymes mentioned above. The physiological meaning of such a wide variety of fatty acid activating enzymes, with overlapping acyl specificities, is not clear.

2. CARNITINE AND ACYL TRANSPORT

In 1955, Fritz discovered that carnitine stimulates the oxidation of long-chain fatty acids in rat liver slices and homogenates. This finding was extended to other tissues, including heart muscle (Fritz *et al.*, 1962). In fact, the stimulating effect of carnitine on palmitate oxidation was found to be greater in heart than in liver mitochondria (Fritz, 1963). The oxidation of fatty acids with a chain length of less than C_{10} is not stimulated by carnitine (Fritz *et al.*, 1962).

Thus far, three enzymes have been described that play a role in acylcarnitine metabolism. These carnitine acyl-transferases catalyze the reaction

$$\text{Acyl-SCoA} + \text{L-carnitine} \rightleftharpoons \text{CoASH} + \text{acyl-L-carnitine} \qquad (21)$$

The apparent equilibrium constant of reactions of this type is near unity (Fritz *et al.*, 1963; Norum, 1964), indicating that the acylcarnitines must be energy-rich compounds.

Purified carnitine acetyltransferase reacts with acyl-CoA esters of chain length C_2 to C_{10}, with decreasing V_{max} at infinite substrate concentration (Chase, 1967). Norum (1964) partially purified a carnitine palmitoyltransferase from calf liver mitochondria that reacts with increasing velocity with acylcarnitine esters from C_2 to C_{16}. The activity at lower chain length was attributed to contamination with carnitine acetyltransferase (Norum, 1964).

Carnitine octanoyltransferase has been demonstrated in mitochondria from several tissues, including rat heart and pigeon breast muscle. This enzyme shows maximal activity with C_6 to C_{10} acyl esters (Solberg, 1971).

In order that the acylcarnitines should function as carriers of acyl units across the mitochondrial inner membrane (Fritz and Yue, 1963), the carnitine acyltransferases must have a double localization, since the substrate for β-oxidation is acyl-CoA. Evidence for a double localization has been presented by Yates and Garland (1970) in the case of carnitine palmitoyltransferase. They found that the activity of the enzyme with acyl-CoA as substrate was greatly increased by sonication of rat liver and also beef heart mitochondria, indicating that the bulk, but not all of the activity was present in a compartment inaccessible to acyl-CoA. West *et al.* (1971) separated two different carnitine palmitoyltransferases from beef liver. These were believed to have different localization in the mitochondria.

Finally, some carnitine acetyltransferase (Makita and Sandborn, 1971) and carnitine palmitoyltransferase (Van Tol and Hülsmann, 1969) activity has been reported to be present in the microsomal fraction.

It is clear that carnitine palmitoyltransferase has a direct function in promoting long-chain acyl transport for fatty acid oxidation. It is probable that the role of carnitine acetyltransferase differs from that of carnitine palmitoyltransferase. The short-chain fatty acids permeate the mitochondrial membrane in the undissociated form (for review, see Klingenberg, 1970a) and activation occurs in the matrix (Scholte *et al.*, 1971). It has been suggested that the function of carnitine acetyltransferase is to act as an acetyl buffer (Greville and Tubbs, 1968) or that it has a CoA-sparing effect (Hülsmann *et al.*, 1964). Another possibility is that the enzyme in heart functions as a carrier of acetyl units for the fatty acid elongation system situated in the outer membrane (see Section IV,C) or for acetylcholine synthesis (Fritz, 1965).

It has been suggested that the "outer" carnitine palmitoyltransferase may be rate-limiting for the oxidation of long-chain fatty acids activated outside the inner mitochondrial membrane (Shepherd *et al.*, 1966). Another factor regulating the rate of oxidation of long-chain fatty acids is the level of carnitine; this is low, for instance, in hearts from neonatal rats and in flight muscle from the bee, a carbohydrate-utilizing insect (Childress *et al.*, 1967).

Our present knowledge based on the scheme of Fritz and Yue (1963) is summarized in Fig. 1.

In muscle, long-chain fatty acids are activated mainly on the outer mitochondrial membrane and to a lesser extent on the microsomes. The acyl-CoA esters are converted to the carnitine esters by an "outer" carnitine palmitoyltransferase, which is located outside the inner membrane (perhaps a part is on the outer surface of the inner membrane and a vari-

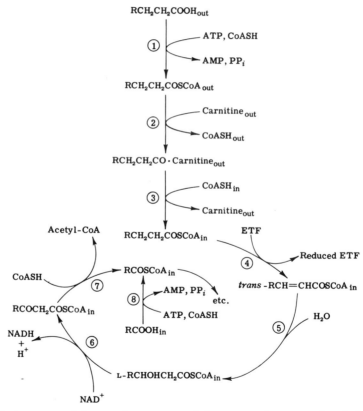

Fig. 1. Activation of fatty acids, transport of the activated compound, and degradation via the β-oxidation spiral. Enzymes: (1) membrane-bound acyl-CoA synthetases, (2) outer carnitine palmitoyl transferase, (3) inner carnitine palmitoyl transferase, (4) acyl-CoA dehydrogenases, (5) enoyl-CoA dehydratases, (6) L-3-hydroxyacyl-CoA dehydrogenase, (7) 3-ketoacyl-CoA thiolase and acetyl-CoA acetyltransferase, (8) soluble acyl-CoA synthetases.

able part in the microsomes). They then react with the "inner" carnitine palmitoyltransferase plus $CoASH_{in}$, yielding $carnitine_{out}$ plus $acyl-CoA_{in}$. The latter is degraded by the enzymes of β-oxidation. This implies that the "inner" carnitine palmitoyltransferase has vectorial properties (Yates and Garland, 1970). An alternative possibility is that acylcarnitine can pass through the inner membrane and that the carnitine produced in the matrix leaves this compartment as acetylcarnitine and/or that carnitine itself slowly diffuses out.

Before the present picture of the organization of fatty acid activation was generally accepted, much controversy about this process existed. To explain the observed effects of atractylate and phosphate and the low intramitochondrial CoASH concentration on palmitate oxidation in mitochondria, it was postulated that palmitate activation would take place in several functional compartments (see Garland *et al.*, 1969b for review). At the moment, it is clear that the apparent complexity of palmitate activation is mainly caused by the following reasons: (a) Atractylate inhibits palmitoyl-CoA synthetase (Alexandre *et al.*, 1969) and carnitine palmitoyltransferase (Skrede and Bremer, 1970); (b) P_i induces a loss of citric acid cycle intermediates (De Jong *et al.*, 1969) and of CoASH (Skrede and Bremer, 1970).

3. THE β-OXIDATION SPIRAL

The early work done by the pioneers on the enzymology of β-oxidation has been reviewed by Lynen (1955). The reactions catalyzed by the Lynen spiral are summarized in Fig. 1. The enzymes involved, which are all mitochondrial, will now be considered.

A. ACYL-CoA DEHYDROGENASES. The acyl-CoA dehydrogenases (for review, see Beinert, 1963) catalyze the general reaction

$$RCH_2CH_2COSCoA + ETF \rightleftharpoons trans\text{-}RCH=CHCOSCoA + \text{reduced ETF} \quad (22)$$

The acyl-CoA dehydrogenases, like their electron acceptor ETF, are flavoproteins with FAD as prosthetic group. In liver mitochondria, three different acyl-CoA dehydrogenases have been described, differing in substrate specificity. The specificity of the corresponding enzymes from heart has not been tested, but it seems clear that there is a butyryl-CoA dehydrogenase acting on C_4 to C_8 acyl-CoA esters, and at least one dehydrogenase active with long-chain acyl-CoA esters.

B. ENOYL-CoA HYDRATASES. These enzymes catalyze the reaction

$$trans\text{-}RCH=CHCOSCoA + H_2O \rightleftharpoons L\text{-}RCHOHCH_2COSCoA \quad (23)$$

One of these enzymes, crotonase, has been extensively studied (for review, see Stern, 1961). It acts on the trans isomers of C_4 to at least C_9 2-ethylenic acyl-CoA compounds, the rate of the reaction decreasing with increasing chain length.

Wit-Peeters and co-workers (1971) have discovered that the enoyl-CoA hydratase responsible for the hydration of *trans*-2-hexadecanoyl-CoA is different from the enzyme that hydrates crotonoyl CoA.

C. L-3-HYDROXYACYL-CoA DEHYDROGENASE. The reaction catalyzed by this enzyme (see Wakil, 1963) is

$$\text{L-RCHOHCH}_2\text{COSCoA} + \text{NAD}^+ \rightleftharpoons \text{RCOCH}_2\text{COSCoA} + \text{NADH} + \text{H}^+ \quad (24)$$

This enzyme has a wide specificity, acting equally well with hydroxy-acyl-CoA esters of chain length C_4 to C_{12}. It is assumed that esters of greater chain length are also attacked by the same enzyme.

D. ACETYL-CoA ACETYLTRANSFERASE AND 3-KETOACYL-CoA THIOLASE. These enzymes were formerly thought to be identical and both were named thiolase (see Hartmann and Lynen, 1961). Stern (1955) purified acetyl-CoA acetyl-transferase from pig heart specific for acetoacetyl-CoA. This enzyme was later crystallized by Gehring *et al.* (1968). The mechanism of the reaction is as follows:

$$\text{CH}_3\text{COCH}_2\text{COSCoA} + \text{enzyme-SH} \rightleftharpoons \text{CH}_3\text{COS-enzyme} + \text{CH}_3\text{COSCoA} \quad (25)$$

$$\text{CH}_3\text{COS-enzyme} + \text{CoA} \rightleftharpoons \text{CH}_3\text{COSCoA} + \text{enzyme-SH} \quad (26)$$

$$\text{Sum: CH}_3\text{COCH}_2\text{COSCoA} + \text{CoASH} \rightleftharpoons \text{CH}_3\text{COSCoA} + \text{CH}_3\text{COSCoA} \quad (27)$$

A different enzyme, 3-ketoacyl-CoA thiolase, was crystallized from beef liver by Seubert *et al.* (1968). The relative activities of this enzyme with the various 3-keto-acyl-CoA esters was found to be $C_4 : C_6 : C_8 : C_{10} : C_{12} : C_{16} = 1 : 5.1 : 4.3 : 3.8 : 3.6 : 3.6$.

E. OVERALL REACTION OF β-OXIDATION. Surprisingly little is known about the interrelationships of the enzymes of the β-oxidation spiral. Some kind of concerted mechanism must be involved, since attempts to detect intermediates have been unsuccessful (see Greville and Tubbs, 1968).

In heart mitochondria, the acetyl-CoA produced by 3-ketoacyl-CoA thiolase and acetyl-CoA acetyltransferase is completely oxidized to carbon dioxide and water. Assuming that the oxidation of 1 mole of NADH and 1 mole of QH_2 via the respiratory chain give rise to 3 and 2 moles of ATP, respectively (see Section III,F), and complete oxidation of

1 mole of stearic acid will give rise to the synthesis of 146 moles of ATP:

$$C_{17}H_{35}COOH + 26O_2 + 146ADP + 146P_i \rightarrow 18CO_2 + 146ATP + 164H_2O \quad (28)$$

F. OXIDATION OF ODD-NUMBERED FATTY ACIDS. When 3-ketopentanoyl-CoA, derived from fatty acids with an odd number of carbon atoms, is cleaved by the last reaction of the β-oxidation spiral, acetyl-CoA plus propionyl-CoA are formed. The latter compound is also formed by the activation of propionate; by the breakdown of the amino acids valine, isoleucine, threonine, tryptophane, and methionine; and by the breakdown of thymine.

Propionyl-CoA is carboxylated to methylmalonyl-CoA-according to the following reactions:

$$\text{Enzyme-biotin} + HCO_3^- \underset{}{\overset{Mg^{2+},K^+}{\rightleftharpoons}} \text{enzyme-biotin-COO}^-$$
$$+ ATP^{4-} \qquad\qquad + ADP^{3-} + P_i^{2-} + H^+ \quad (29)$$

$$\text{Enzyme-biotin-COO}^- \rightleftharpoons \text{enzyme-biotin}$$
$$+ CH_3CH_2COSCoA \qquad\qquad + CH_3CH(COO^-)COSCoA \quad (30)$$

$$\text{Sum: } CH_3CH_2COSCoA + HCO_3^- \rightleftharpoons CH_3CH(COO^-)COSCoA$$
$$+ ATP^{4-} \qquad\qquad + ADP^{3-} + P_i^{2-} + H^+ \quad (31)$$

Propionyl-CoA carboxylase has been crystallized from pig heart (Kaziro *et al.*, 1961) and from beef liver mitochondria (Lane *et al.*, 1960). Both preparations have similar properties. The enzyme is specific for ATP, but propionyl-CoA ($v = 100\%$) can be replaced by acetyl-CoA ($v = 0.7$–1%), butyryl-CoA ($v = 5$–6%), crotonyl-CoA ($v = 3\%$) and valeryl-CoA ($v = 0.5\%$) (Kaziro *et al.*, 1961; Halenz *et al.*, 1962). The enzyme is stimulated by high concentrations of K^+ (Neujahr and Mistry, 1963; Giorgio and Plaut, 1967; Edwards and Keech, 1968). The apparent equilibrium constant for the formation of methylmalonyl-CoA is 5.7 at pH 8.1 and 28°C (Kaziro *et al.*, 1965). The product methylmalonyl-CoA possesses the S-configuration. (Sprecher *et al.*, 1964, 1966; Rétey and Lynen, 1964).

The carboxylated propionyl-CoA carboxylase [Reaction (29)] has been isolated (Halenz *et al.*, 1962). It is possible with the purified enzyme to transfer carbon dioxide from S-methylmalonyl-CoA to butyryl-CoA (Lane and Halenz, 1960) or acetyl-CoA (Scholte, 1970). The relatively low specificity of the enzyme with respect to acyl-CoA could result in the carboxylation of mitochondrial acetyl-CoA to malonyl-CoA. Since malonyl-CoA does not react with carnitine acetyltransferase (Scholte, 1970), it would accumulate within the mitochondria. Therefore, it has been proposed that the function of mitochondrial malonyl-

CoA decarboxylase is to dispose of this unwanted malonyl-CoA (Christ, 1968; Scholte, 1969). In rat liver and guinea pig heart, propionyl-CoA carboxylase and malonyl-CoA decarboxylase are both mitochondrial matrix enzymes (Scholte, 1969; Wit-Peeters *et al.*, 1971). Several cases of inherited lethal propionyl-CoA carboxylase deficiency have been reported (Hommes *et al.*, 1968; Hsia *et al.*, 1969, 1971; Gompertz *et al.*, 1970).

Methylmalonyl-CoA racemase converts S-methylmalonyl-CoA into *R*-methylmalonyl-CoA, which is isomerized by *R*-methylmalonyl-CoA isomerase into succinyl-CoA. The apparent equilibrium constant for the racemase reaction is 1 at 30°C and pH 7.3 (Allen *et al.*, 1963). The enzyme has been purified from sheep liver. Also at 30°C, a slow non-enzymic racemization takes place. The reaction occurs via a shift of the α-hydrogen atom (Mazumder *et al.*, 1962).

R-Methylmalonyl-CoA isomerase, purified from sheep liver by Cannata and co-workers (1965), contains 2 moles cobamide coenzyme per mole of enzyme. It has been demonstrated that the isomerization involves the migration of the —COSCoA group (Eggerer *et al.*, 1960).

G. OXIDATION OF UNSATURATED FATTY ACIDS. Natural unsaturated fatty acids have the cis configuration. They are degraded by the enzymes of the β-oxidation until the *cis*-3-enoyl-CoA or the *cis*-2-enoyl-CoA derivative has been formed.

cis-3-Enoyl-CoA is not a substrate for enoyl-CoA hydratase, but it can be converted into the *trans*-2-enoyl-CoA derivative by the enzyme 3-*cis*-2-*trans*-enoyl-CoA isomerase, which has been purified from sonicated rat liver mitochondria by Stoffel *et al.* (1964). *cis*-2-Enoyl-CoA can be hydrated by the enoyl-CoA hydratases (Stern, 1961; Stoffel and Caesar, 1965) to D-3-hydroxyacyl-CoA. This compound is not a substrate for L-3-hydroxyacyl-CoA dehydrogenase. It is converted into the L-isomer by 3-hydroxyacyl CoA epimerase, first described by Stern (1962), which has been purified from sonicated rat liver mitochondria (Stoffel *et al.*, 1964). Rat liver mitochondria contain higher isomerase and epimerase activities than rat heart mitochondria (Stoffel and Caesar, 1965). The polyunsaturated fatty acids are oxidized by rat liver mitochondria at rates comparable to that of palmitate (Stoffel and Schiefer, 1965).

D. Oxidation of Amino Acids

It has been known for a very long time that muscle tissue can oxidize glutamate and other amino acids very rapidly. It was suggested that during this process ammonia was formed via the transdeamination pathway first suggested by von Euler and co-workers in 1938. Subsequently, however, it was shown (Copenhaver *et al.*, 1950) that heart and skeletal muscle contain only minor amounts of glutamate dehydrogenase. Later it was shown by Borst (1962) that in rat heart mitochondria glutamate is almost quantitatively converted to aspartate. Table II, taken from De Haan (1968), shows the stoicheiometry of glutamate oxidation in rat-heart mitochondria. It is clear that only minor amounts of ammonia are produced under the experimental conditions. Furthermore, the addition of malonate completely suppresses glutamate oxidation, as had already been observed by Borst and Slater (1960). From the data of Gautheron *et al.* (1964), it might be inferred that in pig heart the deamination of glutamate is slightly greater than in rat heart. Pigeon breast muscle is virtually devoid of glutamate dehydrogenase, so that only transamination of glutamate takes place in mitochondria from this tissue (Borst, 1962). In insect flight muscle, no ammonia formation could be measured during flight (Sacktor and Wormser-Shavit, 1966), although Van den Bergh (1962) detected slight glutamate dehydrogenase activity in mitochondria from housefly flight muscle. Gerez and Kirsten (1965) observed ammonia production in muscle tissues poor in mitochondria such as rat skeletal muscle, whereas in mitochondria-rich muscles (like those of pigeon breast muscle), no ammonia was formed. A possible route of ammonia production in sekeletal muscle has recently been demonstrated by Lowenstein and Tornheim (1971) using muscle extracts. This route involves deamination of AMP and the subsequent conversion of the IMP and aspartate into AMP. AMP deaminase has been detected in various types of muscle (Conway and Cook, 1939). Aspartate can be formed from glutamate intramitochondrially.

The physiological role of amino acid breakdown in muscle is uncertain. The extent of oxidation usually observed is too low to be of importance

TABLE II

STOICHIOMETRY OF GLUTAMATE OXIDATION IN RAT HEART MITOCHONDRIA

Additions	Substrate utilization or product formation (μatoms or μmoles/30 min)			
	$-\Delta$ Oxygen	$-\Delta$ Glu	$+\Delta$ Asp	$+\Delta$ NH$_3$
None	3.4	1.4	1.3	0.2
Malonate (20 mM)	0.2	0.2	0.1	0.1
Malate (20 mM)	3.4	2.4	2.6	0.1
Malonate + malate	3.7	2.7	2.5	0

for energy production. Possible roles of amino acid oxidation are in replenishment of Krebs cycle intermediates (Van den Bergh, 1962) or (especially the AMP–IMP cycle) in adjusting the levels of adenine and inosine nucleotides (Lowenstein and Tornheim, 1971).

E. The Respiratory Chain

In the oxidation reactions of the Krebs cycle, the hydrogen atoms (or the electrons derived from them) do not react directly with oxygen, but pass through a series of hydrogen or electron carriers, the respiratory chain. The present picture of the respiratory chain is shown in Fig. 2.

The primary hydrogen acceptor for the oxidation of pyruvate and α-ketoglutarate is the lipoic acid covalently bound to one of the proteins of the corresponding keto–acid dehydrogenase complex. Hydrogens from

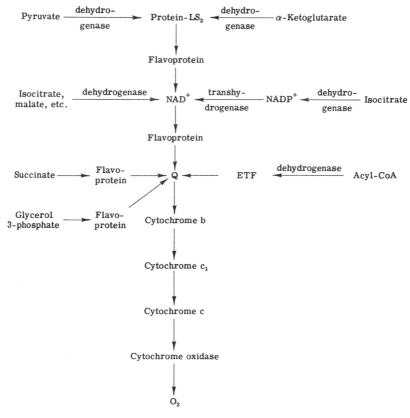

Fig. 2. The respiratory chain in mitochondria.

the reduced lipoic acid are transferred to NAD^+ in a reaction catalyzed by lipoamide dehydrogenase. NAD^+ is also the hydrogen acceptor for the oxidation of a number of substrates, including malate, isocitrate, and L-3-hydroxyacyl-CoA, each oxidoreduction being catalyzed by a specific dehydrogenase.

There are two $NADP^+$-linked dehydrogenases in heart muscle mitochondria catalyzing the oxidative decarboxylation of isocitrate and malate; *threo*-D_s-isocitrate:$NADP^+$ oxidoreductase (decarboxylating) (EC 1.1.1.42) is present both in the mitochondrion (80–85%) and in the extramitochondrial compartment (Goebell and Pette, 1967). Brdiczka and Pette (1971) have shown that in heart muscle, "malic" enzyme (malate:$NADP^+$ oxidoreductase (decarboxylating) (EC 1.1.1.40)) occurs predominantly in the mitochondrial fraction (68–86% of the total cellular activity, depending on the species). The transfer of hydrogens from NADPH to NAD^+ is catalyzed by a transhydrogenase (see Section IV,B).

Another flavoprotein in the respiratory chain transfers hydrogens from NADH to Q:NADH dehydrogenase. Q is also the hydrogen acceptor for the flavoprotein-catalyzed oxidation of succinate and of acyl-CoA. The oxidation of QH_2 by molecular oxygen involves the transfer of electrons rather than hydrogens (or hydride ions), as occurs in the first part of the respiratory chain. Electrons are transferred successively to cytochrome b, cytochrome c_1, and cytochrome c. The enzyme catalyzing the oxidation of reduced cytochrome c by oxygen is cytochrome oxidase; it contains two heme groups, a and a_3, and two copper atoms (Beinert *et al.*, 1970).

Considerable attention has been paid to the occurrence of spectrally distinct species of cytochrome b in the respiratory chain (for reviews, see Slater, 1971; Van Dam and Meijer, 1971). These different species can be detected under different states of energization of the mitochondria.

Several inhibitors of the respiratory chain and of the oxidative phosphorylation coupled with it (Section III,F) are known. Some practical information concerning their application in biochemical research is summarized by Slater (1967).

F. Oxidative Phosphorylation

Each of the oxidative steps involved in the oxidation of a respiratory substrate is coupled with phosphorylation. With one exception only, the phosphorylations are associated with the transfer of reducing equivalents from NADH (or QH_2) to oxygen via the respiratory chain. The exception is the substrate-linked phosphorylation associated with the oxidation of α-ketoglutarate. During the oxidation of this substrate by the α-ketoglutarate dehydrogenase complex succinyl-CoA is formed, which can be utilized for the synthesis of ATP according to the following reactions:

$$\text{Succinyl CoA} + \text{GDP} + \text{P}_i \rightleftharpoons \text{succinate} + \text{CoASH} + \text{GTP} \quad (32)$$

$$\text{GTP} + \text{AMP} \rightleftharpoons \text{GDP} + \text{ADP} \quad (33)$$

$$2\text{ADP} \rightleftharpoons \text{AMP} + \text{ATP} \quad (34)$$

Sum: $\text{Succinyl-CoA} + \text{ADP} + \text{P}_i \rightleftharpoons \text{succinate} + \text{ATP} + \text{CoASH} \quad (35)$

Reaction (32) is catalyzed by succinyl-CoA synthetase, Reaction (33) by nucleoside monophosphatae kinase (Heldt and Schwalbach, 1967), and Reaction (34) by adenylate kinase. Nucleoside diphosphate kinase, which in liver catalyzes the sum of Reactions (33) and (34), is shown to be practically absent in heart and skeletal muscle mitochondria (Klingenberg and Pfaff, 1966).

The substrate-linked phosphorylation associated with the oxidation of α-ketoglutarate is uncoupled by arsenate (Sanadi *et al.*, 1954; see also Needham and Pillai, 1937), but not by uncouplers of respiratory chain phosphorylation (Hunter, 1951; Judah, 1951). It is also insensitive to oligomycin (Chappell and Greville, 1961), the inhibitor of respiratory chain phosphorylation.

Respiratory chain phosphorylation is quantitatively the most important source of ATP in muscle under aerobic conditions. There appear to be three phosphorylation steps associated with the oxidation of NADH by oxygen, known as Sites I, II, and III (Fig. 3).

The respiratory chain and the enzyme systems concerned with respiratory chain phosphorylation are located in the inner membrane of mitochondria. Although the mechanism of substrate-linked phosphorylation is known (see Tager *et al.*, 1969, for a review), that of respiratory chain phosphorylation has not yet been unraveled. Three main hypotheses have been proposed to account for the nature of the primary energy-conserving process in the oxidoreductions of the respiratory chain (for a review, see Slater, 1971).

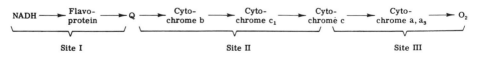

Fig. 3. Sites of ATP synthesis in the respiratory chain.

1. CHEMICAL HYPOTHESIS

This was proposed by Slater in 1953 and is modeled on the mechanism of the substrate-linked phosphorylation. According to this hypothesis, energy is conserved in an energy-rich compound between one of the products of the redox reaction between AH_2 and B and a ligand C [Reaction (36)]. Subsequently, the energy-rich compound designated as A∼C is utilized to synthesize ATP (reaction (37)).

$$AH_2 + B + C \rightleftharpoons A \sim C + BH_2 \tag{36}$$

$$A \sim C + ADP + P_i \rightleftharpoons A + C + ATP \tag{37}$$

$$\text{Sum:} \quad AH_2 + B + ADP + P_i \rightleftharpoons A + BH_2 + ATP \tag{38}$$

This hypothesis explains the fact that the oxidation of a substrate in "tightly coupled" mitochondria is dependent on the presence of ADP and P_i. This phenomenon of respiratory control was first demonstrated clearly by Lardy and Wellman (1952).

A ∼C can be utilized not only for the synthesis of ATP, but also for other energy-requiring processes, such as the uptake of cations (see Section IV,A) and the energy-linked transhydrogenase (Section IV,B).

Reaction (37) is inhibited by oligomycin (see Lardy *et al.*, 1958). The effect of an uncoupler is presumably to cause the hydrolysis of A ∼ C:

$$A \sim C \xrightarrow{\text{uncoupler}} A + C \tag{39}$$

These relationships are illustrated in Fig. 4. Instead of A ∼ C, the more general formulation "energy pressure" can be used.

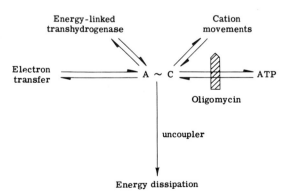

Fig. 4. General formulation of utilization of energy of electron transfer in mitochondria.

It should be noted that all these reactions, except the uncoupler-stimulated dissipation of the energy pressure, are reversible. The utilization of energy for the reversal of the respiratory chain was first observed by Chance (1956a), and it has been demonstrated in several laboratories that the extrusion of cations from mitochondria can be used to synthesize ATP (Cockrell *et al.*, 1967; E. Rossi and Azzone, 1970). Grinius *et al.* (1970), have demonstrated that the energy-linked transhydrogenase is reversible. Extending these observations, Van de Stadt *et al.* (1971) showed that the oxidation of NADPH by NAD+ in submitochondrial particles from beef heart mitochondria can be coupled with the synthesis of ATP. It is also clear from Fig. 4 that: (a) The uncoupler-stimulated ATPase is inhibited by oligomycin. (b) Oligomycin inhibits the utilization of ATP for energy-requiring reactions such as the uptake of cations. (c) The uncoupler allows oxidation to take place ·in the absence of ADP. (d) Oligomycin inhibits the ADP-dependent respiration. (e) The uncoupler releases the inhibition of respiration by oligomycin. (f) Respiration can be stimulated in the absence of ADP by introducing an energy-requiring reaction like the uptake of cations and that this is insensitive to oligomycin.

2. Chemiosmotic Hypothesis

According to this hypothesis (Mitchell, 1961), the primary energy-conserving act is the translocation of protons across the mitochondrial inner membrane, uncompensated by the movement of cations in the opposite direction or of anions in the same direction. The membrane potential built up in this way is utilized to synthesize ATP.

3. Conformational Hypothesis

It is of interest that this hypothesis is based on the belief that muscular contraction is brought about by an effect of ATP on the conformation of myosin. Indeed, Gillis and Maréchal (1969) have reported a reversal of this process—the stretching of previously contracted glycerinated muscle fibers can be coupled to the synthesis of ATP.

According to the conformational hypothesis, then, the primary energy-conservation process in respiratory chain phosphorylation is a confirmational change in a respiratory protein (King *et al.*, 1965; Boyer, 1965). This energy can then be utilized to synthesize ATP. Formally, this hypothesis does not differ in principle from the chemical hypothesis.

The evidence in favor of these different mechanisms has recently been reviewed by Slater (1971) and Van Dam and Meijer (1971). Slater *et al.* (1970) have proposed a mechanism of Site II phosophorylation involving high-energy forms of cytochrome b.

IV. Energy-Consuming Reactions in Muscle Mitochondria

A. *Cation Transport*

1. INTRODUCTION

In the past decade, a vast amount of literature on mitochondrial cation transport has appeared. A number of excellent reviews have covered in detail the ion translocating properties of mitochondria and of submitochondrial particles obtained from intact mitochondria by various treatments (Lehninger *et al.*, 1967; Greville, 1969; Chance and Montal, 1972; Van Dam and Meijer, 1971). Most of the work is done within the framework of the study of oxidative phosphorylation and mitochondrial energy metabolism; therefore, only a general description of mitochondrial ion transport will be given in this review.

Since the early observations that mitochondria contain and are able to retain cations (Slater and Cleland, 1953; Bartley and Davies, 1954), a considerable increase in the knowledge of ion transport has occurred. Cation uptake by mitochondria is an energy-requiring process. Therefore, in analogy with the various hypotheses of oxidative phosphorylation (see Section III,F), two possibilities of energy-driven cation transport are currently being considered (cf. Greville, 1969).

A. THE CHEMICAL HYPOTHESIS. An energy-rich intermediate of oxidative phosphorylation ($A \sim C$ in Fig. 4), generated by electron transport through the respiratory chain or by ATP, catalyzes cation transport in two possible ways, (a) by directly driving a specific cation pump (Rasmussen *et al.*, 1965) or (b) by driving a H^+ pump, which in its turn causes an opposite movement of cations (Chappell and Crofts, 1965).

B. THE CHEMIOSMOTIC HYPOTHESIS. The H^+ gradient generated by oxidoreduction or by ATP is used for the movement of ions (Mitchell and Moyle, 1969). These authors had shown previously (Mitchell and Moyle, 1965) that when mitochondria become energized by a short action of the respiratory chain, they extrude protons. Later it became clear that most of this H^+ movement is compensated for by a movement of cations (mainly Ca^{2+}) in the opposite direction. Similar results were obtained when ATP was used to energize the membrane. However, in particles obtained from mitochondria by sonication, H^+ is taken up upon energization (Mitchell and Moyle, 1965; Chance and Mela, 1967). This is in accordance with the proposal of Lee and Ernster (1966) that the orientation of the membrane in submitochondrial particles is opposite to that of intact mitochondria [see also Racker (1970) and Section V,A].

2. Uptake of Divalent Cations

Mitochondria can take up amounts of Ca^{2+} as high as 2–3 μatoms per milligram of mitochondrial protein (C. S. Rossi and Lehninger, 1963). This unphysiological uptake, which leads to precipitation of calcium phosphates within the mitochondria (Greenawalt *et al.*, 1964), damages the mitochondria considerably, as can be seen from the resulting uncoupling of oxidative phosphorylation. When, however, small amounts of Ca^{2+} are added to mitochondria under conditions of respiratory control (no ADP present) a transient burst of respiration is found that is proportional to the amount of Ca^{2+} added (Chance, 1956b, 1965). This limited calcium uptake is characterized by (a) an opposite movement of H^+ ions, (b) swelling of the mitochondria, (c) a transient splitting of ATP (ATPase) when the process is driven by ATP, and (d) when driven by substrate oxidation, a transient burst in oxygen uptake together with a transient oxidation–reduction cycle of the components of the respiratory chain (see Chance, 1965; Rasmussen *et al.*, 1965; Chance and Mela, 1966; C. S. Rossi *et al.*, 1966).

Much attention has been paid to the ratio of Ca^{2+} taken up to energy-rich intermediate ($A \sim C$) expended, and a value of 2 has been found by various laboratories (see Lehninger *et al.*, 1967). The exact value of this ratio is difficult to estimate, since it has been found that in addition, nonenergy-linked binding of Ca^{2+} to mitochondria, mainly to phospholipids, takes place (Scarpa and Azzi, 1968; Reynafarje and Lehninger, 1969; Jacobus and Brierley, 1969). Also, exchange of Ca^{2+} for intramitochondrial cations can occur (E. J. Harris *et al.*, 1966; Scarpa and Azzone, 1970). Recently, a protein has been obtained from mitochondria that binds Ca^{2+} with an extremely high affinity (Lehninger, 1970, 1971; see however Moore, 1971). It is suggested that this protein is involved in the Ca^{2+} transport over the mitochondrial membrane.

Most of the work on Ca^{2+} uptake has been performed with rat liver mitochondria. Fehmers (1968) obtained the same results with rabbit heart mitochondria. He calculated from his experiments that under optimal conditions the mitochondria in the heart muscle are able to halve the extramitochondrial Ca^{2+} concentration within about 200 msec. Since this applies to an initial concentration of 10^{-5}–10^{-6} M, which lies within the physiological range, Fehmers (1968) concluded that mitochondrial Ca^{2+} uptake contributes considerably to the relaxation of the heart muscle. The same conclusion has been reached by Horn et al. (1969, 1971). In blowfly flight muscle mitochondria Ca^{2+} is taken up in an energy-dependent way, but probably without interaction of a specific Ca^{2+} carrier (Carafoli et al., 1971a,b). The presence of permeant anions that are taken up simultaneously with Ca^{2+} is essential in these mitochondria. Ca^{2+} uptake by blowfly flight muscle mitochondria is not accompanied by a transient stimulation of oxygen uptake (Carafoli et al., 1971; Carafoli and Lehninger, 1971).

The uptake of other divalent cations such as Sr^{2+} (Carafoli et al., 1965; Carafoli, 1965), Mn^{2+}, or Mg^{2+} (Chappell et al., 1962; Brierley, 1963) resembles that of Ca^{2+}, although certain differences exist.

3. Uptake of Monovalent Cations

Isolated mitochondria are relatively impermeable to monovalent cations; K^+ uptake, for instance, is slow (Gamble, 1957). However, it is possible to increase the permeability to K^+ by adding the antibiotic valinomycin to the mitochondria (Moore and Pressman, 1964). In the presence of valinomycin, uptake of K^+ closely resembles that of Ca^{2+}. A detailed review of the work on transport of monovalent ions in mitochondria and submitochondrial particles is given by Van Dam and Meijer (1971).

B. Energy-Linked Transhydrogenase

The transfer of hydrogen atoms between NAD and NADP is catalyzed by NAD(P) transhydrogenase, an enzyme discovered by Kaplan and co-workers (1953). Since the redox potential of the two nicotinamide nucleotides is almost equal (Kaplan et al., 1953; Lee and Ernster, 1964), it is to be expected that at thermodynamic equilibrium intramitochondrial NAD and NADP will be equally reduced. However, Klingenberg and Slenczka (1959) observed that in intact mitochondria, NADP was much more reduced than NAD. This difference was overcome by ADP or uncouplers of oxidative phosphorylation. They postulated the existence of an asymmetric transfer of hydrogen between NADH and NADP⁺ involving energy (Klingenberg and Slenczka, 1959; see also Estabrook and Nissley, 1963). Subsequently, the existence of such an energy-linked hydrogen transfer was demonstrated by Danielson and Ernster

(1963). It is now generally accepted that the energy for the hydrogen transfer

$$NADH + NADP^+ + \sim \rightleftharpoons NAD^+ + NADPH \tag{40}$$

is provided by a high-energy intermediate of oxidative phosphorylation produced by substrate oxidation or by ATP (see Fig. 4). The amount of $\sim$ necessary to drive Reaction (40) is generally thought to be 1 per $NADP^+$ reduced (Danielson and Ernster, 1963; Tager *et al.*, 1967). As mentioned in Section III,F, it has recently been possible to demonstrate the reversibility of Reaction (40) (Grinius *et al.*, 1970; Van de Stadt *et al.*, 1971).

The energy-linked transhydrogenase in submitochondrial particles is strongly inhibited by an antibody against the purified transhydrogenase enzyme, which is not energy-linked (Kawasaki *et al.*, 1964). This is strong evidence in favor of the view that the transhydrogenase enzyme plays a role in the energy-linked transhydrogenase system.

Currently, two hypotheses about the mechanism of the energy-linked transhydrogenase exist. The first involves the formation in the presence of energy of an energized intermediate of NADH (or $NADP^+$) which then reacts with $NADP^+$ (or NADH) (Ernster and Lee, 1964). The alternative hypothesis involves an active, energy-linked transport of one of the reactants (NAD^+) from the transhydrogenase enzyme, thus maintaining a low NAD^+ concentration in the vicinity of the active center of the enzyme, which also leads to virtually complete reduction of $NADP^+$ (Papa and Francavilla, 1967).

The role of the energy-linked transhydrogenase in metabolism is not established. The process is demonstrated in many types of mitochondria with the notable exception of insect flight muscle mitochondria (Donnellan *et al.*, 1970). It might play a role in providing reducing equivalents for synthetic reactions. The lack of a clear physiological function of the energy-linked transhydrogenase has led Krebs and Veech (1969) to "suggest the possibility that the energy-linked transhydrogenation may be an *in vitro* artefact in the sense that it is a reaction which does not occur *in vivo* but only under artificial conditions."

C. Synthesis of Fatty Acids

In liver, three systems for the biosynthesis of fatty acids are present:

1. *The cytosolic system* responsible for the synthesis of the entire fatty acid molecule. This so-called *de novo* system requires malonyl-CoA

as substrate, with acetyl-CoA serving as a sparker molecule, and NADPH as hydrogen donor (see Lynen, 1961).

2. *The microsomal system* elongates acyl-CoA esters, requires malonyl-CoA as substrate and either NADH or NADPH as hydrogen donor (Nugteren, 1965; Guchhait *et al.*, 1966).

3. *The mitochondrial system* also elongates acyl-CoA esters, but requires acetyl-CoA as substrate and NADH as hydrogen donor (see Wit-Peeters, 1969). This system is firmly bound to the inside of the inner membrane (Quagliariello *et al.*, 1968; Wit-Peeters, 1969).

In rat-liver homogenates, these three systems can be measured independently, and their synthesizing capacities have the same order of magnitude. The mitochondrial system, in contrast to the others, is not lowered by fasting (Donaldson *et al.*, 1970).

Fatty acid synthesis in heart mitochondria was discovered by Hülsmann in 1960. The substrate specificity was proven to be the same as described above for the liver mitochondrial system (Whereat *et al.*, 1967; Hull and Whereat, 1967; Christ, 1968; Dahlen and Porter, 1968; Wit-Peeters, 1969). In heart, no fatty acid-synthesizing systems are present in the microsomes or in the cytosol (Wit-Peeters *et al.*, 1970). Also, ATP citrate lyase (Garland and Randle, 1964b; Wit-Peeters *et al.*, 1970) and acetyl-CoA carboxylase (Wit-Peeters *et al.*, 1970) are absent from heart, and the inner membrane of heart mitochondria is relatively impermeable for citrate (compare Section VI). These enzymes and the tricarboxylic acid carrier operate in rat liver and convert intramitochondrial citrate into extramitochondrial malonyl-CoA for extramitochondrial fatty acid synthesis.

Whereat *et al.* (1969) were able to separate the inner and outer membranes of rabbit heart mitochondria and demonstrated synthesizing activity in both membranes. However, the loss of 80% of the fatty acid-synthesizing activity during their isolation procedure detracts somewhat from the conclusiveness of their results. These investigators found elongation activity in the outer membranes and *de novo* activity in the inner membranes. By latency and fractionation studies of guinea pig heart mitochondria, Wit-Peeters and co-workers (1970; Wit-Peeters, 1971) showed that the elongation system is present at the inside of the inner mitochondrial membrane, to which it is firmly bound, but that it also resides (one-third of total) on the outside of the inner membrane or in the outer membrane. They could not detect *de novo* activity in guinea pig heart. Ghosal *et al.* (1969) showed *de novo* activity in isolated rat heart, perfused with [^{14}C]acetate. However, as pointed out by Wit-Peeters (1971), in these experiments contamination by other cell components or cells of a different tissue has not been absolutely excluded. By degradation studies of the fatty acids formed by rat heart mitochondria metabolizing [2-^{14}C]pyruvate, Christ (1968) showed that practically no *de novo* synthesis occurs, but that endogenous or added fatty acids are elongated by no more than two or three C_2 units. He found that most of the radioactivity present in palmitic acid is recovered in the carboxyl carbon atom and that the first six carbon atoms of myristic acid, counted from the methyl end, contained practically no radioactivity. Also, experiments of Dahlen and Porter (1968) with a partially purified preparation of beef heart mitochondria and of Wit-Peeters (1969) with guinea pig heart mitochondria and beef heart muscle preparation are in agreement with the view that heart mitochondria only elongate fatty acids.

The mitochondrial elongation system has been thought to be the reverse of the β-oxidation. By separating guinea pig heart mitochondria into a mixed-membrane fraction, an intermembrane fraction, and a soluble matrix fraction, Wit-Peeters *et al.* (1971) showed that the distribution pattern of none of the enzymes of fatty acid oxidation tested—palmitoyl-CoA dehydrogenase, L-3-hydroxyacyl-CoA dehydrogenase, and the enoyl-CoA hydratases—exactly parallels that of the synthesizing system (compare Section V).

V. Structure of Muscle Mitochondria and Localization of Enzymes within the Mitochondria

A. Morphological Studies

The early history of morphological studies with the light microscope on the structure of muscle mitochondria has recently been reviewed (Weijers, 1971; see also Section I). A combined light microscopic and biochemical study of heart mitochondria was carried out by Cleland and Slater (1953a,b) and Slater and Cleland (1953). The isolated mitochondria clearly showed a membrane, at that time a controversial subject (Harman, 1950; Huennekens and Green, 1950).

The improvements in the technique for preparations of thin sections made it possible to visualize the structure of the interior of the mitochondrion. These investigations started when Palade (1952) and Sjöstrand (1953) independently published electron micrographs of mitochondria in fixed and embedded tissue. Since that time electron microscopic studies of the morphology of mitochondria has been described in many scientific papers. The detailed structural elements seen in electron micrographs of sectioned isolated mitochondria characterize the mitochondria of most known cell types. Going from the outside to the inside of the organelle, the following structural elements are encountered:

1. A smooth outer membrane.
2. An intermembrane space between the inner and outer membrane.
3. An inner membrane, with infoldings, termed "cristae mitochondriales" by Palade (1952). Heart and other muscle mitochondria possess many more cristae than liver mitochondria. As a consequence, the inner membranes of muscle mitochondria have a tubular appearance.
4. A matrix space, which is surrounded by the inner membrane.

In negatively-stained preparations, the inner membrane is found to be studded with small particles, the inner membrane subunits, discovered by Fernández-Morán (1962). These particles, which are confined to the inner membrane (Stoeckenius, 1963; Parsons, 1963), consist of a spherical or hexagonal head (diameter 9.0 nm), attached by means of a stalk (5.0×3.5 nm) to the inner membrane (see also Fernández-Morán *et al.*, 1964). Racker and co-workers (1965; Kagawa and Racker, 1966) identified the inner membrane subunits with the mitochondrial ATPase, which is identical with coupling factor F_1 (Racker and Conover, 1963). This enzyme is thought to be involved in oxidative phosphorylation.

The localization of the inner membrane subunits is a controversial matter (for a discussion, see Stoeckenius, 1970). Most workers in this field believe that the inner membrane subunits are located at the matrix side of the inner membrane. This is supported by images of partially disrupted mitochondria in which the relationship of the submitochondrial structures still can be recognized. When, however, only the intermembrane space is penetrated by the negative stain, the particles are not visible. Submitochondrial vesicles obtained by sonication carry the subunits at the outside. These vesicles are believed to have the "inside-out" configuration, with the intermembrane space inside (Mitchell and Moyle, 1965; Löw, 1966; Lee and Ernster, 1966).

Weijers (1971) has questioned this and argues that several lines of evidence can be brought forward to localize the inner membrane subunits at the intermembrane side of the membrane. Weijers (1971) found that in partly disrupted rat heart mitochondria with highly condensed matrix (Hackenbrock, 1966), the inner-membrane subunits faced the outer membrane. He used the "dense bodies" (Weinbach and von Brand, 1967) and "slender structures" as markers for the matrix side of the inner membrane. Weijers (1971) also proposes, on the basis of both morphological and biochemical evidence, that submitochondrial vesicles obtained by sonication may consist of two types of particles, one type with intermembrane space inside and the other with matrix space inside (see also Chance *et al.*, 1970).

At first sight this view appears to be in contradiction to the observation of Fessenden and Racker (1966) that an antibody to Racker's coupling factor F_1 (ATPase) completely inhibits the ATPase of submitochondrial particles. This observation suggests that all of the ATPase is accessible to the antibody and must therefore be on the outside of the vesicles. However, Souverijn (1971) has pointed out that these particles contain little, if any, adenine nucleotides (Klingenberg, 1967), so that added ATP would not be able to exchange with "internal" adenine nucleotide via the adenine nucleotide translocator (see Section VI) and would therefore be resistant to breakdown by ATPase on the inside of the vesicles (see also Weijers, 1971).

Hall and Crane (1970) using thin sectioning observed a new structural element in the intermembrane space of cristae of beef heart mitochondria that they interpret as a sheet of parallel rods, on the outside of the infoldings of the inner membrane. The diameter of the particles is 5.0 nm.

Morphological changes within intact mitochondria, as a consequence of metabolic changes were first studied with isolated rat liver mitochondria by Hackenbrock (1965, 1966, 1968a,b; Hackenbrock and Kaplan, 1969). He described two main conformations—condensed (with regard to the matrix) and orthodox. The latter conformation can be seen under conditions where lack of ADP or P_i is limiting oxidative phosphorylation (State 4 according to Chance and Williams, 1956). The slow transition from condensed to orthodox after exhaustion of ADP depends upon an intact respiratory chain, but phosphorylation is not required. Addition of ADP to State 4 mitochondria gives a far more rapid transition back to the condensed form.

Morphological transitions in heart mitochondria were studied by Green and co-workers (Penniston *et al.*, 1968; R. A. Harris *et al.*, 1968; Green *et al.*, 1968; Green and McLennan, 1969; Green and Harris, 1969). The three conformations observed by these authors were also thought to be related to the energetic state of the mitochondria. However, Stoner and Sirak (1969) obtained the same conformation changes, believed by Green and co-workers to be essential for oxidative phosphorylation, by allowing swelling to take place. Other investigators also conclude from their experiments that there exists no strict correlation between ultrastructure and metabolic state (Bronk and Jasper, 1968, 1970; Weber *et al.*, 1969; Kuner and Beyer, 1970).

B. Physicochemical Differences between Inner and Outer Membrane

In 1966–1967 several groups of investigators reported techniques for the separation of the inner and outer membrane of liver mitochondria. This was possible because these membranes differ in the following respects:

1. The inner membrane is an osmotic barrier, through which active transport (i.e., energy-dependent transport) or exchange occurs of some ions and passive transport of uncharged lipid-soluble molecules and water (see Papa *et al.*, 1969; Klingenberg, 1970b). The inner membrane of intact mitochondria can shrink and swell, depending upon the osmolarity of the medium (Stoner and Sirak, 1969). The outer membrane is permeable for small molecules and ions (Werkheiser and Bartley, 1957) with a maximal molecular weight of 5000–12,000 (Pfaff *et al.*, 1968). When the inner membrane swells too much, it ruptures the outer membrane (Parsons *et al.*, 1966; Stoner and Sirak, 1969).

2. The outer membrane has a lower buoyant density than the inner membrane, because it contains more phospholipid (Parsons *et al.*, 1966, 1967).

3. The outer membrane contains more cholesterol (Parsons and Yano, 1967). Probably one of the functions of the cholesterol is to promote the fluidity of the lipid part of the membrane and thereby supporting the integrity of the membrane (Graham and Green, 1970). When digitonin is added to intact rat liver mitochondria, it binds to the cholesterol of the outer membrane, which ruptures (Lévy *et al.*, 1966). Much higher digitonin concentrations are required to attack the inner membrane; this probably occurs by the detergent action of digitonin (Schnaitman *et al.*, 1967).

C. Localization of Enzymes within Muscle Mitochondria

1. INTRODUCTION

The following methods for the localization of enzymes within mitochondria are used: (a) separation of inner and outer membranes; (b) separation of the enzymes of the intermembrane space, the matrix space, and the membranes; (c) latency measurements; (d) the use of translocator inhibitors; (e) the use of proteolytic enzymes; and (f) the use of histochemical methods in electron microscopy.

Most localization studies have been done with rat liver mitochondria (for review, see Ernster and Kuylenstierna, 1970). In this section, reference will only be made to work that has been done with heart mito-chondria. When the results are compared with the localization in liver mitochondria, the similarity in localization is striking. This leads us to propose the following general rule: If an enzyme (for which no mito-chondrial isoenzymes exist) occurs both in liver and in heart mitochon-dria, the enzyme has exactly the same localization in both cases.

2. Separation of Inner and Outer Membranes

Parsons and co-workers (1966, 1967) fractionated liver mitochondria by gradient centrifugation after swelling of the mitochondria in P_i, in order to rupture the outer membrane. In this way, they obtained small amounts of pure membrane fractions. This method has, however, not been successfully applied to muscle mitochondria. Also, the digitonin method of Lévy and co-workers (1966, 1967) modified by Schnaitman et al. (1967) fails with heart mitochondria. Oliveira and co-workers (1969) fractionated beef heart mitochondria with the digitonin method, but the outer membrane fraction (monoamine oxidase as marker) was obtained in small yield only and was heavily contaminated with inner membrane fragments. Wit-Peeters, Weijers, and Scholte (see Scholte, 1970) showed that when guinea pig heart mitochondria are treated with digitonin, the outer membranes are ruptured, but most of them remain attached to the inner membrane. Under these conditions, only 2.6% of the mitochondrial creatine kinase was solubilized, which indicates that, in contrast to the current view, this enzyme is membrane bound. They obtained evidence that the enzyme in intact mitochondria is bound to the outside of the inner membrane.

The method of Sottocasa and co-workers (1967) was successfully adapted for the fractionation of rabbit heart mitochondria (Whereat et al., 1969). After swelling, shrinking, sonication, and discontinuous gradient centrifugation, four submitochondrial fractions are obtained—a fraction with solubilized enzymes, an outer membrane fraction (contain-ing 45% of the mitochondrial monoamine oxidase activity purified six-fold), a mixed membrane fraction, and an inner membrane plus matrix fraction (containing 85% of the cytochrome oxidase, purified one- to twofold). They localized fatty acid synthesizing activity in both mem-branes, but obtained a low recovery (20%) (see Section IV,C).

3. Separation of Enzymes of the Intermembrane Space, the Matrix Space, and the Membranes

Pette (1966) and Klingenberg and Pfaff (1966) have applied the technique of differential extraction developed by Delbrück et al. (1959)

to separate enzymes of the intermembrane space, the matrix space, and the membranes, and also to separate extra and intramitochondrial enzymes in homogenates. This method consists of repeated extractions in media of different composition and ionic strength. The method is excellent for the localization of soluble enzymes in the extramitochondrial compartment, the intermembrane space and the matrix, and has been successfully applied for the localization of these enzymes in locust flight muscle and rat heart mitochondria. However, since outer membrane fragments may not be completely sedimented with the centrifugation procedure used, some contamination with outer membrane enzymes may occur. Some soluble Krebs cycle enzymes (malate dehydrogenase, citrate condensing enzyme, isocitrate dehydrogenase) were located in the matrix together with aspartate aminotransferase and glutamate dehydrogenase. Most of cytochrome c, the mitochondrial adenylate kinase, and creatine kinase were recovered in the intermembrane enzyme fraction (Pette, 1966; Klingenberg and Pfaff, 1966). This method was used earlier to fractionate *Locusta migratoria* flight muscle. L-Glycerol-3-phosphate oxidase, NADH oxidase, and succinate dehydrogenase are bound to the membrane fraction, while acetyl-CoA acetyltransferase, alanine aminotransferase, aspartate aminotransferase, citrate synthase, cytochrome c, L-3-hydroxyacyl-CoA dehydrogenase, isocitrate dehydrogenase, and malate dehydrogenase become soluble after the sonic treatment (Beenakkers, 1963; Brosemer *et al.*, 1963).

The method of Delbrück *et al.* (1959) has been modified to ensure that the outer membranes should sediment together with the inner membranes and to accomplish a greater yield of creatine kinase in the intermembrane fraction (Wit-Peeters *et al.*, 1970, 1971; Scholte *et* al., 1971). They showed that in sonicated guinea pig heart mitochondria the matrix enzymes are bound to a different extent to the inner side of the inner membrane. In descending order of tightness were bound: cytochrome c oxidase, the fatty acid elongation system, aspartate aminotransferase, β-hydroxyacyl-CoA dehydrogenase, enoyl-CoA hydratase with crotonoyl-CoA as substrate, palmitoyl-CoA dehydrogenase, malate dehydrogenase, malonyl-CoA decarboxylase, and finally, very loosely bound if at all: propionyl-CoA carboxylase and the acetyl-CoA, malonyl-CoA, propionyl-CoA and butyryl-CoA synthetases. In another partition it was found that the enoyl-CoA hydratase with *trans*-2-hexadecenoyl-CoA as substrate was more loosely bound than the enoyl-CoA hydratase acting with crotonoyl-CoA. It was also found that increased sonication time resulted in an increased release of the matrix enzymes, but that the order of release of the matrix enzymes was not affected. For several matrix enzymes the observed order corresponds closely to that obtained with rat liver mitochondria (Scholte, 1969; Wit-Peeters, 1969).

4. Other Localization Studies

Other localization studies are mentioned in Sections III,C and IV,C. Histochemical studies in electron microscopy were used to identify cyto-chrome c at the outside of the inner membrane of rat heart mitochondria (Seligman *et al.*, 1968; cf. Tyler, 1970). L-Glycerol-3-phosphate dehydro-genase is localized on the outside of the inner membrane of insect flight muscle mitochondria (Donnellan *et al.*, 1970) and skeletal muscle mito-chondria (Klingenberg and Buchholz, 1970).

5. Current Status of Our Knowledge of Localization of Enzymes in Muscle Mitochondria

The localization of enzymes in the different compartments of muscle mitochondria is shown in Table III.

D. *Evidence for Two Types of Mitochondria in Skeletal and Heart Muscle*

Hülsmann and co-workers have described two types of mitochondria in skeletal muscle and in heart muscle (Hülsmann *et al.*, 1968, 1969; Hülsmann, 1970). Mitochondria that were obtained by processing the tissue with a loose homogenizer had a relatively high rate of respiration in the presence of added ATP and Mg^{2+}, due to endogenous ATPase activity, which was largely oligomycin resistant. These mitochondrial preparations also had a relatively high acid phosphatase activity. With a tight homogenizer or treatment with the proteolytic enzyme Nagarse, "deeply localized" mitochondria were obtained with a lower ATPase and a lower acid phosphatase activity.

E. *Criteria for Intactness of Isolated Muscle Mitochondria*

In establishing the morphological intactness of isolated mitochondria, the criterion that must be adopted is that after processing, the mitochon-dria should resemble as closely as possible the structure of the organelle

TABLE III
LOCALIZATION OF ENZYMES IN MUSCLE MITOCHONDRIA[a]

Outer membrane	Aspartate aminotransferase (4,11,13,19)
	Butyryl-CoA synthetase (13)
Fatty acid elongation system (16,17,18)	Carnitine acetyltransferase (2,3)
Monoamine oxidase (9,16)	Carnitine palmitoyl-transferase (20)
Palmitoyl-CoA synthetase (5,10)	Citrate condensing enzyme (4,11)
Rotenone-insensitive NADH–cyto-	Cytochrome c (7,11)
chrome c reductase (16)	Cytochrome c oxidase (9,13,16,18,19)
	Enoyl-CoA hydratase (with crotonoyl-
Outer surface of inner membrane	CoA) (19)
Carnitine palmitoyl transferase (20)[b]	Enoyl-CoA hydratase (with *trans*-2-hexa-
Creatine kinase (12)	decenoyl-CoA) (19)
Cytochrome c (7,11,14,15)	Fatty acid elongation system (17,18,19)
L-Glycerol-3-phosphate dehydrogenase	Glutamate dehydrogenase (11)
(6,8)	L-Hydroxyacyl-CoA dehydrogenase
	(1,19)
Intermembrane space	Isocitrate dehydrogenase (4,11)
Adenylate kinase (7,11)[b]	Malate dehydrogenase (4,7,11,13,18,19)
Carnitine acetyltransferase (2,3,20)	Malonyl-CoA decarboxylase (13,19)
	Palmitoyl-CoA dehydrogenase (19)
Inner membrane plus matrix	Propionyl-CoA carboxylase (13,18,19)
Acetyl-CoA acetyltransferase (1)	Propionyl-CoA synthetase (13)
Acetyl-CoA synthetase (13)	Respiratory chain (4)[c]
Adenylate kinase (7,11)	Succinate dehydrogenase (4)[c]
Alanine aminotransferase (4,11)	

[a] REFERENCES: (1) Beenakkers, 1963; (2) Beenakkers and Klingenberg, 1964; (3) Beenakkers and Henderson, 1967; (4) Brosemer *et al.*, 1963; (5) De Jong and Hülsmann, 1970; (6) Donnellan *et al.*, 1970; (7) Klingenberg and Pfaff, 1966; (8) Klingenberg and Buchholz, 1970; (9) Oliveira *et al.*, 1969; (10) Pande and Blanchaer, 1970; (11) Pette, 1966; (12) Scholte, 1970; (13) Scholte *et al.*, 1971; (14) Seligman *et al.*, 1968; (15) Tyler, 1970; (16) Whereat *et al.*, 1969; (17) Wit-Peeters, 1971; (18) Wit-Peeters *et al.*, 1970; (19) Wit-Peeters *et al.*, 1971; (20) Yates and Garland, 1970.

[b] Has not definitely been established, but inferred from comparison with the localization in rat liver mitochondria.

[c] The Keilin and Hartree heart muscle preparation, which contains purified inner membrane (see above) has a high succinate oxidase activity.

in vivo. The main problem in visualizing the structure in the electron microscope is the fixation, as has been underlined by many authors (see Butler and Judah, 1970).

In the light of present knowledge we can revise the generally accepted biochemical criteria for the intactness of isolated muscle mitochondria (cf. Borst, 1961):

1. The mitochondria should be able to oxidize pyruvate and short-chain fatty acids at optimal rate without added cofactors.

2. They should exhibit respiratory control, i.e., the velocity of oxidation should be dependent upon the presence of ADP and P_i.

3. The oxidation of added reduced cytochrome c should not take place.

4. Rotenone may not lower the oxidation of added NADH in the presence of added cytochrome c.

5. The mitochondria should exhibit a low ATPase activity in the absence of uncoupler.

6. The palmitoyl-CoA synthetase activity should be high.

7. The creatine kinase activity should be high.

VI. Interrelationships between Mitochondrial and Extramitochondrial Processes in Muscle

In an organ with many synthetic functions, like the liver, there is an intimate interrelationship between intra- and extramitochondrial processes. For instance, gluconeogenesis and urea synthesis take place partly in the mitochondria and partly in the extramitochondrial compartment. The function of mitochondria in liver is not only to provide ATP for extramitochondrial synthetic processes, but also to provide reducing equivalents, the carbon skeleton for gluconeogenesis, and the carbon skeleton for extramitochondrial fatty acid synthesis (see Slater et al., 1969).

The situation in muscle is simpler. The most important function of muscle mitochondria is to synthesize ATP. Nevertheless, also in muscle a close cooperation between mitochondria and the rest of the cell exists and a large number of compounds has to cross the inner membrane of the mitochondria in both directions. Substances entering muscle mitochondria are: (a) oxygen; (b) the oxidized end product of glycolysis—pyruvate; (c) the end products of the hydrolysis of lipids, chiefly activated fatty acids (see Section III,C); (d) acetoacetate and β-hydroxybutyrate; (e) glutamate and possibly some other aminoacids; (f) the hydrogen carrier that brings into the mitochondrion the reducing equivalents produced by glycolysis, since the mitochondrial membrane is impermeable to the nicotinamide nucleotides (in muscle tissues other than insect flight muscle, this may be malate together with glutamate, as suggested by Borst (1963); in insect flight muscle, the redox couple responsible for the transport of reducing equivalents has shown to be dihydroxyacetone phosphate and glycerol 3-phosphate (Bücher and Klingenberg, 1958); however, no transport of these compounds through the mitochondrial membrane is necessary, since the mitochondrial glycerol-3-phosphate dehydrogenase is located at the outside of the inner membrane (Klingenberg and Buchholz, 1970); (g) the substrates for oxidative phosphorylation, i.e., P_i and ADP.

Substrates leaving muscle mitochondria are: (a) the endproducts of the mitochondrial oxidations, i.e., carbon dioxide (or bicarbonate), water, and ATP; (b) aspartate; (c) the oxidized form of the redox couple responsible for bringing reducing equivalents from outside the mitochondrion to the inside, i.e., according to the proposal of Borst (1963), aspartate together with α-ketoglutarate.

It should be noted that most of the substances entering or leaving the mitochondrion are anions. Thus, the question arises of how electroneutrality is maintained.

Most of the work on the transport of anions across the inner mitochondrial membrane has been carried out with liver mitochondria. It has become clear since 1964, mostly as a result of the pioneering work of Chappell (see Chappell, 1968), that this transport is brought about by specific carrier systems, or translocators, situated in the inner mitochondrial membrane. These translocators mediate a 1:1 exchange (see Meijer, 1971) of specific anion pairs, so that electroneutrality is maintained. The studies of anion transport in mitochondria have recently been reviewed by Klingenberg (1970a,b) and by Meijer (1971). The translocators may be distinguished not only on the basis of their specificity for the anions exchanged, but also by their sensitivity to inhibitors (see Robinson *et al.*, 1971).

In liver mitochondria, the following translocators have been described: (a) phosphate translocator, mediating a phosphate–hydroxyl exchange; (b) dicarboxylate translocator, mediating a dicarboxylate–phosphate exchange; (c) tricarboxylate translocator, mediating a tricarboxylate–malate or tricarboxylate–tricarboxylate exchange; (d) α-ketoglutarate translocator, mediating an α-ketoglutarate–dicarboxylate exchange; (e) aspartate translocator, first proposed by Azzi *et al.* (1967) on the basis of the stimulation by glutamate of aspartate-dependent mitochondrial reactions, has been shown by Tas and Tager (unpublished observations) to mediate a 1:1 exchange of aspartate for glutamate in submitochondrial preparations derived from beef heart mitochondria; (f) adenine nucleotide translocator mediates the exchange of ADP(ATP) for ATP(ADP); under physiological conditions, it exchanges extramitochondrial ADP for intramitochondrial ATP. A dicarboxylate–dicarboxylate exchange also occurs and is mediated by the dicarboxylate, the tricarboxylate, and the α-ketoglutarate translocators (see Meijer, 1971).

Heart mitochondria appear to contain the translocators mentioned above, with one important exception—it has not been possible to show the presence of the tricarboxylate translocator (Chappell *et al.*, 1968; Meijer, 1971). Also, skeletal muscle mitochondria are devoid of this translocator (Meijer, 1971). In accordance with these observations, England and Robinson (1969) reported that the rate of efflux of citrate from rat heart mitochondria is only 10 nmoles/min per milligram protein at 30°C. At 25°C, the rate is 7 nmoles/min per milligram protein, whereas in rat liver mitochondria this rate is 22 nmoles/min per milligram protein (Meijer, 1971).

Mitochondria from the flight muscle of the housefly are a special case. Under conditions suitable for substrate oxidation (P_i and ADP present), these mitochondria were shown to be practically impermeable to citric acid cycle intermediates (Van den Bergh and Slater, 1962; Van den Bergh, 1964). Later it was observed that this impermeability was only apparent; at low phosphate concentrations, di- and tricarboxylate anions can penetrate the inner membrane (Tulp and Van Dam, 1969; Tulp *et al.*, 1971). Tulp and co-workers demonstrated that the dicarboxylate translocator is strongly inhibited by low concentrations of phosphate.

In the intact flight muscle cell, enough phosphate is present to inhibit the dicarboxylate and tricarboxylate carriers. This might result in prevention of leakage of citric acid cycle intermediates from the mitochondria, as originally postulated by Van den Bergh and Slater (1962).

The question of the transfer of reducing equivalents across the mitochondrial membrane has been mentioned briefly. Although a mechanism for the oxidation of extramitochondrial NADH by mitochondria exists in insect mitochondria in the form of the glycerol 3-phosphate cycle, this cannot occur in mammalian muscle mitochondria, where the activity of the mitochondrial glycerol-3-phosphate dehydrogenase is very low.

Some evidence for the operation of the malate–aspartate shuttle (Borst, 1963) in heart has been presented by LaNoue and Williamson (1971).

A beautiful demonstration of the interrelationship between intra- and extramitochondrial processes on the level of energy supply is provided by the experiments of Harary and Slater (1965) with cultures of beating heart cells. They showed that when the extramitochondrial supply of ATP was cut off by inhibiting glycolysis with iodoacetate, beating continued. If, however, oligomycin was added in addition to iodoacetate, thus inhibiting the mitochondrial generation of ATP, beating stopped. Oligomycin on its own was unable to inhibit beating, showing that the ATP supplied by glycolysis was sufficient to allow beating to occur. Furthermore, they found that addition of the uncoupler 2,4-dinitrophenol, which induces the hydrolysis of all ATP in the cell via the mitochondrial ATPase, inhibited the beating of the cells. When oligomycin was added in addition to 2,4-dinitrophenol so that the uncoupler-induced ATPase was inhibited, beating was observed once more.

Finally, in the past decade it has become clear that mitochondria contain their own unique species of DNA, together with the machinery for replication, transcription, and protein synthesis (for reviews, see Borst and Kroon, 1969; Rabinowitz and Swift, 1970; Ashwell and Work, 1970). Based on measurements of the length of mitochondrial DNA molecules and on the assumption that this DNA is homogeneous, it is concluded that the genetic information of mitochondrial DNA is very limited (Borst, 1970, 1971). In fact the information is not even sufficient to code for the whole protein-synthesizing machinery itself (Borst, 1971). However, it might be that part of this machinery is coded for by the nucleus and that the information of mitochondrial DNA is used for the synthesis of particular mitochondrial proteins (Borst, 1971). Another suggestion, made by Borst (1970, 1971), is that the mitochondrion "started out as a nearly completely autonomous endosymbiotic bacterium" that in the course of evolution has lost nearly all its autonomous functions.

ACKNOWLEDGMENTS

The authors have made grateful use of the chapter by Slater (1960) in the first edition of this treatise in compiling the revised version. In the first edition, Slater (1960) used the word "sarcosome," introduced by Retzius (1890), synonymously with "muscle mitochondrion." The authors now prefer to use the latter term, since this emphasizes the similarity in structure and function between mitochondria from muscle and those from other tissues.

REFERENCES

Aas, M. (1971). *Biochim. Biophys. Acta* **231**, 32.

Aas, M., and Bremer, J. (1968). *Biochim. Biophys. Acta* **164**, 157.

Alexandre, A., Rossi, C. R., Sartorelli, L., and Siliprandi, N. (1969). *FEBS Lett.* **3**, 279.

Allen, S. H. G., Kellermeyer, R., Stjernholm, R., Jacobson, B., and Wood, H. G. (1963). *J. Biol. Chem.* **238**, 1637.

Ashwell, M., and Work, T. S. (1970). *Annu. Rev. Biochem.* **39**, 251.

Atkinson, D. E. (1969). *In* "The Citric Acid Cycle" (J. M. Lowenstein, ed.), p. 137. Dekker, New York.

Azzi, A., Chappell, J. B., and Robinson, B. H. (1967). *Biochem. Biophys. Res. Commun.* **29**, 148.

Bartley, W., and Davies, R. E. (1954). *Biochem. J.* **57**, 37.

Battelli, F., and Stern, L. (1912). *Ergeb. Physiol., Biol. Chem. Exp. Pharmakol.* **12**, 96.

Beenakkers, A. M. T. (1963). *Biochem. Z.* 337, 436.
Beenakkers, A. M. T., and Henderson, P. T. (1967). *Eur. J. Biochem.* 1, 187.
Beenakkers, A. M. T., and Klingenberg, M. (1964). *Biochim. Biophys. Acta* 84, 205.
Beinert, H. (1963). *In* "The Enzymes" (P. D. Boyer, H. Lardy, and K. Myrbäck, eds.), 2nd ed., Vol. 7, p. 447. Academic Press, New York.
Beinert, H., Hartzell, C. R., Van Gelder, B. F., Ganapathy, K., Mason, H. S., and Wharton, D. C. (1970). *J. Biol. Chem.* 245, 225.
Bensley, R. R., and Hoerr, N. (1934). *Anat. Rec.* 60, 449.
Berg, P. (1956). *J. Biol. Chem.* 222, 991.
Bode, C., and Klingenberg, M. (1965). *Biochem. Z.* 341, 271.
Borst, P. (1961). M. D. Thesis, University of Amsterdam, Jacob van Campen, Amsterdam.
Borst, P. (1962). *Biochim. Biophys. Acta* 57, 256.
Borst, P. (1963). *Proc. Int. Congr. Biochem.*, 5th. 1961 Vol. 2, p. 233.
Borst, P. (1970). *In* "Control of Organelle Development" (P. L. Miller, ed.), p. 201. Cambridge Univ. Press, London and New York.
Borst, P. (1971). *In* "Autonomy and Biogenesis of Mitochondria and Chloroplasts" (N. K. Boardman, A. W. Linnane, and R. M. Smillie, eds.), p. 260. North-Holland Publ., Amsterdam.
Borst, P., and Kroon, A. M. (1969). *Int. Rev. Cytol.* 26, 107.
Borst, P., and Slater, E. C. (1960). *Biochim. Biophys. Acta* 41, 170.
Boyer, P. D. (1965). *In* "Oxidases and Related Redox Systems" (T. E. King, H. S. Mason, and M. Morrison, eds.), Vol. 2, p. 994. Wiley, New York.
Boyer, P. D., Mills, R. C., and Fromm, H. J. (1959). *Arch. Biochem. Biophys.* 81, 249.
Brdiczka, D., and Pette, D. (1971). *Eur. J. Biochem.* 19, 546.
Bremer, J. (1965). *Biochim. Biophys. Acta* 104, 581.
Bremer, J. (1966). *Biochim. Biophys. Acta* 116, 1.
Brierley, G. P. (1963). *In* "Energy-linked Functions of Mitochondria" (B. Chance, ed.), p. 237. Academic Press, New York.
Bronk, J. R., and Jasper, D. K. (1968). *J. Cell Biol.* 38, 277.
Bronk, J. R., and Jasper, D. K. (1970). *Biochem. J.* 116, 33P.
Brosemer, R. W., Vogell, W., and Bücher, T. (1963). *Biochem. Z.* 338, 854.
Bücher, T., and Klingenberg, M. (1958). *Angew. Chem.* 70, 552.
Bullard, H. (1913). *Amer. J. Anat.* 14, 1.
Bullard, H. (1916). *Amer. J. Anat.* 19, 1.
Burton, K. (1959). *Biochem. J.* 71, 388.
Butler, W. H., and Judah, J. D. (1970). *J. Cell Biol.* 44, 278.
Cannata, J. J. B., Focesi, A., Jr., Mazumder, R., Warner, R. C., and Ochoa, S. (1965). *J. Biol. Chem.* 240, 3249.
Carafoli, E. (1965). *Biochim. Biophys. Acta* 97, 107.
Carafoli, E., and Lehninger, A. L. (1971). *Biochem. J.* 122, 681.
Carafoli, E., Weiland, S., and Lehninger, A. L. (1965). *Biochim. Biophys. Acta* 97, 88.
Carafoli, E., Hansford, R. G., Sacktor, B., and Lehninger, A. L. (1971). *J. Biol. Chem.* 246, 964.
Chance, B. (1956a). *In* "Enzymes: Units of Biological Structure and Function" (O. H. Gaebler, ed.), p. 447. Academic Press, New York.
Chance, B. (1956b). *Proc. Int. Congr. Biochem.*, 3rd, 1955 p. 300.

Chance, B. (1965). *J. Biol. Chem.* **240**, 2729.

Chance, B., and Mela, L. (1966). *Proc. Nat. Acad. Sci. U.S.* **55**, 1243.

Chance, B., and Mela, L. (1967). *J. Biol. Chem.* **242**, 830.

Chance, B., and Montal, M. (1972). *In* "Current Topics in Membranes and Transport" (F. Bronner and A. Kleinzeller, eds.). Academic Press, New York (in press).

Chance, B., and Williams, G. R. (1956). *Advan. Enzymol.* **17**, 65.

Chance, B., Erecinska, M., and Lee, C.-P. (1970). *Proc. Nat. Acad. Sci. U.S.* **66**, 928.

Chappell, J. B. (1968). *Brit. Med. Bull.* **24**, 150.

Chappell, J. B., and Crofts, A. R. (1965). *Biochem. J.* **95**, 393.

Chappell, J. B., and Greville, G. D. (1961). *Nature (London)* **190**, 502.

Chappell, J. B., and Haarhoff, K. N. (1967). *In* "Biochemistry of Mitochondria" (E. C. Slater, Z. Kaniuga, and L. Wojtczak, eds.), p. 75. Academic Press, New York.

Chappell, J. B., and Perry, S. V. (1953). *Biochem. J.* **55**, 586.

Chappell, J. B., Greville, G. D., and Bicknell, K. E. (1962). *Biochem. J.* **84**, 61P.

Chappell, J. B., Henderson, P. J. F., McGivan, J. D., and Robinson, B. H. (1968). *In* "The Interaction of Drugs and Subcellular Components in Animal Cells" (P. N. Campbell, ed.), p. 71. Churchill, London.

Chase, J. F. A. (1967). *Biochem. J.* **104**, 510.

Chen, R. F., and Plaut, G. W. E. (1963). *Biochemistry* **2**, 1023.

Childress, C. C., Sacktor, B., and Traynor, D. R. (1967). *J. Biol. Chem.* **242**, 754.

Christ, E. J. V. J. (1968). *Biochim. Biophys. Acta* **152**, 50.

Claude, A. (1946). *J. Exp. Med.* **84**, 51.

Cleland, K. W., and Slater, E. C. (1953a). *Quart. J. Microsc. Sci.* **94**, 329.

Cleland, K. W., and Slater, E. C. (1953b). *Biochem. J.* **53**, 547.

Cockrell, R. S., Harris, E. J., and Pressman, B. C. (1967). *Nature (London)* **215**, 1487.

Colowick, S. P., Kaplan, N. O., Estabrook, R. W., and Pullmann, M. E., eds. (1967). "Methods of Enzymology," Vol. 10. Academic Press, New York.

Conway, E. J., and Cook, R. (1939). *Biochem. J.* **33**, 479.

Copenhaver, J. H., McShan, W. H., and Meyer, R. K. (1950). *J. Biol. Chem.* **183**, 73.

Dahlen, J. V., and Porter, J. W. (1968). *Arch. Biochem. Biophys.* **127**, 207.

Danielson, L., and Ernster, L. (1963). *Biochem. Z.* **338**, 188.

Davis, E. J. (1965). *Biochim. Biophys. Acta* **96**, 217.

Davis, E. J. (1967). *Biochim. Biophys. Acta* **143**, 26.

Davis, E. J. (1968). *Biochim. Biophys. Acta* **162**, 1.

De Haan, E. J. (1968). Ph.D. Thesis, University of Amsterdam, Amsterdam.

De Jong, J. W., and Hülsmann, W. C. (1970). *Biochim. Biophys. Acta* **197**, 127.

De Jong, J. W., Hülsmann, W. C., and Meijer, A. J. (1969). *Biochim. Biophys. Acta* **184**, 664.

Delbrück, A., Zebe, E., and Bücher, T. (1959). *Biochem. Z.* **331**, 273.

Donaldson, W. E., Wit-Peeters, E. M., and Scholte, H. R. (1970). *Biochim. Biophys. Acta* **202**, 35.

Donnellan, J. F., Barker, M. D., Wood, J., and Beechey, R. B. (1970). *Biochem. J.* **120**, 467.

Edwards, J. B., and Keech, D. B. (1968). *Biochim. Biophys. Acta* **159**, 167.

Eggerer, H., Stadtman, E. R., Overath, P., and Lynen, F. (1960). Biochem. Z. 333, 1.

England, P. J., and Robinson, B. H. (1969). Biochem. J. 112, 8P.

Ernster, L., and Kuylenstierna, B. (1970). In "Membranes of Mitochondria and Chloroplasts" (E. Racker, ed.), p. 172. Van Nostrand-Reinhold, Princeton, New Jersey.

Ernster, L., and Lee, C.-P. (1964). Annu. Rev. Biochem. 33, 729.

Estabrook, R. W., and Nissley, S. P. (1963). In "Funktionelle und morphologische Organisation der Zelle" (P. Karlson, ed.), p. 119. Springer-Verlag, Berlin and New York.

Fehmers, M. C. O. (1968). M. D. Thesis, University of Amsterdam, Amsterdam.

Fernández-Morán, H. (1962). Circulation 26, 1039.

Fernández-Morán, H., Oda, T., Blair, P., and Green, D. E. (1964). J. Cell Biol. 22, 63.

Fessenden, J. M., and Racker, E. (1966). J. Biol. Chem. 241, 2483.

Fritz, I. B. (1955). Acta Physiol. Scand. 34, 367.

Fritz, I. B. (1963). Advan. Lipid Res. 1, 285.

Fritz, I. B. (1965). In "Recent Research on Carnitine" (G. Wolf, ed.), p. 83. MIT Press, Cambridge, Massachusetts.

Fritz, I. B., and Yue, K. T. N. (1963). J. Lipid Res. 4, 279.

Fritz, I. B., Kaplan, E., and Yue, K. T. N. (1962). Amer. J. Physiol. 202, 117.

Fritz, I. B., Schultz, .S. K., and Srere, P. A. (1963). J. Biol. Chem. 238, 2509.

Gamble, J. L. (1957). J. Biol. Chem. 228, 955.

Garland, P. B., and Randle, P. J. (1964a). Biochem. J. 91, 6C.

Garland, P. B., and Randle, P. J. (1964b). Biochem. J. 93, 678.

Garland, P. B., Newsholme, E. A., and Randle, P. J. (1962). Nature (London) 195, 381.

Garland, P. B., Shepherd, D., Nicholls, D. G., and Ontko, J. (1968). Advan. Enzyme Regul. 6, 1.

Garland, P. B., Shepherd, D., Nicholls, D. G., Yates, D. W., and Light, P. A. (1969a). In "The Citric Acid Cycle" (J. M. Lowenstein, ed.), p. 163. Dekker, New York.

Garland, P. B., Haddock, B. A., and Yates, D. W. (1969b). In "Mitochondria, Structure and Function" (L. Ernster and Z. Drahota, eds.), p. 111. Academic Press, New York.

Gautheron, D., Durand, R., Pialoux, N., and Gaudemer, Y. (1964). Bull. Soc. Chim. Biol. 46, 645.

Gehring, U., Riepertinger, C., and Lynen, F. (1968). Eur. J. Biochem. 6, 264.

Gerez, C., and Kirsten, R. (1965). Biochem. Z. 341, 534.

Ghosal, J., Whitworth, T., and Coniglio, J. G. (1969). Biochim. Biophys. Acta 187, 576.

Gillis, J. M., and Maréchal, G. (1969). Abstr. Int. Biophys. Congr., 3rd, 1969 p. 271.

Giorgio, A. J., and Plaut, G. W. E. (1967). Biochim. Biophys. Acta 139, 487.

Goebell, H., and Klingenberg, M. (1964). Biochem. Z. 340, 411.

Goebell, H., and Pette, D. (1967). Enzymol. Biol. Clin. 8, 161.

Gompertz, D., Bau, D. C. K., Storrs, C. N., Peters, T. J., and Hughes, E. A. (1970). Lancet I, 1140.

Graham, J. M., and Green, C. (1970). Eur. J. Biochem. 12, 58.

Green, D. E. (1936). Biochem. J. 30, 629.

Green, D. E. (1951). Biol. Rev. Cambridge Phil. Soc. 26, 410.

Green, D. E., and Harris, R. A. (1969). *FEBS Lett.* **5**, 241.

Green, D. E., and McLennan, D. H. (1969). *BioScience* **193**, 213.

Green, D. E., Asai, J., Harris, R. A., and Penniston, J. T. (1968). *Arch. Biochem. Biophys.* **125**, 684.

Greenawalt, J. W., Rossi, C. S., and Lehninger, A. L. (1964). *J. Cell Biol.* **23**, 21.

Greville, G. D. (1969). *Curr. Top. Bioenerg.* **3**, 1.

Greville, G. D., and Tubbs, P. K. (1968). *Essays Biochem.* **4**, 155.

Grinius, L. L., Jasaitis, A. A., Kadziauska, Yu.P., Liberman, E. A., Skulachev, V. P., Topali, V. P., Tsofina, L. M., and Vladimirova, M. A. (1970). *Biochim. Biophys. Acta* **216**, 1.

Guchhait, R. B., Putz, G. R., and Porter, J. W. (1966). *Arch. Biochem. Biophys.* **117**, 541.

Hackenbrock, C. R. (1965). *Anat. Rec.* **151**, 356.

Hackenbrock, C. R. (1966). *J. Cell Biol.* **30**, 269.

Hackenbrock, C. R. (1968a). *Proc. Nat. Acad. Sci. U.S.* **61**, 598.

Hackenbrock, C. R. (1968b). *J. Cell Biol.* **37**, 345.

Hackenbrock, C. R., and Kaplan, A. I. (1969). *J. Cell Biol.* **42**, 221.

Halenz, D. R., Feng, J.-Y., Hegre, C. S., and Lane, M. D. (1962). *J. Biol. Chem.* **237**, 2140.

Hall, J. D., and Crane, F. L. (1970). *Exp. Cell Res.* **62**, 480.

Hansford, R. G., and Chappell, J. B. (1968). *Biochem. Biophys. Res. Commun.* **30**, 643.

Harary, I., and Slater, E. C. (1965). *Biochim. Biophys. Acta* **99**, 227.

Harman, J. W. (1950). *Exp. Cell Res.* **1**, 382.

Harman, J. W., and Feigelson, M. (1952). *Exp. Cell Res.* **3**, 47.

Harman, J. W., and Osborne, U. H. (1953). *J. Exp. Med.* **98**, 81.

Harris, E. J., Judah, S. O., and Ahmed, K. (1966). *Curr. Top. Bioenerg.* **1**, 255.

Harris, R. A., Penniston, J. T., Asai, J., and Green, D. E. (1968). *Proc. Nat. Acad. Sci. U.S.* **59**, 830.

Hartmann, G., and Lynen, F. (1961). *In* "The Enzymes" (P. D. Boyer, H. Lardy, and K. Myrbäck, eds.), 2nd ed., Vol. 5, p. 381. Academic Press, New York.

Haslam, R. J. (1966). *Regul. Metab. Processes Mitochondria, Proc. Symp., 1965* p. 108.

Haslam, R. J., and Krebs, H. A. (1963). *Biochem. J.* **86**, 432.

Hathaway, J. A., and Atkinson, D. E. (1963). *J. Biol. Chem.* **238**, 2875.

Hathaway, J. A., and Atkinson, D. E. (1965). *Biochem. Biophys. Res. Commun.* **20**, 661.

Hayakawa, T., and Koike, M. (1967). *J. Biol. Chem.* **242**, 1356.

Hayakawa, T., Muta, H., Hirashima, M., Ide, S., Okabe, K., and Koike, M. (1964). *Biochem. Biophys. Res. Commun.* **17**, 51.

Hayakawa, T., Hirashima, M., Ide, S., Hamada, M., Okabe, K., and Koike, M. (1966). *J. Biol. Chem.* **241**, 4694.

Heldt, H. W., and Schwalbach, K. (1967). *Eur. J. Biochem.* **1**, 199.

Henle, J. (1841). "Allgemeine Anatomie." Voss, Leipzig.

Hogeboom, G. H., Claude, A., and Hotchkiss, R. D. (1946). *J. Biol. Chem.* **165**, 615.

Holmgren, E. (1910). *Arch. Mikrosk. Anat. Entwicklungsmech.* **75**, 240.

Hommes, F. A., Kuipers, J. R. G., Elema, J. D., Jansen, J. F., and Jonxis, J. H. P. (1968). *Pediat. Res.* **2**, 519.

Horn, R. S., Levin, R., and Haugaard, N. (1969). *Biochem. Pharmacol.* **18**, 503.
Horn, R. S., Fyhn, A., and Haugaard, N. (1971). *Biochim. Biophys. Acta* **226**, 459.
Hsia, Y. E., Scully, K. J., and Rosenberg, L. E. (1969). *Lancet* I, 757.
Hsia, Y. E., Scully, K. J., and Rosenberg, L. E. (1971). *J. Clin. Invest.* **50**, 127.
Hülsmann, W. C. (1960). *Biochim. Biophys. Acta* **45**, 623.
Hülsmann, W. C. (1970). *Biochem. J.* **116**, 32P.
Hülsmann, W. C., Siliprandi, D., Ciman, M., and Siliprandi, N. (1964). *Biochim. Biophys. Acta* **93**, 166.
Hülsmann, W. C., De Jong, J. W., and Van Tol, A. (1968). *Biochim. Biophys. Acta* **162**, 292.
Hülsmann, W. C., Meijer, A. E. F. H., Bethlem, J., and Van Wijngaarden, G. K. (1969). *Excerpta Med. Found. Int. Congr. Ser.* **199**, 319.
Huennekens, F. M., and Green, D. E. (1950). *Arch. Biochem.* **27**, 428.
Hull, F. E., and Whereat, A. F. (1967). *J. Biol. Chem.* **22**, 4023.
Hunter, F. E., Jr. (1951). *In* "Phosphorus Metabolism" (W. D. McElroy and B. Glass, eds.), Vol. 1, p. 297. Johns Hopkins Press, Baltimore, Maryland.
Jacobus, W. E., and Brierley, G. P. (1969). *J. Biol. Chem.* **244**, 4995.
Jagannathan, V., and Schweet, R. S. (1952). *J. Biol. Chem.* **196**, 551.
Judah, J. D. (1951). *Biochem. J.* **49**, 271.
Kagawa, Y., and Racker, E. (1966). *J. Biol. Chem.* **241**, 2475.
Kaplan, N. O., Colowick, S. P., and Neufeld, E. F. (1953). *J. Biol. Chem.* **205**, 1.
Kawasaki, T., Satoh, K., and Kaplan, N. O. (1964). *Biochem. Biophys. Res. Commun.* **17**, 648.
Kaziro, Y., Ochoa, S., Warner, R. G., and Chen, J.-Y. (1961). *J. Biol. Chem.* **236**, 1917.
Kaziro, Y., Grossman, A., and Ochoa, S. (1965). *J. Biol. Chem.* **240**, 64.
Keilin, D. (1925). *Proc. Roy. Soc., Ser. B* **98**, 312.
Keilin, D., and Hartree, E. F. (1947). *Biochem. J.* **41**, 503.
King, T. E., Kuboyama, M., and Takemori, S. (1965). *In* "Oxidases and Related Redox Systems" (T. E. King, H. S. Mason, and M. Morrison, eds.), Vol. 2, p. 707. Wiley, New York.
Klingenberg, M. (1967). *In* "Mitochondrial Structure and Compartmentation" (E. Quagliariello *et al.*, eds.), p. 320. Adriatica Editrice, Bari.
Klingenberg, M. (1970a). *Essays Biochem.* **6**, 119.
Klingenberg, M. (1970b). *FEBS Lett.* **6**, 145.
Klingenberg, M., and Buchholz, M. (1970). *Eur. J. Biochem.* **13**, 247.
Klingenberg, M., and Pfaff, E. (1966). *Regul. Metab. Processes Mitochondria, Proc. Symp., 1965* p. 180.
Klingenberg, M., and Slenczka, W. (1959). *Biochem. Z.* **331**, 486.
Knoop, F. (1904). *Beitr. Chim. Physiol. Pathol.* **6**, 150.
Koike, M., Reed, L. J., and Carroll, W. R. (1960). *J. Biol. Chem.* **235**, 1924.
Koike, M., Reed, L. J., and Carroll, W. R. (1963). *J. Biol. Chem.* **238**, 30.
Kölliker, A. (1850). *Mikrosk. Anat. II* **1**, 203.
Kölliker, A. (1857). *Z. Wiss. Zool.* **8**, 311.
Kölliker, A. (1888). *Z. Wiss. Zool.* **47**, 689.
Kornberg, A., and Pricer, W. E. (1953). *J. Biol. Chem.* **204**, 329.
Kosicki, G. W., and Lee, L. P. K. (1966). *J. Biol. Chem.* **241**, 3571.
Krebs, H. A. (1970). *Advan. Enzyme Regul.* **8**, 335.
Krebs, H. A., and Veech, R. L. (1969). *In* "The Energy Level and Metabolic Control in Mitochondria" (S. Papa *et al.*, eds.), p. 329. Adriatica Editrice, Bari.

Kuner, J. M., and Beyer, R. E. (1970). *J. Membrane Biol.* **2,** 71.
Lane, M. D., and Halenz, D. R. (1960). *Biochem. Biophys. Res. Commun.* **2,** 436.
Lane, M. D., Halenz, D. R., Kosow, D. P., and Hegre, C. S. (1960). *J. Biol. Chem.* **235,** 3082.
LaNoue, K. F., and Williamson, J. R. (1971). *Metab., Clin. Exp.* **20,** 119.
Lardy, H. A., and Wellman, H. (1952). *J. Biol. Chem.* **195,** 215.
Lardy, H. A., Johnson, D., and McMurray, W. C. (1958). *Arch. Biochem. Biophys.* **78,** 587.
Lee, C.-P., and Ernster, L. (1964). *Biochim. Biophys. Acta* **81,** 187.
Lee, C.-P., and Ernster, L. (1966). *Regul. Metab. Processes Mitochondria, Proc. Symp., 1965,* p. 218.
Lehninger, A. L. (1970). *Biochem. J.* **119,** 129.
Lehninger, A. L. (1971). *Biochem. Biophys. Res. Commun.* **42,** 312.
Lehninger, A. L., Carafoli, E., and Rossi, C. S. (1967). *Advan. Enzymol.* **29,** 259.
Lévy, M., Toury, R., and André, J. (1966). *C. R. Acad. Sci.* **262,** 1593 and 1766.
Lévy, M., Toury, R., and André, J. (1967). *Biochim. Biophys. Acta* **135,** 599.
Ling, K.-H., Marcus, F., and Lardy, H. A. (1965). *J. Biol. Chem.* **240,** 1893.
Linn, T. C., Pettit, F. H., and Reed, L. J. (1969a). *Proc. Nat. Acad. Sci. U.S.* **62,** 234.
Linn, T. C., Pettit, F. H., Hucho, F., and Reed, L. J. (1969b). *Proc. Nat. Acad. Sci. U.S.* **64,** 227.
Löw, H. (1966). *Regul. Metab. Processes Mitochondria, Proc. Symp., 1965* p. 25.
Lowenstein, J., and Tornheim, K. (1971). *Science* **171,** 397.
Lynen, F. (1955). *Annu. Rev. Biochem.* **24,** 653.
Lynen, F. (1961). *Fed. Proc., Fed. Amer. Soc. Exp. Biol.* **20,** 941.
Mahler, H. R., Wakil, S. J., and Bock, R. M. (1953). *J. Biol. Chem.* **204,** 453.
Makita, T., and Sandborn, E. B. (1971). *Experientia* **27,** 184.
Mansour, T. E. (1963). *J. Biol. Chem.* **238,** 2285.
Mansour, T. E., and Mansour, J. M. (1962). *J. Biol. Chem.* **237,** 629.
Mazumder, R., Sasakawa, T., Kaziro, Y., and Ochoa, S. (1962). *J. Biol. Chem.* **237,** 3065.
Meijer, A. J. (1971). Ph.D. Thesis, University of Amsterdam, Amsterdam.
Mitchell, P. (1961). *Nature (London)* **191,** 144.
Mitchell, P., and Moyle, J. (1965). *Nature (London)* **208,** 147.
Mitchell, P., and Moyle, J. (1969). *Eur. J. Biochem.* **9,** 149.
Moore, C. L. (1971). *Biochem. Biophys. Res. Commun.* **42,** 298.
Moore, C., and Pressman, B. C. (1964). *Biochem. Biophys. Res. Commun.* **15,** 562.
Needham, D. M., and Pillai, R. K. (1937). *Biochem. J.* **31,** 1837.
Neujahr, H. Y., and Mistry, S. P. (1963). *Acta Chem. Scand.* **17,** 1140.
Nicholls, D. G., Shepherd, D., and Garland, P. B. (1967). *Biochem. J.* **103,** 677.
Norum, K. R. (1964). *Biochim. Biophys. Acta* **89,** 95.
Nugteren, D. H. (1965). *Biochim. Biophys. Acta* **106,** 280.
Oliveira, M. M., Weglicki, W. B., Nason, A., and Nair, P. P. (1969). *Biochim. Biophys. Acta* **180,** 98.
Olson, R. E., and Piatnek, D. A. (1959). *Ann. N.Y. Acad. Sci.* **72,** 466.
Palade, G. E. (1952). *Anat. Rec.* **114,** 427.
Pande, S. V., and Blanchaer, M. C. (1970). *Biochim. Biophys. Acta* **202,** 43.
Papa, S., and Francavilla, A. (1967). *In* "Mitochondrial Structure and Compartmentation" (E. Quagliariello *et al.,* eds.), p. 263. Adriatica Editrice, Bari.

Papa, S., Tager, J. M., Quagliariello, E., and Slater, E. C., eds. (1969). "The energy Level and Metabolic Control in Mitochondria." Adriatica Editrice, Bari.

Parmeggiani, A., and Bowman, R. H. (1963). *Biochem. Biophys. Res. Commun.* **12**, 268.

Parmeggiani, A., Luft, J. H., Love, D. S., and Krebs, E. G. (1966). *J. Biol. Chem.* **241**, 4625.

Parsons, D. F. (1963). *Science* **142**, 1176.

Parsons, D. F., and Yano, Y. (1967). *Biochim. Biophys. Acta* **135**, 362.

Parsons, D. F., Williams, G. R., and Chance, B. (1966). *Ann. N.Y. Acad. Sci.* **137**, 643.

Parsons, D. F., Williams, G. R., Thompson, W., Wilson, D., and Chance, B. (1967). *In* "Mitochondrial Structure and Compartmentation" (E. Quagliariello *et al.*, eds.), p. 29. Adriatica Editrice, Bari.

Passonneau, J. V., and Lowry, O. H. (1962). *Biochem. Biophys. Res. Commun.* **7**, 10.

Paul, M. H., and Sperling, E. (1952). *Proc. Soc. Exp. Biol. Med.* **79**, 352.

Penniston, J. T., Harris, R. A., Asai, J., and Green, D. E. (1968). *Proc. Nat. Acad. Sci. U.S.* **59**, 624.

Pette, D. (1966). *Regul. Metab. Processes Mitochondria, Proc. Symp., 1965* p. 28.

Pfaff, E., Klingenberg, M., Ritt, E., and Vogell, W. (1968). *Eur. J. Biochem.* **5**, 222.

Quagliariello, E., Landriscina, C., and Coratelli, P. (1968). *Biochim. Biophys. Acta* **164**, 12.

Rabinowitz, M., and Swift, H. (1970). *Physiol. Rev.* **50**, 376.

Racker, E. (1970). *In* "Membranes of Mitochondria and Chloroplasts" (E. Racker, ed.), p. 127. Van Nostrand-Reinhold, Princeton, New Jersey.

Racker, E., and Conover, T. E. (1963). *Fed. Proc., Fed. Amer. Soc. Exp. Biol.* **22**, 1088.

Racker, E., Tyler, D. D., Estabrook, R. W., Conover, T. E., Parsons, D. F., and Chance, B. (1965). *In* "Oxidases and Related Redox Systems" (T. E. King, H. S. Mason, and H. Morrison, eds.), Vol. 2, p. 1077. Wiley, New York.

Randle, P. J., Newsholme, E. A., and Garland, P. B. (1964). *Biochem. J.* **93**, 652.

Rasmussen, H., Chance, B., and Ogata, E. (1965). *Proc. Nat. Acad. Sci. U.S.* **53**, 1069.

Reed, L. J. (1969). *In* "Current Topics in Cellular Regulation" (B. L. Horecker and E. R. Stadtman, eds.), Vol. 1, p. 233. Academic Press, New York.

Regaud, C. (1909). *C. R. Acad. Sci.* **149**, 426.

Rétey, J., and Lynen, F. (1964). *Biochem. Biophys. Res. Commun.* **16**, 358.

Retzius, G. (1890). *Biol. Untersuch.* [N.S.] **1**, 51.

Reynafarje, B., and Lehninger, A. L. (1969). *J. Biol. Chem.* **244**, 584.

Robinson, B. H., Williams, G. R., Halperin, M. L., and Leznoff, C. C. (1971). *Eur. J. Biochem.* **20**, 65.

Rossi, C. S., and Lehninger, A. L. (1963). *Biochem. Z.* **338**, 698.

Rossi, C. S., Carafoli, E., Drahota, Z., and Lehninger, A. L. (1966). *Regul. Metab. Processes Mitochondria, Proc. Symp., 1965* p. 317.

Rossi, E., and Azzone, G. F. (1970). *Eur. J. Biochem.* **12**, 319.

Sacktor, B. (1955). *J. Biophys. Biochem. Cytol.* **1**, 29.

Sacktor, B., and Childress, C. C. (1967). *Arch. Biochem. Biophys.* **120**, 583.

Sacktor, B., and Cochran, D. G. (1957). *Biochim. Biophys. Acta* **26**, 200.

Sacktor, B., and Hurlbut, E. C. (1966). *J. Biol. Chem.* **241**, 632.

Sacktor, B., and Wormser-Shavit, E. (1966). *J. Biol. Chem.* **241**, 624.

Sanadi, D. R., Gibson, D. M., Ayengar, P., and Ouellet, L. (1954). *Biochim. Biophys. Acta* **13**, 146.

Sanwal, B. D., Zink, M. W., and Stachow, C. S. (1964). *J. Biol. Chem.* **239**, 1597.

Scarpa, A., and Azzi, A. (1968). *Biochim. Biophys. Acta* **150**, 473.

Scarpa, A., and Azzone, G. F. (1970). *Eur. J. Biochem.* **12**, 328.

Schnaitman, C., Erwin, V. G., and Greenawalt, J. W. (1967). *J. Cell Biol.* **32**, 719.

Scholte, H. R. (1969). *Biochim. Biophys. Acta* **178**, 137.

Scholte, H. R. (1970). Ph.D. Thesis, University of Amsterdam, Amsterdam.

Scholte, H. R., Wit-Peeters, E. M., and Bakker, J. C. (1971). *Biochim. Biophys. Acta* **231**, 479.

Schweet, R. S., Katchmann, B., Bock, R. M., and Jagannathan, V. (1952). *J. Biol. Chem.* **196**, 563.

Seligman, A. M., Karnovsky, M. J., Wasserkrug, H. L., and Hanker, J. S. (1968). *J. Cell Biol.* **38**, 1.

Seubert, W., Lamberts, I., Kramer, R., and Ohly, B. (1968). *Biochim. Biophys. Acta* **164**, 498.

Shepherd, D., Yates, D. W., and Garland, P. B. (1966). *Biochem. J.* **98**, 3C.

Shipp, J. C., Opie, L. H., and Challoner, D. (1961). *Nature (London)* **189**, 1018.

Sjöstrand, F. S. (1953). *Nature (London)* **171**, 30.

Skrede, S., and Bremer, J. (1970). *Eur. J. Biochem.* **14**, 465.

Slater, E. C. (1950). *Nature (London)* **166**, 982.

Slater, E. C. (1953). *Nature (London)* **172**, 975.

Slater, E. C. (1960). *In* "The Structure and Function of Muscle (G. H. Bourne, ed.), 1st ed., Vol. 2, p. 105. Academic Press, New York.

Slater, E. C. (1967). *Methods Enzymol.* **10**, 48.

Slater, E. C. (1971). *Quart. Rev. Biophys.* **4**, 35.

Slater, E. C., and Cleland, K. W. (1953). *Biochem. J.* **55**, 566.

Slater, E. C., Tamblyn-Hague, C., and Davis-Van Thienen, W. C. (1965). *Biochim. Biophys. Acta* **96**, 206.

Slater, E. C., Quagliariello, E., Papa, S., and Tager, J. M. (1969). *In* "The Energy Level and Metabolic Control in Mitochondria" (S. Papa *et al.*, eds.), p. 1. Adriatica Editrice, Bari.

Slater, E. C., Lee, C.-P., Berden, J. A., and Wegdam, H. J. (1970). *Nature (London)* **226**, 1248.

Solberg, H. E. (1971). *FEBS Lett.* **12**, 134.

Sottocassa, G. L., Kuylenstierna, B., Ernster, L., and Bergstrand, A. (1967). *J. Cell Biol.* **32**, 448.

Souverijn, J. H. M. (1971). Personal communication.

Sprecher, M., Clark, M. J., and Sprinson, D. B. (1964). *Biochem. Biophys. Res. Commun.* **15**, 581.

Sprecher, M., Clark, M. J., and Sprinson, D. B. (1966). *J. Biol. Chem.* **241**, 872.

Stern, J. R. (1955). *Methods Enzymol.* **1**, 581.

Stern, J. R. (1961). *In* "The Enzymes" (P. D. Boyer, H. Lardy, and K. Myrbäck, eds.)., 2nd ed., Vol. 5, p. 511. Academic Press, New York.

Stern, J. R. (1962). *Methods Enzymol.* **5**, 557.

Stern, J. R., Coon, M. J., Del Campillo, A., and Schneider, M. C. (1956). *J. Biol. Chem.* **221**, 15.

Stoeckenius, W. (1963). *J. Cell Biol.* **17**, 443.
Stoeckenius, W. (1970). *In* "Membranes of Mitochondria and Chloroplasts" (E. Racker, ed.), p. 53. Van Nostrand-Reinhold, Princeton, New Jersey.
Stoffel, W., and Caesar, H. (1965). *Hoppe-Seyler's Z. Physiol. Chem.* **341**, 76.
Stoffel, W., and Schiefer, H-G. (1965). *Hoppe-Seyler's Z. Physiol. Chem.* **341**, 84.
Stoffel, W., Ditzer, R., and Caesar, H. (1964). *Hoppe-Seyler's Z. Physiol. Chem.* **339**, 167.
Stoner, C. D., and Sirak, H. D. (1969). *J. Cell Biol.* **43**, 521.
Struck, E., Ashmore, J., and Wieland, O. (1966). *Advan. Enzyme Regul.* **4**, 219.
Tager, J. M., Groot, G. S. P., Roos, D., Papa, S., and Quagliariello, E. (1967). *In* "Mitochondrial Structure and Compartmentation" (E. Quagliariello *et al.*, eds.), p. 453. Adriatica Editrice, Bari.
Tager, J. M., De Haan, E. J., and Slater, E. C. (1969). *In* "The Citric Acid Cycle" (J. M. Lowenstein, ed.), p. 213. Dekker, New York.
Tulp, A., and Van Berkel, T. J. C. (1972). *In* "Energy Transduction in Respiration and Photosynthesis" (E. C. Slater *et al.*, eds.). Adriatica Editrice, Bari (1971) p. 439.
Tulp, A., and Van Dam, K. (1969). *Biochim. Biophys. Acta* **189**, 337.
Tulp, A., and Van Dam, K. (1970). *FEBS Lett.* **10**, 292.
Tulp, A., Stam, H., and Van Dam, K. (1971). *Biochim. Biophys. Acta* **234**, 301.
Tung, T., Anderson, L., and Lardy, H. A. (1952). *Arch. Biochem. Biophys.* **40**, 194.
Tyler, D. D. (1970). *Biochem. J.* **116**, 30P.
Utter, M. F. (1969). *In* "The Citric Acid Cycle" (J. M. Lowenstein, ed.), p. 249. Dekker, New York.
Utter, M. F., and Scrutton, M. C. (1969). *In* "Current Topics in Cellular Regulation" (B. L. Horecker and E. R. Stadtman, eds.), Vol. 1, p. 253. Academic Press, New York.
Van Dam, K., and Meijer, A. J. (1971). *Annu. Rev. Biochem.* **40**, 115.
Van den Bergh, S. G. (1962). Ph.D. Thesis, University of Amsterdam, Amsterdam.
Van den Bergh, S. G. (1964). *Biochem. J.* **93**, 128.
Van den Bergh, S. G., and Slater, E. C. (1962). *Biochem. J.* **82**, 362.
Van de Stadt, R. J., Nieuwenhuis, F. J. R. M., and Van Dam, K. (1971). *Biochim. Biophys. Acta* **234**, 173.
Van Tol, A., and Hülsmann, W. C. (1969). *Biochim. Biophys. Acta* **189**, 342.
Van Tol, A., and Hülsmann, W. C. (1970). *Biochim. Biophys. Acta* **223**, 416.
von Euler, H., Adler, E., Günther, G., and Das, N. B. (1938). *Hoppe-Seyler's Z. Physiol. Chem.* **254**, 61.
von Jagow, G., Westermann, B., and Wieland, O. (1968). *Eur. J. Biochem.* **3**, 512.
Wakil, S. J. (1963). *In* "The Enzymes" (P. D. Boyer, H. Lardy, and K. Myrbäck, eds.), 2nd ed., Vol. 7, p. 97. Academic Press, New York.
Walter, P., Paetkau, V., and Lardy, H. A. (1966). *J. Biol. Chem.* **241**, 2523.
Warburg, O. (1913). *Pflügers Arch. Gesamte Physiol. Menschen Tiere* **154**, 599.
Watanabe, M. I., and Williams, C. M. (1951). *J. Gen. Physiol.* **34**, 675.
Weber, N. E., Blair, P. V., and Martin, B. (1969). *Biochem. Biophys. Res. Commun.* **36**, 987.
Webster, L. T., Jr. (1963). *J. Biol. Chem.* **238**, 4010.
Webster, L. T., Jr. (1969). *Methods Enzymol.* **13**, 375.

Webster, L. T., Jr., Gerowin, L. D., and Rakita, L. (1965). *J. Biol. Chem.* **240**, 29.

Weijers, P. J. (1971). Ph.D. Thesis, University of Amsterdam, Amsterdam.

Weinbach, E. C., and von Brand, T. (1967). *Biochim. Biophys. Acta* **148**, 256.

Werkheiser, W. C., and Bartley, W. (1957). *Biochem. J.* **66**, 79.

West, D. W., Chase, J. F. A., and Tubbs, P. K. (1971). *Biochem. Biophys. Res. Commun.* **42**, 912.

Whereat, A. F., Hull, F. E., and Orishimo, M. W. (1967). *J. Biol. Chem.* **242**, 4013.

Whereat, A. F., Orishimo, M. W., and Nelson, J. (1969). *J. Biol. Chem.* **244**, 6498.

Wieland, O., and Siess, E. (1970). *Proc. Nat. Acad. Sci. U.S.* **65**, 947.

Wieland, O., and von Jagow-Westermann, B. (1969). *FEBS Lett.* **3**, 271.

Wieland, O., von Jagow-Westermann, B., and Stukowski, B. (1969). *Hoppe-Seyler's Z. Physiol. Chem.* **350**, 329.

Wieland, O., Menahan, L. A., and von Jagow-Westermann, B. (1970). *In* "Metabolic Regulation and Enzyme Action" (A. Sols and S. Grisolia, eds.), p. 77. Academic Press, New York.

Wieland, O., Siess, E., Schulze-Wethmar, F. H., von Funcke, H. G., and Winston, B. (1971). *Arch Biochem. Biophys.* **143**, 593.

Wigglesworth, V. B. (1949). *J. Exp. Biol.* **26**, 150.

Williamson, J. R.(1962). *Biochem. J.* **83**, 377.

Williamson, J. R. (1964). *Biochem. J.* **93**, 97.

Williamson, J. R., and Krebs, H. A. (1961). *Biochem. J.* **80**, 540.

Wit-Peeters, E. M. (1969). *Biochim. Biophys. Acta* **176**, 453.

Wit-Peeters, E. M. (1971). Ph.D. Thesis, University of Amsterdam, Amsterdam.

Wit-Peeters, E. M., Scholte, H. R., and Elenbaas, H. L. (1970). *Biochim. Biophys. Acta* **210**, 360.

Wit-Peeters, E. M., Scholte, H. R., Van den Akker, F., and De Nie, I. (1971). *Biochim. Biophys. Acta* **231**, 23.

Yates, D. W., and Garland, P. B. (1970). *Biochem. J.* **119**, 547.

10

ATP BREAKDOWN FOLLOWING ACTIVATION OF MUSCLE

NANCY A. CURTIN and R. E. DAVIES

I. Historical Introduction

Many compounds have been proposed as the energy source for muscle contraction. About a century ago, it was thought that when muscle contracted, a large molecule—myogen or inogen—inside the muscle took in oxygen and gave off carbon dioxide. This idea, however, was disproved by Fletcher and Hopkins in 1907. They showed that muscle could contract without oxygen, and under these conditions lactic acid was formed. The studies by Parnas (Parnas and Wagner, 1914), Meyerhof (1920), and Hill (1928) further elaborated the lactic acid theory. Their studies showed that the breakdown of glycogen to form lactic acid in muscle under anaerobic conditions was sufficient to provide the energy for muscle contraction. This process was accepted as the energy source.

The lactic acid theory was eventually disproved when Lundsgaard (1930) found that pretreatment with iodoacetate permitted isolated muscle to contract without producing lactic acid. These muscles, however, did produce inorganic phosphate, which came from the breakdown of phosphorylcreatine (Eggleton and Eggleton, 1927; Fiske and SubbaRow, 1929; Nachmansohn, 1928). Although phosphorylcreatine was broken down in the whole muscle, the reaction did not take place in extracts except in the presence of a diffusible cofactor. Lohmann (1934) identified this cofactor as ADP and proposed that phosphorylcreatine is used to reform ATP in the contracting muscle.

In 1939, Engelhardt and Ljubimowa discovered that myosin, the principal structural protein in muscle, is also an ATPase. This further substantiated the idea that the breakdown of ATP was the immediate energy source for contraction. At this time, a series of new approaches involving isolated enzymes, glycerinated muscles, and actomyosin threads was initiated. The investigations of these systems showed that they could "contract" in the presence of ATP (Szent-Györgyi, 1953; H. H. Weber, 1958).

Nevertheless, the direct usage of ATP in intact muscle during contraction could not be demonstrated despite the efforts of many investigators. Fleckenstein *et al.* (1954) found no net change in ATP, ADP, or phosphorylcreatine in frog rectus abdominis muscle that contracted and did work at 0°C. Mommaerts (1954, 1955) did experiments with turtle muscle at 0°C and his measurements of ADP, AMP, creatine, creatinine, and pyruvate did not show a net change of any of these compounds. Experiments by Munch-Petersen (1953) and Lange (1955a,b,c) also left unanswered the question of the immediate chemical events during contraction. An exhaustive search among all the phosphorus-containing compounds was made. The compounds tested included carnosine di- and monophosphates (Davies *et al.*, 1959), all the inosine, guanosine, and cytosine phosphates, phosphoenolpyruvate, and the phosphoglyceric acids (Cain and Davies, 1962a; Cain *et al.*, 1964), and the phosphorus in the acid-insoluble residue of muscle (Seraydarian and Williams, 1960). None of these compounds was found to break down during contraction.

Chance and Connelly (1957) and Jöbsis and Chance (1957) applied very sensitive but indirect spectrophotometric methods to this problem, but the net change in ADP content that was detected was only about 3% of the value required to account for the energy changes. The rest of it could have been reconstituted to ATP before it was possible to detect it.

In 1959, Davies *et al.* found an increase in inorganic phosphate during a single contraction of turtle muscle at 0°C. The amount of inorganic phosphate was consistent with that expected from the amount of work done by the muscle and the estimated free energy of the unidentified "high energy" source of the inorganic phosphate. Mommaerts (1961) also found an increase in inorganic phosphate during brief tetani of frog muscle.

Carlson and Siger (1960) and Mommaerts *et al.* (1962) confirmed and extended Lundsgaard's (1930) observation that phosphorylcreatine is broken down in iodoacetate-treated muscles during a long series of contractions. Phosphorylcreatine usage during single contractions was found by Cain *et al.* (1962). Pretreatment of the muscles with dinitrophenol reduced the resting level of phosphorylcreatine and thus made it possible to detect the small difference between phosphorylcreatine content of the working muscle and the resting muscle. The measurements of phosphorylcreatine breakdown, however, still did not establish the usage of ATP, even though the two reactions were thought to be closely coupled in the contracting muscle. This was the basis of A. V. Hill's first "Challenge to Biochemists" (1950).

What was needed at this point was an inhibitor of the enzyme, ATP:creatine phosphotransferase, which apparently was reconstituting ATP as fast as it was being broken down. In 1959, Kuby and Mahowald showed that 2,4-dinitrofluorobenzene (DNFB) did in fact inhibit the enzyme *in vitro*. DNFB, however, was well known from Sanger's use of it in end-group analysis for the determination of protein structure. From Sanger's work, one would expect this reagent to attack all the free amino groups in muscle proteins and to inhibit much more than just ATP:creatine phosphotransferase. So DNFB was not tried on muscle

TABLE I

CHEMICAL CHANGES DURING SINGLE CONTRACTIONS OF MUSCLE

Production of inorganic phosphate without net change in free creatine during
single isotonic contraction at 0°C of muscles pretreated with DNFB[a]

Type of muscle from *R. pipiens*	Experimental conditions	P$_i$ change (μmole/gm $\pm$ SEM)	Free creatine change (μmole/gm $\pm$ SEM)
Sartorius	untreated	$+0.62 \pm 0.09$	$+0.68 \pm 0.07$
Sartorius	DNFB-treated	$+0.67 \pm 0.11$	$+0.004 \pm 0.32$
Rectus abdominis	untreated	$+0.95 \pm 0.22$	$+0.93 \pm 0.19$
Rectus abdominis	DNFB-treated	$+0.97 \pm 0.17$	$+0.18 \pm 0.16$

Production of inorganic phosphate (P$_i$) and breakdown of ATP during single
isotonic contractions at 0°C of muscles pretreated with DNFB

Type of muscle	Experimental conditions	P$_i$ change (μmole/gm $\pm$ SEM)	ATP change (μmole/gm $\pm$ SEM)
Sartorius	untreated	$+0.80 \pm 0.10$	-0.01 ± 0.12
Sartorius	DNFB-treated	$+0.64 \pm 0.10$	-0.70 ± 0.15

[a] After Infante and Davies (1965).

until two years after Kuby and Mahowald (1959) reported their *in vitro* experiments. Eventually, the conditions for the DNFB treatment of muscle were established, and Cain and Davies (1962b) found that this reagent completely inhibited ATP:creatine phosphotransferase *in vivo* (Table I). The contractile machinery, however, remained intact, and the muscle worked and broke down an energetically equivalent amount of ATP. Murphy (1966) found that both the ATP level and the developed tension decreased when muscles which had been pretreated with DNFB were repeatedly tetanized under isometric conditions. This verified that DNFB treatment did prevent the reformation of ATP that was broken down during contraction *in vivo*.

Since these initial measurements of phosphorylcreatine and ATP usage were made, many different aspects of contraction have been investigated individually. This chapter summarizes these investigations of the processes of activation, isometric tetanus, shortening, positive work, negative work, and relaxation.

It is very interesting that muscles can contract and develop large forces during negative work without the usage of ATP other than that needed for activation. This finding and others raises once again the question of whether ATP breakdown is in fact the most immediate reaction associated with the development of force and mechanical work. The extremely low ATP usage during negative work and other observations support the proposal that the conformational changes of the peptide crossbridge and the making and breaking of hydrogen bonds and hydrophobic interactions are the most immediate source of the energy for contraction.

II. Action Potential

The initial response of a muscle to electrical stimulation is the propagation of an action potential along the surface membrane. Under normal conditions it rapidly initiates the chemical activation involving Ca^{2+} movements which trigger the contractile response—development of tension, shortening, and work. To measure the energy requirements for the action potential alone, it must be "isolated" from these subsequent processes. Cain *et al.* (1962) accomplished this by freezing the muscle just after stimulation so that no shortening or tension development took place. They found that the inorganic phosphate and creatine contents of frog rectus abdominis muscles were changed very little under these conditions. The increase in inorganic phosphate was only 0.08 ± 0.11 μmole/gm.* Klocke *et al.* (1966) found that canine heart muscle conducted normal action potentials but did not shorten or develop tension when these muscles were electrically stimulated after soaking in Ca^{2+}-free solutions. The oxygen used per electrical depolarization was 0.4 μliter/100 gm muscle per pulse. This is equivalent to 0.001 μmole of ATP per gram of muscle per pulse.

Another method that has been used in the study of "noncontractile" processes involves the use of muscle or muscle fibers that have been soaked in hypertonic solutions prior to electrical stimulation. Hodgkin and Horowicz (1957) and Howarth (1958) reported that muscles treated in this way conducted normal action potentials but did not develop tension. The implication that treatment with hypertonic solution prevents Ca^{2+} release upon electrical stimulation has now been proven by Ashley and Ridgeway (1970). They monitored the change in Ca^{2+} levels in the sarcoplasm by measuring the Ca^{2+}-mediated light emission from aequorin that had been injected into barnacle muscle fibers. Electrical stimulation of fibers in hypertonic solutions elicited the normal membrane response, but no Ca^{2+} transient or tension development.

---

* The $\pm$ value represents $\pm$ standard error of the mean.

Several measurements have been made of the chemical changes that accompany electrical stimulation of a muscle that has been pretreated with hypertonic solutions. Cain *et al.* (1962) measured the phosphoryl-creatine usage by frog rectus abdominis muscles stimulated at 12 pulses/sec for up to 3 sec in hypertonic solutions. The muscles had been treated with dinitrophenol to lower their phosphorylcreatine content. The usage of phosphorylcreatine upon stimulation was not significant (0.01 ± 0.01 μmole per gram of muscle per pulse).

In the frog sartorius muscle, however, Infante *et al.* (1964a) did find a significant increase in inorganic phosphate. The muscles were treated with hypertonic solutions and then were stimulated for 1 sec at a rate of 12 pulses/sec. The total amount of inorganic phosphate produced was 0.51 ± 0.13 μmole per gram of muscle, which is equivalent to an increase of 0.042 ± 0.01 μmole per gram of muscle per pulse.

The ATP usage accompanying depolarization of frog sartorius muscle under similar conditions has been measured by Kushmerick *et al.* (1969). The muscles were pretreated with hypertonic solutions and DNFB and then were stimulated at rates of 5, 10, and 15 times per second for 1, 2, and 3 sec. The mean usage of ATP was 0.006 ± 0.008 μmole per gram of muscle per pulse. These small chemical changes accompanying the action potential may supply the energy required for Na^+ pumping. The magnitude of the chemical changes are in the range that would be expected on the basis of the membrane response to electrical stimulation under hypertonic conditions (Bianchi, 1970).

Frog sartorius muscles treated with hypertonic solutions release about 1 mcal of heat per gram of muscle in response to a single stimulation (Hill, 1958a; Homsher *et al.*, 1970). If the conclusion of the experiments by Ashley and Ridgeway (1970) on barnacle muscle in hypertonic solutions also applies to frog muscle, then the heat observed under these conditions does not come from a release of Ca^{2+} ions into the sarcoplasm. It may be that the heat is liberated while Ca^{2+} is released and rapidly rebound to the sarcoplasmic reticulum before the Ca^{2+} level rises in the sarcoplasm.

Although Ca^{2+} movements may be the source of the heat output under hypertonic conditions, there is evidence that suggests that the action potential is not the source of the heat. Homsher *et al.* (1970) found no measurable heat production accompanying the action potential when the T-tubules had been disrupted by treating the muscle with glycerol. There is more heat liberated as a result of electrical stimulation under hypertonic conditions than could be due to the amount of phosphoryl-creatine or ATP breakdown that is found if the enthalpy change for these reactions is −11 kcal/mole (Wilkie, 1968). It may be significant that the heat output was measured for a single stimulation, whereas the chemical changes were measured for a series of stimulations.

III. Activation

A. Heat Output

The term "activation heat" was used by Hill (1949a) to describe the difference (about 1 mcal/gm) between the total heat output during shortening during a preloaded isotonic twitch and the part of the heat specifically associated with shortening. Approximately the same amount of activation heat was liberated by muscles that had been fully shortened by previous stimulation (Hill, 1949a; Gibbs et al., 1966; Chaplain and Pfister, 1970). Under these conditions there is no apparent tension development or shortening by the muscle.

Gibbs et al. (1966) measured an output of heat (1.2 ± 0.3 mcal/gm) which was quantitatively similar to Hill's (1949a) activation heat. This was done by using a "double twitch" technique. The muscles were stimulated twice under isometric conditions, and the period of time between the two stimuli was adjusted so that (1) the heat output from the first stimulus was complete before the second stimulus was given, and (2) the second stimulus was given within the period of mechanical fusion, so only a small amount (if any) additional tension was developed. If there was additional tension, the heat output was adjusted accordingly. The heat output for the first stimulus was taken to be the result of activation and shortening against the series elastic elements and tension development. The additional heat elicited by the second stimulus was interpreted to be activation heat alone.

Gonzalez-Serratos (1966a,b) examined single fiber preparations microscopically and found that relaxation was more complex at this level than it appears to the unaided eye. Although the fiber as a whole remained shortened as it relaxed after stimulation, the individual myofibrils within the fiber elongated and became wavy. This suggested that within a whole muscle, which has been fully shortened by previous stimulation, the myofibrils will shorten as well as become activated by additional stimulation. The heat and chemical changes under fully shortened conditions may reflect both processes—activation and unloaded shortening.

Homsher and Ricchiuti (1969) and Homsher et al. (1970) stretched semitendinosus muscles beyond rest length to a point at which tension was not developed during a twitch. Presumably, the thick and thin filaments were not overlapping and there was no contractile activity during the stimulation. The heat released under these conditions was 1.03 mcal/gm for a single stimulation. It is interesting and significant that the heat released per stimulation turned out to be much smaller (0.54 mcal/gm/stimulation) when a series of ten stimulations at a frequency of one per second were given to the muscle (Homsher *et al.*, 1970). Smith (1970) has measured the heat released by semitendinosus and sartorius muscle under "fully stretched" conditions similar to the conditions used by Homsher *et al.* (1970). Smith reported that 0.57 mcal/gm of heat was released, but it was not clear in the report whether this is the heat for a single stimulation or the average heat per stimulus during a series of stimulations.

Thus, the results of the heat measurements under conditions which prevent tension development and work fall into two groups—(1) those in which a single stimulation was given and the heat output was about 1 mcal/gm, and (2) those in which a series of stimulations was given and the heat output per stimulus was about 0.5 mcal/gm. If the heat were derived entirely from the enthalpy change involved in phosphorylcreatine or ATP breakdown, the expected chemical changes would be 0.1 and 0.06 μmole/gm, respectively. These values are based on an enthalpy of -11 kcal/mole (Wilkie, 1968).

B. Chemical Breakdown

1. TWITCHES

Kushmerick *et al.* (1969) have measured the ATP usage during a series of contraction–relaxation cycles in which the muscles were very lightly loaded (0.35 gm). Consequently, only a small amount of work was done and the complications due to the presence of tension during relaxation (Hill, 1964c) were avoided. The muscles were given a single pulse every 1, 1.2, or 1.3 sec. The total number of pulses ranged from 5 to 31. An average cycle of activity required the use of 0.057 ± 0.006 μmole ATP per gram of muscle. Homsher *et al.* (1970) found essentially the same (0.051 μmole/gm/stimulation) ATP usage by semitendinosus muscle that had been extended beyond rest length to a length at which tension was not developed upon stimulation. The muscles were stimulated 10 times at 1 time per second in this experiment.

The enthalpy change for the amount of chemical change reported by Kushmerick *et al.* (1969) and Homsher *et al.* (1970) would be sufficient to account for the heat output during a series of stimulations (Homsher *et al.*, 1970). Davies (1963) and Homsher *et al.* (1970) have suggested that the activation heat results from the release of Ca^{2+} from binding sites and the ATP usage is associated with the accumulation of Ca^{2+} into the sarcoplasmic reticulum.

This seems to be supported by the results of Chaplain and Pfister (1970). They measured the heat output as described earlier and also monitored in parallel experiments the Ca^{2+} level in the sarcoplasm. This was done by first injecting the muscle with murexide, a compound that changes its light absorption when complexed with Ca^{2+} (Jöbsis and O'Connor, 1966). Then the muscle was stimulated, and the changes in light absorption were measured. The time course of the release and reabsorption of Ca^{2+} was very similar to the time course of the rate of heat output. The ATP usage for a cycle of contraction and relaxation was 0.072 ± 0.008 μmole/gm. This amount of ATP usage is not significantly greater than that measured by Kushmerick *et al.* (1969).

On the other hand, the amount of chemical change that takes place during repetitive stimulation is only about half of what would be required to give rise to the heat output during a single stimulation. The heat output for the initial stimulation in a series is greater apparently than the heat output during successive stimulations. Since none of the chemical experiments was done with a single stimulation, one does not know whether or not all the heat comes from ATP usage and chemical processes directly coupled to it. Table II is a summary of the measured values for chemical breakdown for activation during a twitch.

TABLE II

ATP BREAKDOWN FOR ACTIVATION DURING A SERIES OF TWITCHES

Reference	Experimental conditions	Chemical change	
		Compound	Amount (μmole/gm/stimulation) $\pm$ SEM
Kushmerick *et al.* (1969)	Sartorius *R. pipiens;* 0°C; lightly loaded isotonic twitches	Inorganic phosphate	$+0.057 \pm 0.006$
Homsher *et al.* (1970)	Semitendinosus *R. pipiens;* 0°C; fully lengthened	ATP	-0.051
Chaplain and Pfister (1970)	Sartorius *R. esculenta;* 0°C; fully shortened	ATP	-0.072 ± 0.008

TABLE III

ATP Breakdown for Activation during Tetanus

		Chemical change	
Reference	Experimental conditions	Compound	Amount (μmole/gm/sec) $\pm$ SEM
Infante *et al.* (1964b)	Rectus abdominis *R. pipiens;* 0°C 50% l_0 125% l_0	Phosphorylcreatine	0.0 $\pm$ 0.28 0.0 $\pm$ 0.21
Infante *et al.* (1964a)	Sartorius *R. pipiens;* 0°C 50% l_0 130% l_0	Inorganic phosphate	+0.26 $\pm$ 0.05 +0.26 $\pm$ 0.07
Sandberg and Carlson (1966)	Sartorius *R. pipiens;* 1°C 65% l_0 140% l_0	Phosphorylcreatine	−0.133 $\pm$ 0.011 −0.134 $\pm$ 0.029

2. Tetanus

Infante *et al.* (1964a) measured the production of inorganic phosphate in frog sartorius muscles that were stimulated at very short or very long muscle lengths. Under these conditions no tension was developed when the muscles were stimulated for 1.5 sec at a rate of 12 pulses/sec at 0°C. At 50% l_0* the rate of inorganic phosphate production was 0.26 $\pm$ 0.05 μmole/gm/sec, and at 130% l_0 it was 0.26 $\pm$ 0.07 μmole/gm/sec. Sandberg and Carlson (1966) did a similar experiment, but the duration of stimulation was 20 sec and the rate was 10 pulses/sec. In this case, the rate of phosphorylcreatine usage at 65% l_0 was 0.133 $\pm$ 0.011 μmole/gm/sec, and at 140% l_0 it was 0.134 $\pm$ 0.029 μmole/gm/sec. A significant amount of tension was developed, however, 14 P_0† at 65% l_0, and 22% P_0 at 140% l_0. Thus, part of the chemical breakdown may have been associated with the process of tension development. Table III summarizes these results.

Kushmerick *et al.* (1969) found that the rate of stimulation influenced the ATP breakdown in frog sartorius muscles, which contracted under

---

* The symbol l_0 refers to the rest length of the muscle.

† The symbol P_0 refers to the maximum force developed during contraction under isometric conditions at rest length.

TABLE IV

TABLE IV
INFLUENCE OF RATE OF STIMULATION ON CHEMICAL BREAKDOWN FOR
ACTIVATION DURING TETANUS OF SARTORIUS OF *R. pipiens*

Reference	Experimental conditions	Compound	Chemical change
			Amount/pulse (μmole/gm/pulse) $\pm$ SEM
Sandberg and Carlson (1966)	1°C; 10 pulses/sec	Phosphorylcreatine	-0.0134 ± 0.003
Kushmerick et al. (1969)	0°C; 10 pulses/sec	Inorganic phosphate	$+0.021 \pm 0.010$
Infante et al. (1964a)	0°C; 12 pulses/sec	Inorganic phosphate	$+0.022 \pm 0.006$
Kushmerick et al. (1969)	0°C; 50 pulses/sec	Inorganic phosphate	$+0.0043 \pm 0.0019$

lightly loaded isotonic conditions at 0°C. The stimulation lasted for 1 or 2 sec and during much of this time the muscles were contracting under fully shortened conditions. For stimulation at 10 pulses/sec the ATP breakdown was 0.021 ± 0.01 μmole/gm/pulse, and at 50 pulses/sec, it was only 0.0043 ± 0.0019 μmole/gm/pulse. Table IV compares the chemical breakdown for activation by muscles stimulated tetanically at various rates.

3. ENERGY REQUIREMENTS FOR Ca^{2+} PUMPING *in Vitro*

The problem of identifying the "activation process" has been approached from various other points of view (reviewed by Hasselbach, 1964; Sandow, 1965; Ebashi and Endo, 1968; Winegrad, 1969). The results permit an estimation of the energy requirements for this process *in vivo*.

The activation of the contractile mechanism is due primarily to the influence of Ca^{2+} on troponin (reviewed by Ebashi and Endo, 1968). During the activation–relaxation cycle for skeletal muscle, Ca^{2+} is released from the terminal cisternae of the sarcoplasmic reticulum and then is reaccumulated by the tubular and terminal portions of the reticulum (Winegrad, 1965, 1968, 1970; Costantin et al., 1965; Jöbsis and O'Connor, 1966; Ridgeway and Ashley, 1967; reviewed by Winegrad, 1969).

The amount of Ca^{2+} that moves from the lateral cisternae of the sarcoplasmic reticulum to the interior of the myofibril during contraction *in vivo* is about 0.2 μmole per gram of muscle (Winegrad, 1968, 1970).

This value agrees with the amount 0.125 μmole per gram of muscle, which gives maximal response when injected into muscle (Heilbrunn and Wiercinski, 1947). A. Weber and Herz (1963) found that superprecipitation occurred as a result of the interaction of actomyosin or myofibrils with a quantity of Ca^{2+} equivalent to one or two molecules of Ca^{2+} per myosin molecule. The binding and reaccumulation of Ca^{2+} by vesicles isolated from muscle is rapid enough to account for the rate of relaxation of muscle *in vivo* (Ouishi and Ebashi, 1964). This reaccumulation requires the presence of Mg^{2+} and the splitting of ATP (Ebashi and Lipmann, 1962). The amount of Ca^{2+} transported per molecule of ATP used depends to some extent on the concentration of Ca^{2+} outside the vesicles. One to two moles of Ca^{2+} are transported per mole of ATP split (Hasselbach and Makinose, 1963; Makinose and The, 1965; A. Weber *et al.*, 1966). From the amount of Ca^{2+} released for activation and the ATP needed to reaccumulate this Ca^{2+}, one can calculate an expected ATP usage associated with an activation–relaxation cycle according to this scheme. The amount is 0.1–0.2 μmole ATP per gram of muscle. This is quantitatively consistent with the idea that the heat is due to the release of bound Ca^{2+}. The heat output would represent a dissipation of the free energy from ATP that went into the process of binding the Ca^{2+} (Davies, 1963; Homsher *et al.*, 1970). It is not possible to compare the expected ATP usage for a single cycle of Ca^{2+} pumping with the ATP usage for a single cycle of activation and relaxation *in vivo*, because the measurement for a single cycle *in vivo* has not been made. The amount of ATP usage per cycle during a series of cycles *in vivo* is only about 0.05 μmole/gm, which is less than that expected for Ca^{2+} pumping on the basis of the experiments on isolated systems and other systems just described.

IV. Isometric Contraction

A. Heat Output and Chemical Breakdown during an Isometric Twitch

The heat produced by an untreated sartorius muscle during a single isometric twitch was found to be 3 mcal/gm by Hill (1958b) and 3.7 mcal/gm by Hill and Woledge (1962). The chemical change during an isometric twitch was measured under somewhat different conditions by Carlson and Siger (1960). Each muscle performed a series of isometric twitches, that is, a series of contraction–relaxation cycles. The muscles did not shorten or do external work, but they did develop tension. The developed tension was not the same for all the twitches in each series.

Therefore, rather than use the observed number of twitches, they calculated the equivalent number of "full" twitches. This number was calculated by dividing the observed total tension developed in the series of twitches by the observed maximum tension developed by the muscle. The calculated number of full twitches was assumed to represent the number of complete sets of chemical reactions that occurred in the muscle. This assumes that the tension is directly proportional to the amount of chemical reaction in the muscle and that the amount of chemical reaction is not much affected by the number of electrical stimuli received. The phosphorylcreatine usage per total creatine content in the muscle was measured and plotted against the calculated equivalent number of full twitches. The slope of the line was 8.98 ± 0.18 mmole phosphorylcreatine usage per mole creatine content per twitch. This is equivalent to 0.286 μmole per gram of muscle during one average isometric contraction–relaxation cycle in a series of cycles. It may be important that several seconds elapsed from the end of the last relaxation to the time the muscle was frozen.

Carlson *et al.* (1967) also did experiments involving a series of isometric twitches, but they measured both the heat output and the phosphorylcreatine usage by each muscle. As in the experiment of Carlson and Siger (1960), the results were corrected for the decrease in tension during the series of twitches. The muscles were pretreated with iodoacetate and nitrogen, and they relaxed between twitches and for about 30 sec between the last stimulation and freezing. The heat output during a series of twitches at a frequency of one every 3 sec was 2.76 mcal/gm/twitch. This is lower than the values reported by Hill (1958b) and Hill and Woledge (1962), but the authors suggested that differences in the experimental design (iodoacetate-treated versus untreated muscles and a series of twitches versus a single twitch) may have been responsible for this difference in heat output. The phosphorylcreatine usage was 0.293 μmole/gm/twitch. This agrees with the results of Carlson and Siger (1960).

B. Heat Output and Chemical Breakdown during an Isometric Tetanus

1. Variation with Duration of Contraction

The rate of heat production during isometric tetani of various durations at 0°C was measured by Abbott (1951). The rate decreased during the first few seconds of contraction and then remained constant at about 1.5 mcal/gm/sec. The initial decrease in the rate of heat output was more rapid at a stimulation frequency of 8 pulses/sec than at 1.8 pulses/sec. Abbott states that the shortening of the muscle against the series elastic elements was over during the first 1.5 sec of contraction. This heat associated with shortening could not be the only factor contributing to the initial high rate of heat output.

Aubert (1956) studied the liberation of heat by sartorius muscles of frogs (*Rana temporaria*) during isometric contraction. He measured the rate of heat output at intervals of 0.5 sec during the contraction. The first measurement was made 0.5 sec after the stimulation had started and by this time the tension had reached a constant level. Aubert found that the rate of liberation of maintenance heat could be described mathematically as the sum of two separate rate terms.

$$h = h_a e^{-\alpha t} + h_b \tag{1}$$

where h = rate of total maintenance heat production
$h_a e^{-\alpha t}$ = rate of labile heat production
h_a = constant for the labile heat production
α = time constant for the labile heat
t = duration of the contraction
h_b = rate of stable heat production

The labile rate and the stable rate each varied in a characteristic way as the contraction progressed (Table V). The labile rate was high at first but decreased rapidly in an exponential way. It reached zero after the first 6–8 sec of contraction. The stable heat rate was a constant

TABLE V

RATES OF LABILE, STABLE, AND TOTAL MAINTENANCE HEAT OUTPUT DURING
ISOMETRIC TETANUS OF SARTORIUS MUSCLE OF *Rana temporaria* AT 0°C[a]

Duration of contraction (sec)	Rate of labile heat output (mcal/gm/sec)	Rate of stable heat output (mcal/gm/sec)	Rate of total maintenance heat output (mcal/gm/sec)
0.5	2.44	3.03	5.47
1.0	1.67	3.03	4.70
2.0	0.78	3.03	3.81
3.0	0.20	3.03	3.23
4.0	0.17	3.03	3.20
5.0	0.08	3.03	3.11
6.0	0.04	3.03	3.07
15.0 and above	0.00	3.03	3.03

[a] Calculated from constants given by Aubert (1956) p. 149, Tableau XIIC.

TABLE VI

INFLUENCE OF SOME EXPERIMENTAL CONDITIONS ON RATE OF LABILE AND STABLE
HEAT OUTPUT DURING ISOMETRIC TETANUS OF SARTORIUS MUSCLE OF
Rana temporaria AT 0°C

Reference	Experimental condition	h_a	h_b
Aubert (1956)	Increased frequency of stimulation	Decreased	Same
Aubert (1968)	Repeated tetani	Decreased	Same
Aubert (1956)	Muscle length greater than l_0	Same	Decreased
Aubert (1964)	DNFB-treated	Same	Decreased

during the entire contraction. Because the labile rate was higher during the early part of the contraction the rate of liberation of the total maintenance heat was always greater for shorter contractions (duration less than 8 sec) than it was for longer contractions. This is similar to the findings of Abbott (1951).

Several factors have been found to influence the rate of heat output during isometric contractions. Some of these are listed in Table VI. Infante *et al.* (1964a) measured the ATP usage during single brief (up to 1.5 sec) isometric tetani of sartorius muscles of *Rana pipiens* at 0°C. The DNFB-treated muscles were held at rest length and were stimulated for various periods of time. DNFB effectively inhibited the enzyme ATP:creatine phosphotransferase, preventing the reformation of ATP from ADP and phosphorylcreatine. For each muscle the inorganic phosphate content as well as the ATP content was determined. The amount of ATP used was equal to the inorganic phosphate formed during the contractions. Thus the reformation of ATP from two ADPs by the myokinase* reaction did not proceed significantly during these contractions. It was concluded that the pretreatment with DNFB was not necessary and inorganic phosphate production was taken to be equal to ATP breakdown. Another group of muscles remained untreated and were used in the experiment. For each muscle the progressive development of tension was recorded. When the stimulation began the tension was developed rapidly, and the maximum tension was reached within less than 0.5 sec of the beginning of stimulation. Then the tension remained constant for the rest of the duration of the stimulation. There was some variation in the rates of ATP breakdown (0.62–0.73 μmole/gm/sec) for contractions of durations ranging from 0.56 to 1.5 sec.

From the measurements of the heat production for contractions of

* Myokinase refers to ATP: AMP phosphotransferase.

various durations, one would expect that the rate of heat output during the brief contractions studied by Infante *et al.* (1964a) would be greater than during longer contractions. The chemical changes during longer contractions have been measured in several studies.

Maréchal and Mommaerts (1963) studied the phosphorylcreatine usage by sartorius muscles of *Rana pipiens* during isometric tetani of durations ranging from 3 to 120 sec. By plotting the chemical usage against duration of contraction they calculated an average rate of usage of 0.28 μmole/gm/sec. As the duration of the tetani increased, the tension decreased considerably. The rate of decrease was 0.8% per sec of contraction. When correction was made for this the rate of usage was 0.32 μmole/gm/sec.

Sandberg and Carlson (1966) made a study of the relationship between muscle length and the phosphorylcreatine usage during isometric tetani of 20 sec duration. Their results included a measure of the phosphorylcreatine usage by sartorius muscles of *R. pipiens* during a 20 sec tetanus at rest length. They reported that the rate of usage was 0.26 $\pm$ 0.02 μmole/gm/sec, which is equivalent to a total usage of 5.2 $\pm$ 0.4 μmole/gm.

Carlson *et al.* (1967) measured both the heat output and phosphorylcreatine usage during isometric tetani of two different durations. The muscles were pretreated with iodoacetate and nitrogen and they relaxed for about 30 sec between the last stimulation and freezing. For the set of muscles that were stimulated for 10 sec, the heat output was 39.4 mcal/gm, which is equivalent to an average rate of 3.94 mcal/gm/sec. The chemical usage was 3.35 μmole/gm for 10 sec of stimulation, which is equivalent to an average rate of 0.335 μmole/gm/sec. For the set of muscles that were stimulated for 30 sec, the heat output was 76.3 mcal/gm/30 sec, (an average rate of 2.54 mcal/gm/sec), and the chemical usage was 7.01 μmole/gm/30 sec (average rate 0.234 μmole/gm/sec).

Clearly, for brief contractions, the rate of chemical change is greater than it is for contractions of longer duration. This variation in rate of chemical change parallels that of the rate of heat output during tetanus. The measured values for the chemical breakdown during isometric twitches and tetani of various durations are summarized in Table VII.

At this point, one can speculate as to why the labile heat diminished as contraction progressed (Aubert, 1956). It may be related to the observation that the total heat output per twitch appears to decrease with repeated stimulation (Hill, 1958b; Hill and Woledge, 1962; Carlson *et al.*, 1967), and that the total labile heat itself decreases upon repeated stimulation (Aubert, 1968). It may be that the changes in maintenance

TABLE VII

CHEMICAL BREAKDOWN IN FROG MUSCLES DURING CONTRACTION UNDER ISOMETRIC CONDITIONS

A. Series of twitches at rest length

Reference	Experimental conditions	Chemical change	
		Compound	Amount per twitch (μmole/gm)
Carlson and Siger (1960)	Sartorius *R. pipiens*; 0°C	Phosphorylcreatine	−0.286
Carlson *et al.* (1967)	Sartorius *R. temporaria*; 0°C	Phosphorylcreatine	−0.293

B. Tetani of various durations

Reference	Experimental conditions	Duration of tetanus (sec)	Chemical change	
			Compound measured	Amount per sec (μmole/gm) ± SEM
Infante *et al.* (1964b)	Rectus abdominis *R. pipiens*; 0°C	1	Phosphorylcreatine	−0.19 ± 0.09
		5		−0.18 ± 0.04
Infante *et al.* (1964a)	Sartorius *R. pipiens*; 0°C	1.5	Inorganic phosphate	+0.73 ± 0.07
Carlson *et al.* (1967)	Sartorius *R. temporaria*; 0°C	10	Phosphorylcreatine	−0.335
Sandberg and Carlson (1966)	Sartorius *R. pipiens*; 1°C	20	Phosphorylcreatine	−0.26 ± 0.02
Carlson *et al.* (1967)	Sartorius *R. temporaria*; 0°C	30	Phosphorylcreatine	−0.234
Maréchal and Mommaerts (1963)	Sartorius *R. pipiens*; 0°C	3–120	Phosphorylcreatine	−0.32

heat and metabolism are due to changes in the heat output and metabolism associated with Ca^{2+} ion movements. Homsher *et al.* (1970) have shown that the activation heat output per twitch was lower when repeated twitches were performed by the muscle, than when a single twitch was performed.

It is also possible that a large amount of internal work is done only at the beginning of a tetanus or during the first twitches in a series. The internal work would require ATP usage and would be degraded into heat as the shortened sarcomeres are stretched out by other sarcomeres. Larson (1970) has calculated the average to-and-fro movement (dither) of sarcomeres during isometric contraction from changes in the diffraction pattern produced by shining a laser through a muscle contracting under isometric conditions. At 10°C the internal work was found to amount to from 186 to 370 gm·cm/gm/sec contraction. In these experiments and others (Larson *et al.*, 1968), the internal work was measured only during the first few seconds of contraction. If, in fact, this internal work decreases as the contraction continues, it would explain the observed decrease in the rate of heat output and chemical usage as the duration of contraction increases.

2. RELATIONSHIP OF MUSCLE LENGTH, TENSION, AND CHEMICAL BREAKDOWN

Infante *et al.* (1964b) studied these properties in the rectus abdominis muscle of *R. pipiens* at 0°C. The amount of breakdown of phosphorylcreatine during a 5 sec tetanus was found to depend on muscle length in the same manner as the developed tension did. At the extremes of muscle length that were considered (50% l_0 and 125% l_0), both chemical breakdown and tension development were very small or zero. At 100% l_0 the chemical breakdown and tension development were maximum.

Infante *et al.* (1964a) performed the same type of experiment using the sartorius of *R. pipiens* at 0°C. The production of inorganic phosphate during a 1.5 sec tetanus depended on muscle length and paralleled tension development. However, unlike the rectus abdominis muscle (Infante *et al.*, 1964b), the sartorius did show chemical breakdown at the extremes of muscle length (50% l_0 and 132% l_0) at which the tension development was zero.

Sandberg and Carlson (1966) also studied the variation of chemical breakdown and tension developed by the sartorius of *R. pipiens* at 0°C. The isometric contractions were 20 sec in duration, which was longer than in the study by Infante *et al.* (1964a). Sandberg and Carlson found that the tension at lengths above and below 100% l_0 did not vary in the same way. The tension fell more rapidly at lengths below 100% l_0 than at lengths greater than 100% l_0. The breakdown of phosphorylcreatine, however, decreased in the same manner at muscle lengths above as it did at muscle lengths below 100% l_0. At the extremes of length that were studied (65% l_0 and 140% l_0), the chemical breakdown was the same.

V. Shortening

A. Shortening Heat

Hill's (1938) description of shortening heat was based on experiments that involved the comparison of the rate of heat output during isometric tetanus with that during shortening which occurred after release from the isometric condition. The shortening heat was equal to the increase in heat output during shortening. It was proportional to the distance shortened, but did not appear to be proportional to the load on the muscle. Hill (1938) proposed an equation in which the rate of release of the total energy (work and shortening heat) was related to the load on the muscle.

$$(P + a)v = b(P_0 - P) \tag{2}$$

where P = load on the muscle
P_0 = force developed under isometric conditions
v = velocity of shortening
a = constant for shortening heat
b = constant defining the absolute rate of energy liberation

From this it would seem that the constant for shortening heat, a, could be determined by measuring mechanical properties of the muscle, that is, P_0, and the velocity of shortening under various loads.

Hill (1949a,c) studied the heat output and work during isotonic twitches and proposed that the total energy output during contraction could be described by the equation

$$E = A + ax + W \tag{3}$$

where E = total energy
A = energy for activation
ax = shortening heat
a = constant for shortening heat
x = distance shortened
W = work done by the muscle

Later, Hill (1964a) established that the constant for shortening heat depends on the load on the muscle; this fact was not apparent from earlier studies (Hill, 1938, 1949a). In the recent experiments (Hill, 1964a) muscles were tetanized under isometric conditions until the maximum tension was developed, then they were released and permitted to shorten under isotonic conditions. After shortening, the muscle resumed contraction under isometric conditions. The rate of heat output during shortening under various loads was compared with that during the final isometric part of the contraction. From the comparison, Hill found that the value of the constant for shortening heat could be related to the load on the muscle by the equation

$$\alpha = (0.16 \pm 0.015)P_0 + (0.18 \pm 0.027)P \tag{4}$$

where α = constant for shortening heat
      P_0 = tension under isometric conditions
      P = observed tension

The new symbol, α, was adopted to distinguish the constant for shortening heat from a, which relates force and velocity of shortening in Eq. (2).

B. Chemical Breakdown and Heat Output

An important characteristic of the energy changes associated with shortening was established by Mommaerts *et al.* (1962) and Carlson *et al.* (1963) from experiments involving series of afterloaded isotonic contraction–relaxation cycles. They measured the total energy output as heat plus work (Carlson *et al.*, 1963) or as chemical breakdown (Mommaerts *et al.*, 1962; Carlson *et al.*, 1963) and then determined what fraction was associated with work. The rest of the energy output (the nonwork part) was about the same, regardless of the load on the muscle and the distance shortened. Under their experimental conditions for a contraction–relaxation cycle, the existence of an energy change correlated with shortening could not be established.

Mommaerts *et al.* (1962) suggested that activation and maintenance metabolism may not be independent of the amount of shortening. In this case, it would be possible that as the muscle shortens the maintenance metabolism decreases by an amount equal to the shortening metabolism, so that the summation of the two types of metabolism would not appear to depend on the degree of shortening.

Carlson *et al.* (1963) suggested that their results could be due to the fact that during contraction, the shortening isotonic muscle liberated more heat than the nonshortening isometric muscle. But then, during the relaxation, the shortened muscle liberated less heat than the non-shortened muscle. Thus, the difference between the total heat for isotonic contraction–relaxation cycles and the total heat for isometric cycles was not proportional to the distance shortened by the muscle under isotonic conditions. This idea could apply to the usage of the phosphorylcreatine also.

Hill (1964b) did a set of experiments somewhat like those of Carlson et al. (1963), except that in Hill's experiments, the velocity rather than the load was the independent variable. Toad muscles performed a single twitch against an ergometer, which controlled the velocity of short-ening, and the total energy output was measured. The net heat output (total energy — work) varied with the velocity of shortening in a compli-cated way. The net heat was not simply proportional to the distance shortened. Hill suggested that other processes in addition to the shorten-ing heat, including those occurring during relaxation, were responsible for the variation in heat output with a velocity of shortening.

These results (Mommaerts et al., 1962; Carlson et al., 1963; Hill, 1964b) showed that heat and chemical changes associated with shorten-ing cannot be demonstrated in a straightforward way for a series of contraction–relaxation cycles. The experiments that will be discussed next were designed so that one measures the chemical changes during that part of the contraction in which the muscle is in the process of shortening. It was during the process of shortening of the muscle that Hill (1938) observed the release of shortening heat.

Cain et al. (1962) pretreated frog rectus abdominis muscles with dinitrophenol in order to lower the phosphorylcreatine content so that the small usage of this compound during a single contraction could be detected easily. Unloaded muscles were activated; they shortened to about one-half their rest length and were immediately frozen. When these muscles which had shortened were compared with unstimulated controls, the difference in inorganic phosphate content (0.08 ± 0.20 μmole/gm) and creatine content (0.02 ± 0.2 μmole/gm) indicated that no significant amount of phosphorylcreatine had been used for shortening.

In other experiments on dinitrophenol-treated rectus abdominis mus-cles, measurements of phosphorylcreatine breakdown during either one or two isotonic contractions under various loads showed that the chemi-cal change was associated with work, rather than with shortening. For single contractions of muscles that did the same amount of work but shortened to very different degrees and were frozen before relaxation, the change in phosphorylcreatine was proportional to the work done rather than to the distance shortened (Cain *et al.*, 1962; Infante *et al.*, 1965).

The energy change during shortening has also been measured in terms of ATP breakdown, and has been compared with the heat of shortening calculated from Eq. (4) (Kushmerick *et al.*, 1969). In this experiment frog sartorius muscles performed a series of isotonic twitches and were manually reextended to their initial lengths between twitches. Since they were very lightly loaded, the muscles developed only tiny amounts of tension during shortening and thus the complication due to energy associated with tension during relaxation (Hill, 1964c) was avoided. The total breakdown of ATP per twitch was 0.088 ± 0.009 μmole per gram of muscle. From the amount of shortening observed, the shortening heat expected on the basis of Eq. (4) was calculated. Assuming that the enthalpy of hydrolysis was as high as -10 kcal/mole and that no work was derived from the breakdown of ATP, at least 0.154 ± 0.014 μmole per gram of muscle would be necessary to account for the shortening heat. The observed ATP breakdown was highly significantly lower than this. The discrepancy between the expected and the observed ATP breakdown was 0.066 ± 0.017 μmole/gm/stimulation. Thus, the heat of shortening could not have been degraded free energy from the breakdown of ATP.

Hill (1966) in his "Further Challenge to Biochemists," proposed an experimental design for the measurement of the chemical energy associated with shortening which involved comparing a single isometric contraction with a single, lightly loaded isotonic contraction. This experiment was done by Davies *et al.* (1967) and Kushmerick *et al.* (1969). Both contractions had the same duration and were interrupted by rapid freezing before relaxation or lengthening began. They found that in the isometric case 0.128 ± 0.035 μmole/gm more ATP was broken down than in the isotonic case. On the basis of the calculated shortening heat for isotonic contraction, one would have expected the isotonic case to require the use of 0.229 ± 0.008 μmole/gm more ATP than the isometric case. Thus, the total discrepancy between the observed value of the ATP usage and that expected from the shortening heat was 0.357 ± 0.037 μmole/gm, which is very highly significantly different from zero. The conclusion of this experiment and the preceding one for the series of isotonic shortenings was the same; this is, the shortening heat is not degraded free energy from ATP (Table VIII).

Thus, the amount of phosphorylcreatine and ATP usage with an enthalpy change sufficient to account for the observed shortening heat was not observed during the full contraction–relaxation cycle, nor during the shortening phase of contraction. Shortening heat, nevertheless, is

TABLE VIII

ABSENCE OF ATP BREAKDOWN EQUIVALENT TO SHORTENING HEAT[a]

Experimental conditions	Observed chemical change (μmole/gm) ± SEM	Expected chemical change[b] (μmole/gm) ± SEM	Discrepancy between observed and expected chemical change (μmole/gm) ± SEM	Significance
Series of unloaded isotonic twitches compared to unstimulated controls	+0.088 ± 0.009[c] per twitch	+0.154 ± 0.014 per twitch	−0.066 ± 0.017 per twitch	$P < 0.005$
Brief lightly loaded isotonic tetanus compared to brief isometric tetanus	−0.128 ± 0.035[d]	+0.229 ± 0.008	−0.357 ± 0.037	$P \ll 0.001$

[a] Results from Kushmerick *et al.* (1969) for Sartorius *R. pipiens* at 0°C.

[b] Calculated from shortening heat = $\alpha \times$ distance shortened; where α was determined from Eq. (3).

[c] Inorganic phosphate content (isotonic − unstimulated).

[d] Inorganic phosphate content (isotonic − isometric).

observed during shortening and must be accounted for in some other way besides the degradation of free energy from ATP breakdown.

C. A Theory for the Source of Shortening Heat

The theory of muscle contraction proposed by Davies (1963) includes a mechanism for the production of shortening heat. Since shortening heat is a reversible heat, which can be detected during shortening but not after a cycle of shortening followed by relaxation and lengthening, Davies suggested that it is derived from an entropy change. According to his theory of contraction, shortening heat is released when a small polypeptide, which constitutes a link from each crossbridge on the myosin filaments to the thin actin–troponin filaments, forms hydrogen bonds and changes from a largely extended random coil to a helical conformation. This transition results in a shortening of the polypeptide and a release of heat due to decrease in entropy. The heat from the transformed entropy is much greater than from the degraded free energy. When the polypeptide reextends an amount of heat is absorbed because of the entropy increase. This uptake of heat is equal to that released during shortening of the polypeptide. This cycle of events is repeated many times by many crossbridge polypeptides during a contraction. At any particular time during the time the muscle is shortening there will be a net heat release, since some of the polypeptides have shortened and released heat whereas others will have shortened and reextended, and their net heat productions will be zero. When the muscle has completely relaxed all the crossbridge polypeptides are extended back to their random coil form, and the net heat change from entropy change in this cyclic set of reactions of the crossbridge polypeptides is zero. Thus, for the whole cycle of contraction and relaxation, the shortening heat is not seen.

VI. Positive Work

A. Heat Output

Fenn (1923, 1924) compared the total energy liberated by muscles which shortened and did work with that liberated during isometric contraction. The working muscles contracted under afterloaded isotonic conditions, or against an inertia lever. The total energy output was found by measuring the heat and work output by the muscles. Hill and Woledge (1962) found that Fenn's results required quantitative adjustment, since the heat calibration was incorrect in the original experiments. This, however, did not alter Fenn's original conclusion that the total energy liberated during a contraction–relaxation cycle involving work was greater than that liberated under isometric conditions (Hill, 1965, p. 147). The energy in excess of isometric was greatest for contractions in which the greatest amount of work was done.

As mentioned in Section V,A, Hill (1949a,c) did a series of experiments in which he investigated the heat output during the rising phase of a preloaded twitch. The total energy output during the rising phase of the twitch could be described by Eq. (3). According to this description, the total energy output was increased slightly as the load on the muscle increased just above zero. Then the total energy decreased as the load increased. The heat output, on the other hand, was greatest when there was no load on the muscle during contraction. The heat decreased as the load on the muscle increased. The work was zero when the load on the muscle was either zero or maximum, that is, at P_0. The largest amount of work done was during contractions in which the muscle was loaded with about 40% P_0.

B. *Heat Output and Chemical Breakdown during a Series of Contractions*

Mommaerts *et al.* (1962) did a study of the chemical breakdown during a series of isotonic tetani. As a starting point they assumed that Hill's equation, Eq. (3), for the total energy output under the conditions of Hill's experiments (Hill, 1949a,c) could be transformed into chemical terms and applied to their results. Their representation of the chemical counterpart of Eq. (3) was

$$\Delta PC_{\text{total}} = A_{t,1} + b_s S + b_w W \tag{5}$$

where ΔPC_{total} = total phosphorylcreatine breakdown during a cycle of contraction and relaxation

$\quad\quad A_{t,1}$ = breakdown associated with activation and maintenance

$\quad\quad b_s$ = breakdown associated with shortening a unit distance

$\quad\quad S$ = distance shortened

$\quad\quad b_w$ = breakdown per unit work done

$\quad\quad W$ = amount of work done

Iodoacetate-treated muscles performed a series of twelve isotonic contraction–relaxation cycles, and the amount of free creatine formation was determined. Free creatine formation was taken to be equivalent to phosphorylcreatine breakdown. Multiple regression analysis was done on the results for the free creatine content of resting muscles, and the free creatine content, work done, and distance shortened by the stimulated muscles. There was good correlation between the free creatine content of stimulated muscles and all other parameters except the distance shortened. The correlation in this case was *not* statistically significant. Their analysis of the results was summarized in the equation

$$\Delta PC_{\text{total}} = 0.45 + (0.896 \pm 2.16)S + (0.00257 \pm 0.00068)W \tag{6}$$

where ΔPC_{total} is expressed in μmole/gm/tetanus

    0.45 μmole/gm/tetanus was the breakdown for activation and maintenance

    S is expressed in centimeters per centimeter of muscle

    W is expressed in gram-centimeters per gram

Their interpretation of their results suggested that 390 gm·cm of work, or 9.1 mcal of work, was done for each micromole of phosphorylcreatine breakdown specifically associated with work (work/ΔPC_{total} − 0.45 − 0.896S). This figure suggests that the contractile mechanism was operating at a very high thermodynamic efficiency, but the statistical significance of this interpretation is doubtful. Their own analysis showed that the form of Eq. (3) which they used did not describe their results. Their regression analysis of the experimental results showed that there was no significant correlation between the phosphorylcreatine usage and the distance shortened.

$$\Delta PC_{\text{shortening}} = (0.896 \pm 2.16)S$$

A comparison of the total phosphorylcreatine breakdown with the total amount of work done shows that for each micromole broken down, 107 gm·cm, or 2.52 mcal, of work was done.

Carlson *et al.* (1963) studied the relationship between the total energy liberated as heat and work and the chemical energy change during a series of isotonic twitches. Under the conditions of the experiment, the work that had been done during shortening was degraded into heat as the muscle relaxed and lengthened between twitches. The amount of external work done was measured independently and subtracted from the observed total energy output. The remaining energy output or net heat output was approximately the same for all contraction cycles despite variation in loads and the distances shortened. The results were described by the equation

$$E = A + kW \tag{7}$$

$$= (2.95 \pm 0.07) \text{ mcal/gm/twitch} + (0.971 \pm 0.081)W \tag{8}$$

$$= 125 \text{ gm} \cdot \text{cm/gm/twitch} + 0.971W \tag{9}$$

where E = total energy output
A = average load-independent heat
k = constant
W = work done by the muscle

In Eq. (8), E, A, and W are given in units of millicalories per gram per twitch and in Eq. (9) in units of gram-centimeters per gram per twitch.

As discussed in Section V,B, these results, unlike those of Hill (1949a,c), Eq. (3), did not include a separate term corresponding to energy associated with shortening. Related to this is the fact that for small loads, and thus large amounts of shortening, the total amount of energy liberated for contraction–relaxation cycles (Carlson *et al.*, 1963) was less than the total energy for the rising phase of the twitch (Hill, 1949a,c). Thus for small loads the work represented a larger fraction of the total energy output for the contraction–relaxation cycles than for the rising phase of the twitch.

Carlson *et al.* (1963), in addition to measuring the heat and work, also compared the phosphorylcreatine usage during a series of isotonic twitches with the usage during a series of isometric twitches. The difference between the phosphorylcreatine content of muscle which contracted under isotonic conditions and muscle which contracted under isometric conditions was measured. This quantity will be referred to here as $\Delta PC_{\text{isotonic–isometric}}$. Regression analysis of the $\Delta PC_{\text{isotonic–isometric}}$, the work, and the distance shortened was done. The correlation between the chemical change and the work was significant, but there was no significant correlation between the chemical change and the distance shortened. The total phosphorylcreatine usage during an isotonic twitch could be expressed as

$$\Delta PC_{\text{isotonic}} = (0.235 \pm 0.013) \text{ } \mu\text{mole/gm/twitch} + (0.166 \pm 0.023)W \tag{10}$$

where $\Delta PC_{\text{isotonic}}$ is the phosphorylcreatine breakdown per twitch in micromoles per gram and W is the work done in units of millicalories per gram. Thus the chemical change may be described by the same type of relationship as the total energy output [Eq. (7)], a constant term plus a term that was proportional to work.

If the amount of work done was compared with the total chemical change, then 106 gm·cm or 2.49 mcal of work was done for each micromole of phosphorylcreatine that was broken down. If the work were compared with the portion of the total chemical change that was statistically correlated with the amount of work done, then 256 gm·cm or 6.02 mcal of work would have been done per micromole of phosphorylcreatine broken down.

Fales (1969) repeated Fenn's (1923, 1924) experiment, but measured the heat output with a gradient layer calorimeter rather than with a thermopile. The muscles performed afterloaded isotonic contraction–relaxation cycles. The plot of total energy liberated (heat plus work) against the work done by the muscle was a straight line. The equation for the line was

$$Y = 19.51 \text{ ergs/gm} \times 10^4 + 2.33W \tag{11}$$

$$= 199 \text{ gm} \cdot \text{cm/gm} + 2.33W \tag{12}$$

where Y was the total energy released and W was the external work done by the muscle. The equation has been given in the units ergs per gram $\times 10^4$ used by Fales and also in the units gram-centimeters per gram. At the intercept of the plot, the total energy, 19.51 ergs/gm $\times 10^4$ or 199 gm·cm/gm, was slightly lower than the total energy measured during an isometric twitch, 23.6 ergs/gm $\times 10^4$ or 240 gm·cm/gm, but this difference was not statistically significant. The slope (energy released/work done) of the line was 2.33 which was above the range of values (1–2) given by Fenn (1923). Equation (11) which described Fales's results was similar in form to those of Carlson *et al.* (1963), Eqs. (7)–(10).

C. Chemical Breakdown during a Single Contraction

Cain *et al.* (1962) measured chemical changes corresponding to the work done during a single contraction. The phosphorylcreatine usage and work done were measured for dinitrophenol-treated frog rectus abdominis muscle which performed an isotonic contraction. The results showed that the chemical breakdown was a linear function of the work done. The slope of the line indicated that 110 gm·cm or 2.6 mcal of work was done per micromole of phosphorylcreatine broken down. One reason for the low value of the work/chemical change in this experiment was that the muscles had been pretreated with dinitrophenol, which lowered the phosphorylcreatine content of the muscles. The ratio of free creatine and inorganic phosphate to phosphorylcreatine was high, and this decreases the amount of free energy made available by the breakdown of phosphorylcreatine.

The ATP requirements for work at various rates was measured by Kushmerick and Davies (1969). DNFB-treated frog sartorius muscles were tetanized and they shortened at a constant velocity against an ergometer. During shortening at each velocity, the muscle changed its developed force as the length of the muscle changed. Thus the muscles performed the maximum work possible during a single contraction. In these experiments, the distance and velocity of shortening were preset. Each experimental muscle was stimulated during the entire period of shortening and it was frozen before relaxation took place. The ATP level in the experimental muscle was compared with the level in the control muscle which had not been stimulated. The amount of external work performed was calculated from the average active tension developed by the muscle and the distance the muscle shortened.

The amount of work done per micromole of ATP broken down was found to depend on the velocity of shortening. The work per micromole of ATP breakdown is, by definition, zero when the contraction is isometric (zero shortening) and also when the muscle shortens at maximum velocity (zero tension), since in these two cases no work is done. The value of the work per micromole ATP breakdown was found to be greatest for contractions involving shortening at 2 cm/sec. In this case 245 gm·cm or 5.75 mcal of work was done per micromole of total ATP breakdown. When the ATP breakdown for activation was deducted from the total breakdown, the work per micromole ATP split by the "mechanical generator" itself (Kushmerick *et al.*, 1969) could be calculated. For shortening at 2 cm/sec, it was 323 gm·cm or 7.58 mcal per micromole ATP broken down.

These values of work per micromole ATP were converted to thermodynamic efficiencies by expressing ATP breakdown in terms of the free energy that would be made available by the reaction. The *in vivo* free energy was calculated taking into account the conditions in the sarcoplasm of DNFB-treated anaerobic muscle during contraction (Kushmerick and Davies, 1969, Appendix). The free energy change was found to be -10 kcal/mole ATP. If the free energy had been converted into work with 100% efficiency, then 426 gm·cm of work would have been done for each micromole of ATP split. For shortening at 2 cm/sec, the observed 323 gm·cm of work done per micromole of ATP broken down by the mechanical generator was equivalent to 76% of maximum thermodynamic efficiency. Table IX is a summary of the experiments in which chemical breakdown and work done during contraction have been measured.

TABLE IX

EXTERNAL WORK DONE FOR 1μMOLE OF CHEMICAL CHANGE ($\equiv$ ATP)[a]

Reference	Experimental conditions	Compound measured	Work/chemical change (gm·cm/μmole)	Efficiency (%)	Work/chemical change associated with external work only[b] (gm·cm/μmole)	Efficiency (%)
Mommaerts et al. (1962)	Series of isotonic tetani, iodo-acetate-treated sartorius R. pipiens, 0°C	Phosphorylcreatine	107	25	390	92
Carlson et al. (1963)	Series of isotonic twitches, iodo-acetate-treated, sartorius R. pipiens, 1–2°C	Phosphorylcreatine	106	25	256	60
Cain et al. (1962)	Dinitrophenol-treated rectus abdominis R. pipiens, 0°C	Phosphorylcreatine	110	26	—	—
Kushmerick and Davies (1969)	Single isovelocity (2 cm/sec) tetanus, DNFB-treated sartorius R. pipiens, 0°C	Inorganic phosphate	245	58	323	76

[a] 100% thermodynamic efficiency (external work/$-\Delta F$) with $-\Delta F$ for ATP usage taken as 10 kcal/mole ATP (Kushmerick and Davies, 1969, appendix) $\equiv 426$ gm·cm/μmole ATP.

[b] See text for the different methods used to calculate this quantity for each set of experiments.

VII. Negative Work

A. *Introduction*

If an activated muscle is subjected to a load which is greater than the force which it can develop at its length at the time, the muscle will lengthen (stretch) and do negative work. The word "stretch" is not meant to imply that the muscle must be at a length greater than its rest length. Muscle can do negative work over the same range of lengths at which it performs isometric contractions and contractions involving shortening.

From experience, it seems that lengthening muscle doing negative work requires less metabolic energy than shortening muscle doing positive work. Activities which mainly depend on lengthening muscle, like climbing down a mountain or lowering one's self on a rope, are obviously less tiring than the reverse motions. It might seem that the ease of downward motion is due to the difference in the gravitational potential rather than the difference in energy requirements of the muscle itself. However, for these movements at constant velocity the forces in the muscles must be identical whether the movement is upward or downward. What is different during negative work is the direction of movement of the muscle, and the influence of this factor on the mechanical, thermal, and chemical processes in the muscle will be considered.

B. *Exercise Studies*

More than seventy years ago, the first of a variety of exercise experiments were done, in which the oxygen requirements for positive work (concentric exercise) and for negative work (eccentric exercise) were compared. Chauveau (1896) compared ascending stairs with descending, Zuntz *et al.* (1906) compared walking uphill with walking down, and Johansson (1901) compared lifting weights with lowering them. With these simple methods, they found that negative work required about half as much oxygen as was needed to do an equivalent amount of positive work.

More recently, Abbott *et al.* (1952) in their studies on oxygen consumption used two bicycle ergometers set back to back. One person pedaled in the conventional direction at a set speed, thus driving the pedals of the second bicycle in a backward direction. The person on the second bicycle resisted the backward motion of the pedals strongly enough to maintain an agreed force read on a gauge. Thus both persons exerted the same force through the experiment. Positive work always required more oxygen than negative work. The rate of oxygen consumption remained almost constant for negative work even when the rate of working was increased. Hill (1960) described the experiment as being "enthusiastically received" when it was shown at a *conversazione* of the Royal Society in London. It turned out that the young lady doing negative work very easily exhausted a strong young man who was doing the positive work!

Eccentric and concentric exercises have been compared to determine their long-term effects on the strengthening of muscle and the altering of oxygen consumption (Seliger *et al.,* 1968). One group of trained athletes performed an exercise program consisting of concentric exercises involving lifting dumbbell weights which were 90–95% of the maximum that could be lifted. The other group did eccentric exercises by lowering weights that were 145–150% of the maximum that could be lifted. Both types of exercise increased the muscle strength by the same amount. However, at the beginning of the program the ratio of the oxygen consumption for lifting to that for lowering was 1.87, and at the end of the program it was 1.94. So eccentric exercise was more advantageous than concentric in terms of its influence on oxygen consumption.

C. Isolated Muscle Experiments

1. Mechanical Properties during Stretch—Force and Velocity

In 1938 Hill proposed his "characteristic" equation, which described the hyperbolic force–velocity relationship for muscle

$$v = b(P_0 - P)/(P + a) \tag{13}$$

where v = velocity of shortening
  P_0 = maximum isometric force at l_0
  P = developed force
a and b = constants

Katz (1939) did a series of experiments to determine whether the equation also applied to muscles which were lengthened by a force greater than P_0. When isotonic loads ranging from P_0 to 170% P_0 were imposed on the activated muscles, the observed velocities of lengthening were always much less than those predicted by Hill's equation. This discrepancy increased as the load was increased up to 170% P_0.

Abbott and Aubert (1952) stretched activated toad and dogfish muscles, using an ergometer so that the velocity was constant during the stretch. The three velocities chosen were greater than those used by Katz (1939). The amount of force developed by the muscles was about the same (160% P_0) for these three velocities of stretch. The force–velocity relationship for stretch is not predicted by Hill's equation. At least for the conditions that have been studied, a greater force is developed during stretch than during shortening at the same velocity.

2. PROGRESSIVE DEVELOPMENT OF TENSION DURING STRETCH

Gasser and Hill (1924) recorded the progressive development of tension by tetanized frog sartorius muscles as they were stretched at various velocities. The amount of tension developed at a particular muscle length during stretch was found to depend on the velocity at which the muscle was being stretched. Four distinct patterns were observed. (1) At very slow rates of stretch, the tension at each instant was equal to the isometric tension characteristic for the muscle length at that instant. (2) If the muscle was stretched faster, the tension exceeded the isometric or equilibrium tension expected for the longer length, and then it declined, asymptotically approaching the isometric tension of the longer length. (3) At a higher velocity of stretch, the tension just reached the expected isometric tension for the longer length and was maintained at this level. (4) At the greatest velocity used, the tension quickly reached that characteristic of the longer length, but was not maintained at this level. The tension fell immediately when the stretching ended and was then restored to the equilibrium value for the longer length.

Using small bundles of muscle fibers, Sugi (1969) has done experiments similar to the one Gasser and Hill (1924) did with whole muscles. The two patterns of tension development for the fiber preparations were the same as patterns 2 and 4 reported by Gasser and Hill (1924). At moderate velocities, the tension during the stretch was always greater than the equilibrium value for the instantaneous length of the muscle. After the stretch was completed, the tension decayed exponentially, approaching but not reaching the equilibrium value for the longer length. When the muscle fibers were stretched very rapidly, the tension at the end of the stretch exceeded the equilibrium value for the longer length, but this tension was not maintained. It dropped as soon as the stretch ended and was then redeveloped. Sugi also found that the tension development responded to changes in the velocity during a single stretch.

When the velocity of stretch decreased, the tension was found to decay. On the other hand, if the velocity of stretch was then increased the tension was redeveloped. These studies show that during stretch the amount of tension is not related to muscle length or to velocity of movement in the same way as it is during shortening.

3. HEAT AND NEGATIVE WORK

Fenn (1924) measured the energy changes that occurred when muscles shortened and did positive work and also when they were lengthened while being stimulated. He found that positive work was accompanied by an increase in the energy (heat plus work) liberated by the muscle, and that negative work decreased the output to a value below that for isometric contraction. Since these findings were reported, several other studies on the heat and work changes accompanying these processes have been done.

Abbott *et al.* (1951) measured the heat output by frog and toad sartorius muscles when they were stretched at various times during a twitch or tetanic contraction–relaxation cycle. The total heat output was greater than that for a similar cycle under isometric conditions. For stretches during relaxation, the amount of excess heat output above that for an isometric contraction was numerically equal to the work done on the muscle during stretch. On the other hand, for stretches during the early part of the contraction, the excess heat was numerically less than the work done on the muscle. The authors suggested three possible "mechanisms" which could have lead to these results.

1. If it is assumed that the same (heat-releasing) chemical reactions took place during the contraction with stretch as during contraction under isometric conditions, then it seems that part of the energy put into the muscle as work done on it during stretch had disappeared. The authors suggested that this part of the work was used to reverse the chemical reactions that took place.

2. It is possible that the work was absorbed in some unknown process.

3. It is also possible that the work done on the muscle was degraded into heat but that the chemical reactions which take place during isometric contraction were prevented by the stretch.

Abbott and Aubert (1951) measured the heat output during very slow stretches and during isometric contraction. Compared to the isometric case, the heat output was decreased during stretch and this decrease was linearly proportional to the distance stretched. This was described as negative heat of lengthening by analogy to the output of heat during shortening. The constant for the negative heat of lengthening varied with the load and was numerically greater than the constant for shortening heat.

Hill and Howarth (1959) followed the time course of the heat output and work done on toad muscles during contraction with stretch. The difference between the heat output and the absolute value of the work done on the contractile elements of the muscle reached a negative value during stretch. In other words, more energy was put into the muscle as work done on it than was released by the muscle as heat. If chemical reactions were being reversed, it would imply net synthesis. During relaxation, however, the value returned to a small positive value. If the proposed thermoelastic absorption of heat during the rise of tension during stretch was taken into account, the difference between heat and work was always a positive number.

4. CHEMICAL CHANGES AND NEGATIVE WORK

Aubert and Maréchal (1963) found that more phosphorylcreatine was used during contraction with rapid stretch than during isometric contraction. This result was the opposite of what would be expected on the basis of the interpretations of the heat measurements by Abbott et al. (1951). In later experiments Maréchal (1964a,b) compared the phosphorylcreatine content of resting muscles, muscles contracting under isometric conditions and muscles contracting with slow shortening or lengthening. The chemical usage during contraction with slow stretch was less than the usage for other types of contraction. There was, however, no evidence that net chemical synthesis had occurred during stretch. Maréchal's interpretation was that stretch resulted in an economy in chemical usage during contraction. This conclusion was supported by another set of experiments (Maréchal and Beckers-Bleukx, 1965). They compared the phosphorylcreatine levels in muscles which were stretched slowly during very prolonged contraction with the levels in muscles which were stimulated at a very short length, when little tension was developed.

Infante et al. (1964c) measured the ATP and inorganic phosphate levels in DNFB-treated muscles which remained at rest, performed a contraction under isometric conditions, or were stretched during contraction. A series of different combinations of these conditions was used in order to duplicate the designs of the experiments in which heat had been measured (Hill and Howarth, 1959). The ATP usage by the muscles during stretch was only one half of the usage by the muscles contracting under isometric conditions, even though there was 70% more tension developed during stretch than during isometric conditions. There was never net synthesis of ATP evident from comparison with the level of ATP in resting muscle.

Other experiments (Curtin *et al.*, 1969, 1970) have shown that there was only a small variation in the ATP usage by muscles that were stretched the same distance at velocities ranging from 0.13 l_0/sec up to 2 l_0/sec. At the slowest velocity, the ATP usage was not significantly different from the usage by muscle stimulated under fully shortened or fully lengthened conditions. Under these last two conditions, the main ATP usage is that involved in pumping Ca^{2+}. When the muscles were stretched, the ATP usage was not proportional to the integral of tension and time. The fact that a net synthesis of ATP or phosphorylcreatine was never found when chemical changes were measured makes the idea of the reversal of their chemical reactions during stretch seem unlikely (Abbott *et al.*, 1951; Abbott and Aubert, 1951; Hill and Howarth, 1959). The other suggestions, however, remain possible.

The observed reduction in the ATP usage and the small heat output during stretch is predicted by a theory of muscle contraction (Davies, 1963). According to the theory, a peptide forms the crossbridges between the thick and the thin filaments. The last chemical reaction associated with shortening of muscles is the transition from the extended random coil form of the peptide to the alpha helical form. This transition involves the formation of hydrogen bonds and hydrophobic interactions which release heat and provide the energy for the shortening of the muscle. Then ATP is required to reextend each crossbridge after it has shortened. However, when the muscle is stretched, the transition to the alpha helical form and the hydrogen bond formation is prevented, thus the heat output is reduced. Since the crossbridges do not shorten fully, ATP hydrolysis is not necessary for the reextension of the bridges. Thus, the theory predicts that aside from the ATP required for Ca^{2+} pumping, no ATP is needed for the development of tension during stretch.

VIII. Relaxation

A. *Heat Output*

Fenn (1924) observed that muscles that were loaded and thus lengthened during relaxation liberated more heat than muscles that were freed of their load when they finished shortening and thus remained shortened during relaxation. He concluded that this extra heat was due to the dissipation of the work that had been done during shortening and also heat from processes involved in the lengthening of the loaded muscle during relaxation. This second type of heat is of interest here.

Hill (1949b) found that the heat production during relaxation was influenced by the movement of the muscle and the load on it as it relaxed. When a muscle shortened against an ergometer during a twitch there was no load on the muscle during relaxation and no heat was released after the shortening was over. During an isotonic twitch under a light load, there was no further release of heat after the shortening had finished.

Hill (1964c) also studied relaxation following isometric contraction. Under these conditions relaxation did not involve extensive lengthening of the muscle. Heat output was observed during the exponential decay of tension during normal relaxation following isometric contraction. On the other hand, if the muscle was quickly released to a shorter length as soon as full isometric tension was developed, the tension immediately dropped to zero and no additional heat was liberated by the muscle. The total heat output during a normal isometric contraction–relaxation cycle was 15–20% greater than the heat output during isometric contraction with a quick release during relaxation. Hill concluded that the extra heat liberated during relaxation in the presence of tension was due to a prolongation of the declining active state of the muscle. In other words, the tension acted as a positive feedback mechanism continuing the activation of the muscle.

These studies suggest that the amount of heat output, during relaxation may depend on a number of factors—movement, tension, and muscle length. The fact that any heat was released raised the question of the possible source of the heat. Several studies have been done in which chemical changes during relaxation have been examined.

B. Chemical Breakdown

1. ANTERIOR BYSSUS RETRACTOR MUSCLE

In one such study, the inorganic phosphate production from phosphorylarginine breakdown during the relaxation of the anterior byssus retractor muscle (ABRM) of the mussel *Mytilus edulis* was measured (Minihan and Davies, 1965; Nauss and Davies, 1966). The ABRM developed a prolonged contraction or "catch" in response to activation by direct current or acetylcholine. The muscle relaxed rapidly when it was exposed to 5-hydroxytryptamine (serotonin). Inorganic phosphate contents of muscles that were frozen at peak of contraction and muscles that were contracted and then relaxed were compared. The difference in the inorganic phosphate for experiments at 8°C was 0.19 ± 0.06 μmole/gm, which indicated that phosphorylarginine was broken down during relaxation from the "catch."

This observed usage agreed with the expected usage calculated by making some assumptions and using the experimental results. The number of crossbridges per gram of ABRM muscle has been calculated to be 2.75×10^{16} (Bear, 1944; Bear and Selby, 1956; Elliott, 1964; Lowy *et al.*, 1964). Assuming that one molecule of phosphorylarginine or ATP is cleaved each time a crossbridge completes its operation, 0.05 μmole per gram of muscle would be cleaved during a single complete round of operation by all the crossbridges. During steady lowering of the load during relaxation, new bonds must be formed and broken a few times. Thus, the total usage is probably between 0.05 and 0.15 μmole per gram of muscle. Considering the amount of Ca^{2+} required to activate the muscle and the usage of one phosphorylarginine or ATP per one or two Ca^{2+} ions reaccumulated during relaxation (Makinose and The, 1965), the total breakdown for Ca^{2+} pumping would be expected to be about 0.05 μmole per gram of muscle. Thus, during the relaxation following a "catch," the total phosphorylarginine or ATP usage for crossbridge operation and Ca^{2+} pumping would be expected to be 0.1–0.2 μmole/gm. This compares very well with the observed phosphorylarginine breakdown (Nauss and Davies, 1966).

2. Rectus Abdominis Muscle

Cain *et al.* (1962) determined the difference between the inorganic phosphate content of rectus abdominis muscles (*Rana pipiens*) at 0°C which had contracted and others which had contracted and relaxed. The mean value indicated that inorganic phosphate was produced during relaxation, but the value was not statistically significant (0.08 ± 0.14 μmole/gm).

3. Sartorius Muscle

Infante and Davies (1962) did two sets of experiments using DNFB-treated sartorius muscles of *R. pipiens* at 0°C, and from the results of these experiments, the ATP usage during relaxation can be calculated. In the first experiments one muscle of each pair performed a single isotonic twitch and was frozen when it stopped shortening.

The mean work done was 17.4 ± 1.3 gm·cm/gm. The ATP usage by the muscle during the single twitch was 0.22 ± 0.07 μmole/gm. In another set of experiments, one muscle of each pair performed an isotonic twitch but was not frozen until the muscle had partly relaxed, that is, partly returned to its initial length. ATP usage by the experimental muscle was 0.43 ± 0.07 μmole/gm. By comparing the two sets of experiments, the ATP usage during relaxation following an isotonic twitch in which the muscle did work was found to be 0.21 ± 0.14 μmole/gm.

Despite the difference in the experimental design, the ATP usage just discussed was almost the same as the ATP usage during relaxation following an isotonic tetanus of a very lightly loaded (0.35 gm) muscle. Davies *et al.* (1967) and Kushmerick *et al.* (1969) found an ATP usage of 0.18 ± 0.072 μmole/gm when muscles that were stimulated at 50 pulses/sec for 2 sec and then relaxed for 3 sec were compared with those that were frozen as soon as the stimulation ended.

ATP usage during relaxation following a short isotonic or short isometric tetanus has been studied by Mommaerts and Wallner (1967). In the experiments involving isotonic contractions, the load on the muscle was 4 gm. The muscles received 250 msec of stimulation (5 pulses at 20 pulses/sec) during either isotonic or isometric contraction. The first muscle received 250 msec of stimulation and was frozen 650 msec after the stimulation ended. The second muscle received 250 msec of stimulation and was frozen 50 msec after its stimulation ended. The results for the isotonic and isometric contractions were quite similar and were considered together. The changes in free creatine, phosphorylcreatine, ATP, ADP, AMP, and inorganic phosphate were measured. The total usage determined from the sum of the observed change in ATP plus the change in AMP is 0.03 ± 0.03 μmole/gm. Alternatively, the total ATP usage can be determined directly from the observed change in inorganic phosphate, which was $+0.06 \pm 0.07$ μmole/gm. Considering the standard error of this figure, these results cannot rule out an ATP usage during relaxation. The difference between the ATP usage during relaxation following a tetanus as measured by Mommaerts and Wallner (1967) ($\Delta P_i = +0.06 \pm 0.07$ μmole/gm) and the usage as measured by Kushmerick *et al.* (1969) ($\Delta P_i = +0.18 \pm 0.072$ μmole/gm) is not statistically significant.[*]

It has been proposed that chemical breakdown during relaxation provides the energy required for the reaccumulation of the Ca^{2+} responsible for activation (Davies, 1963). In Section III,B,3, the calculation of the amount of ATP needed for the reaccumulation process in frog sartorius

---

[*] P_i represents inorganic phosphate.

TABLE X
CHEMICAL CHANGE DURING RELAXATION

Reference	Experimental conditions and type of contraction preceding relaxation	Chemical change during relaxation	
		Compound	Amount (μmole/gm) ± SEM
Nauss and Davies (1966)	ABRM *Mytilus edulis*, 8°C, "catch"	Phosphorylarginine	−0.19 ± 0.06
Cain *et al.* (1962)	Rectus abdominis *R. pipiens*, 0°C, single isotonic twitch	Phosphorylcreatine	−0.08 ± 0.14
Infante and Davies (1962)	Sartorius *R. pipiens*, 0°C, single isotonic twitch	ATP	−0.21 ± 0.14
Mommaerts and Wallner (1967)	Sartorius *R. pipiens*, 0°C, isometric or isotonic tetanus	ATP + AMP Inorganic phosphate	−0.03 ± 0.03 +0.06 ± 0.07
Kushmerick *et al.* (1969)	Sartorius *R. pipiens*, 0°C, single isotonic tetanus	ATP	−0.18 ± 0.072

muscle was discussed. On this basis, an ATP breakdown of 0.1–0.2 μmole/gm would be expected to take place during relaxation. The results of experiments by Infante and Davies (1962), Davies *et al.* (1967), Kushmerick *et al.* (1969), and Mommaerts and Wallner (1967) are consistent with the calculated value. The measured values of chemical breakdown during relaxation are summarized in Table X.

IX. Total Energy (Heat + Work) and Chemical Change

Having discussed various aspects of contraction individually, we now turn to the question of whether the breakdown of ATP and phosphoryl-creatine is the source of the total energy (heat + work) output by the muscle. The experiments of Carlson *et al.* (1963) and Wilkie (1968) have shown that for a series of full cycles of contraction and relaxation, the total energy (heat + work) can be accounted for in terms of chemical breakdown. The values of (heat + work)/phosphorylcreatine breakdown were 9.8 ± 0.5 kcal/mole (Carlson *et al.*, 1963) and 11.0 ± 0.23 kcal/mole (Wilkie, 1968). These values are consistent with the expected enthalpy of hydrolysis of phosphorylcreatine.

Several experiments, however, show that the relationship of the total energy (heat + work) to chemical breakdown from moment to moment during contraction is more complex. As discussed in Section V,B, Kushmerick *et al.* (1969) found that shortening heat could not be accounted for in terms of ATP breakdown during shortening under lightly loaded conditions. In experiments involving a single brief isometric tetanus, Gilbert *et al.* (1970) found that a larger amount of heat was released than could be accounted for by the amount of phosphorylcreatine that was broken down if the enthalpy of hydrolysis was −11 kcal/mole.

Kushmerick and Davies (1969) have done experiments which are relevant to the question of whether the free energy from ATP breakdown could supply the total energy (heat + work) output during a contraction involving work. Podolsky (1960) and Huxley (1957) have predicted that, if the free energy from ATP did supply the total energy, then the rate of breakdown of ATP would be minimum for contractions at zero velocity of shortening (isometric contraction) and would increase with increasing velocity of shortening. Kushmerick and Davies (1969), however, found that the rate of ATP breakdown was almost the same for contractions with shortening at low velocities as it was for those shortening at maximum velocity. The free energy of ATP was predominantly associated with the performance of work. This conclusion has been shown by Gilbert and Kushmerick (1970) to hold also for muscles treated with iodoacetate in which phosphorylcreatine was broken down during the performance of work. The experiments on single contractions indicate that more heat was released than can be accounted for by the amount of chemical breakdown that was observed if the enthalpy of hydrolysis was -11 kcal/mole. Wilkie (1960) has pointed out that heat can be derived from two different sources in this case—(1) Degraded free energy, that is, the part of the total free energy that is not converted into work, and (2) transformed entropy. It is possible, but seems unlikely, that the excess observed heat, which is not accounted for by ATP or phosphorylcreatine, may represent degraded free energy from an unknown reaction. It is also possible that the heat is transformed entropy from processes, such as conformational change (Davies, 1963), that accompany known chemical reactions. Since heat changes associated with entropy changes are reversible, this would explain the finding for full cycles of contraction and relaxation (Carlson *et al.*, 1963; Wilkie, 1968). The reason that the observed chemical change accounts for the total heat $+$ work would be that the net heat due to entropy could be nearly zero for the full cycle.

REFERENCES

Abbott, B. C. (1951). *J. Physiol.* (*London*) 112, 438.

Abbott, B. C., and Aubert, X. M. (1951). *Proc. Roy. Soc., Ser. B* 139, 104.

Abbott, B. C., and Aubert, X. M. (1952). *J. Physiol.* (*London*) 117, 77.

Abbott, B. C., Aubert, X. M., and Hill, A. V. (1951). *Proc. Roy. Soc., Ser. B* 139, 86.

Abbott, B. C., Bigland, B., and Ritchie, J. M. (1952). *J. Physiol* (*London*) 117, 380.

Ashley, C. C., and Ridgeway, E. B. (1970). *J. Physiol.* (*London*) 209, 105.

Aubert, X. (1956). "Le couplage énergetique de la contraction musculaire." Editions Arscia, Brussels.

Aubert, X. (1964). *Pfluegers Arch. Gesamte Physiol. Menschen Tiere* 281, 13.

Aubert, X. M. (1968). *In* "Symposium on Muscle, Budapest, 1966" (E. Ernst and F. B. Straub, eds.), pp. 187–189. Akadémiai Kiadó, Budapest.

Aubert, X., and Maréchal, G. (1963). *J. Physiol.* (*Paris*) 55, 186.

Bear, R. S. (1944). *J. Amer. Chem. Soc.* 66, 2043.

Bear, R. S., and Selby, C. C. (1956). *J. Biophys. Biochem. Cytol.* 2, 55.

Bianchi, C. P. (1970). Personal communication.

Cain, D. F., and Davies, R. E. (1962a). *In* "Muscle as a Tissue" (K. Rodahl and S. M. Horvath, eds.), pp. 84–96. McGraw-Hill, New York.

Cain, D. F., and Davies, R. E. (1962b). *Biochem. Biophys. Res. Commun.* 8, 361.

Cain, D. F., Infante, A. A., and Davies, R. E. (1962). *Nature* (*London*) 196, 214.

Cain, D. F., Kushmerick, M. J., and Davies, R. E. (1964). *Biochim. Biophys. Acta* **86**, 81.

Carlson, F. D., and Siger, A. (1960). *J. Gen. Physiol.* **44**, 33.

Carlson, F. D., Hardy, D. J., and Wilkie, D. R. (1963). *J. Gen. Physiol.* **46**, 851.

Carlson, F. D., Hardy, D., and Wilkie, D. R. (1967). *J. Physiol. (London)* **189**, 209.

Chance, B., and Connelly, C. M. (1957). *Nature (London)* **179**, 1235.

Chaplain, R. A., and Pfister, E. (1970). *Experientia* **26**, 505.

Chauveau, A. (1896). *C. R. Acad. Sci.* **122**, 113.

Costantin, L. L., Franzini-Armstrong, C., and Podolsky, R. J. (1965). *Science* **147**, 158.

Curtin, N. A., Drobnis, D. D., Larson, R. E., and Davies, R. E. (1969). *Fed. Proc., Fed. Amer. Soc. Exp. Biol.* **28**, 711.

Curtin, N. A., Svensson, S. M. M., and Davies, R. E. (1970). *Fed. Proc., Fed. Amer. Soc. Exp. Biol.* **29**, 714.

Davies, R. E. (1963). *Nature (London)* **199**, 1068.

Davies, R. E., Cain, D., and Delluva, A. M. (1959). *Ann. N.Y. Acad. Sci.* **81**, 468.

Davies, R. E., Kushmerick, M. J., and Larson, R. E. (1967). *Nature (London)* **214**, 148.

Ebashi, S., and Endo, M. (1968). *Progr. Biophys. Mol. Biol.* **18**, 123.

Ebashi, S., and Lipmann, F. (1962). *J. Cell Biol.* **14**, 389.

Eggleton, P., and Eggleton, G. P. (1927). *J. Physiol. (London)* **63**, 155.

Elliott, G. F. (1964). *J. Mol. Biol.* **10**, 89.

Engelhardt, W. A., and Ljubimowa, M. N. (1939). *Nature (London)* **144**, 668.

Fales, J. T. (1969). *Amer. J. Physiol.* **216**, 1184.

Fenn, W. O. (1923). *J. Physiol. (London)* **58**, 175.

Fenn, W. O. (1924). *J. Physiol. (London)* **58**, 373.

Fiske, C. H., and SubbaRow, Y. (1929). *J. Biol. Chem.* **81**, 629.

Fleckenstein, A., Janke, J., Davies, R. E., and Krebs, H. A. (1954). *Nature (London)* **174**, 1081.

Fletcher, W. M., and Hopkins, F. G. (1907). *J. Physiol. (London)* **35**, 247.

Gasser, H. S., and Hill, A. V. (1924). *Proc. Roy. Soc., Ser. B* **96**, 398.

Gibbs, C. L., Ricchiuti, N. V., and Mommaerts, W. F. H. M. (1966). *J. Gen. Physiol.* **49**, 517.

Gilbert, C., and Kushmerick, M. J. (1970). *J. Physiol. (London)* **210**, 146P.

Gilbert, C., Kretzschmar, M., Wilkie, D. R., and Woledge, R. C. (1970). *J. Physiol. (London)* **207**, 15P.

Gonzalez-Serratos, H. (1966a). *J. Physiol. (London)* **185**, 20P.

Gonzalez-Serratos, H. (1966b). *J. Physiol. (London)* **186**, 17P.

Hasselbach, W. (1964). *Progr. Biophys. Mol. Biol.* **14**, 167.

Hasselbach, W., and Makinose, M. (1963). *Biochem. Z.* **339**, 94.

Heilbrunn, L. V., and Wiercinski, F. J. (1947). *J. Cell. Comp. Physiol.* **29**, 15.

Hill, A. V. (1928). *Proc. Roy. Soc., Ser. B* **103**, 163.

Hill, A. V. (1938). *Proc. Roy. Soc., Ser. B* **126**, 136.

Hill, A. V. (1949a). *Proc. Roy. Soc., Ser. B* **136**, 195.

Hill, A. V. (1949b). *Proc. Roy. Soc., Ser. B* **136**, 211.

Hill, A. V. (1949c). *Proc. Roy. Soc., Ser. B* **136**, 220.

Hill, A. V. (1950). *Proc. Roy. Soc., Ser. B* **137**, 40.

Hill, A. V. (1958a). *Proc. Roy. Soc., Ser. B* **148**, 397.
Hill, A. V. (1958b). *Proc. Roy. Soc., Ser. B* **149**, 58.
Hill, A. V. (1960). *Science* **131**, 897.
Hill, A. V. (1964a). *Proc. Roy. Soc., Ser. B* **159**, 297.
Hill, A. V. (1964b). *Proc. Roy. Soc., Ser. B* **159**, 596.
Hill, A. V. (1964c). *Proc. Roy. Soc., Ser. B* **159**, 589.
Hill, A. V. (1965). "Trails and Trials in Physiology." Williams & Wilkins, Baltimore, Maryland.
Hill, A. V. (1966). *Biochem. Z.* **345**, 1.
Hill, A. V., and Howarth, J. V. (1959). *Proc. Roy. Soc., Ser. B* **151**, 169.
Hill, A. V., and Woledge, R. C. (1962). *J. Physiol. (London)* **162**, 311.
Hodgkin, A. L., and Horowicz, P. (1957). *J. Physiol. (London)* **136**, 17P.
Homsher, E., and Ricchiuti, N. V. (1969). *Physiologist (Washington, D.C.)* **12**, 256.
Homsher, E., Mommaerts, W. F. H. M., Ricchiuti, N. V., and Wallner, A. (1970). *Fed. Proc., Fed. Amer. Soc. Exp. Biol.* **29**, 655.
Howarth, J. V. (1958). *J. Physiol. (London)* **144**, 167.
Huxley, A. F. (1957). *Progr. Biophys. Biophys. Chem.* **7**, 257.
Infante, A. A., and Davies, R. E. (1962). *Biochem. Biophys. Res. Commun.* **9**, 410.
Infante, A. A., and Davies, R. E. (1965). *J. Biol. Chem.* **240**, 3996.
Infante, A. A., Klaupiks, D., and Davies, R. E. (1964a). *Biochim. Biophys. Acta* **88**, 215.
Infante, A. A., Klaupiks, D., and Davies, R. E. (1964b). *Nature (London)* **201**, 620.
Infante, A. A., Klaupiks, D., and Davies, R. E. (1964c). *Science* **144**, 1577.
Infante, A. A., Klaupiks, D., and Davies, R. E. (1965). *Biochim. Biophys. Acta* **94**, 504.
Jöbsis, F. F., and Chance, B. (1957). *Fed. Proc., Fed. Amer. Soc. Exp. Biol.* **16**, 68.
Jöbsis, F. F., and O'Connor, M. J. (1966). *Biochem. Biophys. Res. Commun.* **25**, 246.
Johansson, J. E. (1901). *Skand. Arch. Physiol.* **11**, 273.
Katz, B. (1939). *J. Physiol. (London)* **96**, 45.
Klocke, F. J., Braunwald, E., and Ross, J., Jr. (1966). *Circ. Res.* **18**, 357.
Kuby, S. A., and Mahowald, T. A. (1959). *Fed. Proc., Fed. Amer. Soc. Exp. Biol.* **18**, 267.
Kushmerick, M. J., and Davies, R. E. (1969). *Proc. Roy. Soc., Ser. B* **174**, 315.
Kushmerick, M. J., Larson, R. E., and Davies, R. E. (1969). *Proc. Roy. Soc., Ser. B* **174**, 293.
Lange, G. (1955a). *Biochem. Z.* **326**, 172.
Lange, G. (1955b). *Biochem. Z.* **326**, 225.
Lange, G. (1955c). *Biochem. Z.* **326**, 369.
Larson, R. E. (1970). Ph.D. Dissertation, University of Pennsylvania, Philadelphia.
Larson, R. E., Kushmerick, M. J., Haynes, D. H., and Davies, R. E. (1968). *Biophys. Soc. Abstr., 11th Annu. Meet.*, P. A-8.
Lohmann, K. (1934). *Biochem. Z.* **271**, 264.
Lowy, J., Millman, B. M., and Hanson, J. (1964). *Proc. Roy. Soc., Ser. B* **160**, 525.
Lundsgaard, E. (1930). *Biochem. Z.* **217**, 162.
Makinose, M., and The, R. (1965). *Biochem. Z.* **343**, 383.

Maréchal, G. (1964a). "Le métabolisme de la phosphorylcréatine et de l'adénosine triphosphate durant la contraction musculaire." Editions Arscia, Brussels.

Maréchal, G. (1964b). *Arch. Int. Physiol. Biochim.* **72**, 306.

Maréchal, G., and Beckers-Bleukx, G. (1965). *J. Physiol.* (*Paris*) **57**, 652.

Maréchal, G., and Mommaerts, W. F. H. M. (1963). *Biochim. Biophys. Acta* **70**, 53.

Meyerhof, O. (1920). *Pfluegers Arch. Gesamte Physiol. Menschen Tiere* **182**, 232.

Minihan, K., and Davies, R. E. (1965). *Nature* (*London*) **203**, 1327.

Mommaerts, W. F. H. M. (1954). *Nature* (*London*) **174**, 1083.

Mommaerts, W. F. H. M. (1955). *Amer. J. Physiol.* **182**, 585.

Mommaerts, W. F. H. M. (1961). *Circulation* **24**, 410.

Mommaerts, W. F. H. M., and Wallner, A. (1967). *J. Physiol.* (*London*) **193**, 343.

Mommaerts, W. F. H. M., Seraydarian, K., and Maréchal, G. (1962). *Biochim. Biophys. Acta* **57**, 1.

Munch-Petersen, A. (1953). *Acta Physiol. Scand.* **29**, 202.

Murphy, R. A. (1966). *Amer. J. Physiol.* **211**, 1082.

Nachmansohn, D. (1928). *Biochem. Z.* **196**, 73.

Nauss, K. Minihan, and Davies, R. E. (1966). *Biochem. Z.* **345**, 173.

Ohnishi, T., and Ebashi, S. (1964). *J. Biochem.* (*Tokyo*) **55**, 599.

Parnas, J., and Wagner, R. (1914). *Biochem. Z.* **61**, 387.

Podolsky, R. J. (1960). *In* "The Structure and Function of Muscle" (G. H. Bourne, ed.), Vol. 2, pp. 359–385. Academic Press, New York.

Ridgeway, E. B., and Ashley, C. C. (1967). *Biochem. Biophys. Res. Commun.* **29**, 229.

Sandberg, J. A., and Carlson, F. D. (1966). *Biochem. Z.* **345**, 212.

Sandow, A. (1965). *Pharmacol. Rev.* **17**, 265.

Seliger, V., Dolejš, L., Karas, V., and Pachlopníková, I. (1968). *Int. Z. Angew. Physiol. Einschl. Arbeitsphysiol.* **26**, 227.

Seraydarian, M. W., and Williams, E. B. (1960). *Biochim. Biophys. Acta* **41**, 352.

Smith, I. C. H. (1970). *J. Physiol.* (*London*) **208**, 71P.

Sugi, H. (1969). *Proc. Jap. Acad.* **45**, 413.

Szent-Györgyi, A. (1953). "Chemistry of Muscular Contraction." Academic Press, New York.

Weber, A., and Herz, R. (1963). *J. Biol. Chem.* **238**, 599.

Weber, A., Herz, R., and Reiss, I. (1966). *Biochem. Z.* **345**, 329.

Weber, H. H. (1958). "The Motility of Muscle and Cells." Harvard Univ. Press, Cambridge, Massachusetts.

Wilkie, D. R. (1960). *Progr. Biophys. Mol. Biol.* **10**, 259.

Wilkie, D. R. (1968). *J. Physiol.* (*London*) **195**, 157.

Winegrad, S. (1965). *Fed. Proc., Fed. Amer. Soc. Exp. Biol.* **24**, 1146.

Winegrad, S. (1968). *J. Gen. Physiol.* **51**, 65.

Winegrad, S. (1969). *In* "Mineral Metabolism" (C. L. Comar and F. Bronner, eds.), Vol. 3, pp. 191–233. Academic Press, New York.

Winegrad, S. (1970). *J. Gen. Physiol.* **55**, 77.

Zuntz, N., Loewy, A., Müller, F., and Caspari, W. (1906). "Höhenklima und Bergwanderungen in ihrer Wirkung auf den Menschen. Ergebnisse experimenteller Forschungn. im Hochgebirge und Laboratorium." Deutsches Verlagshaus Bong & Co., Berlin.

AUTHOR INDEX

Numbers in italics refer to the pages on which complete references are listed.

SUBJECT INDEX